STUDENT SOLUTIONS MANUAL TO ACCOMPANY

SECOND EDITION

CHEMISTRY

STANLEY R. RADEL

City College of the City University of New York

MARJORIE H. NAVIDI

Queens College of the City University of New York

Prepared by

ALAN TCHERNOFF

WEST PUBLISHING COMPANY

Minneapolis/St. Paul New York Los Angeles San Francisco

Cover Image © 1993 Joel Gordon Photography

WEST'S COMMITMENT TO THE ENVIRONMENT

In 1906, West Publishing Company began recycling materials left over from the production of books. This began a tradition of efficient and responsible use of resources. Today, up to 95% of our legal books and 70% of our college texts and school texts are printed on recycled, acid-free stock. West also recycles nearly 22 million pounds of scrap paper annually—the equivalent of 181,717 trees. Since the 1960s, West has devised ways to capture and recycle waste inks, solvents, oils, and vapors created in the printing process. We also recycle plastics of all kinds, wood, glass, corrugated cardboard, and batteries, and have eliminated the use of Styrofoam book packaging. We at West are proud of the longevity and the scope of our commitment to the environment.

Production, Prepress, Printing and Binding by West Publishing Company.

610 Opperman Drive
P.O. Box 64526
St. Paul, MN 55164–0526

Printed in the United States of America
01 00 99 98 97 96 95 94 8 7 6 5 4 3 2 1 0

ISBN 0–314–03399–8

Table of Contents

Preface

To The Student:

Once you have completed study of a chapter you should then attempt to work the final exercises. Do not give in to the temptation to look up answers before you complete the problems, because then you will lose the opportunity of discovering for yourself the thought process involved in problem solving and will greatly reduce your confidence level. Because problem solving is critical for the study of chemistry, you will be doing yourself a disservice. Rather, you should complete your problem selection and only then check your solutions against the solutions manual. This process will help you spot weaknesses in your problem solving technique and thereby improve your skills.

I wish to thank Dr. Marjorie Navidi for her assistance in checking this manual. I also wish to thank Dr. David Adams for his checking of the manual. My many talks with Dr. Stanley Radel were enlightening and very helpful. I would also like to thank Denise Bayko of West Educational Publishing for her cooperation and patience. Thanks also goes to Luana Richards, who did and excellent and much appreciated job of copyediting the manuscript. Finally, I would like to thank my wife Rachel and daughter Rebecca for their support.

Dr. Alan Tchernoff

CHAPTER 1

CHEMISTRY AND MEASUREMENT

Solutions To Practice Exercises

PE 1.1 (a) Matter is anything that has mass and occupies space. For example, wood, water, air, and rubber are four different kinds of matter.

(b) Energy is the capacity for doing work. Electrical energy, chemical energy, radiant energy, and nuclear energy are four examples.

PE 1.2 Consult Table 1.2.

PE 1.3 Table 1.3 shows that

(a) 1 dm = 10^{-1} m; therefore, 21 dm = 21 × 10^{-1} m = 2.1 m.

(b) 1 nm = 10^{-9} m; therefore, 410 nm = 410 × 10^{-9} m = 4.10 × 10^{-7} m.

(c) 1 μm = 10^{-6} m; therefore, 0.55 μm = 0.55 × 10^{-6} m = 5.5 × 10^{-7} m.

PE 1.4 (a) 1 kL = 10^3 L = 1000 L.

(b) 1 mL = 10^{-3} L; therefore, 18 mL = 18 × 10^{-3} L = 0.018 L.

(c) 500 cm^3 = 500 mL; therefore, 500 mL = 500 × 10^{-3} L = 0.500 L.

PE 1.5 Counting the number of digits in the measurement gives the number of significant figures.

(a) 65 mL has two digits; two.

(b) 173.4 mL has four digits; four.

(c) 13.2 mL has three digits; three.

(d) 5 mL has one digit; one.

PE 1.6 (a) Rule 3 applies; 0.0060 g has two significant figures.

(b) Rule 2 applies; 7.03000 pm has six significant figures.

(c) Rule 3 applies; 0.000001 s has one significant figure.

(d) Rule 2 applies; 1.800 × 10^{-6} m has four significant figures.

PE 1.7 (a) The measurement 6.8 mL has one decimal place; therefore, the sum will have one decimal place.

6.8	mL
71.3	5 mL
78.1	5 mL

= 78.2 mL after rounding off to one decimal place.

(b) The measurement 3.43 cm has two decimal places; therefore, the difference will have two decimal places.

$$\begin{array}{r|l} 9.22 & 41\text{ cm} \\ -\ 3.43 & \quad\text{cm} \\ \hline 5.79 & 41\text{ cm} \end{array} = 5.79\text{ cm after rounding off to the second decimal place.}$$

(c) The measurement 21.33 g has two decimal places; therefore, the sum will have two decimal places.

$$\begin{array}{r|l} 0.00 & 41\text{ g} \\ +\ 21.33 & \quad\text{g} \\ \hline 21.33 & 41\text{ g} \\ -\ 7.08 & 44\text{ g} \\ \hline 14.24 & 97\text{ g} \end{array} = 14.25\text{ g after rounding off to the second decimal place.}$$

PE 1.8 (a) The number 5.0003 has five significant figures, 2.100 has four significant figures, and 0.0104 has three significant figures; the product will have three significant figures.

$5.0003 \times 2.100 \times 0.0104 = 0.10920 = 0.109$ after rounding off to three significant figures.

(b) Both 0.024 and 0.0015 have two significant figures; the quotient will have two significant figures.

$0.024/0.0015 = 16$

(c) 3.005×10^{-5} has four significant figures while 1.1×10^{6} has two significant figures; the quotient will have two significant figures.

$3.005 \times 10^{-5}/1.1 \times 10^{6} = 2.7318 \times 10^{-11} = 2.7 \times 10^{-11}$ after rounding off to two significant figures.

PE 1.9 The number of objects in a dozen is exactly 12 by definition. The average weight of one orange is 3.10 kg/12 = 0.258 kg. The answer has three significant figures because 3.10 is given to three significant figures.

PE 1.10 We can find out how many liters the tank holds by using the conversion factor 3.785 L/1 gallon from Table 1.1.

$$17.0\text{ gallons} \times \frac{3.785\text{ L}}{1\text{ gallon}} = 64.3\text{ L}$$

PE 1.11 The conversion factor we need is

$$\frac{3\text{ ft}}{1\text{ yard}} \times \frac{12\text{ in}}{1\text{ ft}} \times \frac{2.54\text{ cm}}{1\text{ in}} \times \frac{1\text{ m}}{100\text{ cm}} = 0.9144\text{ m/yard}$$

$$\text{Area} = \text{length} \times \text{width} = 100\text{ yards} \times 53.33\text{ yards} \times \left(\frac{0.9144\text{ m}}{1\text{ yard}}\right)^2 = 4.46 \times 10^3\text{ m}^2$$

PE 1.12

$$\frac{99\text{ miles}}{1\text{ h}} \times \frac{1\text{ h}}{3600\text{ s}} \times \frac{1.609\text{ km}}{1\text{ mile}} \times \frac{1000\text{ m}}{1\text{ km}} = 44\text{ m/s}$$

PE 1.13 (a)

$$\frac{1.00\text{ lb}}{32\text{ pats}} \times \frac{1\text{ kg}}{2.205\text{ lb}} \times \frac{10^3\text{ g}}{1\text{ kg}} = 14.2\text{ g/pat}$$

(b)

$$\frac{14.2\text{ g}}{1\text{ pat}} \times \frac{10^3\text{ mg}}{1\text{ g}} = 1.42 \times 10^4\text{ mg/pat}$$

PE 1.14 (a) For this part we use the conversion factor 5.00 g alloy/3.00 g gold.

$$0.500 \text{ kg gold} \times \frac{10^3 \text{ g}}{1 \text{ kg}} \times \frac{5.00 \text{ g alloy}}{3.00 \text{ g gold}} = 833 \text{ g alloy}$$

(b) For this part we use the conversion factor 2.00 g platinum/5.00 g alloy.

$$833 \text{ g alloy} \times \frac{2.00 \text{ g platinum}}{5.00 \text{ g alloy}} = 333 \text{ g platinum}$$

or by difference, 833 g alloy – 500 g gold = 333 g platinum.

PE 1.15 The mass of an object is equal to the gravitational force divided by the gravitational acceleration.

$$m = \frac{f}{g} = \frac{2.45 \text{ kg} \cdot \text{m/s}^2}{9.81 \text{ m/s}^2} = 0.250 \text{ kg or } 250 \text{ g}$$

PE 1.16

$$d = \frac{m}{V} = \frac{18.00 \text{ g}}{1.325 \text{ mL}} = 13.58 \text{ g/mL}$$

PE 1.17 The density of aluminum = 2.70 g/mL; density of lead = 11.35 g/mL.

$$\text{volume } 100.0 \text{ g aluminum} = 100.0 \text{ g} \times \frac{1 \text{ mL}}{2.70 \text{ g}} = 37.0 \text{ mL or } 37.0 \text{ cm}^3$$

$$\text{volume } 100.0 \text{ g lead} = 100.0 \text{ g} \times \frac{1 \text{ mL}}{11.35 \text{ g}} = 8.811 \text{ mL or } 8.811 \text{ cm}^3$$

Aluminum occupies a little more than four times the volume of an equal mass of lead.

PE 1.18 First we must convert the speed into m/s.

$$v = \frac{80 \text{ km}}{1 \text{ h}} \times \frac{10^3 \text{ m}}{1 \text{ km}} \times \frac{1 \text{ h}}{3600 \text{ s}} = 22 \text{ m/s}$$

Then,

$$E_k = \frac{m \times v^2}{2} = \frac{1680 \text{ kg} \times (22 \text{ m/s})^2}{2} = 4.1 \times 10^5 \text{ kg} \cdot \text{m}^2/\text{s}^2 = 4.1 \times 10^5 \text{ N} \cdot \text{m} = 4.1 \times 10^5 \text{ J} = 4.1 \times 10^2 \text{ kJ}$$

Remember: 1 N = 1 kg•m/s^2; 1 J = 1 N•m.

PE 1.19 (a) Work must be done to lift the book against the gravitational force. Therefore, the potential energy increases.

(b) Work must be done to stretch a rubber band; the energy is stored as potential energy in the rearranged internal structure of the rubber. As the rubber band returns to its original shape, the potential energy converts to kinetic energy and/or heat energy. The potential energy decreases.

(c) Oppositely charged particles attract each other. Work must be done to separate them. Consequently, the potential energy decreases as they come together.

PE 1.20 More. Imagine splitting the 100 mL of water at 95°C into two identical 50-mL portions, each at 95°C. Obviously, they each must contain the same amount of energy and together twice as much energy as another identical 50-mL portion at 95°C. Since the 100-mL portion contains twice as much energy, it can provide more heat energy. In this case, the 100-mL sample will provide twice as much heat energy as the 50-mL sample.

PE 1.21 (a) $1.00 \text{ kg fuel oil} \times \frac{10^3 \text{ g}}{1 \text{ kg}} \times \frac{269 \text{ kcal}}{25.0 \text{ g fuel oil}} = 1.08 \times 10^4 \text{ kcal}$

(b) $1.00 \text{ kg fuel oil} \times \frac{10^3 \text{ g}}{1 \text{ kg}} \times \frac{269 \text{ kcal}}{25.0 \text{ g fuel oil}} \times \frac{4.184 \text{ kJ}}{1 \text{ kcal}} = 4.50 \times 10^4 \text{ kJ}$

Solutions To Final Exercises

1.1

(a) Physicians require a knowledge of the chemical composition of the body and the chemical reactions that take place within the body. Physicians also rely on chemists to supply them with the drugs they use to fight disease.

(b) Geologists are interested in the composition of the earth and its changes, many of which are nothing more than chemical reactions on a grand scale. Chemists help elucidate these processes and do compositional analyses for geologists.

(c) Certain electrical engineers require materials that can withstand the rigors of space and other hostile environments without malfunctioning. Chemists have developed and are continuing to develop such materials.

(d) Growers are continuously trying to increase output and find new uses for their produce. Agricultural chemists develop new, improved, and safer fertilizers, pesticides, and plant growth stimulants in order to increase crop yields. They also develop new uses for the raw materials that can be obtained from crops.

(e) Astronomers are interested in the origin and evolution of the universe. Chemical analysis of stars and other heavenly bodies by spectrophotometry and analysis of objects such as meteors and moon samples can help astronomers unravel this mystery.

(f) The most advanced computer known is the human brain. Biochemists are helping unravel the mysteries of how the brain works. Their results will help computer scientists design superior computers. Chemists also help develop new and improved materials such as semiconductors for the computer industry.

1.3 There are three steps in the scientific method.

(1) The gathering of facts. For example, the effort required to raise two pounds is twice that required to raise one pound.

(2) Construction of laws that summarize these facts. For example, the effort required to lift an object is directly proportional to its mass.

(3) Construction of theories or models of nature that will account for the laws. For example, there exists a force of attraction between any two bodies; this force is directly proportional to the masses of the bodies, and inversely proportional to the square of the distance separating them.

1.5 Refer to Table 1.2 in the textbook.

1.7 A derived unit is a unit made by combining two or more other units.

(a) Speed is distance/time. Therefore it has the units m/s.

(b) Area is length squared. Therefore it has the units m^2.

(c) Volume is length cubed. Therefore it has the units m^3.

(d) Density is mass/volume. Therefore it has the units kg/m^3.

(e) Force is mass × acceleration or work/distance. Therefore it has the units $kg{\bullet}m/s^2$.

(f) Energy is the ability to do work and is force × distance. Therefore it has the units $kg{\bullet}m^2/s^2$.

1.9

(a) The SI unit of length is the meter (m), and since volume is defined as length cubed, the SI unit of volume is the cubic meter (m^3).

(b) The liter is defined as $10^3 \text{ cm}^3 = 10^{-3} \text{ m}^3$. Therefore, a liter (L) is equal to one thousandth (1/1000) of the SI unit of volume (m^3).

1.11 (a) $2.2\ \text{m} \times \frac{1\ \text{cm}}{10^{-2}\ \text{m}} = 2.2 \times 10^{2}\ \text{cm}$ (b) $2.2\ \text{m} \times \frac{1\ \text{km}}{10^{3}\ \text{m}} = 2.2 \times 10^{-3}\ \text{km}$ (c) $2.2\ \text{m} \times \frac{1\ \text{nm}}{10^{-9}\ \text{m}} = 2.2 \times 10^{9}\ \text{nm}$

The meter is the preferred unit since it provides the least cumbersome value to work with.

1.13 (a) Since a millimeter is 10^{-3} m and a centimeter 10^{-2} m, there are ten (10) millimeters in a centimeter.
(b) Since 10 mm = 1 cm, there are a hundred (10^2) square millimeters (mm^2) in a square centimeter (cm^2).
(c) Since 10 mm = 1 cm, there are a thousand (10^3) cubic millimeters (mm^3) in a cubic centimeter (cm^3).

1.15

$$\text{length} = 78\ \text{ft} \times \frac{12\ \text{in}}{1\ \text{ft}} \times \frac{1\ \text{m}}{39.37\ \text{in}} = 23.77\ \text{m} = 24\ \text{m (two significant figures)}$$

$$\text{width} = 36\ \text{ft} \times \frac{12\ \text{in}}{1\ \text{ft}} \times \frac{1\ \text{m}}{39.37\ \text{in}} = 10.97\ \text{m} = 11\ \text{m (two significant figures)}$$

$$\text{area} = \text{length} \times \text{width} = 24\ \text{m} \times 11\ \text{m} = 264\ \text{m}^2 = 2.6 \times 10^2\ \text{m}^2\ \text{(two significant figures)}$$

1.17 The volume of a sphere is given by the formula $V = (4/3)\pi r^3 = (1/6)\pi d^3$, where r is the radius and d the diameter of the sphere.

(a) $V = (1/6) \times 3.14 \times (2.50\ \text{in})^3 \times \frac{(2.54\ \text{cm})^3}{1\ \text{in}^3} = 134\ \text{cm}^3$

(b) $134\ \text{cm}^3 \times \left(\frac{1\ \text{dm}}{10\ \text{cm}}\right)^3 = 0.134\ \text{dm}^3$

(c) $134\ \text{cm}^3 \times \left(\frac{1\ \text{m}}{10^2\ \text{cm}}\right)^3 = 1.34 \times 10^{-4}\ \text{m}^3$

1.19

$$V = (4/3)\pi r^3 = (4/3) \times 3.14 \times (1.18 \times 10^2\ \text{pm})^3 = 6.88 \times 10^6\ \text{pm}^3$$

1.21 (a) Precision is a measure of the reproducibility of an experiment. If one obtains similar results from repeated determinations, then the experiment is said to be precise. Accuracy is directly related to how close the experimental measurement is to the actual value of the quantity being measured. The smaller the error, the more accurate the experiment. Precise measurements are not necessarily accurate.
(b) Weigh the object many times and compare the results. If the results all fall within the calculated experimental error, then the result can be considered precise.
(c) Accuracy depends on the true value of the quantity being measured, which in actual practice is known only to within plus and minus some error. Consequently, the accuracy of a determination depends on the error range. However, if you can employ several independent, and possibly even different, ways of measuring the same quantity, and if the results of these measurements all agree within experimental errors, then the value can be assumed to be accurate. In this case, the object could be weighed on several different scales and the results compared.
(d) It is much easier to evaluate the precision, since it is not always possible to measure a quantity in several independent ways.

1.23 Systematic errors are caused mainly by inaccurate instruments, problems or inaccuracies inherent in the method employed, and personal tendencies in the experimenter's technique and analysis. Random errors are small variations and accidental errors over which the experimenter has little control.
(a) Random. The volume readings should be randomly scattered about the fuzzy markings.
(b) Random. The increases and decreases should cancel each other out.
(c) Systematic. Poor personal technique here leads to each weight being higher than it actually is.

1.25 (b), (c), and (f) are obtained by measurement.
(a) No. The number of eggs is obtained by counting.
(d) No. The women's age is obtained by counting.
(e) No. 1000 mL is equivalent to a liter. This is a defined relationship.

1.27 (a) 5,267,250 red cells/mm^3 = 5.27×10^6 red cells/mm^3
(b) 4,968,667 red cells/mm^3 = 4.97×10^6 red cells/mm^3
(c) 7,000,000 red cells/mm^3 = 7.00×10^6 red cells/mm^3

1.29 (a) 5 (b) 2 (c) 2 (d) 5 (e) 1 (f) 3

1.31 (a) $\frac{3.0 \times 22.4}{1.120} = 60$

(b)
$$\begin{array}{r} 56.85 \\ -\ 9.9 \\ -\ 43.214 \\ \hline 3.736 \end{array}$$
= 3.7 after rounding off to one decimal place.

(c) $\frac{4.38 \times 10^4}{5.1 \times 10^{-3}} = 8.6 \times 10^6$ after rounding off to two significant figures.

(d)
$$\begin{array}{r} 22.22 \\ 2.71828 \\ 2001. \\ +\ 0.03 \\ \hline 2025.96828 \end{array}$$
= 2026 after rounding off to zero decimal places.

1.33 (a)
$$\begin{array}{r} 3.61 \times 10^5 \\ +\ 0.275 \times 10^5 \\ \hline 3.885 \times 10^5 \end{array}$$
$= 3.88 \times 10^5$ after rounding off to two decimal places.

(b) $9.1095 \times 10^{-28} \times 6.022 \times 10^{23} = 5.486 \times 10^{-4}$ after rounding off to four significant figures.

(c)
$$\begin{array}{r} 1.0073 \\ +\ 0.0005486 \\ \hline 1.0078486 \end{array}$$
= 1.0078 rounded off to four decimal places.

(d) $\frac{(7.01 \times 10^7)(3.0 \times 10^{24})}{6.219 \times 10^{-16}} = 3.4 \times 10^{47}$ after rounding off to two significant figures.

1.35 (a) $\frac{1 \text{ km}}{0.6214 \text{ mile}}$ and $\frac{0.6214 \text{ mile}}{1 \text{ km}}$

(b) $\frac{1 \text{ gal}}{3.785 \text{ L}}$ and $\frac{3.785 \text{ L}}{1 \text{ gal}}$

(c) $\frac{453.6 \text{ g}}{1 \text{ lb}}$ and $\frac{1 \text{ lb}}{453.6 \text{ g}}$

(d) $\frac{4.184 \text{ J}}{1 \text{ cal}}$ and $\frac{1 \text{ cal}}{4.184 \text{ J}}$

(e) $\frac{1000 \text{ kg}}{1 \text{ metric ton}}$ and $\frac{1 \text{ metric ton}}{1000 \text{ kg}}$

(f) $\frac{1 \text{ h}}{3600 \text{ s}}$ and $\frac{3600 \text{ s}}{1 \text{ h}}$

1.37 55 miles/h is an exact number.

(a) $$\frac{55\text{ miles}}{1\text{ h}} \times \frac{1.609\text{ km}}{1\text{ mile}} = \frac{88.50\text{ km}}{1\text{ h}}$$

(b) $$\frac{55\text{ miles}}{1\text{ h}} \times \frac{1\text{ h}}{3600\text{ s}} \times \frac{1.609\text{ km}}{1\text{ mile}} \times \frac{10^3\text{ m}}{1\text{ km}} = \frac{24.58\text{ m}}{1\text{ s}}$$

1.39 $$\frac{32\text{ miles}}{1\text{ gal}} \times \frac{1\text{ gal}}{3.785\text{ L}} \times \frac{1.609\text{ km}}{1\text{ mile}} = \frac{14\text{ km}}{1\text{ L}}$$

1.41 $$\frac{0.050\text{ mL}}{1\text{ s}} \times \frac{1\text{ L}}{10^3\text{ mL}} \times \frac{3600\text{ s}}{1\text{ h}} = \frac{0.18\text{ L}}{1\text{ h}}$$

1.43 The weight of an object is the gravitational force exerted on that object. Mass is a measure of the quantity of matter in an object. Weight and mass are related by the gravitational acceleration (g). The relationship is weight = mass × g. The mass is the proportionality constant between the force (weight) and the gravitational acceleration (g).

Comment: Weight and mass are not the same thing, but most scientists use them interchangeably. Some confusion arises because in the English system of units amounts of substance are reported in pounds (lbs), which are weight or force units, whereas in the metric system they are reported in kilograms (kg), which are mass units. Keep in mind that the conversion factors 1 lb = 453.6 g and 1 kg = 2.205 lb are only valid on the surface of the earth, since they are derived using the standard g value for the earth's surface.

1.45 The balance is the instrument used in science for determining the mass of an object. The balance makes use of the fact that the gravitational attraction on objects of equal mass is the same. In other words, it makes use of the fact that equal masses have equal weights. The double-pan balance measures the mass of an object by comparing its mass (placed in one pan of the balance) with that of "weights" of known mass, which are placed on the other pan of the balance.

A single-pan balance measures the mass of an object by comparing its mass (placed on the single pan) against "weights" of known mass within the balance.

A spring scale measures the weight of an object by comparing its deflection to that of known weights whose deflections were recorded. (Some spring scales have both weight and mass calibrations, and these can be used to measure either the weight or mass of the object. However, the mass calibration will strictly hold only for the location where the spring scale was calibrated.)

1.47 (a) $$\text{height} = 66\text{ in} \times \frac{2.54\text{ cm}}{1\text{ in}} \times \frac{1\text{ m}}{10^2\text{ cm}} = 1.7\text{ m}$$

(b) $$\text{mass} = 145\text{ lb} \times \frac{1\text{ kg}}{2.205\text{ lb}} = 65.8\text{ kg}$$

(c) $$\text{weight} = 145\text{ lb} \times \frac{1\text{ kg}}{2.205\text{ lb}} \times \frac{9.81\text{ m}}{1\text{ s}^2} = 645\text{ kg}\bullet\text{m/s}^2 = 645\text{ N}$$

1.49 To solve this problem we must remember that the sum or difference of several measurements can have no more decimal places than the measurement with the least number of decimal places. We also must be familiar with the unit prefixes that expand and contract the size of units. For instance, c = 1/100 = 10^{-2} of a unit, d = 1/10 = 10^{-1} of a unit, and m = 1/1000 = 10^{-3} of a unit. See Table 1.3 for the meaning of the other prefix symbols.

Each mass is changed to grams before adding.

2125.		g
404.	3	g
600.	4	g
15.	25	g
3144.	95	g = 3145 g after rounding off to zero decimal places.

1.51

$$1.00 \text{ L beer} \times \frac{6.0 \text{ g alcohol}}{0.1435 \text{ L beer}} = 42 \text{ g alcohol}$$

1.53

$$55 \text{ mL water} \times \frac{71 \text{ g Epsom salts}}{100 \text{ mL water}} = 39 \text{ g Epsom salts}$$

1.55 Using the conversion factors 1.30 kg magnesium/1 m^3, 10^3 g/1 kg, and 1 $m^3/(100 \text{ cm})^3$, and remembering that 1 mL = 1 cm^3,

$$200 \text{ mL} \times \frac{1 \text{ m}^3}{10^6 \text{ mL}} \times \frac{1.30 \text{ kg magnesium}}{1 \text{ m}^3} \times \frac{10^3 \text{ g}}{1 \text{ kg}} = 0.260 \text{ g magnesium}$$

1.57 Density is defined as mass per unit volume. So,

$$\text{density} = \frac{\text{mass}}{\text{volume}} = \frac{88.9 \text{ g}}{10.5 \text{ mL}} = 8.47 \text{ g/mL}$$

1.59 Rearrangement of the density equation gives mass = density × volume. See Table 1.7 in textbook.

(a) The density of water at 3.98°C is 1.00000 g/mL. The mass of 1.00 L of water at 3.98°C therefore equals

$$1.00 \text{ L} \times \frac{1000.00 \text{ g}}{1 \text{ L}} = 1.00 \times 10^3 \text{ g}$$

(b) The density of ice at 0°C is 0.917 g/mL. The mass of 1.00 L of ice at 0°C therefore equals

$$1.00 \text{ L} \times 917 \text{ g/L} = 917 \text{ g}$$

(c) The density of steam at 100°C and 1.00 atm pressure is 0.000596 g/mL. The mass of 1.00 L of steam at 100°C therefore equals

$$1.00 \text{ L} \times 0.596 \text{ g/L} = 0.596 \text{ g}$$

The mass of 1.00 L of ice at 0°C is 83 g less than that of liquid water at 3.98°C. The more massive liquid can support the less massive ice, i.e., ice floats in water. The mass of 1.00 L of steam at 100°C is 1678 times less than that of liquid water at 3.98°C and 1539 times less than that of ice at 0°C.

1.61 To solve this problem we merely use the appropriate conversion factors.

(a) $$\frac{13.6 \text{ g}}{1 \text{ cm}^3} \times \frac{1 \text{ kg}}{10^3 \text{ g}} \times \frac{10^3 \text{ cm}^3}{1 \text{ dm}^3} = \frac{13.6 \text{ kg}}{1 \text{ dm}^3}$$

(b) $$\frac{13.6 \text{ g}}{1 \text{ cm}^3} \times \frac{1 \text{ lb}}{453.6 \text{ g}} \times \frac{(2.54 \text{ cm})^3}{1 \text{ in}^3} \times \frac{(12 \text{ in})^3}{1 \text{ ft}^3} = \frac{849 \text{ lb}}{1 \text{ ft}^3}$$

1.63

$$\underbrace{1.00 \text{ kg boric acid} \times \frac{10^3 \text{ g}}{1 \text{ kg}} \times \frac{1 \text{ cm}^3}{1.435 \text{ g boric acid}}}_{\text{volume of 1.00 kg of boric acid in cm}^3} \times \underbrace{\frac{1 \text{ L}}{10^3 \text{ cm}^3}}_{\text{converts to liters}} = 0.697 \text{ L}$$

Any bottle holding 0.70 L or more will hold 1.00 kg of boric acid.

1.65 (a) Density of mercury = 13.55 g/cm^3 and volume of cylinder = cross-sectional area × height.

$$\text{mass} = \text{density} \times \text{volume} = \frac{13.55 \text{ g}}{1 \text{ cm}^3} \times 2.00 \text{ cm}^2 \times 76.0 \text{ cm} = 2.06 \times 10^3 \text{ g mercury}$$

(b) volume = mass/density = 2.06×10^3 g/0.99823 g/mL = 2.06×10^3 mL or 2.06 L water at 20° C

1.67

$$\text{specific gravity} = \frac{\text{density of substance}}{\text{density of water}} = \frac{0.7893 \text{ g/mL}}{0.99823 \text{ g/mL}} = 0.7907$$

1.69 Work is the product of a force (f) and the distance (d) through which it operates, $w = f \times d$. Work is a form of energy exchange. It is the energy exchanged or transferred when a force (f) moves an object a distance (d) from its original position.

The joule (J) or newton-meter (N•m) is the SI unit of work and energy. The newton is the SI unit of force, $1 \text{ N} = 1 \text{ kg•m/s}^2$. The calorie is an alternative energy unit often used by chemists, 1 cal = 4.184 J.

1.71 Kinetic energy is the energy an object possesses because it is in motion. It only depends on how fast an object is moving at any instant and doesn't depend on its position. Potential energy is the energy an object possesses because of its position or the positions of its parts relative to each other. The potential energy of an object only depends on the position of the object and not on how fast it is moving. So, kinetic energy depends only on the speed and potential energy only on the position.

The SI unit of energy is the newton-meter (N•m) or joule (J).

1.73 Matter has potential energy due to its position, condition, or composition. If work must be performed to get it into a certain position or condition, then its potential energy increases. On the other hand, if energy is released on changing its position or condition, then its potential energy decreases.

(a) Potential energy increases. Positive charges repel each other, and therefore work must be done to push them together.

(b) Potential energy decreases. Negative charges repel each other. Work must be performed to bring them together. Therefore, the potential energy decreases as they fly apart.

(c) Potential energy increases. Opposite charges attract each other. When separating them, work must be performed to overcome this attractive force. Therefore, the potential energy increases as they are separated.

1.75 (a) $w = f \times d = 15.0 \text{ N} \times 500 \text{ cm} \times \dfrac{1 \text{ m}}{10^2 \text{ cm}} = 75.0 \text{ N•m} = 75.0 \text{ J}$

(b) $w = 75.0 \text{ J} \times \dfrac{1 \text{ cal}}{4.184 \text{ J}} = 17.9 \text{ cal}$

1.77 Kinetic energy is the energy of motion and is defined as

$$E_k = \frac{\text{mass} \times v^2}{2}$$

(a) To solve this problem we first have to convert pounds to kilograms and miles per hour to meters per second.

$$0.64\text{ lb} \times \frac{453.6\text{ g}}{1\text{ lb}} \times \frac{1\text{ kg}}{10^3\text{ g}} = 0.29\text{ kg}$$

$$\frac{88\text{ miles}}{1\text{ h}} \times \frac{1609\text{ m}}{1\text{ mile}} \times \frac{1\text{ h}}{3600\text{ s}} = 39\text{ m/s}$$

$$E_k = \frac{0.29\text{ kg} \times (39\text{ m/s})^2}{2} = 2.2 \times 10^2\text{ J}$$

(b) $$E_k = \frac{5.31 \times 10^{-26}\text{ kg} \times (500\text{ m/s})^2}{2} = 6.64 \times 10^{-21}\text{ J}$$

1.79

$$v = \sqrt{2E_k / m}$$

$$v = \sqrt{\frac{2 \times 1.1 \times 10^{-19}\text{ J}}{9.1 \times 10^{-31}\text{ kg}}} = \sqrt{0.2418 \times 10^{12}\text{ m}^2/\text{s}^2} = 4.9 \times 10^5\text{ m/s}$$

1.81

$$1.00\text{ g water} \times \frac{40.67\text{ kJ}}{18.0\text{ g water}} \times \frac{10^3\text{ J}}{1\text{ kJ}} \times \frac{1\text{ cal}}{4.184\text{ J}} = 5.40 \times 10^2\text{ cal}$$

1.83

°F = 32 + 1.8°C = 32 + 1.8(1772) = 3222°F

1.85

The scientific method is based on the reproducibility of experimental results, which means that you must be working in an area that can be quantitatively examined and described, i.e., you must be working in a scientific area or science. If your results cannot be expressed quantitatively, then there would be no way of knowing whether the result of an experiment agrees with your theoretical model. You could never tell whether your model is working, needs revision, or should be rejected. Also, any law you do propose would be of little value, since the law would only provide you with vague and unmeasureable correlations and predictions. Consequently, the scientific method is of value only in science.

1.87

Distance is a continuous variable, which means that it can take on an infinite number of values and, in fact, can take on an infinite number of values between any two points. Distance is not limited to a discrete number of values. Therefore, the possibility that we will ever be able to determine the exact value by measurement is highly unlikely, if not impossible.

1.89

It implies that the size of the components are about 10^{-9} m in size. The components would be closest to atoms in size. They would be about five times larger than atoms. Protons are much smaller, and human hairs and blood cells are much bigger, in size.

1.91

There are 1968/2 = 984 leaves. The thickness is

$$\frac{5.74\text{ cm}}{984\text{ leaves}} = 5.83 \times 10^{-3}\text{ cm/leaf.}$$

(a) $$5.83 \times 10^{-3}\text{ cm} \times \frac{10\text{ mm}}{1\text{ cm}} = 5.83 \times 10^{-2}\text{ mm}$$

(b) $$5.83 \times 10^{-3}\text{ cm} \times \frac{1\text{ m}}{10^2\text{ cm}} \times \frac{1\ \mu\text{m}}{10^{-6}\text{ m}} = 58.3\ \mu\text{m}$$

1.93 Each 5-mL teaspoonful contains 4 mg phenobarbital, 65 mg theophylline, and 12 mg ephedrine hydrochloride, giving us the conversion factors:

$$\frac{4\text{ mg phenobarbital}}{5\text{ mL}};\ \frac{65\text{ mg theophylline}}{5\text{ mL}};\ \frac{12\text{ mg ephedrine hydrochloride}}{5\text{ mL}}$$

The label recommends one teaspoonful four times daily per 60 lb of body weight, giving us the conversion factor:

$$\frac{20\text{ mL}}{1\text{ day}\times 60\text{ lb}}$$

Combining these conversion factors we get

$$\frac{16\text{ mg phenobarbital}}{1\text{ day}\times 60\text{ lb}};\ \frac{260\text{ mg theophylline}}{1\text{ day}\times 60\text{ lb}};\ \frac{48\text{ mg ephedrine hydrochloride}}{1\text{ day}\times 60\text{ lb}}$$

(a) $$1\text{ day}\times 90\text{ lb}\times\frac{16\text{ mg phenobarbital}}{1\text{ day}\times 60\text{ lb}} = 24\text{ mg phenobarbital}$$

$$1\text{ day}\times 90\text{ lb}\times\frac{260\text{ mg theophylline}}{1\text{ day}\times 60\text{ lb}} = 390\text{ mg theophylline}$$

$$1\text{ day}\times 90\text{ lb}\times\frac{48\text{ mg ephedrine hydrochloride}}{1\text{ day}\times 60\text{ lb}} = 72\text{ mg ephedrine hydrochloride}$$

(b) $$24\text{ mg phenobarbital}\times\frac{1\text{ grain}}{64.80\text{ mg}} = 0.37\text{ grains phenobarbital}$$

(c) $$\frac{44\text{ mL}}{\text{bottle}}\times\frac{4\text{ mg phenobarbital}}{5\text{ mL}}\times\frac{1\text{ day}}{24\text{ mg phenobarbital}} = 1.5\text{ days/bottle}$$

1.95 The volume of a sphere is given by the formula $V = (4/3)\pi r^3 = (1/6)\pi d^3$ and the circumference of a circle is given by $C = 2\pi r = \pi d$, where r is the radius and d is the diameter. So, $d = C/\pi = 30\text{ in}/3.14$ and

$$d = \frac{\text{mass}}{\text{volume}} = \frac{21\text{ oz}\times\frac{1\text{ lb}}{16\text{ oz}}\times\frac{453.6\text{ g}}{1\text{ lb}}}{(1/6)\times 3.14\times(30\text{ in}/3.14)^3\times(2.54\text{ cm}/1\text{ in})^3} = \frac{595.35\text{ g}}{7479.19\text{ cm}^3} = 8.0\times 10^{-2}\text{ g/cm}^3$$

1.97 The densities of copper and gold are 8.96 g/mL and 19.32 g/mL, respectively (Table 1.7). Using the rearranged density equation mass = density × volume we get

$$\text{mass gold} = \text{density gold}\times\text{volume gold} = \frac{19.32\text{ g gold}}{1\text{ mL}}\times\underbrace{\frac{100.0\text{ g copper}}{8.96\text{ g copper/mL}}}_{\substack{\text{volume 100.0 g copper}\\ =\text{ volume of gold}}} = 216\text{ g gold}$$

1.99 Using the rearranged density equation mass = density × volume we get

mass ethanol = 0.7893 g/mL × 50.0 mL

mass soln = 0.7893 g/mL × 50.0 mL + 0.99823 g/mL × 100.0 mL

$$\text{and ethanol }\% = \frac{\text{mass ethanol}\times 100\%}{\text{mass soln}} = \frac{39.465\text{ g}\times 100\%}{139.288\text{ g}} = 28.3\%$$

1.101 To calculate the amount of work it takes to lift the lemon, we must first calculate the magnitude of the downward gravitational force acting on the lemon, i.e., the weight of the lemon. From Equation 1.1 of the text we have f = m × g, where g is the gravitational acceleration, 9.80665 m/s^2.

$$f = 85\ \text{g} \times \frac{1\ \text{kg}}{10^3\ \text{g}} \times 9.80665\ \text{m/s}^2 = 0.83\ \text{N}$$

We can now calculate the work required to lift the lemon.

$$w = f \times d = 0.83\ \text{N} \times 4\ \text{ft} \times \frac{12\ \text{in}}{1\ \text{ft}} \times \frac{1\ \text{m}}{39.37\ \text{in}} = 1\ \text{J}$$

Comment: The value of g is actually a variable of the distance from the center of the earth. For small changes in that distance the g value can be treated as a constant.

1.103 Fixed Temperatures for Defining Scales

	Freezing point of water	Boiling point of water	
Celsius scale	0°	100°	divided into 100 equal intervals
Fahrenheit scale	32°	212°	divided into 180 equal intervals

So, 100 Celsius degrees equals 180 Fahrenheit degrees, and a degree Fahrenheit is 100/180 or 5/9 of a degree Celsius. This ratio of 100/180 = 5/9 between the size of the Celsius and Fahrenheit degrees holds for all temperature ranges. Consequently, for any other range starting from the fixed freezing point of water, we have

$$\frac{\text{Celsius range}}{\text{Fahrenheit range}} = \frac{°\text{C} - 0}{°\text{F} - 32} = \frac{5}{9}$$

Solving this equation for °C and then for °F gives °C = 5/9 (°F – 32) = 1.8(°F – 32) and °F = 9/5°C + 32 = 1.8°C + 32.

CHAPTER 2

ATOMS, MOLECULES, AND IONS

Solutions To Practice Exercises

PE 2.1 See Table 2.1 in text.

PE 2.2 (a) The mass number (A) is equal to the total number of protons and neutrons in the nucleus of an atom. A = number of protons + number of neutrons. A = 26 + 30 = 56.

(b) The symbol for iron is Fe. To obtain the atomic symbol of a given isotope, attach the mass number A as a left superscript and the atomic number Z as a left subscript to the symbol. The atomic symbol for this iron isotope is therefore

$^{56}_{26}Fe$

PE 2.3 The symbol for chlorine is Cl. The number of protons = Z = 17. Mass number A = number of protons + number of neutrons = 17 + 18 = 35. The atomic symbol is therefore

$^{35}_{17}Cl^{-}$

The – sign means that the neutral chlorine atom has gained an additional negative charge, i.e., an additional electron, and is now a negative ion, the chloride ion.

PE 2.4 Let x = fractional abundance of lithium-6 and $(1 - x)$ = fractional abundance of lithium-7. Then $6.01512x + 7.01600(1 - x) = 6.941$ from the definition of the chemical atomic weight.

$6.01512x + 7.01600 - 7.01600x = 6.941$

$1.00088x = 0.075$

$x = 0.075/1.00088 = 0.075$ and $(1 - x) = 1 - 0.075 = 0.925$

The percent abundance is equal to 100% times the fractional abundance. Hence, percent abundance of lithium-6 = 7.5%; percent abundance of lithium-7 = 92.5%.

PE 2.5 The charge of one hydrogen ion (proton) is 1.6022×10^{-19} C. One mole of hydrogen ions contains Avogadro's number of hydrogen ions; i.e., 1 mol of hydrogen ions = 6.0221×10^{23} H^+ ions. The charge of 1 mol of hydrogen ions is

$$6.0221 \times 10^{23}\ H^+ \text{ions} \times \frac{1.6022 \times 10^{-19}\ C}{1\ H^+\ \text{ion}} = 9.6486 \times 10^4\ C$$

PE 2.6

$$1.675 \times 10^{-24}\ g \times \frac{6.022 \times 10^{23}\ u}{1\ g} = 1.009\ u$$

PE 2.7 The atomic weights of magnesium and gold are 24.30 u and 197.0 u, respectively (see inside front cover). Hence, their molar masses in grams are 24.30 g/mol and 197.0 g/mol, respectively.

PE 2.8 The atomic weight of calcium to four significant figures is 40.08 u; therefore, the mass of 1 mol of calcium is 40.08 g. The mass in grams of 1.55 mol of calcium is

$$1.55\ \text{mol Ca} \times \frac{40.08\ g}{1\ \text{mol Ca}} = 62.1\ g$$

PE 2.9 The atomic weight of helium is 4.003 u, and its molar mass is 4.003 g/mol. Using conversion factors:

$$2.00 \times 10^{21}\ \text{helium atoms} \times \frac{4.003\ u}{1\ \text{helium atom}} \times \frac{1\ g}{6.022 \times 10^{23}\ u} = 1.33 \times 10^{-2}\ g$$

This problem could also be solved by using the definition of a mole. One mole of any substance contains Avogadro's number of particles = 6.022×10^{23} particles. The mass of 1 mol, called the molar mass, in grams is numerically equal to the atomic weight in atomic mass units. So,

1 mol of helium atoms = 6.022×10^{23} helium atoms = 4.003 g

$$2.00 \times 10^{21}\ \text{helium atoms} \times \frac{4.003\ g}{6.022 \times 10^{23}\ \text{helium atoms}} = 1.33 \times 10^{-2}\ g$$

PE 2.10 (a) Sodium chloride: (1) white crystalline solid (clear when in the form of large crystals); (2) tastes salty; (3) soluble in water; (4) noncombustible; (5) melts at high temperatures (801°C); (6) insoluble in nonpolar solvents, e.g., benzene.

(b) Sucrose: (1) white crystalline solid; (2) tastes sweet; (3) soluble in water; (4) undergoes combustion in air to form CO_2 and H_2O; (5) melts (185°C) to form a syrupy liquid; (6) insoluble in nonpolar solvents, e.g., benzene.

PE 2.11 (a) Mixture; milk is a complex mixture of many substances. It contains water, fats, sugars, etc.

(b) Pure substance; CO_2 is a compound always containing 27.3% carbon and 72.7% oxygen by mass.

(c) Mixture; steel is a mixture of iron with various small percentages of carbon. Certain steels have other metals mixed in as well to produce specific properties.

(d) Pure substance; sucrose is a compound composed of carbon, hydrogen, and oxygen combined in a fixed ratio.

(e) Mixture; wood is the hard fibrous substance beneath the bark in the stems and branches of trees and shrubs; its contents vary widely.

(f) Mixture; blood is a very complex mixture containing hundreds of different substances (red cells, white cells, serum, glucose, etc.).

PE 2.12 (a) In CH_3COOH there are two carbon atoms, four hydrogen atoms, and two oxygen atoms.

(b) The dimethylamine molecule, $(CH_3)_2NH$, contains one nitrogen atom, one hydrogen atom, and two (CH_3) groups, each containing one carbon atom and three hydrogen atoms. In total, there are one nitrogen atom, two carbon atoms, and seven hydrogen atoms.

PE 2.13 (a) Eleven elements are gases at room temperature: helium (He), neon (Ne), argon (Ar), krypton (Kr), xenon (Xe), radon (Rn), hydrogen (H_2), nitrogen (N_2), oxygen (O_2), fluorine (F_2), and chlorine (Cl_2).

(b) Two elements are liquids at room temperature: bromine (Br_2) and mercury (Hg).

PE 2.14 (a) One mole of calcium chloride ($CaCl_2$) contains 1 mol of calcium ions (Ca^{2+}) and 2 mol of chloride ions (Cl^-), so 3.5 mol of calcium chloride will contain $3.5 \times 1 = 3.5$ mol of Ca^{2+} and $3.5 \times 2 = 7.0$ mol of Cl^-.

(b) Each mole of ions contains 6.022×10^{23} ions.

$$3.5 \text{ mol } Ca^{2+} \times \frac{6.022 \times 10^{23} \text{ } Ca^{2+} \text{ ions}}{1 \text{ mol } Ca^{2+}} = 2.1 \times 10^{24} \text{ } Ca^{2+} \text{ ions}$$

$$7.0 \text{ mol } Cl^- \times \frac{6.022 \times 10^{23} \text{ } Cl^- \text{ ions}}{1 \text{ mol } Cl^-} = 4.2 \times 10^{24} \text{ } Cl^- \text{ ions}$$

PE 2.15 The atom ratio is the same as the mole ratio, and the formula reflects the atom ratio. The atom ratio is three magnesium ions for every two nitride ions. The formula is Mg_3N_2.

PE 2.16 (a) One mole of C_4H_{10} contains 4 mol of C atoms and 10 mol of H atoms. The molar masses of C and H are 12.01 g/mol and 1.008 g/mol, respectively. The total mass is

$$\begin{aligned} \text{C: } 4 \text{ mol} \times 12.01 \text{ g/mol} &= 48.04 \text{ g} \\ \text{H: } 10 \text{ mol} \times 1.008 \text{ g/mol} &= \underline{10.08 \text{ g}} \\ \text{Total molar mass} &= 58.12 \text{ g} \end{aligned}$$

(b) One mole of butane contains 6.022×10^{23} molecules and has a mass of 58.12 g. The mass of one molecule is

$$\frac{58.12 \text{ g}}{1 \text{ mol butane}} \times \frac{1 \text{ mol butane}}{6.022 \times 10^{23} \text{ molecules}} = 9.651 \times 10^{-23} \text{ g / molecule}$$

PE 2.17 (a) The molar masses of sulfur and oxygen are 32.07 g/mol and 16.00 g/mol, respectively. The molar mass of sulfur dioxide is

$$\begin{aligned} \text{S: } 1 \text{ mol} \times 32.07 \text{ g/mol} &= 32.07 \text{ g} \\ \text{O: } 2 \text{ mol} \times 16.00 \text{ g/mol} &= \underline{32.00 \text{ g}} \\ \text{Total molar mass} &= 64.07 \text{ g} \end{aligned}$$

The number of moles of SO_2 is

$$1000 \text{ g } SO_2 \times \frac{1 \text{ mol } SO_2}{64.07 \text{ g } SO_2} = 15.61 \text{ mol } SO_2$$

(b) $$15.61 \text{ mol } SO_2 \times \frac{6.022 \times 10^{23} \text{ molecules}}{1 \text{ mol } SO_2} = 9.400 \times 10^{24} \text{ molecules}$$

PE 2.18 (a) From Table 2.6 we see that the sodium ion is Na^+ and the carbonate ion is CO_3^{2-}. A total charge of zero is obtained from two sodium ions ($2 \times 1+ = 2+$) and one carbonate ion ($1 \times 2- = 2-$). The formula is $(Na^+)_2CO_3^{2-}$ or Na_2CO_3.

(b) From Table 2.6 we see that the magnesium ion is Mg^{2+} and the chloride ion is Cl^-. A total charge of zero is obtained from one magnesium ion ($1 \times 2+ = 2+$) and two chloride ions ($2 \times 1- = 2-$). The formula is $Mg^{2+}(Cl^-)_2$ or $MgCl_2$.

(c) From Table 2.6 we see that the iron(III) ion is Fe^{3+} and the bromide ion is Br^-. A total charge of zero is obtained from one iron(III) ion ($1 \times 3+ = 3+$) and three bromide ions ($3 \times 1- = 3-$). The formula is $Fe^{3+}(Br^-)_3$ or $FeBr_3$.

PE 2.19 (a) Potassium ion (b) calcium ion (c) gallium ion.

PE 2.20 (a) Cu^+, copper(I) ion; Cu^{2+}, copper(II) ion. (b) Cr^{2+}, chromium(II) ion; Cr^{3+}, chromium(III) ion.

PE 2.21 (a) Bromide ion (b) nitride ion (c) oxide ion.

PE 2.22 (a) BrO_3^- (b) BrO^- (c) BrO_4^-.

PE 2.23 (a) barium chloride (b) calcium sulfide (c) magnesium nitride
(d) iron(II) sulfate (e) potassium cyanide (f) aluminum nitrate

PE 2.24 (a) $FeCl_2$ (b) Na_2S (c) $K_2Cr_2O_7$ (d) $Mg_3(PO_4)_2$ (e) $Fe_2(SO_4)_3$

Comment: Remember that you must maintain electrical neutrality, i.e., the positive and negative charges must add up to zero.

PE 2.25 (a) sulfur dioxide (b) sulfur trioxide (c) phosphorus trichloride
(d) phosphorus pentachloride (e) uranium hexafluoride (f) carbon tetrachloride

PE 2.26 (a) NH_3 (b) CH_4 (c) HBr (d) BrF_5 (e) CS_2 (f) N_2O_4 (g) IBr

PE 2.27 (a) $CaCO_3$ (b) $NaHCO_3$ (c) CO_2 (solid) (d) SiO_2 (e) CaO (f) N_2O (g) NaOH

Solutions To Final Exercises

2.1 (a) Cathode ray beams always bend toward positively charged objects, showing that they carry a negative charge. Hence, he concluded that electrons are negatively charged.

(b) Thomson observed that the properties of electron beams are independent of the cathode composition and, therefore, concluded that electrons must be basic to every kind of matter.

2.3 Positive rays are formed when the rapidly moving electrons collide with gaseous atoms, causing some of the gaseous atoms to break up into positive ions and electrons. The direction of the positive rays is opposite to that of the cathode rays. The positive rays move toward the cathode or negative electrode.

2.5 (a) The force increases. The equations describing the magnitude of the force experienced by a charge q moving in a magnetic field of strength H are

$$\text{magnetic force} = H \times q \times v$$

$$\text{magnetic force} = \frac{m \times v^2}{r}$$

where v is the velocity or speed of the charged particle. From either equation we see that the magnetic force increases with increasing speed.

(b) The force remains the same. The force exerted on a charge q by an electric field of strength E is given by the equation electric force = E × q.

From this equation we see that the electric force is independent of the velocity of the charged particle.

2.7 A narrow beam of alpha (α) particles is aimed at a thin metal foil (gold, platinum, silver, etc.). A zinc sulfide fluorescent screen is used to detect the alpha particles after they are scattered by the foil.

The observers didn't expect to find any alpha particles deflected through large angles, since they had in mind Thomson's pudding model for the atom. When they performed the experiment, they were surprised to find that some alpha particles were indeed deflected through rather large angles and that a few were even going backward toward the source.

(a) The deflection of alpha particles through large angles was interpreted as a head-on or near head-on collision of the α particle with a small, massive, positively charged core or nucleus.

(b) The fact that most of the α particles were not appreciably deflected was interpreted to mean that most of the atom was empty space, since α particles not encountering a nucleus would continue in nearly straight paths.

2.9

$$m = \frac{3.2044 \times 10^{-19}\ \text{C}}{4.8224 \times 10^{4}\ \text{C/g}} = 6.6448 \times 10^{-24}\ \text{g}$$

2.11 (a) The mass number is the total number of protons and neutrons in the nucleus of an atom, while the atomic number is just the number of protons in the nucleus of an atom.

(b) The atomic number is equal to the number of protons in the nucleus and the number of electrons in the neutral atom.

2.13 Isotopes are atoms of the same element that have different masses. They are atoms with the same atomic number but different mass numbers.

The isotopes of hydrogen differ from each other in the number of neutrons in their nuclei. Hydrogen-1 has no neutrons in its nucleus; hydrogen-2 (deuterium) has one neutron in its nucleus; and hydrogen-3 (tritium) has two neutrons in its nucleus.

2.15 There are eighteen (18) groups and seven (7) periods in the periodic table.

2.17 (a) iron (Fe); mercury (Hg); sodium (Na); nitrogen (N); phosphorus (P); calcium (Ca); sulfur (S); copper (Cu).

(b) Pb (lead); Ag (silver); K (potassium); Au (gold); Ne (neon); Zn (zinc); Ni (nickel).

2.19 (a) Hydrogen, helium, nitrogen, oxygen, fluorine, neon, chlorine, argon, krypton, xenon, and radon. There are eleven elements that are gases at 25°C.

(b) Bromine and mercury.

2.21 Atomic number Z = number of protons = number of electrons in the neutral atom. Mass number A = Z + number of neutrons.

Symbol	Z	A	Protons	Neutrons	Electrons
Zn	30	64	30	34	30
I	53	127	53	74	53
Eu	63	153	63	90	63

2.23 The symbol for a specific atom or ion is

${}^{A}_{Z}X^{\text{charge}}$, where A = mass number and Z = atomic number

(a) $^{208}_{82}Pb^{2+}$

Since Z = 82 there are 82 protons. There are A – Z = 208 – 82 = 126 neutrons. Since there is a charge of 2+ on the ion there must be 2 fewer electrons than protons; 82 – 2 = 80 electrons.

Answer: 82 protons, 126 neutrons, and 80 electrons.

(b) $^{14}_{7}N^{3-}$

Since Z = 7 there are 7 protons. There are 14 – 7 = 7 neutrons. Since there is a charge of 3– on the ion there must be 3 more electrons than protons; 7 + 3 = 10 electrons.

Answer: 7 protons, 7 neutrons, and 10 electrons.

2.25 Z = 13. A = Z + number of neutrons = 13 + 14 = 27. Charge = number of protons – number of electrons = 13 – 10 = 3 or 3+. (The charge is positive if the number of protons > number of electrons; negative if the number of protons < number of electrons.) The element corresponding to Z = 13 is aluminum Al. The complete symbol for the particle is therefore

$^{27}_{13}Al^{3+}$

2.27 Z = 17 for chlorine (Cl). Therefore chlorine-37 has 17 protons, A – Z = 37 – 17 = 20 neutrons, and 17 electrons.

Z = 18 for argon (Ar). Therefore argon-37 has 18 protons, A – Z = 37 – 18 = 19 neutrons, and 18 electrons.

Z = 16 for sulfur (S). Therefore sulfur-37 has 16 protons, A – Z = 37 – 16 = 21 neutrons, and 16 electrons.

2.29 (a) 12 u; exact (b) 1 u (c) 1 u (d) 4 u

2.31 Chemists use a weighted average mass because the isotopes of an element have different abundances. A number of elements have one or two isotopes that are in much greater abundance than any of the others. A simple average in which each isotope is assigned an equal abundance value would give an inaccurate and useless value for the chemical atomic weight.

2.33 A constant magnetic field (H) causes accelerated charged particles to move in circular paths. The radius (r) of these paths depends on the accelerating voltage (V) and mass/charge ratio of the ion (see Equation 2.6). By adjusting V, ions with different m/q ratios focus to the same, preset radius where a detector is positioned. By measuring H, V, r, and the ion current, the mass and fractional abundance of the ion can be calculated (see Equation 2.6).

2.35 From the table on the inside front cover we see that beryllium has four protons in its nucleus. Each proton and neutron contributes approximately 1.0 u to the mass of the isotope. The mass contributed by the electrons is quite small and can be ignored in this case. Therefore, the protons contribute 4.0 u to the mass and the rest is contributed by the neutrons.

Hence, the 10.0135-u isotope has 10.0 u – 4.0 u = 6.0 u = 6 neutrons, and the 9.01218-u isotope has 9.0 u – 4.0 u = 5.0 u = 5 neutrons in the nucleus.

2.37

$$\text{total mass of coins} = 250 \text{ dimes} \times \frac{2.27 \text{ g}}{1 \text{ dime}} + 45 \text{ quarters} \times \frac{5.67 \text{ g}}{1 \text{ quarter}} = 823 \text{ g}$$

$$\text{average mass of a coin} = \frac{823 \text{ g}}{295 \text{ coins}} = 2.79 \text{ g/coin}$$

This average is a weighted average. A simple average confers equal abundance for all types of items. A weighted average takes into account the different abundances and multiplies each weight by the abundance.

A chemical atomic weight is a weighted average. The contribution to the average mass from each isotope is in proportion to its abundance.

Comment: If abundance values are accurate, a weighted average gives the same result as an average where the contribution from each of the n individual objects is summed up and the total is divided by the total number of objects n. In chemistry the number of objects is much too large for such an averaging process, and therefore a weighted average must be taken.

2.39 The chemical atomic weight is found by multiplying each isotopic mass by its fractional abundance and adding the products. The fractional abundance equals the percentage abundance/100%.

$$0.7870 \times 23.98504 \text{ u} = 18.88 \text{ u}$$
$$0.1013 \times 24.98584 \text{ u} = 2.531 \text{ u}$$
$$0.1117 \times 25.98259 \text{ u} = \underline{2.902 \text{ u}}$$
$$\text{Chemical atomic weight Mg} = 24.31 \text{ u}$$

2.41 The fractional abundances must add up to 1, and the weighted average must equal the chemical atomic weight. Let x be the fractional abundance of chlorine-35 and $(1 - x)$ the fractional abundance of chlorine-37. Then

$$34.968852x + 36.965903(1 - x) = 35.4527$$

$$1.997051x = 1.5132$$

$$x = 0.75772 \text{ and } (1 - x) = 1 - 0.75772 = 0.24228$$

Answer: percent abundance chlorine-35 = 75.772%
percent abundance chlorine-37 = 24.228%

2.43 The sum of the mass of each isotope times its fractional abundance equals 63.546 u.

$$(0.6917)(62.939598 \text{ u}) + (0.3083)(\text{mass copper-65}) = 63.546 \text{ u}$$

$$\text{mass copper-65} = \frac{20.011 \text{ u}}{0.3083} = 64.91 \text{ u}$$

2.45 (a) Even very small amounts of matter contain an enormous number of atoms. The use of a small counting unit like the dozen would force us to use large and unwieldy numbers in our calculations, making them difficult and impractical.
(b) The mole is such a large number, 6.022×10^{23} particles or objects, that counting a small number of objects in moles would require the use of very small, unwieldy numbers and factors. Using these very small numbers would make counting difficult and unmanageable.

2.47 Use the table on the inside front cover.
(a) Silver (Ag); 107.87 u/atom; 107.87 g/mol.
(b) Mercury (Hg); 200.59 u/atom; 200.59 g/mol.
(c) Radium (Ra); 226.0 u/atom; 226.0 g/mol.

2.49 (a) The molar mass of aluminum is 26.98 g/mol.

$$2.55 \text{ mol Al} \times \frac{26.98 \text{ g}}{1 \text{ mol Al}} = 68.8 \text{ g}$$

(b) One aluminum atom has a mass of 26.98 u. The mass in grams is

$$26.98 \text{ u} \times \frac{1 \text{ g}}{6.022 \times 10^{23} \text{ u}} = 4.480 \times 10^{-23} \text{ g}$$

2.51 (a) One mole of iron contains 6.022×10^{23} atoms. The number of atoms in 0.100 mol is

$$0.100 \text{ mol} \times \frac{6.022 \times 10^{23} \text{ atoms}}{1 \text{ mol}} = 6.02 \times 10^{22} \text{ atoms}$$

(b) The molar mass of gold is 197.0 g/mol. The number of atoms in 1.00 kg is

$$1.00 \text{ kg} \times \frac{10^3 \text{ g}}{1 \text{ kg}} \times \frac{1 \text{ mol}}{197.0 \text{ g}} \times \frac{6.022 \times 10^{23} \text{ atoms}}{1 \text{ mol}} = 3.06 \times 10^{24} \text{ atoms}$$

(c) The molar mass of mercury is 200.6 g/mol. The number of atoms in 0.250 L is

$$0.250 \text{ L} \times \frac{10^3 \text{ mL}}{1 \text{ L}} \times \frac{13.55 \text{ g}}{1 \text{ mL}} \times \frac{1 \text{ mol}}{200.6 \text{ g}} \times \frac{6.022 \times 10^{23} \text{ atoms}}{1 \text{ mol}} = 1.02 \times 10^{25} \text{ atoms}$$

2.53 (a) Carbon-12 has Z = 6, A = 12. Therefore, it has 6 protons, 6 neutrons, and 6 electrons in each atom. The number of mol of protons in 2.35 mmol carbon-12 is

$$2.35 \text{ mmol carbon-12} \times \frac{1 \text{ mol}}{10^3 \text{ mmol}} \times \frac{6 \text{ mol protons}}{1 \text{ mol carbon-12}} = 0.0141 \text{ mol protons}$$

(b) and (c) Likewise, there are 0.0141 mol neutrons and 0.0141 mol electrons.

2.55 Hydrogen is a colorless, odorless, and tasteless gas under normal or everyday conditions. It is the lightest substance known. Hydrogen doesn't support combustion but is combustible, i.e., it burns in air or oxygen. Oxygen is a colorless, odorless, and tasteless gas under normal conditions. Oxygen supports combustion. Carbon exists in several forms under normal conditions, diamond and graphite being the forms most well known. Diamond is a colorless, odorless, and tasteless crystal. Diamond is brittle, but it is the hardest substance known in that it will scratch, but cannot be scratched by, any other substance. Graphite is a soft solid with a grayish black metallic luster and a slippery, greasy feel. Graphite is a conductor of electricity. Diamond and graphite are both insoluble in water. Sucrose is a white powdery crystalline solid. It is soluble in water. Sucrose is combustible and has a pleasant sweet taste.

2.57 An element contains only atoms of the same atomic number (Z), while a compound contains atoms that have different atomic numbers. In other words, a compound is composed of atoms from at least two elements.

2.59 A molecule is any electrically neutral distinct group consisting of a specific, unvarying, number of atoms bonded together.

No. Some substances, the noble gases, consist of separate independent atoms. Ionic solids, like sodium chloride, consist of ion aggregates of indefinite size. Metallic elements, like iron and gold, consist of atom aggregates of indefinite size. Other substances, like carbon and quartz (silicon dioxide), consist of atoms linked together in three-dimensional arrangements of indefinite extent.

2.61 Allotropes are two or more forms of an element that exist in the same physical state (solid, liquid, or gas).
(a) diamond and graphite (b) oxygen (O_2) and ozone (O_3) (c) white phosphorus (P_4) and red phosphorus

2.63 The noble gases in order of increasing atomic number are: He, Ne, Ar, Kr, Xe, and Rn.

The noble gases are found in Group 8A (Group 18).

The noble gases exist as separate atoms. Therefore the noble gases do not consist of molecules, which must contain at least two atoms bonded together.

2.65 (a) $CaCl_2$. There are 1 calcium atom and 2 chlorine atoms.
(b) $Mg(OH)_2$; count OH twice. There is 1 magnesium atom, 2 oxygen atoms, and 2 hydrogen atoms.
(c) $Al(NH_4)(SO_4)_2$; count SO_4 twice. There is 1 aluminum atom, 1 nitrogen atom, 4 hydrogen atoms, 2 sulfur atoms, and 8 oxygen atoms.
(d) $CH_3(CH_2)_2CH_3$; count CH_2 twice. There are 4 carbon atoms and 10 hydrogen atoms.

2.67 The formulas for molecular elements have a subscript showing the number of atoms in the molecule. The formulas of nonmolecular elements are simply their atomic symbols.
(a) Yes. Hydrogen consists of diatomic molecules; H_2.
(b) Yes. Oxygen consists of diatomic molecules; O_2.
(c) Yes. White phosphorus consists of four-atom molecules; P_4.
(d) No. Argon consists of separate atoms; Ar.
(e) No. Copper is a metallic element; Cu.
(f) Yes. Bromine consists of diatomic molecules; Br_2.

2.69 There are eleven elements that are gases at 25°C. H_2, He, N_2, O_2, F_2, Ne, Cl_2, Ar, Kr, Xe, and Rn.

2.71 Molecular compounds have molecular formulas with subscripts showing the numbers of each kind of atom in the molecule. Nonmolecular compounds have formulas with subscripts that reflect the fixed ratios of the elements in the compound.
(a) C_2H_6O or C_2H_5OH (b) $C_2H_4O_2$ or $HC_2H_3O_2$ or CH_3COOH (c) H_2O (d) H_2O_2 (e) CO (f) NH_3.

2.73 (a) True. Each H_2O formula unit contains two hydrogen atoms, therefore one mole of H_2O formula units contains two moles of hydrogen atoms.
(b) False. One mole of H_2O contains only one mole of oxygen atoms, while one mole of oxygen molecules (O_2) contains two moles of oxygen atoms.
(c) False. One mole of NH_3 contains only three moles of hydrogen atoms, while three moles of hydrogen molecules (H_2) contains six moles of hydrogen atoms.
(d) False. One mole of NH_3 contains only one mole of nitrogen atoms, while one mole of nitrogen molecules (N_2) contains two moles of nitrogen atoms.

2.75 (a) $1.00 \text{ mol } P_4 \times \dfrac{4 \text{ mol P atoms}}{1 \text{ mol } P_4} \times \dfrac{6.022 \times 10^{23} \text{ atoms}}{1 \text{ mol atoms}} = 2.41 \times 10^{24} \text{ P atoms}$

(b) The number of moles of atoms of an element in a mole of substance equals the number of atoms of the element in the formula of the substance.

P_4 has 4 atoms of phosphorus in its formula. Therefore, 1.00 mol of P_4 contains 4.00 mol of P.

2.77 (a) One mole of ammonia (NH_3) contains 3 mol of hydrogen atoms. Therefore, 2.50 mol of ammonia will contain $2.50 \times 3 = 7.50$ mol of hydrogen atoms.

(b) $2.50 \text{ mol } NH_3 \times \dfrac{3 \text{ mol H atoms}}{1 \text{ mol } NH_3} \times \dfrac{6.022 \times 10^{23} \text{ atoms}}{1 \text{ mol atoms}} = 4.52 \times 10^{24} \text{ H atoms}$

2.79 The mole ratio of the elements in a compound is the same as the atom ratio in the formula of the compound; hence, the formula of the compound will contain three iron atoms and four oxygen atoms. The formula is Fe_3O_4.

2.81 The molar mass of a compound is the sum of the molar masses of the atoms in its formula.

(a) One mole of $C_6H_{12}O_6$ contains 6 mol of carbon atoms, 12 mol of hydrogen atoms, and 6 mol of oxygen atoms. The molar masses of carbon, hydrogen, and oxygen are 12.01 g/mol, 1.008 g/mol, and 16.00 g/mol, respectively. The total mass is

Carbon:	6 mol × 12.01 g/mol	=	72.06 g
Hydrogen:	12 mol × 1.008 g/mol	=	12.096 g
Oxygen:	6 mol × 16.00 g/mol	=	96.00 g
Total molar mass		=	180.156 g = 180.16 g

(b) One mole of sulfur trioxide contains 1 mol of sulfur atoms and 3 mol of oxygen atoms. The molar masses of sulfur and oxygen are 32.07 g/mol and 16.00 g/mol, respectively. The total mass is

Sulfur:	1 mol × 32.07 g/mol	=	32.07 g
Oxygen:	3 mol × 16.00 g/mol	=	48.00 g
Total molar mass		=	80.07 g

2.83 (a) One mole of silicon carbide contains 1 mol of silicon atoms and 1 mol of carbon atoms. The molar masses of Si and C are 28.09 g/mol and 12.01 g/mol, respectively. The total mass is

Si:	1 mol × 28.09 g/mol	=	28.09 g
C:	1 mol × 12.01 g/mol	=	12.01 g
Total molar mass		=	40.10 g

(b) One mole of $NaPO_3$ contains 1 mol of sodium atoms, 1 mol of phosphorus atoms, and 3 mol of oxygen atoms. The molar masses of Na, P, and O are 22.99 g/mol, 30.97 g/mol, and 16.00 g/mol, respectively. The total mass is

Na:	1 mol × 22.99 g/mol	=	22.99 g
P:	1 mol × 30.97 g/mol	=	30.97 g
O:	3 mol × 16.00 g/mol	=	48.00 g
Total molar mass		=	101.96 g

(c) One mole of Al_2O_3 contains 2 mol of aluminum atoms and 3 mol of oxygen atoms. The molar masses of Al and O are 26.98 g/mol and 16.00 g/mol, respectively. The total mass is

Al:	2 mol × 26.98 g/mol	=	53.96 g
O:	3 mol × 16.00 g/mol	=	48.00 g
Total molar mass		=	101.96 g

(d) One mole of isopropyl alcohol contains 3 mol of C atoms, 8 mol of H atoms and 1 mol of O atoms. The molar masses of C, H, and O are 12.01 g/mol, 1.008 g/mol, and 16.00 g/mol, respectively. The total mass is

Carbon:	3 mol × 12.01 g/mol	=	36.03 g
Hydrogen:	8 mol × 1.008 g/mol	=	8.064 g
Oxygen:	1 mol × 16.00 g/mol	=	16.00 g
Total molar mass		=	60.094 g = 60.09 g

2.85 (a) The molar mass of SiC was found to be 40.10 g/mol in Exercise 2.83 (a).

$$1.50 \text{ mol SiC} \times \frac{40.10 \text{ g}}{1 \text{ mol SiC}} = 60.2 \text{ g}$$

(b) The molar mass of Al_2O_3 was found to be 101.96 g/mol in Exercise 2.83 (c).

$$6.45 \text{ mol } Al_2O_3 \times \frac{101.96 \text{ g}}{1 \text{ mol } Al_2O_3} = 658 \text{ g}$$

2.87 The molar masses of H and O are 1.008 g/mol and 16.00 g/mol, respectively. The molar mass of H_2O is 2×1.008 g/mol + 16.00 g/mol = 18.02 g/mol.

(a) $$20.0 \text{ g } H_2O \times \frac{1 \text{ mol } H_2O}{18.02 \text{ g } H_2O} = 1.11 \text{ mol } H_2O$$

(b) $$20.0 \text{ g } H_2O \times \frac{1 \text{ mol } H_2O}{18.02 \text{ g } H_2O} \times \frac{6.022 \times 10^{23} \text{ molecules } H_2O}{1 \text{ mol } H_2O} = 6.68 \times 10^{23} \text{ molecules } H_2O$$

2.89 (a) One mole of NaCl contains 1 mol of Na^+ ions and 1 mol of Cl^- ions, so 2.50 mol will contain $2.50 \times 1 =$ 2.50 mol Na^+ ions and $2.50 \times 1 = 2.50$ mol Cl^- ions.

(b) $$2.50 \text{ mol } Na^+ \text{ ions} \times \frac{6.022 \times 10^{23} \text{ } Na^+ \text{ ions}}{1 \text{ mol } Na^+ \text{ ions}} = 1.51 \times 10^{24} \text{ } Na^+ \text{ ions}$$

Likewise, there are 1.51×10^{24} Cl^- ions.

2.91 The molar masses of Na, O, H, and N are 22.99 g/mol, 16.00 g/mol, 1.008 g/mol, and 14.01 g/mol, respectively. The molar mass of NaOH is 22.99 g + 16.00 g + 1.008 g = 40.00 g. The molar mass of HNO_3 is 1.008 g + 14.01 g + (3×16.00 g) = 63.02 g. The calculation will proceed as follows: g HNO_3 $\rightarrow$ mol HNO_3 $\rightarrow$ mol NaOH $\rightarrow$ g NaOH.

$$126 \text{ g } HNO_3 \times \frac{1 \text{ mol } HNO_3}{63.02 \text{ g } HNO_3} \times \frac{1 \text{ mol NaOH}}{1 \text{ mol } HNO_3} \times \frac{40.00 \text{ g NaOH}}{1 \text{ mol NaOH}} = 80.0 \text{ g NaOH}$$

2.93 Ionic compounds when melted (liquid) conduct electricity, while covalent compounds in the liquid state do not. A simple experimental arrangement of an electric lamp connected in series with a battery, a switch, and a pair of platinum electrodes to dip into the liquid under investigation could be used to distinguish between ionic and covalent compounds. If the liquid tested is a conductor of electricity, the lamp filament will glow when the switch is closed.

(a) If the electrodes are lowered into a beaker of methyl chloride, the filament will not glow. Methyl chloride is a nonconductor and, therefore, a covalent compound.

(b) If the electrodes are lowered into melted (liquid) KF, the filament will light up. KF is a conductor and, therefore, an ionic compound.

Comment: Certain substances are covalent in their pure states and ionic in solution. For instance, sulfuric acid (H_2SO_4) in its pure liquid state doesn't conduct appreciable electricity, while an aqueous solution of sulfuric acid does. In its pure liquid state sulfuric acid is covalent, while in aqueous solution it is ionic. Most covalent compounds do not ionize (do not break up into ions or ionic parts) in solution. Ionization is discussed in subsequent chapters of the text.

2.95 Cesium (Cs) is in Group 1A. Group 1A elements form 1+ ions.
Strontium (Sr) is in Group 2A. Group 2A elements form 2+ ions.
Bromine (Br) is in Group 7A. Group 7A elements form 1– ions.

(a) To maintain electrical neutrality in the binary ionic compound formed from cesium and bromine, we need a 1-to-1 relationship between the cesium and bromine ions. The formula is therefore CsBr.

(b) To maintain electrical neutrality in the binary ionic compound formed from strontium and bromine, we need two bromine ions for every strontium ion. The formula is therefore $SrBr_2$.

2.97 Referring to Table 2.6 and making sure that we maintain electrical neutrality we get: (a) $KMnO_4$, (b) $(NH_4)_2CO_3$, (c) Hg_2Cl_2, (d) $Hg(NO_3)_2$, (e) NaSCN, and (f) $Mg(ClO_4)_2$.

2.99 (a) Pb^{2+} is the lead ion; (b) Sn^{2+} is the tin(II) ion (stannous ion is its informal name); (c) Fe^{3+} is the iron(III) ion (ferric ion is its informal name); (d) NH_4^+ is the ammonium ion; (e) Cu^{2+} is the copper(II) ion (cupric ion is its informal name); (f) Hg_2^{2+} is the mercury(I) ion (mercurous ion is its informal name).

2.101 The name of an ionic compound consists of the cation name followed by the anion name.
(a) potassium dichromate; (b) iron(II) sulfate; (c) potassium permanganate; (d) mercury(I) chloride; (e) potassium oxalate; (f) sodium chlorate.

2.103 (a) Hydrogen chloride, (b) hydrogen sulfide; (c) water (officially recognized common name), (d) dinitrogen pentoxide, (e) iodine monobromide or iodine bromide (iodine and bromine form several binary molecular compounds; the prefix "mono" is quite often omitted), and (f) disulfur difluoride (sulfur and fluorine form several binary molecular compounds).

2.105 See diagrams in textbook.

2.107 Both the Rutherford and Thomson atoms are distinct neutral entities of the same size. In both of these atoms, there exist a number of electrons and some form of offsetting positive charge such that the whole atom is neutral. The differences between them are in the way the mass, the positive charge, and the electrons are distributed.

In the Thomson atom, the positive charge is distributed uniformly throughout a sphere and the electrons are embedded in the positive charge like raisins in pudding. The mass of the atom is distributed uniformly throughout the sphere.

In the Rutherford atom, the positive charge and almost all the mass are concentrated in a dense central core called the nucleus, which is very small compared to the total size of an atom. The Rutherford atom is mostly empty space, while the Thomson atom has very little or none. In the Rutherford atom, the electrons reside outside the nucleus in the remainder (practically all) of the atom's volume; they are separated or away from the protons.

2.109 All atoms of a given element have the same number of protons in their nucleus, i.e., they have the same atomic number Z, and the same number of electrons surrounding the nucleus.

Atoms of a given element can differ in their masses, i.e., they can have different mass numbers (different A values). Some atoms may have more neutrons in their nucleus than others. The atoms with more neutrons have a greater atomic mass.

2.111 To solve this problem we have to calculate the factor or multiplier needed to make the nuclear diameter equal to the diameter of a U.S. penny.

The nuclear diameter is 2×10^{-15} m, the diameter of a U.S. penny is 1.9 cm = 0.019 m, and the factor is

2×10^{-15} m $\times x = 0.019$ m

$$x = \frac{0.019 \text{ m}}{2 \times 10^{-15} \text{m}} = 1 \times 10^{13}$$

The atomic diameter is 2×10^{-10} m. If it is expanded by the same factor, i.e., by the same proportion as the nuclear diameter, it would equal 2×10^{-10} m $\times 1 \times 10^{13} = 2 \times 10^{3}$ m or 2 km.

2.113 The percent of the total atomic volume occupied by the nucleus is

$$\frac{V_{\text{nucleus}}}{V_{\text{atom}}} \times 100\% = \frac{(4/3)\pi r^3_{\text{nucleus}}}{(4/3)\pi r^3_{\text{atom}}} \times 100\% = \frac{r^3_{\text{nucleus}}}{r^3_{\text{atom}}} \times 100\%$$

$$= \frac{\left(10^{-15}\text{ m}\right)^3}{\left(10^{-10}\text{ m}\right)^3} \times 100\% = \frac{10^{-45}\text{ m}^3}{10^{-30}\text{ m}^3} \times 100\% = 10^{-13}\%$$

2.115 The fractional abundances must add up to 1. Let x stand for the fractional abundance of the fifth isotope. Then

$0.4889 + 0.2781 + 0.0411 + 0.0062 + x = 1$

and $x = 0.1857$

The percent abundance of the fifth isotope is $0.1857 \times 100\% = 18.57\%$. The sum of the mass of each isotope times its fractional abundance equals 65.39 u. Let x stand for the mass of the fifth isotope.

$(0.4889)(63.9291 \text{ u}) + (0.2781)(65.9260 \text{ u}) + (0.0411)(66.9721 \text{ u}) + (0.0062)(69.9253 \text{ u}) + (0.1857)\,x$
$= 65.39$ u

$$x = \frac{12.61 \text{ u}}{0.1857} = 67.91 \text{ u}$$

The isotopic mass of the fifth isotope is 67.91 u. The closest whole number to the isotopic mass is 68. The name of the isotope should be zinc-68.

2.117 (a) $1.00 \text{ mol electrons} \times \dfrac{6.022 \times 10^{23} \text{electrons}}{1 \text{ mol electrons}} \times \dfrac{1.6022 \times 10^{-19}\text{C}}{1 \text{ electron}} = 9.65 \times 10^{4}\text{C}$

(b) $1.00 \text{ mol alpha particles} \times \dfrac{6.022 \times 10^{23} \text{alpha particles}}{1 \text{ mol alpha particles}} \times \dfrac{3.2044 \times 10^{-19}\text{C}}{1 \text{ alpha particle}} = 1.93 \times 10^{5}\text{C}$

2.119 (a) The red blood cells can be removed from the blood plasma by centrifugation. Whole blood is put in a test tube and placed in the centrifuge. After centrifugation the plasma rests on top of the red blood cells and can be decanted.

(b) Water can be removed from air by use of a dehumidifier. The water vapor condenses on the cold dehumidifier coils and is collected as liquid water in a suitable, removable container.

(c) Oxygen and nitrogen can be separated by boiling since they boil at different temperatures. First the oxygen and nitrogen mixture is liquefied. Then the temperature is raised until the lower boiling component (nitrogen) starts boiling off, leaving the higher boiling component (oxygen) behind in the liquid state.

2.121 The molar masses of the elements needed for this problem are C, 12.01 g; H, 1.008 g; and O, 16.00 g. The molar mass of glucose is $6 \times 12.01 \text{ g} + 12 \times 1.008 \text{ g} + 6 \times 16.00 \text{ g} = 180.16$ g.

The number of moles of glucose in 1.00 L of solution is

$$1.00 \text{ L solution} \times \frac{5.51 \text{ g}}{100 \text{ mL solution}} \times \frac{10^3 \text{ mL}}{1 \text{ L}} \times \frac{1 \text{ mol glucose}}{180.16 \text{ g}} = 0.306 \text{ mol glucose}$$

2.123 To solve this problem you have to calculate the molar mass of benzene. The molar mass of benzene (C_6H_6) is $6 \times 12.01 \text{ g} + 6 \times 1.008 \text{ g} = 78.12 \text{ g}$. The volume of 1.00 mol of benzene is

$$1.00 \text{ mol benzene} \times \frac{78.12 \text{ g}}{1 \text{ mol benzene}} \times \frac{1 \text{ mL}}{0.879 \text{ g}} = 88.9 \text{ mL}$$

CHAPTER 3

CHEMICAL REACTIONS, EQUATIONS, AND STOICHIOMETRY

Solutions To Practice Exercises

PE 3.1 (a) Chemical reaction; souring is caused by microorganisms that break down the proteins and sugars in milk, forming new substances.
(b) Physical change; no new substances are formed.
(c) Chemical change; the interaction of the moisture, carbon dioxide, and sulphur oxides present in air with the copper in bronze causes the bronze statue to turn green; new substances are formed.

PE 3.2 The reactants are the original substances in a chemical reaction: methane and oxygen are the reactants. The products are the new substances formed: water and carbon dioxide are the products.

PE 3.3 (a) See Table 2.4 and Table 1.7.
(b) Any chemical reaction in which water is a reactant is a chemical property of water. Several have been mentioned in the text.

PE 3.4 (a) Any 100-g sample of water contains 88.8 g of oxygen and 11.2 g of hydrogen. The fraction of oxygen in water is 88.8 g/100 g = 0.888. The percentages of oxygen and hydrogen in water are, therefore, percent oxygen = 0.888 × 100% = 88.8% ; percent hydrogen = 100% – 88.8% = 11.2%.
(b) Any 100-g sample of NaCl contains 39.3 g of Na and 60.7 g of Cl. The fractions of Na and Cl in NaCl are 0.393 and 0.607, respectively. The percentages of Na and Cl in NaCl are, therefore, 0.393 × 100% = 39.3% Na and 0.607 × 100% = 60.7% Cl.

PE 3.5 The molar mass of chromite ($FeCr_2O_4$) is 223.84 g/mol. The percentages of Fe, Cr, and O are

$$\text{percent Fe} = \frac{55.85 \text{ g}}{223.84 \text{ g}} \times 100\% = 24.95\%$$

$$\text{percent Cr} = \frac{103.99 \text{ g}}{223.84 \text{ g}} \times 100\% = 46.46\%$$

$$\text{percent O} = \frac{64.00 \text{ g}}{223.84 \text{ g}} \times 100\% = 28.59\%$$

PE 3.6 The atomic weight of sulfur is 32.07 u. The molecular weight divided by the atomic weight is 256.5 u/32.07 u = 8.00. Therefore, there must be eight atoms of sulfur in one molecule of sulfur. The formula is S_8.

PE 3.7 The atomic weights of S and H are 32.07 u and 1.008 u, respectively. Since the molecular weight of hydrogen sulfide is 34.08 u, the hydrogen sulfide molecule can only contain one sulfur atom. The remaining mass, 34.08 – 32.07 = 2.01 u, is due to hydrogen. Since each hydrogen is 1.008 u, there must be two atoms of hydrogen in the molecule. The formula is H_2S.

PE 3.8 (a) The molecular formula $C_6H_{12}O_6$ has an atom ratio of 6:12:6. Dividing by 6 gives the simplest atom ratio of 1:2:1. The empirical formula is therefore CH_2O.

(b) The molecular formula $Hg_2(NO_3)_2$ has an atom ratio of 2:2:6. Dividing by 2 gives the simplest atom ratio of 1:1:3. The empirical formula is therefore $HgNO_3$.

PE 3.9 A 100-g sample of chloroform would contain 89.10 g of chlorine, 0.84 g of hydrogen, and 100 – 89.10 – 0.84 = 10.06 g of carbon. The number of moles of each element in 100 g is

$$89.10 \text{ g Cl} \times \frac{1 \text{ mol Cl atoms}}{35.45 \text{ g Cl}} = 2.51 \text{ mol Cl atoms}$$

$$0.84 \text{ g H} \times \frac{1 \text{ mol H atoms}}{1.008 \text{ g H}} = 0.83 \text{ mol H atoms}$$

$$10.06 \text{ g C} \times \frac{1 \text{ mol C atoms}}{12.01 \text{ g C}} = 0.84 \text{ mol C atoms}$$

By inspection we see that there are 3 mol of Cl atoms for every 1 mol of H atoms and 1 mol of C atoms; the ratio is 3:1:1. The atom ratio in the formula is the same as the mole ratio. The formula is $CHCl_3$.

PE 3.10 The sample contained 8.98 g of Al and 25.0 g – 8.98 g = 16.0 g of S. The number of moles of each element in the sample is

$$8.98 \text{ g Al} \times \frac{1 \text{ mol Al atoms}}{26.98 \text{ g Al}} = 0.333 \text{ mol Al atoms}$$

$$16.0 \text{ g S} \times \frac{1 \text{ mol S atoms}}{32.07 \text{ g S}} = 0.499 \text{ mol S atoms}$$

To find the simplest whole-number mole ratio we have to divide through by the number of moles of Al.

$$\text{Al:} \quad \frac{0.333}{0.333} = 1.00 \qquad 1.00 \times 2 = 2$$

$$\text{S:} \quad \frac{0.499}{0.333} = 1.50 \qquad 1.50 \times 2 = 3$$

We then multiply by 2 to obtain a whole number ratio. Therefore, there are 2 mol of Al for every 3 mol of S, and the formula is Al_2S_3.

PE 3.11 One mole of CH_2 units has a mass of 12.01 + (2 × 1.008) = 14.03 g. One mole of molecules has a molar mass of 56.10 g. If we divide the molar mass by 14.03 g we get 56.10 g/14.03 g = 4.0, so 1 mol of molecules contains 4 mol of CH_2 units. The molecular formula is $(CH_2)_4$ or C_4H_8.

PE 3.12 The numbers of moles of nitrogen and selenium in the molar mass are

$$\text{N:} \quad 371.9 \text{ g compound} \times \frac{15.07 \text{ g N}}{100 \text{ g compound}} \times \frac{1 \text{ mol N atoms}}{14.01 \text{ g N}} = 4.00 \text{ mol N atoms}$$

$$\text{Se:} \quad 371.9 \text{ g compound} \times \frac{84.93 \text{ g Se}}{100 \text{ g compound}} \times \frac{1 \text{ mol Se atoms}}{78.96 \text{ g Se}} = 4.00 \text{ mol Se atoms}$$

The atom ratio is the same as the mole ratio, and the molecular formula is N_4Se_4.

PE 3.13 (a) To calculate the empirical formula, we have to calculate the number of moles of each element present in the Vitamin A sample. We know from Example 3.12 that there are 24.02 mg of C, 3.024 mg of H, and 1.60 mg of O in 28.64 mg of Vitamin A. The number of moles of each element is

$$24.02 \text{ mg C} \times \frac{1 \text{ g}}{10^3 \text{ mg}} \times \frac{1 \text{ mol C atoms}}{12.01 \text{ g C}} = 0.0020 \text{ mol C atoms}$$

$$3.024 \text{ mg H} \times \frac{1 \text{ g}}{10^3 \text{ mg}} \times \frac{1 \text{ mol H atoms}}{1.008 \text{ g H}} = 0.0030 \text{ mol H atoms}$$

$$1.60 \text{ mg O} \times \frac{1 \text{ g}}{10^3 \text{ mg}} \times \frac{1 \text{ mol O atoms}}{16.00 \text{ g O}} = 0.00010 \text{ mol O atoms}$$

To find the simplest whole-number ratio we will divide through by the number of moles of O. This gives us 20 mol of C for every 30 mol of H and 1 mol of O; the ratio is 20:30:1. The empirical formula is $C_{20}H_{30}O$.

(b) To find the molecular formula, we have to determine how many empirical unit masses there are in the molar mass of the compound. One mole of $C_{20}H_{30}O$ units has a mass of $(20 \times 12.011) + (30 \times 1.00794) + 15.9994 = 286.5$ g. The molar mass of Vitamin A is 286.5 g, so the molecular formula is the same as the empirical formula. The molecular formula is $C_{20}H_{30}O$.

PE 3.14 (a) To get the percentage composition, we need the mass of each element in the original sample. After combustion the carbon is all in the carbon dioxide and the hydrogen is all in the water. The masses of C and H in the sample are

$$\text{C:} \quad 81.36 \text{ mg CO}_2 \times \frac{12.01 \text{ mg C}}{44.01 \text{ mg CO}_2} = 22.20 \text{ mg C}$$

$$\text{H:} \quad 24.98 \text{ mg H}_2\text{O} \times \frac{2.016 \text{ mg H}}{18.02 \text{ mg H}_2\text{O}} = 2.795 \text{ mg H}$$

The percentages are C: $\frac{22.20 \text{ mg}}{25.00 \text{ mg}} \times 100\% = 88.80\%$; H: $\frac{2.795 \text{ mg}}{25.00 \text{ mg}} \times 100\% = 11.18\%$

(b) When both the molar mass and the percentage composition are known, the molecular formula can be determined directly by calculating the number of moles of each element present in the molar mass. The atom ratio in the formula is the same as the mole ratio. The number of moles of C and H in 1 mol of compound is

$$54.09 \text{ g compound} \times \frac{88.80 \text{ g C}}{100 \text{ g compound}} \times \frac{1 \text{ mol C atoms}}{12.01 \text{ g C}} = 4.00 \text{ mol C atoms}$$

$$54.09 \text{ g compound} \times \frac{11.18 \text{ g H}}{100 \text{ g compound}} \times \frac{1 \text{ mol H atoms}}{1.008 \text{ g H}} = 6.00 \text{ mol H atoms}$$

In 1 mol of compound there are 4 mol of C atoms and 6 mol of H atoms. The formula is C_4H_6.

PE 3.15 Step 1. ethylene + oxygen → carbon dioxide + water
(reactants) (products)

Step 2. $C_2H_4(g) + O_2(g) \rightarrow CO_2(g) + H_2O(g)$

Step 3. Balance the equation. $C_2H_4(g) + 3O_2(g) \rightarrow 2CO_2(g) + 2H_2O(g)$

PE 3.16 The unbalanced equation is $SO_2(g) + O_2(g) \rightarrow SO_3(g)$. There is one S atom in SO_2 on the left and one S atom in SO_3 on the right, so the S atoms are balanced. Now balance the oxygen. There are four oxygen atoms on the left and only three on the right. We may not change the coefficient of SO_3, because the S atoms are already balanced. Instead, we change the coefficient in front of O_2 to give one less oxygen on the left side.

$SO_2(g) + \frac{1}{2}O_2(g) \rightarrow SO_3(g)$

Whole-number coefficients in the same ratio are obtained by multiplying each coefficient by two: $2SO_2(g) + O_2(g) \rightarrow 2SO_3(g)$.

PE 3.17 The ions involved in this reaction are: potassium, K^+; chromate, CrO_4^{2-}; silver, Ag^+; and nitrate, NO_3^-.

potassium chromate + silver nitrate → silver chromate + potassium nitrate
$K_2CrO_4(aq) + AgNO_3(aq) \rightarrow Ag_2CrO_4(s) + KNO_3(aq)$

The CrO_4^{2-} and NO_3^- ions remain intact in this reaction and can be treated as units for the purposes of balancing the equation. Two K^+ ions on the left require two KNO_3 formula units on the right, and two Ag^+ ions on the right require two $AgNO_3$ formula units on the left. The balanced equation is $K_2CrO_4(aq) + 2AgNO_3(aq) \rightarrow Ag_2CrO_4(s) + 2KNO_3(aq)$.

PE 3.18 (a) Nitrogen and hydrogen exist as diatomic molecular gases under normal conditions.

$N_2(g) + 3H_2(g) \rightarrow 2NH_3(g)$

(b) $BaO_2(s) \rightarrow BaO(s) + O_2(g)$ (unbalanced)

$BaO_2(s) \rightarrow BaO(s) + \frac{1}{2}O_2(g)$ (balanced)

$2BaO_2(s) \rightarrow 2BaO(s) + O_2(g)$ (balanced–integral coefficients)

PE 3.19 $F_2(g) + H_2O(l) \rightarrow O_2(g) + HF(aq)$ (unbalanced)

$F_2(g) + H_2O(l) \rightarrow \frac{1}{2}O_2(g) + 2HF(aq)$ (balanced)

$2F_2(g) + 2H_2O(l) \rightarrow O_2(g) + 4HF(aq)$ (balanced–integral coefficients)

PE 3.20 The unbalanced equation is $C_4H_4S(l) + O_2(g) \rightarrow CO_2(g) + H_2O(g) + SO_2(g)$.

Balancing gives $C_4H_4S(l) + 6O_2(g) \rightarrow 4CO_2(g) + 2H_2O(g) + SO_2(g)$.

PE 3.21 $2Al(s) + Fe_2O_3(s) \rightarrow 2Fe(l) + Al_2O_3(l)$

Aluminum (Al) combines with oxygen, hence it is oxidized. Iron(III) oxide (Fe_2O_3) loses oxygen, hence it is reduced.

PE 3.22 $C_3H_8(g) + 5O_2(g) \rightarrow 3CO_2(g) + 4H_2O(g)$

From the balanced equation we see that 4 mol of H_2O are produced for every mole of C_3H_8 consumed. The conversion factor 4 mol H_2O/1 mol C_3H_8 gives

$$2.00 \text{ mol } C_3H_8 \times \frac{4 \text{ mol } H_2O}{1 \text{ mol } C_3H_8} = 8.00 \text{ mol } H_2O$$

PE 3.23 $SnO_2(s) + 2C(s) \rightarrow Sn(s) + 2CO(g)$
50.0 kg kg = ?

The steps in the calculation are: g SnO_2 → mol SnO_2 → mol Sn → kg Sn.

$$50.0 \times 10^3 \text{ g } SnO_2 \times \frac{1 \text{ mol } SnO_2}{150.7 \text{ g } SnO_2} \times \frac{1 \text{ mol Sn}}{1 \text{ mol } SnO_2} \times \frac{118.7 \text{ g Sn}}{1 \text{ mol Sn}} = 39.4 \times 10^3 \text{ g Sn} = 39.4 \text{ kg Sn}$$

PE 3.24 $Mg(s) + 2HCl(aq) \rightarrow MgCl_2(aq) + H_2(g)$
g = ? 2.25 mol

The steps in the calculation are: mol HCl → mol Mg → g Mg.

$$2.25 \text{ mol HCl} \times \frac{1 \text{ mol Mg}}{2 \text{ mol HCl}} \times \frac{24.30 \text{ g Mg}}{1 \text{ mol Mg}} = 27.3 \text{ g Mg}$$

PE 3.25 $2KClO_3(s) \rightarrow 2KCl(s) + 3O_2(g)$
g = ? 5.00 g

The steps in the calculation are: g O_2 → mol O_2 → mol $KClO_3$ → g $KClO_3$.

$$5.00 \text{ g } O_2 \times \frac{1 \text{ mol } O_2}{32.00 \text{ g } O_2} \times \frac{2 \text{ mol } KClO_3}{3 \text{ mol } O_2} \times \frac{122.5 \text{ g } KClO_3}{1 \text{ mol } KClO_3} = 12.8 \text{ g } KClO_3$$

PE 3.26 The equation for the reaction is $P_4(s) + 5O_2(g) \rightarrow P_4O_{10}(s)$

(a) The equation states that 5 mol of oxygen reacts with 1 mol of phosphorus. Therefore, 1.00 mol of oxygen will react with 0.20 mol of phosphorus. There is 1.00 mol of phosphorus in the reaction mixture; hence, excess phosphorus is present, and oxygen is limiting.

(b) 1.00 mol O_2 consumes 0.20 mol P_4. Originally there was 1.00 mol P_4. After the reaction there will be 1.00 mol – 0.20 mol = 0.80 mol P_4 remaining.

PE 3.27 (a) The equation for the reaction is

$3CaCl_2(aq) + 2Na_3PO_4(aq) \rightarrow Ca_3(PO_4)_2(s) + 6NaCl(aq)$
50.0 g 100.0 g

To solve this problem we first have to convert the known masses to moles. The number of moles of each is

$$50.0 \text{ g } CaCl_2 \times \frac{1 \text{ mol } CaCl_2}{111.0 \text{ g } CaCl_2} = 0.450 \text{ mol } CaCl_2$$

$$100.0 \text{ g } Na_3PO_4 \times \frac{1 \text{ mol } Na_3PO_4}{163.94 \text{ g } Na_3PO_4} = 0.610 \text{ mol } Na_3PO_4$$

From the equation we see that 3 mol of $CaCl_2$ are needed for every 2 mol of Na_3PO_4. Since the mixture contains fewer moles of $CaCl_2$ than Na_3PO_4, $CaCl_2$ must be the limiting reactant and Na_3PO_4 the excess reactant.

(b) The number of grams of Na_3PO_4 consumed is

$$0.450 \text{ mol } CaCl_2 \times \frac{2 \text{ mol } Na_3PO_4}{3 \text{ mol } CaCl_2} \times \frac{163.94 \text{ g } Na_3PO_4}{1 \text{ mol } Na_3PO_4} = 49.2 \text{ g } Na_3PO_4$$

Since there was originally 100.0 g of Na_3PO_4, there are 100.0 g – 49.2 g = 50.8 g Na_3PO_4 remaining.

PE 3.28

$$\begin{array}{ccc} S(s) + & O_2(g) \rightarrow & SO_2(g) \\ 100.0\text{ g} & 200.0\text{ g} & \text{g} = ? \end{array}$$

The numbers of moles of S and O_2 are

$$100.0 \text{ g S} \times \frac{1 \text{ mol S}}{32.07 \text{ g S}} = 3.118 \text{ mol S}$$

$$200.0 \text{ g } O_2 \times \frac{1 \text{ mol } O_2}{32.00 \text{ g } O_2} = 6.25 \text{ mol } O_2$$

From the equation we see that 1 mol of S reacts with 1 mol of O_2. Since we have 6.25 mol O_2 and only 3.118 mol S, S is the limiting reactant. The mass of SO_2 formed is calculated as follows: mol S $\rightarrow$ mol SO_2 $\rightarrow$ g SO_2.

$$3.118 \text{ mol S} \times \frac{1 \text{ mol } SO_2}{1 \text{ mol S}} \times \frac{64.065 \text{ g } SO_2}{1 \text{ mol } SO_2} = 199.8 \text{ g } SO_2$$

PE 3.29

$$\begin{array}{ll} 2Cu(s) + S(s) \rightarrow & Cu_2S(s) \\ 99.8\text{ g} & \text{g} = ? \end{array}$$

The theoretical yield of copper(I) sulfide (Cu_2S) is calculated in the usual way from the balanced equation. The steps in the calculation are: g Cu $\rightarrow$ mol Cu $\rightarrow$ mol Cu_2S $\rightarrow$ g Cu_2S.

$$99.8 \text{ g Cu} \times \frac{1 \text{ mol Cu}}{63.55 \text{ g Cu}} \times \frac{1 \text{ mol } Cu_2S}{2 \text{ mol Cu}} \times \frac{159.2 \text{ g } Cu_2S}{1 \text{ mol } Cu_2S} = 125 \text{ g } Cu_2S$$

The percent yield is

$$\text{percent yield} = \frac{\text{actual yield}}{\text{theoretical yield}} \times 100\% = \frac{50.0 \text{ g}}{125 \text{ g}} \times 100\% = 40.0\%$$

Solutions To Final Exercises

3.1

(a) Physical change; no new substances are formed. The toaster filament returns to its original state on cooling.
(b) Chemical change; the brown material consists of new substances that result from the breakdown (decomposition) of the apple molecules by oxygen and microbes.
(c) Chemical change; most greases are organic compounds that are combustible and form $CO_2(g)$ and $H_2O(g)$ when they undergo combustion. New substances are formed.
(d) Physical change; the grease is just dissolved away by the cleaning fluid. No new substances are formed.

3.3

A chemical property is any chemical reaction that a substance can undergo. A physical property is any property other than a chemical property.
(a) Physical property; no new substances are formed.
(b) Chemical property; new substances are formed; therefore a chemical reaction has taken place.
(c) Chemical property; rusting is the oxidation of iron during exposure to air and moisture.
(d) Physical property; the state of iron changes from solid to the liquid, but no new substances are formed.

3.5

(a) The law of constant composition states that the elemental composition by mass of a given compound is the same for all samples of the compound. Atomic theory assumes that the atoms combine in fixed ratios when they form compounds. If the kinds and amounts of atoms in a compound are indeed fixed, then the percent by mass of each element composing the compound is fixed and will be the same for all samples of the compound.

(b) The law of conservation of mass states that there is no measurable change in total mass during a chemical reaction. A chemical reaction occurs when one or more substances are changed into one or more new substances. Atomic theory assumes that the creation of a new substance involves just the rearrangement of the atoms present, not the creation or destruction of atoms. Therefore, the total mass before and after the reaction will be the same, since the same atoms are present before and after the reaction.

3.7 (a) Using the law of conservation of mass, mass oxygen = mass mercury oxide – mass mercury $= 7.00\text{ g} - 6.48\text{ g} = 0.52\text{ g}$.

(b) The percentage composition of a compound is the percent of each element by mass.

$$\text{percent Hg} = \frac{6.48\text{ g}}{7.00\text{ g}} \times 100\% = 92.6\%$$

$$\text{percent O} = \frac{0.52\text{ g}}{7.00\text{ g}} \times 100\% = 7.4\%$$

3.9

$$\text{percent iron} = \frac{3.05\text{ g}}{4.36\text{ g}} \times 100\% = 70.0\%$$

percent oxygen = 100% – percent iron = 100% – 70.0% = 30.0%

3.11 One mole of Cu_2S contains 2 mol of Cu atoms and 1 mol of S atoms; therefore, the number of grams of copper per gram of sulfur in Cu_2S is

$$\frac{127.1\text{ g Cu}}{32.07\text{ g S}} = \frac{3.963\text{ g Cu}}{1\text{ g S}}$$

Similarly, the number of grams of copper per gram of sulfur in CuS is

$$\frac{63.55\text{ g Cu}}{32.07\text{ g S}} = \frac{1.982\text{ g Cu}}{1\text{ g S}}$$

The law of multiple proportions states that the masses of one element combined with a fixed mass of another element are in the ratio of small whole numbers. Using the two results we just calculated, we get the following ratio between the copper masses combined with 1 g of S

$$\frac{3.963\text{ g}}{1.982\text{ g}} = \frac{2.00}{1.00} = \frac{2}{1}$$

This is a ratio of small whole numbers; therefore, it illustrates the law of multiple proportions.

3.13 (a) The molar mass of oxalic acid ($H_2C_2O_4$) is $(2 \times 1.008) + (2 \times 12.01) + (4 \times 16.00) = 90.04$ g/mol.

$$\text{percent carbon} = \frac{24.02\text{ g}}{90.04\text{ g}} \times 100\% = 26.68\%$$

$$\text{percent oxygen} = \frac{64.00\text{ g}}{90.04\text{ g}} \times 100\% = 71.08\%$$

$$\text{percent hydrogen} = \frac{2.016\text{ g}}{90.04\text{ g}} \times 100\% = 2.239\%$$

(b) The molar mass of SnF_2 is 118.7 + (2 × 19.00) = 156.7 g/mol.

$$\text{percent tin} = \frac{118.7\text{ g}}{156.7\text{ g}} \times 100\% = 75.75\%$$

$$\text{percent fluorine} = \frac{38.00\text{ g}}{156.7\text{ g}} \times 100\% = 24.25\%$$

(c) The molar mass of acetone (C_3H_6O) is (3 × 12.01) + (6 × 1.008) + 16.00 = 58.08 g/mol.

$$\text{percent carbon} = \frac{36.03\text{ g}}{58.08\text{ g}} \times 100\% = 62.04\%$$

$$\text{percent hydrogen} = \frac{6.048\text{ g}}{58.08\text{ g}} \times 100\% = 10.41\%$$

$$\text{percent oxygen} = \frac{16.00\text{ g}}{58.08\text{ g}} \times 100\% = 27.55\%$$

(d) The molar mass of phosgene ($COCl_2$) is 12.01 + 16.00 + (2 × 35.45) = 98.91 g/mol.

$$\text{percent carbon} = \frac{12.01\text{ g}}{98.91\text{ g}} \times 100\% = 12.14\%$$

$$\text{percent oxygen} = \frac{16.00\text{ g}}{98.91\text{ g}} \times 100\% = 16.18\%$$

$$\text{percent chlorine} = \frac{70.90\text{ g}}{98.91\text{ g}} \times 100\% = 71.68\%$$

3.15 The molecular formula gives the actual number of atoms of each element in the molecule, while the empirical formula only gives the simplest whole-number ratio of the atoms in a compound.

These formulas will be the same whenever the molar mass of the compound equals the mass of a mole of empirical formula units.

3.17 An empirical formula gives the simplest whole-number ratio of the atoms in a compound.

(a) NH_2 (b) CH (c) S

3.19 The atomic weight of boron is 10.81 u. The number of boron atoms in a boron molecule is 129.7 u/10.81 u = 12. The formula of the boron molecule is therefore B_{12}.

3.21 (a) The masses of arsenic and sulfur in the sample are 12.18 g and 20.00 – 12.18 = 7.82 g, respectively. The numbers of moles of arsenic and sulfur are

$$12.18\text{ g As} \times \frac{1\text{ mol As atoms}}{74.92\text{ g As}} = 0.1626\text{ mol As atoms}$$

$$7.82\text{ g S} \times \frac{1\text{ mol S atoms}}{32.07\text{ g S}} = 0.244\text{ mol S atoms}$$

To find the simplest whole-number mole ratio, we divide through by the number of moles of arsenic. This gives us 1 mol As to 0.244/0.1626 = 1.50 mol S. To obtain a whole-number ratio we multiply by 2. There are 2 mol of arsenic for every 3 mol of S, and the empirical formula is As_2S_3.

(b) No. To determine if this is the molecular formula, you must know the molecular weight or molar mass of the molecule (or compound).

3.23 The masses of the elements are given, 6.58 g nitrogen and 1.42 g hydrogen. The numbers of moles of nitrogen and hydrogen are

$$6.58 \text{ g N} \times \frac{1 \text{ mol N atoms}}{14.01 \text{ g N}} = 0.470 \text{ mol N atoms}$$

$$1.42 \text{ g H} \times \frac{1 \text{ mol H atoms}}{1.008 \text{ g H}} = 1.41 \text{ mol H atoms}$$

To find the simplest whole-number ratio we divide through by the number of moles of N. This gives us a nitrogen-to-hydrogen ratio of 1:3. The empirical formula is NH_3.

3.25 (a) A 100-g sample of the compound would contain 43.18 g K, 39.15 g Cl, and 17.67 g O. The numbers of moles of each element in a 100-g sample are

$$43.18 \text{ g K} \times \frac{1 \text{ mol K atoms}}{39.10 \text{ g K}} = 1.104 \text{ mol K atoms}$$

$$39.15 \text{ g Cl} \times \frac{1 \text{ mol Cl atoms}}{35.45 \text{ g Cl}} = 1.104 \text{ mol Cl atoms}$$

$$17.67 \text{ g O} \times \frac{1 \text{ mol O atoms}}{16.00 \text{ g O}} = 1.104 \text{ mol O atoms}$$

From inspection we see that there is 1 mol of each element. The empirical formula is KClO.

(b) A 100-g sample of the compound would contain 31.90 g K, 28.93 g Cl, and 39.17 g O. The numbers of moles of each element in 100 g of the compound are

$$31.90 \text{ g K} \times \frac{1 \text{ mol K atoms}}{39.10 \text{ g K}} = 0.8159 \text{ mol K atoms} \qquad \frac{0.8159}{0.8159} = 1.00$$

$$28.93 \text{ g Cl} \times \frac{1 \text{ mol Cl atoms}}{35.45 \text{ g Cl}} = 0.8161 \text{ mol Cl atoms} \qquad \frac{0.8161}{0.8159} = 1.00$$

$$39.17 \text{ g O} \times \frac{1 \text{ mol O atoms}}{16.00 \text{ g O}} = 2.448 \text{ mol O atoms} \qquad \frac{2.448}{0.8159} = 3.00$$

After dividing through by the number of moles of K, we see that the resulting numbers are 1, 1, and 3. There is 1 mol of K for every 1 mol of Cl and 3 mol of O. The empirical formula is $KClO_3$.

(c) A 100-g sample of the compound would contain 28.22 g K, 25.59 g Cl, and 46.19 g O. The numbers of moles of each element in 100 g of the compound are

$$28.22 \text{ g K} \times \frac{1 \text{ mol K atoms}}{39.10 \text{ g K}} = 0.7217 \text{ mol K atoms} \qquad \frac{0.7217}{0.7217} = 1.00$$

$$25.59 \text{ g Cl} \times \frac{1 \text{ mol Cl atoms}}{35.45 \text{ g Cl}} = 0.7218 \text{ mol Cl atoms} \qquad \frac{0.7218}{0.7217} = 1.00$$

$$46.19 \text{ g O} \times \frac{1 \text{ mol O atoms}}{16.00 \text{ g O}} = 2.887 \text{ mol O atoms} \qquad \frac{2.887}{0.7217} = 4.00$$

After dividing through by the number of moles of K, we see that the resulting ratio is 1:1:4. The empirical formula is $KClO_4$.

3.27 The masses of carbon and hydrogen in the molar mass are 56.1 g × 0.856 = 48.0 g of carbon and 56.1 g × 0.144 = 8.08 g of hydrogen. The number of moles of carbon and hydrogen are

$$48.0 \text{ g C} \times \frac{1 \text{ mol C atoms}}{12.01 \text{ g C}} = 4.00 \text{ mol C atoms}$$

$$8.08 \text{ g H} \times \frac{1 \text{ mol H atoms}}{1.008 \text{ g H}} = 8.02 \text{ mol H atoms}$$

In 1 mol of the hydrocarbon there are 4 mol of C and 8 mol of H. The atom ratio is the same as the mole ratio, and the molecular formula is C_4H_8.

3.29 The numbers of moles of carbon, oxygen, and hydrogen in the molar mass are

$$\text{C:} \quad 194 \text{ g acid} \times \frac{37.12 \text{ g C}}{100 \text{ g acid}} \times \frac{1 \text{ mol C atoms}}{12.01 \text{ g C}} = 6.00 \text{ mol C atoms}$$

$$\text{O:} \quad 194 \text{ g acid} \times \frac{57.69 \text{ g O}}{100 \text{ g acid}} \times \frac{1 \text{ mol O atoms}}{16.00 \text{ g O}} = 6.99 \text{ mol O atoms}$$

$$\text{H:} \quad 194 \text{ g acid} \times \frac{5.188 \text{ g H}}{100 \text{ g acid}} \times \frac{1 \text{ mol H atoms}}{1.008 \text{ g H}} = 9.98 \text{ mol H atoms}$$

The molecular formula is $C_6H_{10}O_7$.

3.31 (a) The masses of lead and oxygen in the 2.500-g sample are 2.266 g lead and 2.500 – 2.266 = 0.234 g oxygen. The numbers of moles of lead and oxygen in the molar mass are

$$\text{Pb:} \quad 685.6 \text{ g red lead} \times \frac{2.266 \text{ g Pb}}{2.500 \text{ g red lead}} \times \frac{1 \text{ mol Pb atoms}}{207.2 \text{ g Pb}} = 3.00 \text{ mol Pb atoms}$$

$$\text{O:} \quad 685.6 \text{ g red lead} \times \frac{0.234 \text{ g O}}{2.500 \text{ g red lead}} \times \frac{1 \text{ mol O atoms}}{16.00 \text{ g O}} = 4.01 \text{ mol O atoms}$$

The formula is Pb_3O_4.

(b) No. If the formula calculated from the molar mass and composition fractions turns out to be an integral multiple of the simpler empirical formula, then the compound must be molecular. In all other situations the information is insufficient to enable you to distinguish whether the compound is a molecular or a nonmolecular substance.

3.33 (a) After combustion of the styrene, all the carbon is in the CO_2 and all the hydrogen is in the H_2O. The masses of C and H in the sample are

$$\text{C:} \quad 33.80 \text{ mg } CO_2 \times \frac{12.01 \text{ mg C}}{44.01 \text{ mg } CO_2} = 9.224 \text{ mg C}$$

$$\text{H:} \quad 6.92 \text{ mg } H_2O \times \frac{2.016 \text{ mg H}}{18.02 \text{ mg } H_2O} = 0.774 \text{ mg H}$$

The percentage composition of styrene is

$$\text{C: } \frac{9.224 \text{ mg}}{10.00 \text{ mg}} \times 100\% = 92.24\%; \quad \text{H: } \frac{0.774 \text{ mg}}{10.00 \text{ mg}} \times 100\% = 7.74\%$$

(b) The number of moles of C and H in the sample are

$$\text{C: } 9.224 \text{ mg C} \times \frac{1 \text{ g}}{10^3 \text{ mg}} \times \frac{1 \text{ mol C atoms}}{12.01 \text{ g C}} = 7.68 \times 10^{-4} \text{ mol C atoms}$$

$$\text{H: } 0.774 \text{ mg H} \times \frac{1 \text{ g}}{10^3 \text{ mg}} \times \frac{1 \text{ mol H atoms}}{1.008 \text{ g H}} = 7.68 \times 10^{-4} \text{ mol H atoms}$$

From inspection we see that there is a carbon-to-hydrogen ratio of 1:1. The empirical formula is CH.

3.35 After combustion of the rocket fuel, all the carbon is in the CO_2 and all the hydrogen is in the H_2O. The masses of C, H, and N in the sample are

$$\text{C: } 1.048 \text{ g } CO_2 \times \frac{12.01 \text{ g C}}{44.01 \text{ g } CO_2} = 0.2860 \text{ g C}$$

$$\text{H: } 0.8573 \text{ g } H_2O \times \frac{2.016 \text{ g H}}{18.02 \text{ g } H_2O} = 0.09591 \text{ g H}$$

$$\text{N: } 0.7148 \text{ g} - 0.2860 \text{ g} - 0.09591 \text{ g} = 0.3329 \text{ g N}$$

The numbers of moles of C, H, and N in the molar mass are

$$\text{C: } 60.10 \text{ g compound} \times \frac{0.2860 \text{ g C}}{0.7148 \text{ g compound}} \times \frac{1 \text{ mol C atoms}}{12.01 \text{ g C}} = 2.00 \text{ mol C atoms}$$

$$\text{H: } 60.10 \text{ g compound} \times \frac{0.09591 \text{ g H}}{0.7148 \text{ g compound}} \times \frac{1 \text{ mol H atoms}}{1.008 \text{ g H}} = 8.00 \text{ mol H atoms}$$

$$\text{N: } 60.10 \text{ g compound} \times \frac{0.3329 \text{ g N}}{0.7148 \text{ g compound}} \times \frac{1 \text{ mol N atoms}}{14.01 \text{ g N}} = 2.00 \text{ mol N atoms}$$

The molecular formula is $C_2H_8N_2$.

3.37 The phrase *balanced equation* means that you have the same number of each kind of atom on both sides of the chemical equation.

A chemical reaction just involves the rearrangement of atoms; no atoms are created or destroyed. Therefore, there must be the same number of each kind of atom before and after the reaction. Balancing a chemical reaction ensures that there is the same number of each kind of atom before and after the reaction, i.e., the same number of each kind of atom on both sides of the arrow. Unbalanced equations do not comply with the law of conservation of mass; balanced equations do.

3.39 (a) sugar $\rightarrow$ carbon + water
(b) hydrochloric acid + sodium hydroxide $\rightarrow$ sodium chloride + water
(c) carbon dioxide + water $\rightarrow$ glucose + oxygen

3.41 (a) $C_6H_6(l) + O_2(g) \rightarrow CO_2(g) + H_2O(g)$ (unbalanced)
$C_6H_6(l) + 7\frac{1}{2}O_2(g) \rightarrow 6CO_2(g) + 3H_2O(g)$ (balanced)
$2C_6H_6(l) + 15O_2(g) \rightarrow 12CO_2(g) + 6H_2O(g)$ (balanced–integral coefficients)

(b) $ZnS(s) + O_2(g) \rightarrow ZnO(s) + SO_2(g)$ (unbalanced)
$ZnS(s) + \frac{3}{2}O_2(g) \rightarrow ZnO(s) + SO_2(g)$ (balanced)
$2ZnS(s) + 3O_2(g) \rightarrow 2ZnO(s) + 2SO_2(g)$ (balanced–integral coefficients)

(c) $FeO(s) + O_2(g) \rightarrow Fe_2O_3(s)$ (unbalanced)
$2FeO(s) + \frac{1}{2}O_2(g) \rightarrow Fe_2O_3(s)$ (balanced)
$4FeO(s) + O_2(g) \rightarrow 2Fe_2O_3(s)$ (balanced–integral coefficients)

(d) $CS_2(l) + O_2(g) \rightarrow CO_2(g) + SO_2(g)$ (unbalanced)

$CS_2(l) + 3O_2(g) \rightarrow CO_2(g) + 2SO_2(g)$ (balanced)

3.43 The balanced equations are

(a) $P_4O_{10}(s) + 6H_2O(l) \rightarrow 4H_3PO_4(aq)$

(b) $XeF_6(s) + 3H_2O(l) \rightarrow XeO_3(s) + 6HF(g)$

(c) $2K(s) + 2H_2O(l) \rightarrow 2KOH(aq) + H_2(g)$

(d) $PCl_5(s) + 4H_2O(l) \rightarrow H_3PO_4(aq) + 5HCl(aq)$

(e) $PBr_3(l) + 3H_2O(l) \rightarrow H_3PO_3(aq) + 3HBr(aq)$

3.45 (a) $C_6H_{12}O_6(aq) \rightarrow C_2H_5OH(aq) + CO_2(g)$ (unbalanced)
$C_6H_{12}O_6(aq) \rightarrow 2C_2H_5OH(aq) + 2CO_2(g)$ (balanced)

(b) $6CO_2(g) + 6H_2O(l) \rightarrow C_6H_{12}O_6(aq) + 6O_2(g)$ (balanced)

3.47 (a) A combination reaction is one in which two or more simpler substances combine to form a compound. Three examples are

$H_2(g) + Cl_2(g) \rightarrow 2HCl(g)$

$Na_2O(s) + SO_2(g) \rightarrow Na_2SO_3(s)$

$KOH(s) + CO_2(g) \rightarrow KHCO_3(s)$

(b) A decomposition reaction is one in which a compound breaks down into simpler substances that may or may not be elements. Three examples are

$NH_4NO_3(s) \xrightarrow{heat} N_2O(g) + 2H_2O(g)$

$MgCO_3(s) \xrightarrow{heat} MgO(s) + CO_2(g)$

$2Pb(NO_3)_2 \xrightarrow{heat} 2PbO(s) + 4NO_2(g) + O_2(g)$

3.49 Carbon monoxide (CO) is most likely to be formed at very high combustion temperatures.

3.51 (a) $Ca(s) + 2H_2O(l) \rightarrow H_2(g) + Ca(OH)_2(aq)$. Calcium is displacing hydrogen from water; displacement reaction.
(b) $2AgBr(s) \rightarrow 2Ag(s) + Br_2(l)$. Silver bromide is breaking down into its elements; decomposition reaction.
(c) $PbO(s) + C(s) \rightarrow Pb(l) + CO(g)$. Carbon is taking the place of lead in the compound; displacement reaction.
(d) $2K(s) + Cl_2(g) \rightarrow 2KCl(s)$. Potassium chloride is formed from its elements; combination reaction.

3.53 (a) In unlimited oxygen the products are CO_2 and H_2O.

$C_8H_{18}(l) + O_2(g) \rightarrow CO_2(g) + H_2O(g)$ (unbalanced)

$C_8H_{18}(l) + 12\frac{1}{2}O_2(g) \rightarrow 8CO_2(g) + 9H_2O(g)$ (balanced)

$2C_8H_{18}(l) + 25O_2(g) \rightarrow 16CO_2(g) + 18H_2O(g)$ (balanced–integral coefficients)

(b) $CH_3CHOHCH_3$ can be rewritten as C_3H_8O.

$C_3H_8O(l) + 5O_2(g) \rightarrow 3CO_2(g) + 4H_2O(g)$ (balanced)

(c) $C_6H_{12}O_6(s) + O_2(g) \rightarrow CO_2(g) + H_2O(g)$ (unbalanced)

$C_6H_{12}O_6(s) + 6O_2(g) \rightarrow 6CO_2(g) + 6H_2O(g)$ (balanced)

(d) CH_3COCH_3 can be rewritten as C_3H_6O.

$C_3H_6O(l) + 4O_2(g) \rightarrow 3CO_2(g) + 3H_2O(g)$ (balanced)

3.55 (a) $C(s) + O_2(g) \rightarrow CO_2(g)$ (excess oxygen)

(b) $2C(s) + O_2(g) \rightarrow 2CO(g)$ (limited oxygen)

3.57 (a) $GeO_2(s) + 2C(s) \rightarrow Ge(s) + 2CO(g)$

(b) During a reaction, a substance that combines with oxygen is oxidized; a substance that loses oxygen is reduced. Carbon is oxidized; GeO_2 is reduced.

3.59 (a) For the blast furnace reduction

$2C(s) + O_2(g) \xrightarrow{\Delta} 2CO(g)$

and $Fe_2O_3(s) + 3CO(g) \xrightarrow{\Delta} 2Fe(l) + 3CO_2(g)$

(b) $SnO_2(s) + 2C(s) \xrightarrow{\Delta} Sn(l) + 2CO(g)$

(c) For the blast furnace reduction

$2C(s) + O_2(g) \xrightarrow{\Delta} 2CO(g)$

and $Fe_3O_4(s) + 4CO(g) \xrightarrow{\Delta} 3Fe(l) + 4CO_2(g)$

(d) $2ZnS(s) + 3O_2(g) \xrightarrow{\Delta} 2ZnO(s) + 2SO_2(g)$

$ZnO(s) + C(s) \xrightarrow{\Delta} Zn(g) + CO(g)$

3.61 (a) $2KClO_3(s) \rightarrow 2KCl(s) + 3O_2(g)$

(b) $0.50 \text{ mol } KClO_3 \times \dfrac{2 \text{ mol KCl}}{2 \text{ mol } KClO_3} = 0.50 \text{ mol KCl}$

$0.50 \text{ mol } KClO_3 \times \dfrac{3 \text{ mol } O_2}{2 \text{ mol } KClO_3} = 0.75 \text{ mol } O_2$

(c) $0.50 \text{ mol KCl} \times \dfrac{74.55 \text{ g KCl}}{1 \text{ mol KCl}} = 37 \text{ g KCl}$

$0.75 \text{ mol } O_2 \times \dfrac{32.00 \text{ g } O_2}{1 \text{ mol } O_2} = 24 \text{ g } O_2$

3.63

$$4NH_3(g) + 5O_2(g) \rightarrow 4NO(g) + 6H_2O(g)$$
75.0 g g = ?

The steps in the calculation are: $g\ NH_3 \rightarrow mol\ NH_3 \rightarrow mol\ O_2 \rightarrow g\ O_2$.

$$75.0\ g\ NH_3 \times \frac{1\ mol\ NH_3}{17.03\ g\ NH_3} \times \frac{5\ mol\ O_2}{4\ mol\ NH_3} \times \frac{32.00\ g\ O_2}{1\ mol\ O_2} = 176\ g\ O_2$$

3.65

$$C_6H_{12}O_6(s) + 6O_2(g) \rightarrow 6CO_2(g) + 6H_2O(g)$$
15.0 g mol = ? g = ?

(a) The steps in the calculation are: g glucose → mol glucose → mol CO_2.

$$15.0\ g\ glucose \times \frac{1\ mol\ glucose}{180.2\ g\ glucose} \times \frac{6\ mol\ CO_2}{1\ mol\ glucose} = 0.499\ mol\ CO_2$$

(b) The steps in the calculation are: g glucose → mol glucose → mol H_2O → g H_2O.

$$15.0\ g\ glucose \times \frac{1\ mol\ glucose}{180.2\ g\ glucose} \times \frac{6\ mol\ H_2O}{1\ mol\ glucose} \times \frac{18.02\ g\ H_2O}{1\ mol\ H_2O} = 9.00\ g\ H_2O$$

3.67

$$2SO_2(g) + O_2(g) \rightarrow 2SO_3(g)$$
500 g g = ?

The steps in the calculation are: $g\ SO_2 \rightarrow mol\ SO_2 \rightarrow mol\ O_2 \rightarrow g\ O_2$.

$$500\ g\ SO_2 \times \frac{1\ mol\ SO_2}{64.06\ g\ SO_2} \times \frac{1\ mol\ O_2}{2\ mol\ SO_2} \times \frac{32.00\ g\ O_2}{1\ mol\ O_2} = 125\ g\ O_2$$

3.69

The molar mass of CH_2O is 12.01 + (2 × 1.008) + 16.00 = 30.03 g/mol. The mass of CH_2O that contains 85.5 g C is

$$85.5\ g\ C \times \frac{30.03\ g\ CH_2O}{12.01\ g\ C} = 214\ g\ CH_2O$$

3.71

The number of grams of Ag in 31.96 g of AgCl is

$$31.96\ g\ AgCl \times \frac{107.87\ g\ Ag}{143.32\ g\ AgCl} = 24.05\ g\ Ag$$

The percentage of silver in the coin is $\%\ Ag = \frac{24.05\ g}{26.73\ g} \times 100\% = 89.97\%$

3.73

(a) The candle is the limiting reactant; the air (oxygen) is the excess reactant.
(b) The candle is the excess reactant; the air (oxygen) is the limiting reactant.

3.75

$$3Cl_2(g) + 6KOH(aq) \rightarrow 5KCl(aq) + KClO_3(aq) + 3H_2O(l)$$
6.00 mol 8.00 mol

(a) The number of moles of KOH needed to react with 6.00 mol Cl_2 is

$$6.00\ mol\ Cl_2 \times \frac{6\ mol\ KOH}{3\ mol\ Cl_2} = 12.0\ mol\ KOH$$

Since there are only 8.00 mol KOH in the original mixture, KOH is the limiting reagent.

(b) The number of moles of $KClO_3$ that can be formed from 8.00 mol of KOH is

$$8.00 \text{ mol KOH} \times \frac{1 \text{ mol KClO}_3}{6 \text{ mol KOH}} = 1.33 \text{ mol KClO}_3$$

(c) The number of moles of Cl_2 consumed by 8.00 mol KOH is

$$8.00 \text{ mol KOH} \times \frac{3 \text{ mol Cl}_2}{6 \text{ mol KOH}} = 4.00 \text{ mol Cl}_2$$

Originally there was 6.00 mol Cl_2. After the reaction there will be 6.00 mol – 4.00 mol = 2.00 mol Cl_2 remaining.

3.77

$$\text{P}_4\text{O}_{10}(s) + 6\text{H}_2\text{O}(l) \rightarrow 4\text{H}_3\text{PO}_4(aq)$$

30.0 g   75.0 g   g = ?

To find the limiting reactant we convert the known masses to moles.

$$30.0 \text{ g P}_4\text{O}_{10} \times \frac{1 \text{ mol P}_4\text{O}_{10}}{283.9 \text{ g P}_4\text{O}_{10}} = 0.106 \text{ mol P}_4\text{O}_{10}$$

$$75.0 \text{ g H}_2\text{O} \times \frac{1 \text{ mol H}_2\text{O}}{18.02 \text{ g H}_2\text{O}} = 4.16 \text{ mol H}_2\text{O}$$

Since 0.106 mol P_4O_{10} only needs 0.636 mol of H_2O to completely react, P_4O_{10} is the limiting reactant. The number of grams of H_3PO_4 formed is (g P_4O_{10} → mol P_4O_{10} → mol H_3PO_4 → g H_3PO_4)

$$30.0 \text{ g P}_4\text{O}_{10} \times \frac{1 \text{ mol P}_4\text{O}_{10}}{283.9 \text{ g P}_4\text{O}_{10}} \times \frac{4 \text{ mol H}_3\text{PO}_4}{1 \text{ mol P}_4\text{O}_{10}} \times \frac{98.00 \text{ g H}_3\text{PO}_4}{1 \text{ mol H}_3\text{PO}_4} = 41.4 \text{ g H}_3\text{PO}_4$$

3.79

$$2\text{H}_2\text{S}(g) + 3\text{O}_2(g) \rightarrow 2\text{SO}_2(g) + 2\text{H}_2\text{O}(g)$$

17.0 g   200 g

To find the limiting reactant we convert the known masses to moles.

$$17.0 \text{ g H}_2\text{S} \times \frac{1 \text{ mol H}_2\text{S}}{34.08 \text{ g H}_2\text{S}} = 0.499 \text{ mol H}_2\text{S}$$

$$200 \text{ g O}_2 \times \frac{1 \text{ mol O}_2}{32.00 \text{ g O}_2} = 6.25 \text{ mol O}_2$$

From inspection of the reaction equation we see that H_2S is the limiting reactant. The moles of each of the four gases in the final mixture are:

H_2S: totally consumed; zero moles.

$$\text{SO}_2\text{: } 0.499 \text{ mol H}_2\text{S} \times \frac{2 \text{ mol SO}_2}{2 \text{ mol H}_2\text{S}} = 0.499 \text{ mol SO}_2$$

$$\text{H}_2\text{O: } 0.499 \text{ mol H}_2\text{S} \times \frac{2 \text{ mol H}_2\text{O}}{2 \text{ mol H}_2\text{S}} = 0.499 \text{ mol H}_2\text{O}$$

$$\text{O}_2\text{: } 6.25 \text{ mol O}_2 - \left(0.499 \text{ mol H}_2\text{S} \times \frac{3 \text{ mol O}_2}{2 \text{ mol H}_2\text{S}}\right) = 5.50 \text{ mol O}_2$$

3.81

(a) $$2\text{NaCl}(aq) + 2\text{H}_2\text{O}(l) \rightarrow 2\text{NaOH}(aq) + \text{H}_2(g) + \text{Cl}_2(g)$$

125 kg   kg = ?

The steps in the calculation of the theoretical yield are:
kg NaCl → g NaCl → mol NaCl → mol NaOH → kg NaOH.

$$125 \text{ kg NaCl} \times \frac{10^3 \text{ g}}{1 \text{ kg}} \times \frac{1 \text{ mol NaCl}}{58.44 \text{ g NaCl}} \times \frac{2 \text{ mol NaOH}}{2 \text{ mol NaCl}} \times \frac{40.00 \text{ g NaOH}}{1 \text{ mol NaOH}} = 85.6 \text{ kg NaOH}$$

(b) $\text{percent yield} = \frac{55.4 \text{ kg}}{85.6 \text{ kg}} \times 100\% = 64.7\%$

3.83

$$\text{percent yield} = \frac{\text{actual yield}}{\text{theoretical yield}} \times 100\%$$

Therefore, in order to get an actual yield of 25.0 g Freon-12 with a percent yield of 72.0%, we would need a theoretical yield of

$$\text{theoretical yield} = \frac{\text{actual yield}}{\text{percent yield}} \times 100\% = 25.0 \text{ g} \times \frac{100\%}{72.0\%} = 34.7 \text{ g Freon-12}$$

$$3CCl_4(l) + 2SbF_3(s) \rightarrow 3CCl_2F_2(g) + 2SbCl_3(s)$$
g = ?   34.7 g

The theoretical mass of SbF_3 needed to produce 34.7 g of Freon-12 is calculated as follows:
g CCl_2F_2 → mol CCl_2F_2 → mol SbF_3 → g SbF_3.

$$34.7 \text{ g } CCl_2F_2 \times \frac{1 \text{ mol } CCl_2F_2}{120.9 \text{ g } CCl_2F_2} \times \frac{2 \text{ mol } SbF_3}{3 \text{ mol } CCl_2F_2} \times \frac{178.7 \text{ g } SbF_3}{1 \text{ mol } SbF_3} = 34.2 \text{ g } SbF_3$$

Answer: 34.2 g SbF_3 are needed in order to produce 25.0 g CCl_2F_2.

3.85

$$CH_3OH(l) + CO(g) \rightarrow CH_3COOH(l)$$
105 g   75.0 g   g = ?

(a) To find the limiting reactant we convert the known masses to moles.

$$105 \text{ g } CH_3OH \times \frac{1 \text{ mol } CH_3OH}{32.04 \text{ g } CH_3OH} = 3.28 \text{ mol } CH_3OH$$

$$75.0 \text{ g CO} \times \frac{1 \text{ mol CO}}{28.01 \text{ g CO}} = 2.68 \text{ mol CO}$$

From inspection of the reaction equation we see that CO is the limiting reactant. The theoretical yield is calculated in the usual manner (g CO → mol CO → mol CH_3COOH → g CH_3COOH).

$$75.0 \text{ g CO} \times \frac{1 \text{ mol CO}}{28.01 \text{ g CO}} \times \frac{1 \text{ mol } CH_3COOH}{1 \text{ mol CO}} \times \frac{60.05 \text{ g } CH_3COOH}{1 \text{ mol } CH_3COOH} = 161 \text{ g } CH_3COOH$$

(b) $\text{actual yield} = \text{theoretical yield} \times \frac{\text{percent yield}}{100\%} = 161 \text{ g} \times \frac{88.0\%}{100\%} = 142 \text{ g}$

3.87 One must examine the material after the change to see if any new substances were formed. If no new substances were formed, then the change was physical. If new substances have appeared, then the change was a chemical reaction.

3.89 The law of multiple proportions states that in different compounds containing the same elements, the masses of one element combined with a fixed mass of the other element are in the ratio of small whole numbers.

(a) The numbers of grams of Ti per gram of Br in $TiBr_2$ and $TiBr_4$ are:

$$TiBr_2: \frac{47.88\text{ g Ti}}{159.8\text{ g Br}} = \frac{0.300\text{ g Ti}}{1\text{ g Br}}; \quad TiBr_4: \frac{47.88\text{ g Ti}}{319.6\text{ g Br}} = \frac{0.150\text{ g Ti}}{1\text{ g Br}}$$

The ratio of the Ti masses combined with 1 g of Br is 0.300 g/0.150 g = 2/1.

(b) The numbers of grams of Ag per gram of O in $AgIO_3$ and $AgIO_4$ are:

$$AgIO_3: \frac{107.9\text{ g Ag}}{48.00\text{ g O}} = \frac{2.25\text{ g Ag}}{1\text{ g O}}; \quad AgIO_4: \frac{107.9\text{ g Ag}}{64.00\text{ g O}} = \frac{1.69\text{ g Ag}}{1\text{ g O}}$$

The ratio of the Ag masses combined with 1 g O is 2.25 g/1.69 g = 1.33/1.00 = 4/3.

(c) The numbers of grams of N per gram of O in NO and N_2O_3 are:

$$NO: \frac{14.01\text{ g N}}{16.00\text{ g O}} = \frac{0.876\text{ g N}}{1\text{ g O}}; \quad N_2O_3: \frac{28.01\text{ g N}}{48.00\text{ g O}} = \frac{0.584\text{ g N}}{1\text{ g O}}$$

The ratio of the N masses combined with 1 g O is 0.876 g/0.584 g = 1.50/1.00 = 3/2.

(d) The numbers of grams of N per gram of O in N_2O_5 and N_2O_4 are:

$$N_2O_5: \frac{28.01\text{ g N}}{80.00\text{ g O}} = \frac{0.350\text{ g N}}{1\text{ g O}}; \quad N_2O_4: \frac{28.01\text{ g N}}{64.00\text{ g O}} = \frac{0.438\text{ g N}}{1\text{ g O}}$$

The ratio of the N masses combined with 1 g O is 0.350 g/0.438 g = 1.00/1.25 = 4/5.

3.91 The ores of many metals are either oxides or are converted to oxides for easier treatment. The metals are obtained by removing the oxygen, that is, by reduction.

3.93 (a) Mercury(II) oxide; $2HgO(s) \xrightarrow{\text{heat}} 2Hg(l) + O_2(g)$

(b) Water; $2H_2O(l) \xrightarrow{\text{electric current}} 2H_2(g) + O_2(g)$

3.95 The number of iron atoms in a hemoglobin molecule is

$$64{,}500\text{ u hemoglobin} \times \frac{0.35\text{ u}}{100\text{ u hemoglobin}} \times \frac{1\text{ Fe atom}}{55.85\text{ u Fe}} = 4\text{ Fe atoms}$$

3.97 After combustion of the compound, all the carbon is in the CO_2 and all the hydrogen is in the H_2O. The numbers of moles of C and H are

$$C: \quad 6.60\text{ g }CO_2 \times \frac{12.01\text{ g C}}{44.01\text{ g }CO_2} \times \frac{1\text{ mol C atoms}}{12.01\text{ g C}} = 0.150\text{ mol C atoms}$$

$$H: \quad 4.10\text{ g }H_2O \times \frac{2.016\text{ g H}}{18.02\text{ g }H_2O} \times \frac{1\text{ mol H atoms}}{1.008\text{ g H}} = 0.455\text{ mol H atoms}$$

Dividing through by the number of moles of C gives the simple whole-number ratio 3 hydrogen:1 carbon. The empirical formula is CH_3.

3.99 The percentage composition of A is Cu: 79.9% and O: 20.1%. The percentage composition of B is

$$Cu: 100\% - 20.1\% = 79.9\% \text{ and } O: \frac{0.201\text{ g}}{1.00\text{ g}} \times 100\% = 20.1\%$$

The percentage composition of C is

$$\text{Cu: } \frac{12.8\text{ g}}{16.0\text{ g}} \times 100\% = 80\% \text{ and O: } 100\% - 80\% = 20\%$$

The percentage compositions of A, B, and C are all the same; therefore, they have the same empirical formula. Barring the existence of two binary copper-oxygen compounds with the same empirical formula, A, B, and C are the same compound.

3.101 (a) $5.00\text{ kg SiO}_2 \times \dfrac{32.00\text{ kg O}_2}{60.08\text{ kg SiO}_2} = 2.66\text{ kg O}_2$

(b) $5.00\text{ kg CaCO}_3 \times \dfrac{48.00\text{ kg O}_2}{100.1\text{ kg CaCO}_3} = 2.40\text{ kg O}_2$

3.103 $H_2SO_4(aq) + 2NH_3(g) \rightarrow (NH_4)_2SO_4(aq)$

g = ?

The calculation of the number of grams of ammonia follows the scheme:
V × d (of acid mixture) → g acid mixture → g H_2SO_4 → mol H_2SO_4 → mol NH_3 → g NH_3.

$$35.5\text{ mL} \times \frac{1.84\text{ g acid mixture}}{1\text{ mL}} \times \frac{98.0\text{ g H}_2\text{SO}_4}{100\text{ g acid mixture}} \times \frac{1\text{ mol H}_2\text{SO}_4}{98.08\text{ g H}_2\text{SO}_4} \times \frac{2\text{ mol NH}_3}{1\text{ mol H}_2\text{SO}_4} \times \frac{17.03\text{ g NH}_3}{1\text{ mol NH}_3}$$

$= 22.2\text{ g NH}_3$

3.105 $2NaIO_3(aq) + 5NaHSO_3(aq) + 3NaCl(aq) \rightarrow I_2(s) + 5Na_2SO_4(aq) + 3HCl(aq) + H_2O(l)$

1.00 kg

The number of kilograms of Chile saltpeter needed is calculated as follows: g I_2 → mol I_2 → theoretical mol $NaIO_3$ needed → theoretical g $NaIO_3$ needed → actual g $NaIO_3$ needed → kg Chile saltpeter needed.

$$1.00 \times 10^3\text{ g I}_2 \times \frac{1\text{ mol I}_2}{253.8\text{ g I}_2} \times \frac{2\text{ mol NaIO}_3}{1\text{ mol I}_2} \times \frac{197.9\text{ g NaIO}_3}{1\text{ mol NaIO}_3} \times \frac{100\%}{82\%} \times \frac{100\text{ g Chile saltpeter}}{0.20\text{ g NaIO}_3}$$

$= 9.5 \times 10^5\text{ g} = 9.5 \times 10^2\text{ kg Chile saltpeter}$

3.107 Hematite reduction: $Fe_2O_3(s) + C(s) \rightarrow 2Fe(l) + 3CO(g)$

1000 kg kg = ?

The molar mass of Fe_2O_3 is 159.7 g/mol, of which 111.7 g is Fe. The ratio of Fe to Fe_2O_3 is therefore 111.7 g Fe/159.7 g Fe_2O_3 or 111.7 kg Fe/159.7 kg Fe_2O_3. The road map for obtaining the actual mass of iron from hematite is: kg Fe_2O_3 → theoretical kg Fe → actual kg Fe.

$$1000\text{ kg Fe}_2\text{O}_3 \times \frac{111.7\text{ kg Fe}}{159.7\text{ kg Fe}_2\text{O}_3} \times \frac{78\%}{100\%} = 5.5 \times 10^2\text{ kg Fe}$$

Magnetite reduction: $Fe_3O_4(s) + 4C(s) \rightarrow 3Fe(l) + 4CO(g)$

1000 kg kg = ?

The molar mass of Fe_3O_4 is 231.5 g/mol, of which 167.5 g is Fe. The ratio of Fe to Fe_3O_4 is therefore 167.5 g Fe/231.5 g Fe_3O_4 or 167.5 kg Fe/231.5 kg Fe_3O_4. The road map for obtaining the actual mass of iron from magnetite is: kg Fe_3O_4 → theoretical kg Fe → actual kg Fe.

$$1000 \text{ kg } Fe_3O_4 \times \frac{167.5 \text{ kg Fe}}{231.5 \text{ kg } Fe_3O_4} \times \frac{72\%}{100\%} = 5.2 \times 10^2 \text{ kg Fe}$$

Answer: Hematite yields the greatest amount of iron per metric ton.

3.109

$$MgCO_3(s) \rightarrow MgO(s) + CO_2(g)$$

x g $\quad$ $\text{grams} = \dfrac{x \text{ g} \times \text{molar mass MgO}}{\text{molar mass } MgCO_3}$

$$CaCO_3(s) \rightarrow CaO(s) + CO_2(g)$$

$(24.00 - x)$ g $\quad$ $\text{grams} = \dfrac{(24.00 - x) \text{ g} \times \text{molar mass CaO}}{\text{molar mass } CaCO_3}$

Let x equal the number of grams of $MgCO_3$ in the sample. The molar masses of $MgCO_3$, $CaCO_3$, MgO, and CaO are 84.31 g/mol, 100.1 g/mol, 40.30 g/mol, and 56.08 g/mol, respectively. From the equations we see that 1 mol of MgO is formed for each mol of $MgCO_3$ consumed, and 1 mol of CaO is formed for each mol of $CaCO_3$ consumed. Setting the grams of MgO and CaO formed equal to 12.00 gives the following equation to be solved for x.

$$\frac{40.30\,x}{84.31} + \frac{56.08(24.00 - x)}{100.1} = 12.00$$

$$0.4780x + 13.45 - 0.5602x = 12.00$$

$$0.0822x = 1.45$$

$$x = 17.6 \text{ g } MgCO_3$$

Answer: The mass of $MgCO_3$ in the original mixture is 17.6 g.

CHAPTER 4

SOLUTIONS AND SOLUTION STOICHIOMETRY

Solutions To Practice Exercises

PE 4.1 (a) $Al_2(SO_4)_3(s) \rightarrow 2Al^{3+}(aq) + 3SO_4^{2-}(aq)$
(b) $HI(aq) \rightarrow H^+(aq) + I^-(aq)$
(c) $HNO_2(aq) \rightleftharpoons H^+(aq) + NO_2^-(aq)$

PE 4.2 $Al_2(SO_4)_3(aq) + 6NaOH(aq) \rightarrow 2Al(OH)_3(s) + 3Na_2SO_4(aq)$
$Al^{3+}(aq) + 3OH^-(aq) \rightarrow Al(OH)_3(s)$

PE 4.3 (a) Sodium sulfite is soluble (Rule 2)
(b) Silver phosphate is sparingly soluble (Rule 6)
(c) Barium nitrate is soluble (Rule 1)
(d) Sodium phosphate is soluble (Rule 2)

PE 4.4 (a) When lead nitrate ($PbNO_3$) and potassium bromide (KBr) are mixed in aqueous solution, the ions initially present are Pb^{2+}, NO_3^-, K^+, and Br^-. The possible products of a precipitation reaction are lead bromide ($PbBr_2$) and potassium nitrate (KNO_3). The solubility rules state that $PbBr_2$ is sparingly soluble (Rule 3) and KNO_3 is soluble (Rule 2). A precipitate of $PbBr_2$ will form.

$Pb(NO_3)_2(aq) + 2KBr(aq) \rightarrow PbBr_2(s) + 2KNO_3(aq)$
$Pb^{2+}(aq) + 2Br^-(aq) \rightarrow PbBr_2(s)$

(b) After mixing $Na_2S(aq)$ and $ZnCl_2(aq)$, the ions momentarily present are Na^+, S^{2-}, Zn^{2+}, and Cl^-. The possible products of a precipitation reaction are NaCl and ZnS. NaCl is soluble (Rule 2); ZnS is sparingly soluble (Rule 6). A precipitate of ZnS will form.

$Na_2S(aq) + ZnCl_2(aq) \rightarrow ZnS(s) + 2NaCl(aq)$
$S^{2-}(aq) + Zn^{2+}(aq) \rightarrow ZnS(s)$

PE 4.5 Silver phosphate (Ag_3PO_4) is sparingly soluble (Rule 6). Silver nitrate ($AgNO_3$) is soluble (Rule 1); sodium phosphate (Na_3PO_4) is soluble (Rule 2). They can be used as reactants.

$3AgNO_3(aq) + Na_3PO_4(aq) \rightarrow Ag_3PO_4(s) + 3NaNO_3(aq)$

Sodium nitrate ($NaNO_3$) is soluble (Rules 1 and 2).

PE 4.6 All the chloride ions (Cl^-) initially in the solution are now in the form of AgCl. The steps in the calculation are: 0.3050 g AgCl $\rightarrow$ g Cl^- $\rightarrow$ % Cl^-.

One mole of AgCl (143.32 g) contains 1 mol of Cl (35.453 g). The mass of chloride in 0.3050 g AgCl is

$$0.3050 \text{ g AgCl} \times \frac{35.453 \text{ g Cl}^-}{143.32 \text{ g AgCl}} = 0.07545 \text{ g Cl}^-$$

The percent of chloride in the original mixture is

$$\frac{\text{mass of Cl}^-}{\text{mass of sample}} \times 100\% = \frac{0.07545 \text{ g}}{0.2501 \text{ g}} \times 100\% = 30.17\%$$

PE 4.7 (a) $Fe(OH)_3(s) + 3HNO_3(aq) \rightarrow Fe(NO_3)_3(aq) + 3H_2O(l)$
(b) $Mg(OH)_2(s) + H_2SO_4(aq) \rightarrow MgSO_4(aq) + 2H_2O(l)$

PE 4.8 (a) $2NaOH(aq) + H_3PO_4(aq) \rightarrow Na_2HPO_4(aq) + 2H_2O(l)$
(b) We will use KOH for the supply of K^+ ions and H_2SO_4 for the supply of HSO_4^- ions.

$KOH(aq) + H_2SO_4(aq) \rightarrow KHSO_4(aq) + H_2O(l)$

PE 4.9 (a) $NaCN(s) + HCl(aq) \rightarrow NaCl(aq) + HCN(g)$
(b) $NH_4HCO_3(s) + HCl(aq) \rightarrow NH_4Cl(aq) + H_2O(l) + CO_2(g)$
(c) $MgCO_3(s) + 2HCl(aq) \rightarrow MgCl_2(aq) + H_2O(l) + CO_2(g)$
(d) $NaHSO_3(s) + HCl(aq) \rightarrow NaCl(aq) + H_2O(l) + SO_2(g)$

PE 4.10 $Mg(s) + H_2SO_4(aq) \rightarrow MgSO_4(aq) + H_2(g)$
$Mg(s) + 2H^+(aq) \rightarrow Mg^{2+}(aq) + H_2(g)$

PE 4.11 (a) Magnesium (Mg) is above silver (Ag) in the activity series and will displace silver from aqueous $AgNO_3$. Magnesium will dissolve according to the equation

$Mg(s) + 2AgNO_3(aq) \rightarrow Mg(NO_3)_2(aq) + 2Ag(s)$

(b) Lead (Pb) is above copper (Cu) in the activity series and will displace copper from aqueous $CuSO_4$. It will react according to the equation

$Pb(s) + CuSO_4(aq) \rightarrow PbSO_4(s) + Cu(s)$

PE 4.12

$$750 \text{ mL soln} \times \frac{1.19 \text{ g soln}}{1 \text{ mL soln}} \times \frac{26.00 \text{ g H}_2\text{SO}_4}{100 \text{ g soln}} = 232 \text{ g H}_2\text{SO}_4$$

PE 4.13 A dioxin level of 50 ppt by mass means 50 g of dioxin in one trillion (1.0×10^{12}) grams of fish. The amount of dioxin in 100 g of fish is

$$100 \text{ g fish} \times \frac{50 \text{ g dioxin}}{1.0 \times 10^{12} \text{ g fish}} = 5.0 \times 10^{-9} \text{ g dioxin}$$

PE 4.14

$$\text{molarity} = \frac{\text{number of moles of KNO}_3}{\text{volume of solution in liters}} = \frac{1.22 \text{ mol KNO}_3}{0.750 \text{ L}} = 1.63 \text{ mol KNO}_3/\text{L}$$

The solution is 1.63 M in KNO_3.

PE 4.15 The steps in the calculation are: g $AgNO_3$/mL → g $AgNO_3$/L → mol $AgNO_3$/L → M $AgNO_3$.

$$\text{molarity} = \frac{5.25 \text{ g AgNO}_3}{125 \text{ mL}} \times \frac{10^3 \text{ mL}}{1 \text{ L}} \times \frac{1 \text{ mol AgNO}_3}{169.87 \text{ g AgNO}_3} = 0.247 \text{ mol AgNO}_3/\text{L} = 0.247 \text{ M AgNO}_3$$

PE 4.16 The calculation will proceed as follows: g soln/mL → g soln/L → g HNO_3/L → mol HNO_3/L → M HNO_3.

$$\frac{1.424 \text{ g soln}}{1 \text{ mL}} \times \frac{1000 \text{ mL}}{1 \text{ L}} \times \frac{70.9 \text{ g HNO}_3}{100 \text{ g soln}} \times \frac{1 \text{ mol HNO}_3}{63.01 \text{ g HNO}_3} = 16.0 \text{ mol HNO}_3/\text{L} = 16.0 \text{ M HNO}_3$$

The solution is 16.0 M in HNO_3.

PE 4.17 The milliliters of 1.50 M HNO_3 required is

$$0.250 \text{ mol HNO}_3 \times \frac{1 \text{ L}}{1.50 \text{ mol HNO}_3} \times \frac{10^3 \text{ mL}}{1 \text{ L}} = 167 \text{ mL}$$

PE 4.18 The calculation will proceed as follows: V × M (of $NaNO_3$) → mol $NaNO_3$ → g $NaNO_3$.

$$0.250 \text{ L} \times \frac{1.45 \text{ mol NaNO}_3}{1 \text{ L}} \times \frac{84.99 \text{ g NaNO}_3}{1 \text{ mol NaNO}_3} = 30.8 \text{ g NaNO}_3$$

PE 4.19 The number of grams of Na_2CO_3 needed is

$$100.0 \text{ mL} \times \frac{0.250 \text{ mol Na}_2\text{CO}_3}{1000 \text{ mL}} \times \frac{106.0 \text{ g Na}_2\text{CO}_3}{1 \text{ mol Na}_2\text{CO}_3} = 2.65 \text{ g Na}_2\text{CO}_3$$

Put 2.65 g Na_2CO_3 in a 100-mL volumetric flask. Dissolve and bring solution up to mark.

PE 4.20 One mole of $Fe_2(SO_4)_3$ dissociates into 2 mol of Fe^{3+} and 3 mol of SO_4^{2-}. A 0.750 M $Fe_2(SO_4)_3$ solution will be 2 × 0.750 M = 1.50 M in Fe^{3+} and 3 × 0.750 M = 2.25 M in SO_4^{2-}. The number of moles of SO_4^{2-} and Fe^{3+} in 1.00 L of solution is

(a) $1.00 \text{ L} \times \frac{2.25 \text{ mol SO}_4^{2-}}{1 \text{ L}} = 2.25 \text{ mol SO}_4^{2-}$

(b) $1.00 \text{ L} \times \frac{2.50 \text{ mol Fe}^{3+}}{1 \text{ L}} = 1.50 \text{ mol Fe}^{3+}$

PE 4.21 The total number of moles of chloride ion (Cl^-) is the sum of the moles of chloride ion in each of the separate solutions.

$$0.250 \text{ L MgCl}_2 \times \frac{0.125 \text{ mol MgCl}_2}{1 \text{ L}} \times \frac{2 \text{ mol Cl}^-}{1 \text{ mol MgCl}_2} + 0.800 \text{ L} \times \frac{0.350 \text{ mol FeCl}_3}{1 \text{ L}} \times \frac{3 \text{ mol Cl}^-}{1 \text{ mol FeCl}_3}$$
$$= 0.902 \text{ mol Cl}^-$$

The volume of the mixed solution = 0.250 L + 0.800 L = 1.050 L. The molarity of the chloride ion is

$$\text{molarity} = \frac{0.902 \text{ mol Cl}^-}{1.050 \text{ L}} = 0.859 \text{ mol Cl}^-/\text{L}$$

The solution is 0.859 M in chloride ion.

PE 4.22 (a) CH_3COOH, (b) HCl, (c) H_2SO_4, (d) HNO_3, and (e) $NH_3(aq)$.

PE 4.23 $V_C = ?$; $M_C = 12$ M; $V_D = 1.0$ L; $M_D = 0.10$ M

$$V_C \times M_C = V_D \times M_D$$

$$V_C = \frac{1000 \text{ mL} \times 0.10 \text{ mmol/mL}}{12 \text{ mmol/mL}} = 8.3 \text{ mL}$$

PE 4.24 The balanced equation for the neutralization is

$H_2SO_4(aq) + 2NaOH(aq) \rightarrow Na_2SO_4(aq) + 2H_2O(l)$
M = ?

The steps in the calculation are: V × M (of NaOH) → mmol NaOH → mmol H_2SO_4 → molarity of H_2SO_4 solution.

$$40.0 \text{ mL} \times \frac{0.750 \text{ mmol NaOH}}{1 \text{ mL}} \times \frac{1 \text{ mmol } H_2SO_4}{2 \text{ mmol NaOH}} \times \frac{1}{10.0 \text{ mL}} = \frac{1.50 \text{ mmol } H_2SO_4}{1 \text{ mL}} = 1.50 \text{ M } H_2SO_4$$

PE 4.25 The balanced equation is

$NaHCO_3(s) + HNO_3(aq) \rightarrow NaNO_3(aq) + H_2O(l) + CO_2(g)$
5.23 g mL = ?

The steps in the calculation are: g $NaHCO_3$ → mol $NaHCO_3$ → mol HNO_3 → mL HNO_3.

$$5.23 \text{ g } NaHCO_3 \times \frac{1 \text{ mol } NaHCO_3}{84.01 \text{ g } NaHCO_3} \times \frac{1 \text{ mol } HNO_3}{1 \text{ mol } NaHCO_3} \times \frac{1 \text{ L } HNO_3}{1.25 \text{ mol } HNO_3} = 0.0498 \text{ L } HNO_3$$

$$= 49.8 \text{ mL } HNO_3$$

PE 4.26 The balanced equation is $BaCl_2(aq) + H_2SO_4(aq) \rightarrow BaSO_4(s) + 2HCl(aq)$
5.55 g mL = ?

The steps in the calculation are: g $BaCl_2$ → mol $BaCl_2$ → mol H_2SO_4 → mL H_2SO_4.

$$5.55 \text{ g } BaCl_2 \times \frac{1 \text{ mol } BaCl_2}{208.2 \text{ g } BaCl_2} \times \frac{1 \text{ mol } H_2SO_4}{1 \text{ mol } BaCl_2} \times \frac{1 \text{ L } H_2SO_4}{2.00 \text{ mol } H_2SO_4} = 0.0133 \text{ L } H_2SO_4 = 13.3 \text{ mL } H_2SO_4$$

PE 4.27 The balanced equation is $H_3PO_4(aq) + 3KOH(aq) \rightarrow K_3PO_4(aq) + 3H_2O(l)$
M = ?

The steps in the calculation are: V × M (of KOH) → mmol KOH → mmol H_3PO_4 → molarity H_3PO_4 solution.

$$25.54 \text{ mL} \times \frac{2.111 \text{ mmol KOH}}{1 \text{ mL}} \times \frac{1 \text{ mmol } H_3PO_4}{3 \text{ mmol KOH}} \times \frac{1}{40.00 \text{ mL}} = \frac{0.4493 \text{ mmol } H_3PO_4}{1 \text{ mL}} = 0.4493 \text{ M } H_3PO_4$$

PE 4.28 The balanced equation for the reaction is

$Na_2CO_3(s) + 2HCl(aq) \rightarrow 2NaCl(aq) + H_2O(l) + CO_2(g)$
g = ? 33.75 mL of
0.4150 M

The steps in the calculation are: V × M (of HCl) → mol HCl → mol Na_2CO_3 → g Na_2CO_3 → % Na_2CO_3.

$$0.03375\ \text{L} \times \frac{0.4150\ \text{mol HCl}}{1\ \text{L}} \times \frac{1\ \text{mol Na}_2\text{CO}_3}{2\ \text{mol HCl}} \times \frac{106.0\ \text{g Na}_2\text{CO}_3}{1\ \text{mol Na}_2\text{CO}_3} = 0.7423\ \text{g Na}_2\text{CO}_3$$

The percent by mass of Na_2CO_3 in the sample is

$$\%\ \text{Na}_2\text{CO}_3 = \frac{0.7423\ \text{g}}{2.9929\ \text{g}} \times 100\% = 24.80\%$$

Solutions To Final Exercises

4.1

(a) Vinegar is a homogeneous mixture. It is a clear solution of acetic acid and water.
(b) Clean air is a homogeneous mixture. It is a clear uniform mixture of several gaseous substances.
(c) Chicken soup is a heterogeneous mixture. Separate pieces can be seen floating in the soup.
(d) 14-karat gold is a homogeneous mixture. It is usually a solid solution of copper in gold.
(e) Smog is a heterogeneous mixture. Smog is misty, indicating large suspended particles.
(f) Gasoline is a homogeneous mixture. It is a clear solution of several organic substances.

4.3

(a) clean air, (b) a solution of sodium chloride, (c) a mercury amalgam like mercury in silver, and (d) hydrogen in platinum.

4.5

(a) The crystal will dissolve and disappear into the solution. An unsaturated solution contains less than the equilibrium amount of solute.
(b) The crystal will cause no change in the solution. It will fall to the bottom of the solution container and remain there (no apparent change will occur to it). A saturated solution cannot dissolve any more solute.
(c) Crystallization of all excess solute in solution will occur. A supersaturated solution holds more solute than the equilibrium amount. This excess solute exists in a metastable state that is usually broken or disrupted by a seed crystal.

4.7

Ions in solution conduct electricity; the more ions present, the greater the conductivity. A strong electrolyte is completely or almost completely ionized in aqueous solution. Therefore, the solution of a soluble strong electrolyte will have a relatively high conductivity. A weak electrolyte is a molecular substance that ionizes only to a limited extent in water, and its solution will have a relatively low conductivity. A nonelectrolyte does not form ions in aqueous solution; it therefore cannot conduct current and has a zero conductivity. Hence, we can distinguish between strong electrolytes, weak electrolytes, and nonelectrolytes simply by measuring their conductivity. See Figure 4.7 in the textbook.

4.9

(a) $K_2CO_3(s) \rightarrow 2K^+(aq) + CO_3^{2-}(aq)$
(b) $NaHCO_3(s) \rightarrow Na^+(aq) + HCO_3^-(aq)$
(c) $FeSO_4(s) \rightarrow Fe^{2+}(aq) + SO_4^{2-}(aq)$
(d) $Ba(OH)_2(s) \rightarrow Ba^{2+}(aq) + 2OH^-(aq)$
(e) $CaCl_2(s) \rightarrow Ca^{2+}(aq) + 2Cl^-(aq)$
(f) $Na_3PO_4(s) \rightarrow 3Na^+(aq) + PO_4^{3-}(aq)$

4.11

A metathesis reaction is a reaction in which atoms or ions exchange partners. In aqueous solution all the ions act as independent entities. The ions in this state do not belong to any particular compound. Therefore, if all the ions remained in solution there would be no actual exchange taking place. Hence, a precipitate or molecular compound must be produced for a metathesis reaction between ions in aqueous solution to occur.

4.13

(a) $Pb(NO_3)_2$ is soluble (Rule 1); (b) Li_2CO_3 is soluble (Rule 2); (c) Hg_2Cl_2 is sparingly soluble (Rule 3); (d) $Mg(OH)_2$ is sparingly soluble (Rule 5); (e) $FeSO_4$ is soluble (Rule 4); and (f) $AlPO_4$ is sparingly soluble (Rule 6).

4.15 Yes. At the moment of mixing the ions present are 0.250 mol Sr^{2+}, 0.500 mol NO_3^-, 0.500 mol Na^+, and 0.500 mol OH^- in 2.00 L of water. Since the formula of strontium hydroxide is $Sr(OH)_2$, the 0.500 mol OH^- and 0.250 mol Sr^{2+} form 0.250 mol $Sr(OH)_2$. The number of grams of $Sr(OH)_2$ per 100 mL of water at the instant of mixing is therefore

$$\frac{0.250 \text{ mol Sr(OH)}_2}{2.00 \text{ L}} \times \frac{121.6 \text{ g Sr(OH)}_2}{1 \text{ mol Sr(OH)}_2} \times \frac{0.100 \text{ L}}{100 \text{ mL}} = \frac{1.52 \text{ g Sr(OH)}_2}{100 \text{ mL}}$$

Since the amount present is 3.7 times greater than the amount needed for saturation, a precipitate of $Sr(OH)_2$ will form.

4.17 $MgCl_2(aq) + 2AgNO_3(aq) \rightarrow 2AgCl(s) + Mg(NO_3)_2(aq)$
g = ? 0.03500 g

The steps in the calculation are: g AgCl $\rightarrow$ mol AgCl $\rightarrow$ mol $MgCl_2$ $\rightarrow$ g $MgCl_2$ $\rightarrow$ % $MgCl_2$.

$$0.03500 \text{ g AgCl} \times \frac{1 \text{ mol AgCl}}{143.32 \text{ g AgCl}} \times \frac{1 \text{ mol MgCl}_2}{2 \text{ mol AgCl}} \times \frac{95.210 \text{ g MgCl}_2}{1 \text{ mol MgCl}_2} = 0.01163 \text{ g MgCl}_2$$

$$\% \text{ MgCl}_2 = \frac{\text{mass MgCl}_2}{\text{mass sample}} \times 100\% = \frac{0.01163 \text{ g}}{0.3250 \text{ g}} \times 100\% = 3.578\%$$

4.19 The mass of chromium in 0.2748 g of Cr_2O_3 is

$$0.2748 \text{ g Cr}_2\text{O}_3 \times \frac{104.0 \text{ g Cr}}{152.0 \text{ g Cr}_2\text{O}_3} = 0.1880 \text{ g Cr}$$

$$\% \text{ chromium} = \frac{\text{mass chromium}}{\text{mass sample}} \times 100\% = \frac{0.1880 \text{ g}}{1.4312 \text{ g}} \times 100\% = 13.14\%$$

4.21
(a) Acids ionize in water to form hydrogen ions, $H^+(aq)$.
(b) Bases produce hydroxide ions, $OH^-(aq)$, in aqueous solution.
(c) The molecular compound water; the reaction is $H^+(aq) + OH^-(aq) \rightarrow H_2O(l)$.

4.23
(a) $H_2SO_3(aq) \rightleftharpoons H^+(aq) + HSO_3^-(aq)$ and to a much smaller extent
$HSO_3^-(aq) \rightleftharpoons H^+(aq) + SO_3^{2-}(aq)$

(b) $H_2CO_3(aq) \rightleftharpoons H^+(aq) + HCO_3^-(aq)$ and to a much smaller extent
$HCO_3^-(aq) \rightleftharpoons H^+(aq) + CO_3^{2-}(aq)$

(c) $HCN(aq) \rightleftharpoons H^+(aq) + CN^-(aq)$

4.25
(a) $Ba(OH)_2(aq) + 2HNO_3(aq) \rightarrow Ba(NO_3)_2(aq) + 2H_2O(l)$
(b) $3LiOH(aq) + H_3PO_4(aq) \rightarrow Li_3PO_4(aq) + 3H_2O(l)$
(c) $Cd(OH)_2(s) + H_2S(aq) \rightarrow CdS(s) + 2H_2O(l)$
(d) $2Fe(OH)_3(s) + 3H_2SO_4(aq) \rightarrow Fe_2(SO_4)_3(aq) + 6H_2O(l)$

4.27 When you treat the salt of a weak acid with a strong acid, you form the weak acid. The equations that follow are just typical examples of the many possible reactions you could use to answer this question. The net ionic equation is the molecular equation with the spectator ions removed.

(a) 1. $CaCO_3(s) + 2HCl(aq) \rightarrow CaCl_2(aq) + H_2O(l) + CO_2(g)$
$CaCO_3(s) + 2H^+(aq) \rightarrow Ca^{2+}(aq) + H_2O(l) + CO_2(g)$
2. $Na_2CO_3(aq) + 2HCl(aq) \rightarrow 2NaCl(aq) + H_2O(l) + CO_2(g)$
$CO_3^{2-}(aq) + 2H^+(aq) \rightarrow H_2O(l) + CO_2(g)$

(b) 1. $Na_2SO_3(aq) + 2HCl(aq) \rightarrow 2NaCl(aq) + H_2O(l) + SO_2(g)$
$SO_3^{2-}(aq) + 2H^+(aq) \rightarrow H_2O(l) + SO_2(g)$
2. $MgSO_3(s) + H_2SO_4(aq) \rightarrow MgSO_4(aq) + H_2O(l) + SO_2(g)$
$MgSO_3(s) + 2H^+(aq) \rightarrow Mg^{2+}(aq) + H_2O(l) + SO_2(g)$

(c) 1. $ZnS(s) + 2HCl(aq) \rightarrow ZnCl_2(aq) + H_2S(g)$
$ZnS(s) + 2H^+(aq) \rightarrow Zn^{2+}(aq) + H_2S(g)$
2. $Na_2S(aq) + 2HNO_3(aq) \rightarrow 2NaNO_3(aq) + H_2S(g)$
$S^{2-}(aq) + 2H^+(aq) \rightarrow H_2S(g)$

(d) 1. $NaCN(aq) + HCl(aq) \rightarrow NaCl(aq) + HCN(g)$
$CN^-(aq) + H^+(aq) \rightarrow HCN(g)$
2. $2LiCN(aq) + H_2SO_4(aq) \rightarrow Li_2SO_4(aq) + 2HCN(g)$
$CN^-(aq) + H^+(aq) \rightarrow HCN(g)$

4.29 Ammonia gas (NH_3) is formed. Salts containing ammonium ion (NH_4^+) give off ammonia gas (NH_3 (g)) when treated with strong concentrated base. The net ionic equation is

$NH_4^+(aq) + OH^-(aq) \rightarrow NH_3(g) + H_2O(l)$

4.31 (a) Iron is the more active and hydrogen the less active element, since iron displaces hydrogen from hydrochloric acid.
(b) Aluminum is the more active and copper the less active element, since aluminum displaces copper from an aqueous solution of copper sulfate.

4.33 (a) $2Cr(s) + 6HCl(aq) \rightarrow 3H_2(g) + 2CrCl_3(aq)$
(b) $2K(s) + 2H_2O(l) \rightarrow H_2(g) + 2KOH(aq)$
(c) $Zn(s) + CuSO_4(aq) \rightarrow ZnSO_4(aq) + Cu(s)$
(d) $Mg(s) + 2H_2O(g) \rightarrow H_2(g) + Mg(OH)_2(s)$
(e) $2Al(s) + 3H_2SO_4(aq) \rightarrow 3H_2(g) + Al_2(SO_4)_3(aq)$
(f) $Fe(s) + 2AgNO_3(aq) \rightarrow 2Ag(s) + Fe(NO_3)_2(aq)$

4.35 (a) $2Al(s) + 3Pb(NO_3)_2(aq) \rightarrow 2Al(NO_3)_3(aq) + 3Pb(s)$
$2Al(s) + 3Pb^{2+}(aq) \rightarrow 2Al^{3+}(aq) + 3Pb(s)$

(b) No. Lead is below aluminum in the activity series and therefore cannot displace aluminum from its compounds.

4.37 (a) The solubility of DDT expressed in parts per thousand means the number of grams of DDT per 10^3 g of water. This can be calculated from the given solubility as follows:

$$\frac{5.0 \times 10^{-5}\,\text{g DDT}}{10^2\,\text{g}\,H_2O} \times \frac{10}{10} = \frac{5.0 \times 10^{-4}\,\text{g DDT}}{10^3\,\text{g}\,H_2O}$$

Answer: Solubility is 5.0×10^{-4} parts per thousand by mass.

(b) The solubility expressed in parts per million means the number of grams of DDT per 10^6 g of water. This can be calculated from the given solubility as follows:

$$\frac{5.0\times10^{-5}\text{g DDT}}{10^2\text{g H}_2\text{O}}\times\frac{10^4}{10^4}=\frac{5.0\times10^{-1}\text{g DDT}}{10^6\text{ g H}_2\text{O}}$$

Answer: Solubility is 0.50 ppm by mass.

4.39 (mL soln → g soln → mg soln → mg H_2O_2)

$$10.0\text{ mL soln}\times\frac{1.0\text{ g soln}}{1\text{ mL soln}}\times\frac{10^3\text{mg}}{1\text{ g}}\times\frac{3.0\text{ mg H}_2\text{O}_2}{100\text{ mg soln}}=3.0\times10^2\text{mg H}_2\text{O}_2$$

4.41 The number of grams of ethanol in 1.50 L of solution is calculated as follows: V × density → g solution → g ethanol.

$$1.50\text{ L soln}\times\frac{10^3\text{mL}}{1\text{ L}}\times\frac{0.9687\text{ g soln}}{1\text{ mL soln}}\times\frac{20.0\text{ g ethanol}}{100.0\text{ g soln}}=291\text{ g ethanol}$$

4.43 (a) $\text{mass of solution}=\dfrac{\text{mass of solute}\times100\%}{\text{percent by mass}}=\dfrac{100\text{ g}\times100\%}{12.0\%}=833\text{ g of glucose solution}$

(b) mass water = mass solution – mass glucose = 833 g – 100 g = 733 g

(c) $\text{volume}=\dfrac{\text{mass}}{\text{density}}=\dfrac{833\text{ g}}{1.05\text{ g/mL}}=793\text{ mL}$

(d) Place the solid glucose in a 1-L volumetric cylinder calibrated in 1-mL increments. Dissolve the glucose in some water and then bring the solution up to the 793-mL mark with additional water.

4.45 You get the number of moles of solute contained in that volume of solution.

molarity = moles of solute/liters of solution, therefore moles of solute = molarity × liters of solution.

4.47 $\text{molarity}=\dfrac{\text{number of moles of solute}}{\text{volume of solution in liters}}$

(a) $\text{molarity}=\dfrac{45.0\text{ g NaCl}}{0.250\text{ L}}\times\dfrac{1\text{ mol NaCl}}{58.44\text{ g NaCl}}=3.08\text{ mol NaCl/L}$

The solution is 3.08 M in NaCl.

(b) $\text{molarity}=\dfrac{40.0\text{ g H}_2\text{SO}_4}{2.00\text{ L}}\times\dfrac{1\text{ mol H}_2\text{SO}_4}{98.08\text{ g H}_2\text{SO}_4}=0.204\text{ mol H}_2\text{SO}_4\text{/L}$

The solution is 0.204 M in H_2SO_4.

(c) $\text{molarity}=\dfrac{2.50\text{ g Ba(OH)}_2}{0.325\text{ L}}\times\dfrac{1\text{ mol Ba(OH)}_2}{171.3\text{ g Ba(OH)}_2}=4.49\times10^{-2}\text{mol Ba(OH)}_2\text{/L}$

The solution is 4.49×10^{-2} M in $Ba(OH)_2$.

(d) $\text{molarity} = \dfrac{5.00 \text{ g KNO}_3}{0.525 \text{ L}} \times \dfrac{1 \text{ mol KNO}_3}{101.1 \text{ g KNO}_3} = 9.42 \times 10^{-2} \text{ mol KNO}_3\text{/L}$

The solution is 9.42×10^{-2} M in KNO_3.

4.49 (d × 10^3 mL/L → g soln/L → g CH_3COOH/L → mol CH_3COOH/L → M CH_3COOH)

$$\frac{1.05 \text{ g soln}}{1 \text{ mL}} \times \frac{10^3 \text{ mL}}{1 \text{ L}} \times \frac{99.5 \text{ g CH}_3\text{COOH}}{100 \text{ g soln}} \times \frac{1 \text{ mol CH}_3\text{COOH}}{60.05 \text{ g CH}_3\text{COOH}} = \frac{17.4 \text{ mol CH}_3\text{COOH}}{1 \text{ L}}$$

$= 17.4 \text{ M CH}_3\text{COOH}$

4.51 moles of solute = molarity × volume of solution in liters

(a) 3.25 mol NaCl/L × 1.50 L = 4.88 mol NaCl

(b) 1.50×10^{-3} mol $(NH_4)_2SO_4$/L × 0.180 L = 2.70×10^{-4} mol $(NH_4)_2SO_4$

(c) 6.50 mol $Ba(NO_3)_2$/L × 0.0250 L = 0.162 mol $Ba(NO_3)_2$

4.53 number of moles = V (L) × M (mol/L)

(a) 0.350 L × 0.100 mol NaOH/L = 3.50×10^{-2} mol NaOH

(b) 1.75 L × 2.00×10^{-5} mol KBr/L = 3.50×10^{-5} mol KBr

(c) 0.0352 L × 0.555 mol $Al_2(SO_4)_3$/L = 0.0195 mol $Al_2(SO_4)_3$ = 1.95×10^{-2} mol $Al_2(SO_4)_3$

4.55 2.00 L of a 1.00 M aqueous glucose solution would contain 2.00 mol of glucose or 2.00 mol × 180.2 g glucose/1 mol glucose = 360 g of glucose. Put 360 g of glucose in a 2-L volumetric flask. Dissolve the glucose in water. Add enough additional water to bring the solution up to the 2-L mark.

4.57 (a) $0.30 \text{ mol HCl} \times \dfrac{1.0 \text{ L}}{0.15 \text{ mol HCl}} = 2.0 \text{ L} = 2.0 \times 10^3 \text{ mL}$

(b) $0.30 \text{ g HCl} \times \dfrac{1 \text{ mol HCl}}{36.46 \text{ g HCl}} \times \dfrac{1.0 \text{ L}}{0.15 \text{ mol HCl}} \times \dfrac{10^3 \text{ mL}}{1 \text{ L}} = 55 \text{ mL}$

4.59 The steps in the calculation are: V × M (of NaF) → mol NaF → mg NaF.

$$0.250 \text{ L} \times \frac{2.0 \times 10^{-5} \text{ mol NaF}}{1 \text{ L}} \times \frac{41.99 \text{ g NaF}}{1 \text{ mol NaF}} = 2.1 \times 10^{-4} \text{ g NaF} = 0.21 \text{ mg NaF}$$

4.61 (a) The steps in the calculation are:
density (g/mL) → g H_2SO_4 soln/L → g H_2SO_4/L → mol H_2SO_4/L → M H_2SO_4.

$$\frac{1.29 \text{ g soln}}{1 \text{ mL}} \times \frac{1000 \text{ mL}}{1 \text{ L}} \times \frac{38.0 \text{ g H}_2\text{SO}_4}{100 \text{ g soln}} \times \frac{1 \text{ mol H}_2\text{SO}_4}{98.08 \text{ g H}_2\text{SO}_4} = 5.00 \text{ mol H}_2\text{SO}_4\text{/L} = 5.00 \text{ M H}_2\text{SO}_4$$

(b) number of mmol = V (mL) × M (mmol/mL) = 50 mL × 5.00 mmol H_2SO_4/mL = 2.5×10^2 mmol H_2SO_4

4.63 (a) 3.0 M KNO_3; 3.0 M K^+ and 3.0 M NO_3^-
(b) 0.55 M $BaCl_2$; 0.55 M Ba^{2+} and 2 × 0.55 = 1.1 M Cl^-
(c) 0.75 M $NaHCO_3$; 0.75 M Na^+ and 0.75 M HCO_3^-
(d) 0.75 M $Al_2(SO_4)_3$; 2 × 0.75 = 1.5 M Al^{3+} and 3 × 0.75 = 2.2 M SO_4^{2-} (rounded from 2.25 M)

(e) 0.25 M Na_3PO_4; 3 × 0.25 = 0.75 M Na^+ and 0.25 M PO_4^{3-}
(f) 1.0 M $Hg_2(NO_3)_2$; 1.0 M Hg_2^{2+} and 2 × 1.0 = 2.0 M NO_3^-

4.65 (a) The steps in the calculation are: V (L) × M (mol/L) → mol OH^- → mol NaOH → g NaOH.

$$0.750 \text{ L soln} \times \frac{1.50 \text{ mol OH}^-}{1 \text{ L soln}} \times \frac{1 \text{ mol NaOH}}{1 \text{ mol OH}^-} \times \frac{40.00 \text{ g NaOH}}{1 \text{ mol NaOH}} = 45.0 \text{ g NaOH}$$

(b) One mole of NaOH provides one mole of OH^-, therefore their molarities will be the same.

Answer: The solution is 1.50 M in NaOH.

4.67 The total volume of the mixed solution is 0.200 L + 0.150 L = 0.350 L.

(a) $$0.200 \text{ L} \times \frac{0.120 \text{ mol Cr}_2(\text{SO}_4)_3}{1 \text{ L}} \times \frac{2 \text{ mol Cr}^{3+}}{1 \text{ mol Cr}_2(\text{SO}_4)_3} \times \frac{1}{0.350 \text{ L}} = 0.137 \text{ M Cr}^{3+}$$

(b) $$0.150 \text{ L} \times \frac{0.100 \text{ mol Na}_2\text{SO}_4}{1 \text{ L}} \times \frac{2 \text{ mol Na}^+}{1 \text{ mol Na}_2\text{SO}_4} \times \frac{1}{0.350 \text{ L}} = 0.0857 \text{ M Na}^+$$

(c) The amount of sulfate ion in the $Cr_2(SO_4)_3$ solution is

$$0.200 \text{ L} \times \frac{0.120 \text{ mol Cr}_2(\text{SO}_4)_3}{1 \text{ L}} \times \frac{3 \text{ mol SO}_4^{2-}}{1 \text{ mol Cr}_2(\text{SO}_4)_3} = 0.0720 \text{ mol SO}_4^{2-}$$

The amount of sulfate ion in the Na_2SO_4 solution is

$$0.150 \text{ L} \times \frac{0.100 \text{ mol Na}_2\text{SO}_4}{1 \text{ L}} \times \frac{1 \text{ mol SO}_4^{2-}}{1 \text{ mol Na}_2\text{SO}_4} = 0.0150 \text{ mol SO}_4^{2-}$$

The total sulfate ion in the mixed solution is 0.0720 mol + 0.0150 mol = 0.0870 mol SO_4^{2-}.

The final molarity of the sulfate ion is

$$\text{molarity} = \frac{0.0870 \text{ mol SO}_4^{2-}}{0.350 \text{ L}} = 0.249 \text{ mol SO}_4^{2-}/\text{L} = 0.249 \text{ M SO}_4^{2-}$$

4.69 $V_C \times M_C$ = moles substance in conc. soln
$V_D \times M_D$ = moles substance in dilute soln

The number of moles does not change upon dilution. Equating these two expressions gives the dilution formula.

$$V_C \times M_C = V_D \times M_D$$

4.71 The dilution formula is $V_C \times M_C = V_D \times M_D$. Remember that molarity(M) = mol/L = mmol/mL. The volume of the concentrated solution (stock solution) needed to make the dilute solution is

(a) $$V_C = \frac{V_D \times M_D}{M_C} = \frac{320 \text{ mL} \times 1.5 \text{ M}}{16 \text{ M}} = 30 \text{ mL}$$

(b) $$V_C = \frac{250 \text{ mL} \times 3.0 \text{ M}}{15 \text{ M}} = 50 \text{ mL}$$

(c) $$V_C = \frac{1.4 \times 10^3 \text{ mL} \times 2.8 \text{ M}}{17 \text{ M}} = 2.3 \times 10^2 \text{ mL}$$

4.73 (a) The steps in the calculation are: d (g/mL) soln → g soln/L → g HCl/L → mol HCl/L → M HCl.

$$\frac{1.19 \text{ g soln}}{1 \text{ mL}} \times \frac{1000 \text{ mL}}{1 \text{ L}} \times \frac{37.0 \text{ g HCl}}{100 \text{ g soln}} \times \frac{1 \text{ mol HCl}}{36.46 \text{ g HCl}} = 12.1 \text{ mol HCl/L} = 12.1 \text{ M HCl}$$

(b) $V_C \times M_C = V_D \times M_D$

$$V_C = \frac{250 \text{ mL} \times 0.500 \text{ mmol/mL}}{12.1 \text{ mmol/mL}} = 10.3 \text{ mL}$$

4.75 The dilute solution is made by removing 152 mL of concentrated ammonia from a beaker into which some of the concentrated ammonia was poured and adding it slowly, with constant mixing, to a volume of water (say, approximately 300 mL) in a 750-mL volumetric flask. Enough additional water is then added to make a final volume of 750 mL. A 1000-mL graduated cylinder could alternatively be used.

4.77 The steps in the titration are:

Step 1. Using either a pipet or a buret, add a known quantity of standard sodium hydroxide solution to an appropriate size beaker or Erlenmeyer flask. Water can be added to the flask as well.

Step 2. Add an appropriate indicator, such as phenolphthalein or methyl red, to the standard sodium hydroxide solution in the flask.

Step 3. Place the unknown sulfuric acid solution in a buret and record the initial volume mark of acid solution.

Step 4. Add unknown sulfuric acid solution to the standard sodium hydroxide solution until the endpoint is reached, as signaled by a change in the indicator.

Step 5. Record the final volume marking of acid solution in the buret. The volume of acid solution used equals the final volume mark minus the initial volume mark. Desired values can now be calculated.

See Figure 4.14 in the textbook.

Comment: Buret volume markings are in reverse. When a buret is full, it reads zero volume. These markings are easier to use, since it is only the difference between the initial and final volumes (that is, the volume of solution required for complete reaction) that is important.

4.79 $H_2SO_4(aq) + 2NaOH(aq) \rightarrow Na_2SO_4(aq) + 2H_2O(l)$

(a) The steps in the calculation are: V × M (of NaOH) → mol NaOH → mol H_2SO_4 → M H_2SO_4.

$$0.04000 \text{ L} \times \frac{0.1000 \text{ mol NaOH}}{1 \text{ L}} \times \frac{1 \text{ mol } H_2SO_4}{2 \text{ mol NaOH}} = 2.000 \times 10^{-3} \text{ mol } H_2SO_4$$

$$\text{molarity A} = \frac{\text{moles solute}}{\text{liter soln}} = \frac{2.000 \times 10^{-3} \text{ mol } H_2SO_4}{4.000 \times 10^{-2} \text{ L}} = 5.000 \times 10^{-2} \text{ mol } H_2SO_4\text{/L} =$$

5.000×10^{-2} M H_2SO_4

(b) mol $H_2SO_4 = 2.000 \times 10^{-3}$ mol; see (a)

$$\text{molarity B} = \frac{2.000 \times 10^{-3} \text{ mol } H_2SO_4}{2.000 \times 10^{-2} \text{ L}} = \frac{0.1000 \text{ mol } H_2SO_4}{1 \text{ L}} = 0.1000 \text{ M } H_2SO_4$$

(c) $$0.01532 \text{ L} \times \frac{0.1000 \text{ mol NaOH}}{1 \text{ L}} \times \frac{1 \text{ mol } H_2SO_4}{2 \text{ mol NaOH}} = 7.660 \times 10^{-4} \text{ mol } H_2SO_4$$

$$\text{molarity C} = \frac{7.660 \times 10^{-4}\ \text{mol H}_2\text{SO}_4}{3.064 \times 10^{-2}\ \text{L}} = 2.500 \times 10^{-2}\ \text{mol H}_2\text{SO}_4\text{/L} = 2.500 \times 10^{-2}\ \text{M H}_2\text{SO}_4$$

(d) $$\text{molarity D} = 0.03064\ \text{L} \times \frac{0.1000\ \text{mol NaOH}}{1\ \text{L}} \times \frac{1\ \text{mol H}_2\text{SO}_4}{2\ \text{mol NaOH}} \times \frac{1}{0.01532\ \text{L}} = 0.1000\ \text{M H}_2\text{SO}_4$$

4.81 $NaOH(aq) + HCl(aq) \rightarrow NaCl(aq) + H_2O(l)$

The steps in the calculation are: $V \times M$ (of NaOH) $\rightarrow$ mol NaOH $\rightarrow$ mol HCl $\rightarrow$ M HCl.

$$0.04010\ \text{L} \times \frac{0.09850\ \text{mol NaOH}}{1\ \text{L}} \times \frac{1\ \text{mol HCl}}{1\ \text{mol NaOH}} = 3.950 \times 10^{-3}\ \text{mol HCl}$$

$$\text{molarity} = \frac{3.950 \times 10^{-3}\ \text{mol HCl}}{4.500 \times 10^{-2}\ \text{L}} = 8.778 \times 10^{-2}\ \text{mol HCl/L} = 8.778 \times 10^{-2}\ \text{M HCl}$$

4.83 The steps in the calculation are: $V \times M$ (of Co^{2+}) $\rightarrow$ mol Co^{2+} $\rightarrow$ g Co $\rightarrow$ % Co.

$$0.0500\ \text{L} \times \frac{0.071\ \text{mol Co}^{2+}}{1\ \text{L}} \times \frac{58.93\ \text{g Co}}{1\ \text{mol Co}^{2+}} = 0.21\ \text{g Co}$$

$$\%\ \text{Co} = \frac{\text{mass Co}}{\text{mass sample}} \times 100\% = \frac{0.21\ \text{g}}{2.122\ \text{g}} \times 100\% = 9.9\%$$

4.85 $CaCO_3(s) + 2HCl(aq) \rightarrow CaCl_2(aq) + CO_2(g) + H_2O(l)$
1.110 g M = ?

The steps in the calculation are: g $CaCO_3$ $\rightarrow$ mol $CaCO_3$ $\rightarrow$ mol HCl $\rightarrow$ M HCl.

$$1.110\ \text{g CaCO}_3 \times \frac{1\ \text{mol CaCO}_3}{100.09\ \text{g CaCO}_3} \times \frac{2\ \text{mol HCl}}{1\ \text{mol CaCO}_3} = 0.02218\ \text{mol HCl}$$

The molarity of the HCl solution is

$$\text{molarity HCl} = \frac{0.02218\ \text{mol HCl}}{0.03545\ \text{L}} = \frac{0.6257\ \text{mol HCl}}{1\ \text{L}} = 0.6257\ \text{M HCl}$$

4.87 $Ca(HCO_3)_2(aq) + Ca(OH)_2(aq) \rightarrow 2CaCO_3(s) + 2H_2O(l)$
800 L of g = ?
1.0×10^{-4} M

(a) The steps in the calculation are: $V \times M$ (of $Ca(HCO_3)_2$) $\rightarrow$ mol $Ca(HCO_3)_2$ $\rightarrow$ mol $Ca(OH)_2$ $\rightarrow$ g $Ca(OH)_2$.

$$800\ \text{L} \times \frac{1.0 \times 10^{-4}\ \text{mol Ca(HCO}_3)_2}{1\ \text{L}} \times \frac{1\ \text{mol Ca(OH)}_2}{1\ \text{mol Ca(HCO}_3)_2} \times \frac{74.09\ \text{g Ca(OH)}_2}{1\ \text{mol Ca(OH)}_2} = 5.9\ \text{g Ca(OH)}_2$$

(b) No. Because the excess $Ca(OH)_2$ would make the water alkaline, and alkaline solutions can be very corrosive.

4.89 $MgCO_3{\cdot}CaCO_3(s) + 4HCl(aq) \rightarrow MgCl_2(aq) + CaCl_2(aq) + 2CO_2(g) + 2H_2O(l)$

The steps in the calculation are:
$V \times M$ (of HCl) $\rightarrow$ mol HCl $\rightarrow$ mol dolomite $\rightarrow$ g dolomite $\rightarrow$ % dolomite.

$$0.0500\ \text{L} \times \frac{0.260\ \text{mol HCl}}{1\ \text{L}} \times \frac{1\ \text{mol dolomite}}{4\ \text{mol HCl}} \times \frac{184.4\ \text{g dolomite}}{1\ \text{mol dolomite}} = 0.599\ \text{g dolomite}$$

$$\%\ \text{dolomite} = \frac{\text{mass dolomite}}{\text{mass sample}} \times 100\% = \frac{0.599\ \text{g}}{1.00\ \text{g}} \times 100\% = 59.9\%$$

4.91 A solution is a mixture of two or more substances uniformly dispersed as separate atoms, molecules, or ions rather than as larger aggregates. A solution is sometimes referred to as a homogeneous mixture. A heterogeneous mixture is one in which the components are not uniformly mixed on the molecular level. One or more of the components exists in aggregates whose sizes are much larger than atomic or molecular dimensions.

True liquid and gaseous solutions are clear, and the mixed substances do not separate or settle upon standing. Heterogeneous liquid and gaseous mixtures can be turbid or cloudy and the components often settle upon standing.

Yes. If one or more of the substances absorbs light, the mixture will be colored.
No. Cloudiness is caused by the presence of particles much larger than atomic or molecular size.

4.93 If the solution is saturated, then 50 mL of water will contain no more than

$$50\ \text{mL} \times \frac{79.4\ \text{g thiosulfate}}{100\ \text{mL}} = 39.7\ \text{g thiosulfate}$$

This supersaturated solution contains 500 g – 39.7 g = 460 g of excess thiosulfate in a metastable state. The addition of a crystal of sodium thiosulfate should cause the crystallization of this 460 g of excess thiosulfate.

4.95 Yes. A strong electrolyte is any substance that is completely or almost completely ionized in aqueous solution. However, if the solubility of the substance is low, only a low concentration of dissolved ions will exist in solution, even though the substance is a strong electrolyte.

4.97 Zinc loses electrons more easily. Zinc is above iron in the activity series and therefore is the more active element. The fact that zinc is more active than iron means that it parts with its electrons more easily than iron.

4.99 True. Ionic compounds completely dissociate when they dissolve. Therefore, a mole of compound provides a number of moles of each ion equal to the number of each ion in the formula. Consequently, each ion's molarity will be either the same or some integral whole-number multiple of the compound's molarity.

4.101

(a) $NaHCO_3(s) + HCl(aq) \rightarrow NaCl(aq) + H_2O(l) + CO_2(g)$
$NaHCO_3(s) + H^+(aq) \rightarrow Na^+(aq) + H_2O(l) + CO_2(g)$

(b) $CaCO_3(s) + 2HCl(aq) \rightarrow CaCl_2(aq) + H_2O(l) + CO_2(g)$
$CaCO_3(s) + 2H^+(aq) \rightarrow Ca^{2+}(aq) + H_2O(l) + CO_2(g)$

(c) $Mg(OH)_2(s) + 2HCl(aq) \rightarrow MgCl_2(aq) + 2H_2O(l)$
$Mg(OH)_2(s) + 2H^+(aq) \rightarrow Mg^{2+}(aq) + 2H_2O(l)$

4.103 The molar mass of cholesterol is 386.7 g/mol. The number of millimoles of cholesterol in 100 mL of serum is

$$\frac{205\ \text{mg cholesterol}}{100\ \text{mL}} \times \frac{1\ \text{mmol cholesterol}}{386.7\ \text{mg cholesterol}} = \frac{0.530\ \text{mmol cholesterol}}{100\ \text{mL}}$$

The serum cholesterol concentration in millimoles per liter is

$$\frac{0.530 \text{ mmol cholesterol}}{100 \text{ mL}} \times \frac{1000 \text{ mL}}{1 \text{ L}} = \frac{5.30 \text{ mmol cholesterol}}{1 \text{ L}}$$

The serum cholesterol concentration in micromoles per liter is

$$\frac{5.30 \text{ mmol cholesterol}}{1 \text{ L}} \times \frac{10^3 \text{ micromoles}}{1 \text{ mmol}} = \frac{5.30 \times 10^3 \text{ micromoles cholesterol}}{1 \text{ L}}$$

4.105 When a solution is diluted to twelve times its original volume, its concentration becomes one-twelfth of what it originally was. If the new concentration is 9.09 mg/100 mL, then the old concentration was twelve times that, or

12×9.09 mg/100 mL = 109 mg/100 mL

Answer: The original serum specimen concentration was 109 mg/100 mL.

Comment: An alternative way to solve this problem is to use the dilution formula $V_C \times C_C = V_D \times C_D$ from Exercise 4.70.

4.107 1 km = 10^3 m and 1 m = 10^2 cm; so 1 km = 10^5 cm and 1 km^3 = 10^{15} cm^3.

(a) $1.37 \times 10^9 \text{ km}^3 \times \frac{10^{15} \text{ cm}^3}{1 \text{ km}^3} \times \frac{4 \times 10^{-6} \text{ mg gold}}{10^3 \text{ cm}^3} \times \frac{1 \text{ g}}{10^3 \text{ mg}} = 5 \times 10^{12} \text{ g gold}$

(b) At \$400 per troy ounce (gold is weighed in the troy unit system; 1 oz troy = 1.09714 oz avior.)

$$\frac{\$400}{1 \text{ oz troy}} \times \frac{1 \text{ oz troy}}{1.097 \text{ oz}} \times \frac{1 \text{ oz}}{28.35 \text{ g}} \times 5 \times 10^{12} \text{ g} = 6 \times 10^{13} \text{ dollars}$$

4.109 All of the original Fe in the ore is now in 3.766 g of iron(III) oxide (Fe_2O_3). The mass of Fe can be found using the mass fraction of Fe in Fe_2O_3.

$$\text{mass Fe} = 3.766 \text{ g Fe}_2\text{O}_3 \times \frac{111.7 \text{ g Fe}}{159.7 \text{ g Fe}_2\text{O}_3} = 2.634 \text{ g Fe}$$

The percent by mass of iron in the ore is

$$\% \text{ Fe} = \frac{\text{mass iron}}{\text{mass sample ore}} \times 100\% = \frac{2.634 \text{ g}}{9.240 \text{ g}} \times 100\% = 28.51\%$$

4.111 No. The order (higher to lower) of the activities of the elements involved is Mg, Fe, Sn, H, Ag, and Pt. Only elements above a given element in the activity series can displace that element. Elements below a given element cannot displace it. Tin can displace hydrogen but not iron. Silver and platinum cannot displace either hydrogen or iron. Magnesium can displace both hydrogen and iron. Therefore, the wire that dissolved in HCl solution but not in $FeCl_2$ solution must be Sn. The other wire that didn't dissolve in either solution could be either Ag or Pt; there is not enough information to tell which one it is.

4.113 (a) proof = 2 × % by volume of ethanol = 2 × 12.5 = 25

Answer: The wine is 25 proof.

(b) You would need

$$150 \text{ mg} \times \frac{1 \text{ oz}}{20 \text{ mg}} = 7.5 \text{ oz}$$

of 90-proof whiskey in order to achieve a concentration of 150 mg per 100 mL of blood. The number of scotch highballs needed to supply 7.5 oz of whiskey would be

$$7.5 \text{ oz whiskey} \times \frac{1 \text{ scotch highball}}{1.5 \text{ oz whiskey}} = 5 \text{ scotch highballs}$$

Answer: A 70-kg man drinking five scotch highballs will have an alcohol concentration of 150 mg/100 mL blood.

4.115

$$1.00 \text{ g NaCl} \times \frac{1 \text{ mol NaCl}}{58.44 \text{ g NaCl}} = 0.0171 \text{ mol NaCl}$$

In order for the two solutions to be identical, 0.0171 mol of each of the other salts would be needed. The masses of the other salts needed are:

$$0.0171 \text{ mol } KNO_3 \times \frac{101.10 \text{ g } KNO_3}{1 \text{ mol } KNO_3} = 1.73 \text{ g } KNO_3$$

$$0.0171 \text{ mol } NaNO_3 \times \frac{84.99 \text{ g } NaNO_3}{1 \text{ mol } NaNO_3} = 1.45 \text{ g } NaNO_3$$

$$0.0171 \text{ mol KCl} \times \frac{74.55 \text{ g KCl}}{1 \text{ mol KCl}} = 1.27 \text{ g KCl}$$

4.117

$BaCl_2(aq) + H_2SO_4(aq) \rightarrow BaSO_4(s) + 2HCl(aq)$

Initially there are 0.1200 L × 0.150 mol H_2SO_4/L = 0.0180 mol H_2SO_4 and 4.10 g $BaCl_2$ × 1 mol $BaCl_2$/208.2 g $BaCl_2$ = 0.0197 mol $BaCl_2$ in 0.2000 L of solution. Since H_2SO_4 is the limiting reagent, 0.0180 mol of $BaSO_4$ forms (precipitates), leaving no sulfate ion (SO_4^{2-}) in solution, $[SO_4^{2-}] = 0$, and 0.0017 mol $BaCl_2$ unreacted.

$$\text{mass } BaSO_4 = 0.0180 \text{ mol } BaSO_4 \times \frac{233.4 \text{ g}}{1 \text{ mol } BaSO_4} = 4.20 \text{ g}$$

$$[Ba^{2+}] = \frac{0.0017 \text{ mol}}{0.2000 \text{ L}} = 8.5 \times 10^{-3} \text{ M}; \ [Cl^-] = \frac{2 \times 0.0197 \text{ mol}}{0.2000 \text{ L}} = 0.197 \text{ M};$$

$$[H^+] = \frac{2 \times 0.0180 \text{ mol}}{0.2000 \text{ L}} = 0.180 \text{ M}$$

CHAPTER 5

GASES AND THEIR PROPERTIES

Solutions To Practice Exercises

PE 5.1 The weight of the cylindrical mercury column is equal to the volume of the cylinder × density of mercury × g.

$$\text{weight} = 76.0\ \text{cm} \times 1.00\ \text{cm}^2 \times 13.6\ \text{g/cm}^3 \times 9.81\ \text{m/s}^2 \times \frac{1\ \text{kg}}{10^3\ \text{g}} = 10.1\ \text{kg}\bullet\text{m/s}^2 = 10.1\ \text{N}$$

The pressure exerted by the cylindrical mercury column is

$$\text{pressure} = \frac{\text{force}}{\text{area}} = \frac{10.1\ \text{N}}{1.00\ \text{cm}^2} \times \left(\frac{10^2\ \text{cm}}{1\ \text{m}}\right)^2 = 1.01 \times 10^5\ \text{N/m}^2$$

PE 5.2 (a) $1.04\ \text{atm} \times \dfrac{760\ \text{torr}}{1\ \text{atm}} = 790\ \text{torr}$ (b) $750\ \text{torr} \times \dfrac{1\ \text{atm}}{760\ \text{torr}} = 0.987\ \text{atm}$

PE 5.3 (a) 1 mm Hg corresponds to a pressure of 1 torr, so 120 mm Hg corresponds to a pressure of 120 torr.
(b) 120 torr × 1 atm/760 torr = 0.158 atm
(c) 120 torr × 1.01325 bar/760 torr = 0.160 bar

PE 5.4 The pressure of the gas is equal to the sum of the atmospheric pressure, P_{atm} = 770 torr, and the pressure exerted by 100 mm of mercury. Hence,

$P_{gas} = P_{atm} + 100\ \text{torr} = 770\ \text{torr} + 100\ \text{torr} = 870\ \text{torr}$

PE 5.5 The initial volume is 15.0 mL. Decreasing the pressure to one-fourth of its original value will increase the volume to four times its original value. The final volume will be 15.0 mL × 4 = 60.0 mL.

PE 5.6 The initial gas pressure is 155 kPa. The volume decreases by a factor of 18.5/25.9. Hence, the gas pressure increases by the inverse of this factor. The final pressure will be 155 kPa × 25.9/18.5 = 217 kPa.

PE 5.7 $P_i = 0.880$ atm, $V_i = 28.9$ mL; $V_f = 40.5$ mL, $P_f = ?$

$$P_f = P_i \times \frac{V_i}{V_f} = 0.880 \text{ atm} \times \frac{28.9 \text{ mL}}{40.5 \text{ mL}} = 0.628 \text{ atm}$$

PE 5.8 (a) Increase. A decrease in pressure will cause the volume to increase.
(b) The factor must be greater than unity; the factor is 780 torr/760 torr.

PE 5.9 (a) Decrease. The pressure varies inversely with the volume. If the volume increases, the pressure must decrease.
(b) The pressure decreases, so the volume factor must be less than unity; the factor is 4.0 L/6.0 L.

PE 5.10 (a) $T = 273.15 + C = 273.15 - 15 = 258$ K
(b) $173 = 273.15 + C$; $C = 173 - 273.15 = -100$°C

PE 5.11 (a) $V_i = 1.46$ L, $T_i = 273 + 20 = 293$ K; $V_f = 1.22$ L, $T_f = ?$

$$T_f = T_i \times \frac{V_f}{V_i} = 293 \text{ K} \times \frac{1.22 \text{ L}}{1.46 \text{ L}} = 245 \text{ K}$$

(b) $245 = 273 + C$; $C = 245 - 273 = -28$°C

PE 5.12 (a) Increase. The volume is directly proportional to the Kelvin temperature. If the temperature increases, the volume must increase.
(b) The initial and final Kelvin temperatures are $T_i = 273 - 83 = 190$ K and $T_f = 273 + 30 = 303$ K. The volume increases, so the temperature factor must be greater than unity; the factor is 303 K/190 K.

PE 5.13 The decrease in temperature from 75°C = 348 K to –10°C = 263 K tends to decrease the volume; the temperature factor will be less than unity, or 263 K/348 K. The increase in pressure from 620 torr to 745 torr also tends to decrease the volume; the pressure factor will be less than unity, or 620 torr/745 torr.

$$V_f = V_i \times \text{temperature factor} \times \text{pressure factor} = 55.0 \text{ mL} \times \frac{263 \text{ K}}{348 \text{ K}} \times \frac{620 \text{ torr}}{745 \text{ torr}} = 34.6 \text{ mL}$$

PE 5.14 The density increases as the temperature goes down from 303 K to 273 K; therefore, the temperature factor must be greater than unity, or 303 K/273 K. The density increases as the pressure goes up from 750 torr to 760 torr; therefore, the pressure factor must be greater than unity, or 760 torr/750 torr.

$$d_f = d_i \times \text{temperature factor} \times \text{pressure factor} = 0.635 \text{ g/L} \times \frac{303 \text{ K}}{273 \text{ K}} \times \frac{760 \text{ torr}}{750 \text{ torr}} = 0.714 \text{ g/L}$$

PE 5.15 The molar volume (volume per mole) of any gas is approximately 22.4 L at standard conditions, 1.00 atm and 0°C = 273 K. The volume increases as the temperature goes up from 273 K to 373 K; therefore, the temperature factor must be greater than unity, or 373 K/273 K. The volume increases as the pressure goes down from 760 torr to 740 torr; therefore, the pressure factor must be greater than unity, or 760 torr/740 torr.

$$22.4 \text{ L/mol} \times \frac{373 \text{ K}}{273 \text{ K}} \times \frac{760 \text{ torr}}{740 \text{ torr}} = 31.4 \text{ L/mol}$$

PE 5.16 The volume and number of moles of gas are fixed. The pressure decreases from 10.0 atm to 9.75 atm. The temperature will decrease by the same factor:

$$T = 273 \text{ K} \times \frac{9.75 \text{ atm}}{10.0 \text{ atm}} = 266 \text{ K} = -7^\circ\text{C}$$

PE 5.17 Example 5.12: P_i = 740 torr, V_i = 127 mL, T_i = 273 K; P_f = 760 torr, T_f = 303 K, V_f = ?

$$V_f = V_i \times \frac{P_i}{P_f} \times \frac{T_f}{T_i} = 127 \text{ mL} \times \frac{740 \text{ torr}}{760 \text{ torr}} \times \frac{303 \text{ K}}{273 \text{ K}} = 137 \text{ mL}$$

Practice Exercise 5.13: P_i = 620 torr, V_i = 55.0 mL, T_i = 348 K; P_f = 745 torr, T_f = 263 K, V_f = ?

$$V_f = 55.0 \text{ mL} \times \frac{620 \text{ torr}}{745 \text{ torr}} \times \frac{263 \text{ K}}{348 \text{ K}} = 34.6 \text{ mL}$$

PE 5.18 V = 0.250 L, T = 298 K, and n = 0.250 g × 1 mol/32.00 g = 7.81×10^{-3} mol.

$$P = \frac{nRT}{V} = \frac{7.81 \times 10^{-3} \text{ mol} \times 0.0821 \text{ L•atm/mol•K} \times 298 \text{ K}}{0.250 \text{ L}} = 0.764 \text{ atm}$$

PE 5.19 V = 0.100 L, T = 301 K, and P = 770 torr × 1 atm/760 torr = 1.01 atm.

$$n = \frac{PV}{RT} = \frac{1.01 \text{ atm} \times 0.100 \text{ L}}{0.0821 \text{ L•atm/mol•K} \times 301 \text{ K}} = 4.09 \times 10^{-3} \text{ mol}$$

Thus, 4.09×10^{-3} mol of the unknown gas has a mass of 0.360 g. The mass of 1 mol (the molar mass) is 0.360 g/4.09×10^{-3} mol = 88.0 g/mol.

PE 5.20 The molar mass of methane is 12.01 + (4 × 1.008) = 16.04 g/mol. P = 0.750 atm, T = 273 K, and $\mathcal{M}$ = 16.04 g/mol.

$$d = \frac{\mathcal{M}P}{RT} = \frac{16.04 \text{ g/mol} \times 0.750 \text{ atm}}{0.0821 \text{ L•atm/mol•K} \times 273 \text{ K}} = 0.537 \text{ g/L}$$

PE 5.21 $CH_4(g) + 2O_2(g) \rightarrow CO_2(g) + 2H_2O(g)$
0.250 mol L = ?

From the equation we see that 0.250 mol CO_2 will be produced from 0.250 mol CH_4.
T = 293 K, P = 1.00 atm.

$$V = \frac{nRT}{P} = \frac{0.250 \text{ mol} \times 0.821 \text{ L•atm/mol•K} \times 293 \text{ K}}{1.00 \text{ atm}} = 6.01 \text{ L}$$

PE 5.22 $2H_2O(l) \rightarrow 2H_2(g) + O_2(g)$

(a) The coefficients in the equation tell us that for every milliliter of oxygen produced there will be 2 mL of hydrogen produced. If 45.0 mL of hydrogen is produced, then the number of milliliters of oxygen produced at the same time is 45.0 mL H_2 × 1 mL O_2/2 mL H_2 = 22.5 mL O_2.

(b) No. This water is liquid, and the rule only applies to gaseous reactants and products.

PE 5.23 (a) $P_{total} = P_{hydrogen} + P_{water\ vapor}$

$P_{hydrogen} = P_{total} - P_{water\ vapor}$ = 740 torr – 23.8 torr = 716 torr

(b) $X_{hydrogen} = \frac{P_{hydrogen}}{P_{total}} = \frac{716 \text{ torr}}{740 \text{ torr}} = 0.968$

(c) mole percent hydrogen = 100% × $X_{hydrogen}$ = 100% × 0.968 = 96.8%

PE 5.24 Table 5.3 shows that the vapor pressure of water at 15°C is 12.8 torr. The partial pressure of hydrogen is obtained from Dalton's law:

$$P_{hydrogen} = P_{total} - P_{water} = 770 \text{ torr} - 12.8 \text{ torr} = 757 \text{ torr}$$

PE 5.25 $2HgO(s) \rightarrow 2Hg(l) + O_2(g)$

The vapor pressure of water is 18.65 torr at 21°C. The partial pressure of oxygen is

$$P_{oxygen} = P_{total} - P_{water} = 765 \text{ torr} - 18.65 \text{ torr} = 746 \text{ torr} \times \frac{1 \text{ atm}}{760 \text{ torr}} = 0.982 \text{ atm}$$

V = 0.250 L, T = 294 K. Substituting these values into the ideal gas law gives the number of moles of oxygen produced:

$$n_{oxygen} = \frac{PV}{RT} = \frac{0.982 \text{ atm} \times 0.250 \text{ L}}{0.0821 \text{ L}\bullet\text{atm/mol}\bullet\text{K} \times 294 \text{ K}} = 0.0102 \text{ mol}$$

The remainder of the calculation is: mol $O_2 \rightarrow$ mol HgO $\rightarrow$ g HgO.

$$0.0102 \text{ mol } O_2 \times \frac{2 \text{ mol HgO}}{1 \text{ mol } O_2} \times \frac{216.6 \text{ g HgO}}{1 \text{ mol HgO}} = 4.42 \text{ g HgO}$$

PE 5.26 $r_{CO_2} \times \sqrt{\mathcal{M}_{CO_2}} = r_x \times \sqrt{\mathcal{M}_x}$, $r_x = 1.66 \times r_{CO_2}$, and $\mathcal{M}_{CO_2} = 44.01$ g/mol.

$$\sqrt{\mathcal{M}_x} = \frac{r_{CO_2}}{1.66 r_{CO_2}} \times \sqrt{44.01 \text{ g/mol}}$$

$$\mathcal{M}_x = \left(\frac{1}{1.66}\right)^2 \times 44.01 \text{ g/mol} = 16.0 \text{ g/mol}$$

PE 5.27 150 g of N_2 is equal to $150 \text{ g } N_2 \times \frac{1 \text{ mol } N_2}{28.01 \text{ g } N_2} = 5.36 \text{ mol } N_2$.

The kinetic energy in joules of 1 mol of N_2 at 25°C is

$E_k = (3/2)RT = (3/2) \times 8.314 \text{ J/mol}\bullet\text{K} \times 298 \text{ K} = 3.72 \times 10^3 \text{ J/mol}$.

The kinetic energy of 150 g of N_2 is therefore 5.36 mol × 3.72×10^3 J/mol = 1.99×10^4 J.

PE 5.28 For O_2, $\mathcal{M}$ = 32.00 g/mol = 0.03200 kg/mol. T = 25°C = 298 K.

$$u_{rms} = \sqrt{\frac{3RT}{\mathcal{M}}} = \sqrt{\frac{3 \times 8.314 \text{ kg}\bullet\text{m}^2/\text{s}^2\bullet\text{mol}\bullet\text{K} \times 298 \text{ K}}{0.03200 \text{ kg/mol}}} = 482 \text{ m/s}$$

PE 5.29 Attractive forces lower the PV product; therefore, from the figure the order of increasing intermolecular attraction should be H_2, N_2, O_2, and CO_2.

PE 5.30 V = 0.750 L, T = 293 K, n = 1.10 mol, a = 3.667 $L^2\bullet$atm/mol^2, and b = 0.04081 L/mol.

(a) $$P = \frac{nRT}{V} = \frac{1.10 \text{ mol} \times 0.0821 \text{ L}\bullet\text{atm/mol}\bullet\text{K} \times 293 \text{ K}}{0.750 \text{ L}} = 35.3 \text{ atm}$$

(b) $$P = \frac{nRT}{V - nb} - \frac{an^2}{V^2}$$

$$P = \frac{1.10 \text{ mol} \times 0.0821 \text{ L•atm/mol•K} \times 293 \text{ K}}{0.750 \text{ L} - (1.10 \text{ mol} \times 0.04081 \text{ L/mol})} - \frac{3.667 \text{ L}^2\text{•atm/mol}^2 \times (1.10 \text{ mol})^2}{(0.750 \text{ L})^2}$$

$$= 37.5 \text{ atm} - 7.89 \text{ atm} = 29.6 \text{ atm}$$

PE 5.31 (a) The proportionality constant *a* is a measure of the intrinsic strength of the attractive forces in the gas. The larger the value, the greater the attractive forces in the gas. In order of increasing attractive force: He, H_2, O_2, N_2, CO_2, and HCl.

(b) The magnitude of *b* is a measure of the size of the molecules making up the gas. In order of increasing molecular size: He, H_2, O_2, N_2, HCl, and CO_2.

Solutions To Final Exercises

5.1 (a) 1. Finely divided particles suspended in a liquid are seen to be in continuous random motion. This observation is evidence that the much smaller liquid molecules themselves are in incessant and random motion. This phenomenon is called *Brownian motion*.
2. The spontaneous spreading of one substance through another, called *diffusion*, is evidence that the molecules are in continuous motion. It is the continuous random motion of the molecules that causes one substance's molecules to intermingle with those of another.
3. Gases have neither a definite volume nor a definite shape, but expand and completely fill any container. This is evidence that the gas molecules are in continuous random motion. The continuous random motion of the gas molecules brings them to every part of the container.
4. Gases exert continuous pressure equally in all directions. This is evidence that the gas molecules are in incessant random motion, since they are bombarding each square centimeter of the container walls equally.
5. The experimentally observed relationships between temperature and the rates of many molecular processes are explainable in terms of molecular motion. For instance, liquids flow more freely when hot, and the rate of diffusion increases with increasing temperature.

(b) 1. The pressure of gases in a confined space increases with increasing temperature. The molecules move more rapidly at the higher temperatures. They collide with the container walls more frequently and with a greater force and therefore exert a greater pressure on the container walls.
2. The rate of diffusion increases with temperature, showing that the molecules are moving faster.
3. Liquids flow more easily when hot, showing that the cohesion of the molecules is disrupted by the increased molecular motion.

5.3 (a) See the textbook (Section 5.1).

(b) *Thermal* means having to do with heat. When the temperature of a gas, liquid, or solid increases due to the absorption of heat energy, the motional or kinetic energies, i.e., the translational, rotational, and vibrational energy, of the atoms and molecules increases and the random motions of the atoms and molecules speed up. Likewise, when the temperature decreases due to a release of heat energy, the random motions of the atoms and molecules slow down. Since the random motion is directly related to the heat energy a substance absorbs or releases, it is often referred to as "thermal motion."

5.5 1. See Demonstration 5.1. The air is evacuated from a can. The can collapses, showing that the atmosphere exerts pressure.
2. See Figure 5.4. A tube filled with mercury is inverted into a dish of mercury. Most of the mercury remains in the tube; showing that the atmosphere exerts pressure.
3. A liquid is heated, but it doesn't start boiling until a certain temperature is reached. The liquid doesn't boil until its vapor pressure equals the atmospheric pressure; showing that the atmosphere exerts pressure.

5.7 (a) Greater than. The pressure of the atmosphere is the same in all directions. Pressure is defined as a force per unit area. The surface area of a basketball is greater than that of a tennis ball. Since the pressure is the same on both surfaces, the total force on the basketball must be greater.

(b) Equal to. While the basketball is larger than the tennis ball, the force per unit area on their surfaces is the same. Therefore, the pressure exerted by the atmosphere on them is the same.

5.9 pressure = force/area, 1 Pa = 1 N/m^2, and 1 N = 1 kg•m/s^2.

weight dime = $m \times g = 2.27 \times 10^{-3}$ kg $\times$ 9.81 m/s^2

$$\text{area dime} = \pi r^2 = \frac{\pi d^2}{4} = \frac{3.14(0.0179\ \text{m})^2}{4}$$

$$\text{pressure} = \frac{\text{force}}{\text{area}} = \frac{2.27 \times 10^{-3}\ \text{kg} \times 9.81\ \text{m/s}^2}{3.14(0.0179\ \text{m})^2/4} = 88.5\ \text{N/m}^2 = 88.5\ \text{Pa}$$

5.11 (a) 745 mm Hg = 745 torr; 745 torr $\times$ 1 atm/760 torr = 0.980 atm

(b) 1 Pa = 1N/m^2; 745 torr $\times$ 1.01325 $\times$ 10^5 Pa/760 torr = 9.93 $\times$ 10^4 Pa

(c) 1 bar = 10^5 Pa; 745 torr $\times$ 1.01325 bar/760 torr = 0.993 bar

5.13 (a) 2.00 $\times$ 10^5 Pa $\times$ 1 atm/1.01325 $\times$ 10^5 Pa = 1.97 atm

(b) 1.97 atm $\times$ 760 torr/atm = 1.50 $\times$ 10^3 torr

5.15 The sea-level pressure on earth is approximately 1 atm = 760 torr. One torr supports a column of mercury one millimeter high, 1 torr = 1 mm Hg (exactly only at 0°C). The surface pressure on Mars is

P_{Mars} = 0.0060 $\times$ 760 torr = 4.6 torr

Answer: The Martian atmosphere will support a column of mercury 4.6 mm high.

5.17

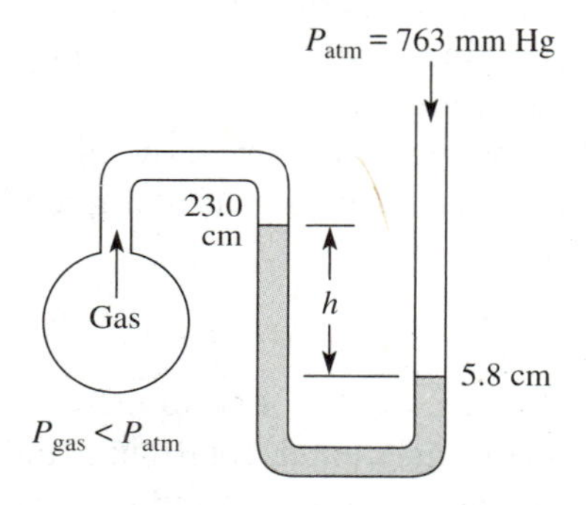

(a) $P_{gas} = P_{atm} - P_{Hg}$; P_{Hg} = 23.0 cm – 5.8 cm = 172 mm = 172 torr

P_{gas} = 763 torr – 172 torr = 591 torr

(b) Every one torr increase in gas pressure will be accompanied by an increase in the height of the mercury on the open side by one-half millimeter and a decrease in the mercury level on the closed side by one-half millimeter. The increase in gas pressure is 805 torr – 591 torr = 214 torr. Therefore the open end level will rise by 214 mm/2 = 107 mm = 10.7 cm and the closed end will fall by 10.7 cm.

Answer: open end = 5.8 cm + 10.7 cm = 16.5 cm
closed end = 23.0 cm – 10.7 cm = 12.3 cm

5.19 Pressure, volume, temperature, and number of moles.

5.21 (a) The volume decreases with increasing pressure at constant temperature.
(b) The volume decreases with decreasing temperature at constant pressure.
(c) The volume increases with increasing number of moles at constant temperature and pressure.

5.23 (a) The temperature remains constant; the pressure increases from 380 torr to 2.00 atm × 760 torr/1 atm = 1520 torr. Since the volume decreases, the pressure factor must be less than unity, or 380 torr/1520 torr.

$V_f = V_i \times$ pressure factor = 150 mL × 380 torr/1520 torr = 37.5 mL

(b) The pressure increases from 380 torr to 800 torr; the pressure factor is 380 torr/800 torr.

$V_f =$ 150 mL × 380 torr/800 torr = 71.2 mL

(c) The pressure increases from 380 torr to 1.50 bar × 760 torr/1.01325 bar = 1.13×10^3 torr; the pressure factor is 380 torr/1.13×10^3 torr.

$V_f =$ 150 mL × 380 torr/1.13×10^3 torr = 50.4 mL

(d) The pressure increases from 380 torr to 200 kPa × 760 torr/1.01325×10^2 kPa = 1500 torr; the pressure factor is 380 torr/1500 torr.

$V_f =$ 150 mL × 380 torr/1500 torr = 38.0 mL

5.25 Since the pressure decreases from 2.1 atm to 1.0 atm, the bubble increases by a factor of 2.1 atm/1.0 atm. The volume of the bubble is originally

$V_{bubble} = (4/3)\pi r^3_{bubble} = (4/3) \times 3.1416 \times (1.5\text{ cm})^3 = 14\text{ cm}^3 = 14\text{ mL}$

$V_f = V_i \times$ pressure factor = 14 mL × 2.1 atm/1.0 atm = 29 mL

5.27 The Kelvin degree and the Celsius degree are the same size. The Celsius temperature can be converted to a Kelvin temperature by adding 273.15. For general purposes, 273 is sufficiently accurate. So, T = 273.15 + C for accurate or detailed calculations; T = 273 + C for general purpose calculations.

(a) T = 273 + 37 = 310 K (b) T = 273 + 0 = 273 K (c) T = 273 + 100 = 373 K (d) T = 273 + 25 = 298 K

5.29 (a) Assuming constant pressure, we can use Charles's law: $V_i/T_i = V_f/T_f$.

T_i = 298.2 K, T_f = 373 K, V_i = 0.525 L, and V_f = ?

$$V_f = V_i \times \frac{T_f}{T_i} = 0.525\text{ L} \times \frac{373\text{ K}}{298.2\text{ K}} = 0.657\text{ L}$$

(b) V_f = 0.525 L × 249.6 K/298.2 K = 0.439 L (c) V_f = 0.525 L × 200 K/298.2 K = 0.352 L

(d) V_f = 0.525 L × 375 K/298.2 K = 0.660 L

5.31 (a) The temperature remains constant; the pressure decreases from 760 torr to 700 torr. Since the volume increases, the pressure factor must be greater than unity, or 760 torr/700 torr.
$V_f = V_i \times$ pressure factor = 2.5 L × 760 torr/700 torr = 2.7 L

(b) The pressure remains constant; the temperature decreases from 273 K to 253 K. Since the volume decreases, the temperature factor must be less than unity, or 253 K/273 K.

$V_f = V_i \times$ temperature factor = 2.5 L × 253 K/273 K = 2.3 L

5.33 (a) Assuming constant temperature and pressure, we can use Avogadro's law: $V_i/n_i = V_f/n_f$.

n_i = 0.0446 mol, n_f = 0.112 mol, V_i = 1.0 L, and V_f = ?

$$V_f = V_i \times \frac{n_f}{n_i} = 1.0\ L \times \frac{0.112\ mol}{0.0446\ mol} = 2.5\ L$$

(b) $$n_f = n_i \times \frac{V_f}{V_i} = 0.0446\ mol \times \frac{0.800\ L}{1.0\ L} = 0.036\ mol$$

5.35 (a) The temperature goes up from 273 K to 323 K; therefore, the temperature factor must be less than unity, or 273 K/323 K. The pressure goes down from 760 torr to 700 torr; therefore, the pressure factor must be less than unity, or 700 torr/760 torr.

$$d_f = d_i \times \text{temperature factor} \times \text{pressure factor} = 1.29\ g/L \times \frac{273\ K}{323\ K} \times \frac{700\ torr}{760\ torr} = 1.00\ g/L$$

(b) The temperature factor must be greater than unity, or 273 K/248 K. The pressure factor must be greater than unity, or 1.10 atm/1.00 atm.

$$d_f = 1.29\ g/L \times \frac{273\ K}{248\ K} \times \frac{1.10\ atm}{1.00\ atm} = 1.56\ g/L$$

5.37 (a) An ideal gas is one that obeys the equation PV = nRT under all conditions.
(b) Real gases behave most nearly like ideal gases at low pressures and high temperatures.

5.39 The ideal gas law is PV = nRT, where R is a constant.
(a) When n and T are also constant, then PV = constant (Boyle's law).
(b) When n and P are also constant, then V = (nR/P) × T = constant × T (Charles's law).
(c) When P and T are also constant; then V = (RT/P) × n = constant × n (Avogadro's law).
(d) When n and V are also constant, then P = (nR/V) × T = constant × T (Amontons' law).

5.41 See similar plots in the textbook (Figure 5.7). Both plots are hyperbolas of the form xy = k. The hyperbolas have different values of the constant k.

5.43 (a) Assuming constant volume, we can use Amontons' law: $P_i/T_i = P_f/T_f$.

T_i = 273 K, T_f = 318 K, P_i = 850 torr, and P_f = ?

$$P_f = P_i \times \frac{T_f}{T_i} = 850\ torr \times \frac{318\ K}{273\ K} = 990\ torr$$

(b) P_f = 850 torr × 83 K/273 K = 258 torr

(c) P_f = 850 torr × 200 K/273 K = 623 torr

(d) P_f = 850 torr × 323 K/273 K = 1.01 × 10^3 torr

5.45
$$\frac{P_i V_i}{T_i} = \frac{P_f V_f}{T_f}$$

P_i = 105 kPa, V_i = 100 L, T_i = 298 K; V_f = 10.0 L, T_f = 273 K, and P_f = ? The final pressure of the N_2 gas is

$$P_f = P_i \times \frac{V_i}{V_f} \times \frac{T_f}{T_i} = 105\ kPa \times \frac{100\ L}{10.0\ L} \times \frac{273\ K}{298\ K} = 962\ kPa$$

5.47 (a) At 0°C and 1.00 atm, 1 mol of oxygen occupies 22.4 L. Therefore, 10.0 g of O_2 will occupy

$$10.0\text{ g O}_2 \times \frac{1\text{ mol O}_2}{32.00\text{ g O}_2} \times \frac{22.4\text{ L}}{1\text{ mol O}_2} = 7.00\text{ L}$$

(b) The temperature decreases from 273 K to 255 K; therefore, the temperature factor must be less than unity, or 255 K/273 K. The pressure increases from 760 torr to 775 torr; therefore, the pressure factor must be less than unity, or 760 torr/775 torr. The volume at –18°C and 775 torr is

$$V_f = 7.00\text{ L} \times \frac{255\text{ K}}{273\text{ K}} \times \frac{760\text{ torr}}{775\text{ torr}} = 6.41\text{ L}$$

Comment: You could also use the ideal gas law (PV = nRT) to calculate the volume in parts (a) and (b).

5.49 PV = nRT and n = PV/RT.

(a) After inhalation: T = 310 K, P = 1.0 atm, and V = 3.5 L.

$$n = \frac{1.0\text{ atm} \times 3.5\text{ L}}{0.0821\text{ L•atm/mol•K} \times 310\text{ K}} = 0.14\text{ mol air}$$

(b) After exhalation: T = 310 K, P = 1.0 atm, and V = 3.0 L.

$$n = 0.14\text{ mol air} \times \frac{3.0\text{ L}}{3.5\text{ L}} = 0.12\text{ mol air}$$

5.51 (a) One mole of any gas occupies approximately 22.4 L at STP. Therefore, its approximate molar mass is

$$\frac{5.86\text{ g}}{1\text{ L}} \times \frac{22.4\text{ L}}{1\text{ mol}} = 131\text{ g/mol}$$

(b) The gaseous element is xenon.

5.53 The mass of the gas is: mass gas = mass filled tank – mass empty tank = 10.673 kg – 10.505 kg = 0.168 kg or 168 g.

The number of moles in the tank is 42.0 mol (from Example 5.17). Therefore, the molar mass of the gas is 168 g/42.0 mol = 4.00 g/mol.

5.55 (a) $\mathcal{M}$ = 153.8 g/mol, T = 293 K, and P = 780 torr × 1 atm/760 torr = 1.03 atm.

$$d = \frac{\mathcal{M}P}{RT} = \frac{153.8\text{ g/mol} \times 1.03\text{ atm}}{0.0821\text{ L•atm/mol•K} \times 293\text{ K}} = 6.59\text{ g/L}$$

(b) $\mathcal{M}$ = 44.01 g/mol, T = 310 K, and P = 1.10 atm.

$$d = \frac{\mathcal{M}P}{RT} = \frac{44.01\text{ g/mol} \times 1.10\text{ atm}}{0.0821\text{ L•atm/mol•K} \times 310\text{ K}} = 1.90\text{ g/L}$$

5.57 (a) V = 0.600 L, P = 427 torr × 1 atm/760 torr = 0.562 atm, T = 343 K.

$$n = \frac{PV}{RT} = \frac{0.562\text{ atm} \times 0.600\text{ L}}{0.0821\text{ L•atm/mol•K} \times 343\text{ K}} = 0.0120\text{ mol}$$

The molar mass is 1.430 g/0.0120 mol = 119 g/mol.

(b) The number of moles of C, H, and Cl in the molar mass is

$$\text{C:}\quad 119 \text{ g compound} \times \frac{10.1 \text{ g C}}{100 \text{ g compound}} \times \frac{1 \text{ mol C atoms}}{12.01 \text{ g C}} = 1.00 \text{ mol C atoms}$$

$$\text{H:}\quad 119 \text{ g compound} \times \frac{0.84 \text{ g H}}{100 \text{ g compound}} \times \frac{1 \text{ mol H atoms}}{1.008 \text{ g H}} = 0.992 \text{ mol H atoms}$$

$$\text{Cl:}\quad 119 \text{ g compound} \times \frac{89.1 \text{ g Cl}}{100 \text{ g compound}} \times \frac{1 \text{ mol Cl atoms}}{35.45 \text{ g Cl}} = 2.99 \text{ mol Cl atoms}$$

There is 1 mol of C atoms and 1 mol of H atoms for every 3 mol of Cl atoms. The atom ratio is the same as the mole ratio. The formula is $CHCl_3$.

5.59

$$2CH_3OH(l) + 3O_2(g) \rightarrow 2CO_2(g) + 4H_2O(g)$$

5.00 g   mL = ?

The steps in the calculation are: g $CH_3OH \rightarrow$ mol $CH_3OH \rightarrow$ mol $O_2 \rightarrow$ mL O_2.

$$5.00 \text{ g } CH_3OH \times \frac{1 \text{ mol } CH_3OH}{32.04 \text{ g } CH_3OH} \times \frac{3 \text{ mol } O_2}{2 \text{ mol } CH_3OH} = 0.234 \text{ mol } O_2$$

The number of milliliters occupied by 0.234 mol O_2 at 0°C and 1.00 atm is

$0.234 \text{ mol } O_2 \times 22.4 \text{ L}/1 \text{ mol } O_2 = 5.24 \text{ L or } 5.24 \times 10^3 \text{ mL } O_2$

Answer: 5.24×10^3 mL of O_2 measured at 0°C and 1.00 atm.

5.61

$$B_2H_6(g) + 3O_2(g) \rightarrow B_2O_3(s) + 3H_2O(g)$$

The number of moles of B_2H_6 under these conditions (V = 3.00 L, T = 293 K, P = 1.01 atm) is:

$$n = \frac{PV}{RT} = \frac{1.01 \text{ atm} \times 3.00 \text{ L}}{0.0821 \text{ L•atm/mol•K} \times 293 \text{ K}} = 0.126 \text{ mol } B_2H_6$$

The rest of the steps in the calculation are: mol $B_2H_6 \rightarrow$ mol $B_2O_3 \rightarrow$ g B_2O_3.

$$0.126 \text{ mol } B_2H_6 \times \frac{1 \text{ mol } B_2O_3}{1 \text{ mol } B_2H_6} \times \frac{69.62 \text{ g } B_2O_3}{1 \text{ mol } B_2O_3} = 8.77 \text{ g } B_2O_3$$

5.63

$$(NH_2)_2CO(aq) + H_2O(l) \rightarrow CO_2(g) + 2NH_3(g)$$

5.00 g   mL = ?

(a) The steps in the calculation are: g urea $\rightarrow$ mol urea $\rightarrow$ mol $NH_3 \rightarrow$ mL NH_3.

$$5.00 \text{ g urea} \times \frac{1 \text{ mol urea}}{60.06 \text{ g urea}} \times \frac{2 \text{ mol } NH_3}{1 \text{ mol urea}} = 0.167 \text{ mol } NH_3$$

The volume of NH_3 produced at T = 310 K and P = 755 torr × 1 atm/760 torr = 0.993 atm is

$$V = \frac{0.167 \text{ mol} \times 0.0821 \text{ L•atm/mol•K} \times 310 \text{ K}}{0.993 \text{ atm}} = 4.28 \text{ L or } 4.28 \times 10^3 \text{ mL } NH_3$$

(b) The coefficients tell us that 2 L of $NH_3(g)$ will be produced for every 1 L of $CO_2(g)$ produced. The volume of CO_2 produced is

$$4.28 \text{ L } NH_3 \times \frac{1 \text{ L } CO_2}{2 \text{ L } NH_3} = 2.14 \text{ L or } 2.14 \times 10^3 \text{ mL } CO_2$$

Answer: 4.28×10^3 mL NH_3 and 2.14×10^3 mL CO_2 for a total of 6.42×10^3 mL of gaseous products.

5.65

$N_2(g) + 3H_2(g) \rightarrow 2NH_3(g)$

The coefficients in the equation tell us that 2 L of $NH_3(g)$ will be produced from 1 L of $N_2(g)$. The volume of N_2 needed to produce 60 L of NH_3 is

$$60 \text{ L } NH_3 \times \frac{1 \text{ L } N_2}{2 \text{ L } NH_3} = 30 \text{ L } N_2$$

The coefficients in the equation tell us that 3 L of $H_2(g)$ are required for each liter of $N_2(g)$. The number of liters of H_2 required is

$$30 \text{ L } N_2 \times \frac{3 \text{ L } H_2}{1 \text{ L } N_2} = 90 \text{ L } H_2$$

5.67

Yes. Since the mole is just a counting unit, the mole fraction of a component, $X_C = n_C/n_T$, is equivalent to the number of component molecules/total number of molecules. Therefore, the mole fraction and the molecule fraction are identical.

5.69

See the derivation in the textbook (Section 5.6).

5.71

Effusion is the passage of gas molecules through small openings. Diffusion is the spontaneous spreading of one substance through another. Effusion requires the presence of some sort of barrier to the free movement of the gas molecules. Diffusion causes the mixing of molecules of different gases through random motion and collisions, so that the gaseous mixture becomes homogeneous.

5.73

(a) $X_{He} = \dfrac{n_{He}}{n_T} = \dfrac{1.50 \text{ mol}}{4.75 \text{ mol}} = 0.316$; $X_{CO_2} = 1 - X_{He} = 1 - 0.316 = 0.684$

(b) mole percent = 100% × mole fraction

mole % helium = 0.316 × 100% = 31.6%
mole % carbon dioxide = 0.684 × 100% = 68.4%

(c) $P_i = P_T \times X_i$
$P_{He} = P_T \times X_{He} = 720 \text{ torr} \times 0.316 = 228 \text{ torr}$
$P_{CO_2} = P_T \times X_{CO_2} = 720 \text{ torr} \times 0.684 = 492 \text{ torr}$

5.75

(a) $P_T = P_{NOCl} + P_{NO} + P_{Cl_2} = 0.730 \text{ atm} + 0.270 \text{ atm} + 0.135 \text{ atm} = 1.135 \text{ atm}$

(b) $X_{NOCl} = P_{NOCl}/P_T = 0.730 \text{ atm}/1.135 \text{ atm} = 0.643$

$X_{NO} = P_{NO}/P_T = 0.270 \text{ atm}/1.135 \text{ atm} = 0.238$

$X_{Cl_2} = 1 - X_{NOCl} - X_{NO} - 1 - 0.643 - 0.238 = 0.119$

5.77

(a) $P_{H_2} = P_{I_2} = \dfrac{0.044 \text{ mol} \times 0.0821 \text{ L•atm/mol•K} \times 731 \text{ K}}{10.0 \text{ L}} = 0.26 \text{ atm}$

$$P_{HI} = \frac{0.312 \text{ mol} \times 0.0821 \text{ L•atm/mol•K} \times 731 \text{ K}}{10.0 \text{ L}} = 1.87 \text{ atm}$$

(b) $P_T = P_{H_2} + P_{I_2} + P_{HI} = 0.26 \text{ atm} + 0.26 \text{ atm} + 1.87 \text{ atm} = 2.39 \text{ atm}$

Comment: Using $n_T = 0.400$ mol, V = 10.0 L, and T = 731 K in the ideal gas equation gives $P_T = 2.40$ atm. The difference in the results is due to rounding at each step in the calculation.

5.79 (a) The gas sample consists of nitrogen and water vapor. The vapor pressure of water at 25°C is 23.8 torr. The partial pressure of N_2 is

$$P_T = P_{N_2} + P_{water}\ ;\ P_{N_2} = P_T - P_{water} = 740\ \text{torr} - 23.8\ \text{torr} = 716\ \text{torr}$$

(b) The mole percent of nitrogen is

$$\text{mol \% nitrogen} = X_{N_2} \times 100\% = (P_{N_2}/P_T) \times 100\%$$

$$\text{mol \% nitrogen} = \frac{716\ \text{torr}}{740\ \text{torr}} \times 100\% = 96.8\%$$

5.81 $CaH_2(s) + 2H_2O(l) \rightarrow 2H_2(g) + Ca(OH)_2(aq)$

$$P_{H_2} = P_T - P_{water} = 750\ \text{torr} - 22.4\ \text{torr} = 728\ \text{torr} \times 1\ \text{atm}/760\ \text{torr} = 0.958\ \text{atm}$$

The number of moles of hydrogen produced is

$$n_{H_2} = \frac{PV}{RT} = \frac{0.958\ \text{atm} \times 1.00\ \text{L}}{0.0821\ \text{L}\bullet\text{atm/mol}\bullet\text{K} \times 297\ \text{K}} = 0.0393\ \text{mol}$$

The remaining steps in the calculation are: mol $H_2 \rightarrow$ mol $CaH_2 \rightarrow$ g CaH_2.

$$0.0393\ \text{mol}\ H_2 \times \frac{1\ \text{mol}\ CaH_2}{2\ \text{mol}\ H_2} \times \frac{42.09\ \text{g}\ CaH_2}{1\ \text{mol}\ CaH_2} = 0.827\ \text{g}\ CaH_2$$

5.83 Light or less massive gases effuse more rapidly than heavy or more massive gases. The mathematical relationship between the rates of effusion of two different gases is given by Graham's law:

$$r_A \times \sqrt{\mathcal{M}_A} = r_B \times \sqrt{\mathcal{M}_B}\ .$$

(a) The molar masses of N_2O and O_2 are 44.01 g/mol and 32.00 g/mol, respectively. Therefore, O_2 effuses more rapidly than N_2O.

(b) To compare the rates, we use Graham's law.

$$r_{oxygen} \times \sqrt{32.00\ \text{g/mol}} = r_{nitrous\ oxide} \times \sqrt{44.01\ \text{g/mol}}$$

$$r_{oxygen} = r_{nitrous\ oxide} \times \sqrt{\frac{44.01}{32.00}} = 1.17 \times r_{nitrous\ oxide}$$

Oxygen should effuse 1.17 times more rapidly than nitrous oxide.

5.85 $r_{Ne} \times \sqrt{\mathcal{M}_{Ne}} = r_x \times \sqrt{\mathcal{M}_x}$, r_{Ne} = volume/26.7 s, and r_x = volume/38.5 s.

$$\mathcal{M}_x = \left(\frac{r_{Ne}}{r_x}\right)^2 \times \mathcal{M}_{Ne} = \left(\frac{38.5}{26.7}\right)^2 \times 20.18\ \text{g/mol} = 42.0\ \text{g/mol}$$

5.87 (a) Greater than. All gases have the same molar kinetic energy at the same temperature. Since there is an approximate nitrogen to oxygen mole ratio of 4:1 in air, the nitrogen molecules should possess a total of four times more kinetic energy than the oxygen molecules.

(b) Greater than. At constant temperature, the root mean square speed varies inversely with the square root of the molar mass; $u_{rms} = \sqrt{3RT/\mathcal{M}} = k \times \sqrt{1/\mathcal{M}}$, where $k = \sqrt{3RT}$. The larger the mass, the smaller the u_{rms}. The molar masses of N_2 and O_2 are 28.01 g/mol and 32.00 g/mol, respectively. Since $\mathcal{M}_{nitrogen} < \mathcal{M}_{oxygen}$, the u_{rms} of the nitrogen molecules is greater than that of the oxygen molecules.

5.89 See Figure 5.19 in the textbook.

As the temperature increases, the values of the most probable speed and u_{rms} increase and the distribution curve flattens out. The flattening of the maximum means that the molecules are more evenly distributed among the various speeds. At higher temperatures there is a significant increase in the number of molecules with greater speeds.

As the temperature decreases, the values of the most probable speed and u_{rms} decrease and the distribution curve narrows and peaks higher. The heightening of the maximum means that a larger fraction of the molecules have speeds around the most probable speed and very few molecules have speeds much greater than the most probable speed.

5.91 (a) $\text{average speed} = \dfrac{375\text{ m/s} + 420\text{ m/s} + 480\text{ m/s}}{3} = 425\text{ m/s}$

$$\text{mean square speed} = \frac{(375\text{ m/s})^2 + (420\text{ m/s})^2 + (480\text{ m/s})^2}{3} = 1.82 \times 10^5\text{ m}^2/\text{s}^2$$

$$\text{root mean square speed} = \sqrt{1.82 \times 10^5\text{ m}^2/\text{s}^2} = 427\text{ m/s}$$

(b) The mean square speed $\left(\overline{u^2}\right)$ is obtained by finding the squares of the speeds of all the molecules, adding these squares together, and taking the average. The square of the average speed $\left(\bar{u}^2\right)$ is obtained by evaluating the average speed of all the molecules and then squaring the result. Since $\overline{u^2}$ is not numerically equal to $\bar{u}^2$, the root mean square speed $\left(\sqrt{\overline{u^2}}\right)$ is not equal to the average speed $\left(\bar{u} = \sqrt{\bar{u}^2}\right)$.

5.93 $u_{rms} = \sqrt{3RT/\mathcal{M}}$

(a) For He, $\mathcal{M} = 4.003$ g/mol $= 0.004003$ kg/mol, T = 298 K, and R = 8.314 kg•m²/s²•mol•K.

$$u_{rms} = \sqrt{\frac{3 \times 8.314\text{ kg}\cdot\text{m}^2/\text{s}^2\cdot\text{mol}\cdot\text{K} \times 298\text{ K}}{0.004003\text{ kg/mol}}} = 1.36 \times 10^3\text{ m/s}$$

(b) For UF_6, $\mathcal{M} = 352.0$ g/mol $= 0.3520$ kg/mol.

$$u_{rms} = \sqrt{\frac{3 \times 8.314\text{ kg}\cdot\text{m}^2/\text{s}^2\cdot\text{mol}\cdot\text{K} \times 298\text{ K}}{0.3520\text{ kg/mol}}} = 145\text{ m/s}$$

5.95 Increase. $u_{rms} = \sqrt{3RT/\mathcal{M}}$. If T is replaced by 2T we get

$$u_{rms}(\text{at 2T}) = \sqrt{\frac{3R \times 2T}{\mathcal{M}}} = \sqrt{2} \times \sqrt{3RT/\mathcal{M}} = \sqrt{2} \times u_{rms}\text{ (at T)}$$

$$u_{rms}\text{ (at 2T)} = 1.414\ u_{rms}\text{ (at T)}$$

5.97 Ideal gas molecules have zero volume (point particles) and no attractive interactions. Real gas molecules have weak attractive interactions and small but definite volumes. As the temperature decreases and the pressure increases, the speed of the gas molecules and the average distance between them drops. The volume occupied by the molecules themselves is no longer negligible compared to the total volume. The intermolecular forces are greater when the molecules are closer together and have a greater effect on the slower moving molecules. Therefore, under conditions of low temperature and high pressure, the real gas molecules act less like ideal gas molecules and the gas becomes less ideal.

5.99 (a) It decreases. Attractive forces reduce either P, or V, or both and cause P × V to be less than the ideal value of nRT. Therefore, the ratio PV/nRT will decrease with increasing intermolecular attractions (all other factors being constant or negligible).

(b) The answer to this question has two parts.

1. In the pressure region where the effect of molecular attractions predominates over the effect of volume of molecules, PV/nRT decreases. In this region, increasing pressure forces the gas molecules closer together. The closer the gas molecules get (up to a point), the stronger the attractive forces experienced by the gas molecules become, since the strength of the force increases as the distance between the gas molecules decreases. Attractive forces cause PV to be less than the ideal value of nRT. Therefore, the ratio PV/nRT will decrease with increasing pressure (increasing attractive forces) in this region.

2. In the pressure region where the effect of volume of molecules predominates over the effect of molecular attractions, PV/nRT increases. In this region, increasing pressure also forces the gas molecules closer together. However, in this region the effect of volume of molecules predominates, and the effect increases as the pressure increases, since the volume decreases with increasing pressure and the molecular volume fraction of the total volume increases. The effect of volume of molecules causes PV to be greater than the ideal value of nRT. Therefore, the ratio PV/nRT will increase with increasing pressure.

(c) Same answer as (b), since increasing the density of a gas is the same as increasing the pressure.

5.101 (a) Real gases have attractive interactions, i.e., there are attractive forces between their molecules. When expanding, work must be performed to overcome these attractive forces. The energy to perform the work comes from the kinetic energy of the gas. According to the kinetic molecular theory of gases, when the average kinetic energy of the gas molecules decreases, the temperature drops. Therefore, gases cool (their temperatures drop) upon expansion.

(b) When a tire is inflated, the pressure of the incoming air pushes back the existing air. This physical moving of the air by a force is a form of work (work = force × distance = pressure × volume change); sometimes called pressure-volume work or PV work. The energy used to perform the work winds up as kinetic energy of the gas. According to the kinetic molecular theory of gases, when the average kinetic energy of the gas molecules increases, the temperature rises. Therefore, the gas warms up (temperature rises) and the tire becomes warm by heat conduction when air is pumped in.

Also, real gases have attractive forces and as more gas molecules that attract each other come closer together, as when the tire is being inflated, their potential energy decreases and the energy appears as kinetic energy. So, the average kinetic energy of the gas molecules increases, which means the temperature of the gas rises, and the tire warms up.

These effects wouldn't occur if the gases were ideal, because ideal gas molecules have no attractive interactions and zero volume (a gas moving into a vacuum performs no work).

5.103 (a) n = 1.00 mol, V = 5.00 L, T = 473 K, and

$$P = \frac{nRT}{V} = \frac{1.00 \text{ mol} \times 0.0821 \text{ L•atm/mol•K} \times 473 \text{ K}}{5.00 \text{ L}} = 7.77 \text{ atm}$$

(b) $P = \dfrac{nRT}{V - nb} - \dfrac{an^2}{V^2}$; $a = 5.464 \text{ L}^2\text{•atm/mol}^2$ and $b = 0.03049$ L/mol

$$P = \frac{1.00 \text{ mol} \times 0.0821 \text{ L•atm/mol•K} \times 473 \text{ K}}{5.00 \text{ L} - (1.00 \text{ mol} \times 0.03049 \text{ L/mol})} - \frac{5.464 \text{ L}^2\text{•atm/mol}^2 \times (1.00 \text{ mol})^2}{(5.00 \text{ L})^2} = 7.81 \text{ atm} - 0.219 \text{ atm}$$

$= 7.59$ atm

The van der Waals equation is more accurate than the ideal gas law; therefore, the van der Waals value of 7.59 atm should be closer to the actual pressure.

5.105 According to the kinetic theory of gases, the attractive forces between gas molecules are weak and usually negligible. Furthermore, the gas molecules are in rapid constant random motion and are separated by distances that are large compared to their own size, except at low temperatures and high pressures. Therefore, at normal temperatures and pressures, the attractive forces are too weak to overcome the molecular motion and the relatively large intermolecular distances to form molecular aggregates that are large or massive enough for the relatively weak gravitational force to pull down to the earth.

5.107 One mole of any gas (pure or mixed) occupies approximately 22.4 L at STP. Therefore, the average molar mass of air is approximately 1.2929 g/L × 22.4 L/mol = 29.0 g/mol.

5.109 Assuming constant temperature, we can use Boyle's law: $P_i \times V_i = P_f \times V_f$.
$V_i = 200$ mL, $V_f = 6.60$ mL, $P_f = 30$ torr, and $P_i = ?$

$$P_i = P_f \times \frac{V_f}{V_i} = 30 \text{ torr} \times \frac{6.60 \text{ mL}}{200 \text{ mL}} = 0.99 \text{ torr}$$

5.111 The molar volume of an ideal gas is given by the equation

$$\text{molar volume ideal gas} = \frac{V}{n} = \frac{RT}{P}$$

At standard conditions, T = 273 K and P = 1.00 atm. The molar volume in liters of an ideal gas at standard conditions is

$$\frac{V}{n} = \frac{0.0821 \text{ L•atm/mol•K} \times 273 \text{ K}}{1.00 \text{ atm}} = 22.4 \text{ L/mol}$$

This result is consistent with Table 5.1. The molar volumes at STP of the real gases listed in Table 5.1 are all fairly close to 22.4 L/mol.

5.113 The volume of the air in the flask before heating is ($V_i = 200$ mL, $T_i = 273$ K; $T_f = 293$ K, $V_f = ?$)

$$V_f = V_i \times \text{temperature factor} = 200 \text{ mL} \times \frac{293 \text{ K}}{273 \text{ K}} = 215 \text{ mL}$$

The volume of the air in the flask after heating is ($V_i = 200$ mL, $T_i = 273$ K; $T_f = 298$ K, $V_f = ?$)

$$V_f = 200 \text{ mL} \times \frac{298 \text{ K}}{273 \text{ K}} = 218 \text{ mL}$$

The increase in volume in going from 20°C to 25°C is

$$V_{25°C} - V_{20°C} = 218 \text{ mL} - 215 \text{ mL} = 3 \text{ mL} = 3 \text{ cm}^3$$

The volume of any section of the tube is $V = h \times$ cross-sectional area $= h \times \pi d^2/4$; h is the amount the liquid will rise in the tube when the air volume increase by 3 cm^3.

$h \times \pi d^2/4 = h \times 3.14 \times (0.69 \text{ cm})^2/4 = h \times 0.37 \text{ cm}^2 = 3 \text{ cm}^3$

$h = 3 \text{ cm}^3/0.37 \text{ cm}^2 = 8 \text{ cm}$

5.115 The mass of C and H in the sample is

$$\text{mass C} = 20.5 \text{ mg } CO_2 \times \frac{12.01 \text{ mg C}}{44.01 \text{ mg } CO_2} = 5.59 \text{ mg C}$$

mass H = mass sample – mass C = 6.54 mg – 5.59 mg = 0.95 mg H

The molar mass of the compound is (d = 1.215 g/L, T = 298 K, and P = 800 torr × 1 atm/760 torr = 1.05 atm):

$$\mathcal{M} = \frac{dRT}{P} = \frac{1.215 \text{ g/L} \times 0.0821 \text{ L·atm/mol·K} \times 298 \text{ K}}{1.05 \text{ atm}} = 28.3 \text{ g/mol}$$

The number of moles of C and H in the molar mass is

$$\text{C: } 28.3 \text{ g compound} \times \frac{5.59 \text{ g C}}{6.54 \text{ g compound}} \times \frac{1 \text{ mol C atom}}{12.01 \text{ g C}} = 2.01 \text{ mol C atoms}$$

$$\text{H: } 28.3 \text{ g compound} \times \frac{0.95 \text{ g H}}{6.54 \text{ g compound}} \times \frac{1 \text{ mol H atoms}}{1.008 \text{ g H}} = 4.1 \text{ mol H atoms}$$

There are 2 mol of C atoms and 4 mol of H atoms in 1 mol of compound. The formula is C_2H_4.

5.117 $2LiOH(s) + CO_2(g) \rightarrow Li_2CO_3(s) + H_2O(l)$
g = ?

The steps in the calculation are: L CO_2 → mol CO_2 → mol LiOH → g LiOH.

The number of moles of CO_2 exhaled every 24 hours can be calculated using the ideal gas equation (V = 450 L, T = 303 K, P = 1.00 atm).

$$n_{CO_2} = \frac{1.00 \text{ atm} \times 450 \text{ L}}{0.0821 \text{ L·atm/mol·K} \times 303 \text{ K}} = 18.1 \text{ mol}$$

In six days the number of moles of CO_2 exhaled would be 6 × 18.1 = 109 mol. The minimum weight of LiOH needed to react with 109 mol CO_2 is

$$109 \text{ mol } CO_2 \times \frac{2 \text{ mol LiOH}}{1 \text{ mol } CO_2} \times \frac{23.95 \text{ g LiOH}}{1 \text{ mol LiOH}} = 5.22 \times 10^3 \text{ g or } 5.22 \text{ kg LiOH}$$

5.119 $NaHCO_3(aq) + HCl(aq) \rightarrow NaCl(aq) + H_2O(l) + CO_2(g)$
520 mg Na L = ?

According to the equation, for every mole of sodium consumed in the form of $NaHCO_3$, 1 mol CO_2 gas is evolved. The steps in the calculation are: g Na → mol Na → mol CO_2 →L CO_2.

$$0.520 \text{ g Na} \times \frac{1 \text{ mol Na}}{22.99 \text{ g Na}} \times \frac{1 \text{ mol } CO_2}{1 \text{ mol Na}} = 0.0226 \text{ mol } CO_2$$

The volume of CO_2 at T = 310 K and P = 1.00 atm can be calculated using the ideal gas equation.

$$V = \frac{nRT}{P} = \frac{0.0226 \text{ mol} \times 0.0821 \text{ L·atm/mol·K} \times 310 \text{ K}}{1.00 \text{ atm}} = 0.575 \text{ L}$$

5.121 Yes. Avogadro's hypothesis states that equal volumes of different gases at the same temperature and pressure contain the same number of moles. Gay-Lussac's law of combining volumes could then be restated in the following way: "At a given temperature and pressure, the moles of gases consumed and produced in a chemical reaction are in the ratio of small whole numbers." This result is in accord with previously drawn conclusions and observations about chemical reactions. Since the law of combining volumes in this form gives correct results, it can be inferred that Avogadro's hypothesis must also be correct. Therefore, the law of combining volumes does support Avogadro's hypothesis.

5.123 (a) The partial pressure P_i of a gas can be calculated from the total pressure P_T using the equation

$$P_i = P_T X_i = P_T \times \frac{\text{mol } \% \; i}{100\%}$$

At a total pressure of 1.00 atm the partial pressures would be

$$P_{\text{cyclopropane}} = 1.00 \text{ atm} \times 0.40 = 0.40 \text{ atm}; \; P_{\text{oxygen}} = 1.00 \text{ atm} \times 0.20 = 0.20 \text{ atm}$$

$$P_{\text{helium}} = 1.00 \text{ atm} - 0.40 \text{ atm} - 0.20 \text{ atm} = 0.40 \text{ atm}$$

(b) The average molar mass is the mass of gas that contains 1 mol of gas molecules. Each gas contributes to the average molar mass in proportion to its mole fraction.

$$\mathcal{M}_{\text{average}} = \sum_i \mathcal{M}_i \times X_i$$

$$\mathcal{M}_{\text{average}} = \mathcal{M}_{\text{cyclopropane}} \times X_{\text{cyclopropane}} + \mathcal{M}_{\text{oxygen}} \times X_{\text{oxygen}} + \mathcal{M}_{\text{helium}} \times X_{\text{helium}}$$
$$= 42.08 \text{ g/mol} \times 0.40 + 32.00 \text{ g/mol} \times 0.20 + 4.003 \text{ g/mol} \times 0.40 = 25 \text{ g/mol}$$

5.125

$$r_x \times \sqrt{\mathcal{M}_x} = r_{\text{He}} \times \sqrt{\mathcal{M}_{\text{He}}}; \; r_{\text{He}} = 2.7 r_x$$

$$\mathcal{M}_x = \left(\frac{r_{\text{He}}}{r_x}\right)^2 \times \mathcal{M}_{\text{He}} = \left(\frac{2.7 r_x}{r_x}\right)^2 \times \mathcal{M}_{\text{He}} = (2.7)^2 \times 4.003 \text{ g/mol} = 29 \text{ g/mol}$$

After combustion of the compound, all the carbon is in the CO_2 and all the hydrogen is in the H_2O. The numbers of moles of C and H are

$$\text{C: } 3.30 \times 10^{-3} \text{ g } CO_2 \times \frac{12.01 \text{ g C}}{44.01 \text{ g } CO_2} \times \frac{1 \text{ mol C atoms}}{12.01 \text{ g C}} = 7.50 \times 10^{-5} \text{ mol C atoms}$$

$$\text{H: } 2.05 \times 10^{-3} \text{ g } H_2O \times \frac{2.016 \text{ g H}}{18.02 \text{ g } H_2O} \times \frac{1 \text{ mol H atoms}}{1.008 \text{ g H}} = 2.28 \times 10^{-4} \text{ mol H atoms}$$

Dividing through by the number of moles of C gives the simple whole-number ratio 3 hydrogen:1 carbon. The empirical formula is CH_3.

The molar mass of the empirical formula is 15.03 g/mol. The number of empirical formula units in the molar mass is 29 g/15.03 g = 1.9 or 2 to the nearest whole-number. The molecular formula is C_2H_6.

5.127 (a) $1 \text{ J} = 1 \text{ N•m}$, $1 \text{ L} = 10^3 \text{ cm}^3 = 10^{-3} \text{ m}^3$, and $1 \text{ atm} = 1.01325 \times 10^5 \text{ N/m}^2$.

$$1.00 \text{ L•atm} \times \frac{10^{-3} \text{ m}^3}{1 \text{ L}} \times \frac{1.01325 \times 10^5 \text{ N}}{1 \text{ atm•m}^2} = 1.01 \times 10^2 \text{ N•m} = 1.01 \times 10^2 \text{ J}$$

(b) $$R = \frac{0.0821 \text{ L•atm}}{1 \text{ mol•K}} \times \frac{101 \text{ J}}{1.00 \text{ L•atm}} = 8.29 \text{ J/mol•K}$$

5.129 $E_k = (3/2)RT$ and $P = 2nE_k/3V$. For a confined gas V is fixed. If the temperature doubles, the molar kinetic energy doubles.

E_k (at 2T) = (3/2)R × 2T = 2 × (3/2)RT = 2 × E_k (at T)

If the molar kinetic energy doubles, the pressure doubles.

P (at 2T) = $2nE_k$ (at 2T)/3V = 2n × $2E_k$ (at T)/3V = 2 × $2nE_k$ (at T)/3V = 2 × P (at T)

Therefore, doubling the Kelvin temperature doubles the pressure in accordance with Charles's law.

5.131 (a) V = 50.0 L, T = 273 K, n = 31.25 mol, and

$$P = \frac{nRT}{V} = \frac{31.25\text{ mol} \times 0.0821\text{ L•atm/mol•K} \times 273\text{ K}}{50.0\text{ L}} = 14.0\text{ atm}$$

(b) $P = \frac{nRT}{V - nb} - \frac{an^2}{V^2}$; $a = 1.360\text{ L}^2\text{•atm/mol}^2$ and $b = 0.03183$ L/mol.

$$P = \frac{31.25\text{ mol} \times 0.0821\text{ L•atm/mol•K} \times 273\text{ K}}{50.0\text{ L} - (31.25\text{ mol} \times 0.03183\text{ L/mol})} - \frac{1.360\text{ L}^2\text{•atm/mol}^2 \times (31.25\text{ mol})^2}{(50.0\text{ L})^2}$$

= 14.3 atm – 0.531 atm = 13.8 atm

5.133 $Ce + \frac{x}{2}O_2 \rightarrow CeO_x$

moles Ce: $2.8\text{ g Ce} \times \frac{1\text{ mol Ce}}{140.1\text{ g Ce}} = 0.020\text{ mol Ce}$

initial moles O_2: $n_i = \frac{P_iV_i}{RT_i} = \frac{1.50\text{ atm} \times 1.0\text{ L}}{0.0821\text{ L•atm/mol•K} \times 298\text{ K}} = 0.061\text{ mol } O_2$

final moles O_2: $n_f = n_i \times$ pressure factor $= 0.061\text{ mol } O_2 \times \frac{1.00\text{ atm}}{1.50\text{ atm}} = 0.041\text{ mol } O_2$

moles O_2 consumed in reaction = 0.061 mol – 0.041 mol = 0.020 mol
So, 1 mol of Ce consumes 1 mol of O_2 in the reaction and, therefore, $(x/2) = 1$ and $x = 2$.

CHAPTER 6

THERMOCHEMISTRY

Solutions To Practice Exercises

PE 6.1 (a) system loses energy; $q = -10$ kJ. (b) system gains energy; $q = +5$ kJ.

PE 6.2 (a) system loses energy; $w = -500$ J. (b) system gains energy; $w = +20$ kJ.

PE 6.3 The liquid water absorbs 2600 J of heat; $q = +2600$ J. The steam does 170 J of work; $w = -170$ J. $\Delta E = q + w = 2600 \text{ J} - 170 \text{ J} = +2430$ J.

PE 6.4 (a) For the formation of NH_3, ΔH is negative and, therefore, the formation is exothermic.
(b) For the formation of NO_2, ΔH is positive and, therefore, the formation is endothermic.

PE 6.5 (a) $q = \Delta H$, q is negative because heat is being given off. Therefore, ΔH is negative.
(b) Since the system is colder than the surroundings, heat will flow into the system. q is positive and, therefore, ΔH is positive.

PE 6.6 The number of moles of methane burned can be calculated using the ideal gas equation. $V = 10.0$ L, $T = 298$ K, $P = 1.00$ atm, and

$$n = \frac{PV}{RT} = \frac{1.00 \text{ atm} \times 10.0 \text{ L}}{0.0821 \text{ L·atm/mol·K} \times 298 \text{ K}} = 0.409 \text{ mol } CH_4$$

The amount of heat given off by the combustion of 0.409 mol CH_4 is

$$0.409 \text{ mol } CH_4 \times \frac{802.3 \text{ kJ}}{1 \text{ mol } CH_4} = 328 \text{ kJ}$$

PE 6.7 To find the number of moles in 1.00 L: V × d (of compound) → g compound → mol compound.

$$1.00 \times 10^3 \text{ mL} \times \frac{0.692 \text{ g}}{1 \text{ mL}} \times \frac{1 \text{ mol}}{114.23 \text{ g}} = 6.06 \text{ mol}$$

The number of kilojoules of heat given off during the combustion is

$6.06 \text{ mol} \times 5455.6 \text{ kJ/1 mol} = 3.31 \times 10^4 \text{ kJ}$.

PE 6.8 (a) Fe_2O_3 forms from iron and oxygen. Iron and rust are solids and oxygen is a gas at 1 atm and 25°C.

$$2Fe(s) + 3/2\ O_2(g) \rightarrow Fe_2O_3(s) \quad \Delta H^\circ_f = -824.2 \text{ kJ}$$

(b) The sign of the enthalpy change shows that 824.2 kJ are given off when 2 mol of Fe reacts with oxygen. The steps in the calculation are: 50.0 kg Fe → mol Fe → kJ.

$$5.00 \times 10^4 \text{ g Fe} \times \frac{1 \text{ mol Fe}}{55.85 \text{ g Fe}} \times \frac{824.2 \text{ kJ}}{2.00 \text{ mol Fe}} = 3.69 \times 10^5 \text{ kJ}$$

PE 6.9 The required reaction (3) can be regarded as the sum of reactions 1 and 2.

1. $SO_2(g) \rightarrow S(s) + O_2(g) \quad \Delta H^\circ_1 = +297 \text{ kJ}$
2. $S(s) + 3/2\ O_2(g) \rightarrow SO_3(g) \quad \Delta H^\circ_2 = -396 \text{ kJ}$
3. $SO_2(g) + 1/2\ O_2(g) \rightarrow SO_3(g) \quad \Delta H^\circ_3 = ?$

By Hess's law

$$\Delta H^\circ_1 + \Delta H^\circ_2 = \Delta H^\circ_3$$

$$\Delta H^\circ_3 = 297 \text{ kJ} - 396 \text{ kJ} = -99 \text{ kJ}$$

PE 6.10 The combustion reactions are

		ΔH°
(1) $CH_3COOH(l) + 2O_2(g)$	$\rightarrow 2CO_2(g) + 2H_2O(l)$	–874.2 kJ
(2) $H_2(g) + 1/2\ O_2(g)$	$\rightarrow H_2O(l)$	–285.8 kJ
(3) $C(graphite) + O_2(g)$	$\rightarrow CO_2(g)$	–393.5 kJ

The thermochemical equation for the formation of CH_3COOH is

$$(4)\ 2C(graphite) + 2H_2(g) + O_2(g) \rightarrow CH_3COOH(l) \quad \Delta H^\circ_f = ?$$

Reversing and multiplying as needed gives us the steps that add up to 4.

		ΔH°
(1a) $2CO_2(g) + 2H_2O(l)$	$\rightarrow CH_3COOH(l) + 2O_2(g)$	+874.2 kJ
(2a) $2H_2(g) + O_2(g)$	$\rightarrow 2H_2O(l)$	2 × –285.8 kJ
(3a) $2C(graphite) + 2O_2(g)$	$\rightarrow 2CO_2(g)$	2 × –393.5 kJ

The enthalpy change for equation 4 is obtained by adding the enthalpy changes of the three steps:
$\Delta H^\circ_f = 874.2 \text{ kJ} - 2 \times 285.8 \text{ kJ} - 2 \times 393.5 \text{ kJ} = -484.4 \text{ kJ}$

PE 6.11

	$FeCr_2O_4(s)$ +	$4C(graphite) \rightarrow$	$Fe(s)$ +	$2Cr(s)$ +	$4CO(g)$	$\Delta H^\circ = +988.4 \text{ kJ}$
ΔH°_f:	?	4×0 kJ	1×0 kJ	2×0 kJ	4×-110.5 kJ	

$\Delta H^\circ = 1 \times 0 \text{ kJ} + (2 \times 0 \text{ kJ}) + (4 \times -110.5 \text{ kJ}) - \Delta H^\circ_f\ FeCr_2O_4 - (4 \times 0 \text{ kJ}) = 988.4 \text{ kJ}$

$\Delta H^\circ_f\ FeCr_2O_4 = -988.4 \text{ kJ} - 442.0 \text{ kJ} = -1430.4 \text{ kJ}$

Answer: The enthalpy of formation of $FeCr_2O_4$ is $\Delta H_f^\circ = -1430.4$ kJ/mol.

PE 6.12 The molar mass of chloroform ($CHCl_3$) is 119.38 g/mol.

molar heat capacity = specific heat capacity × molar mass

$$\text{specific heat capacity} = \frac{\text{molar heat capacity}}{\text{molar mass}} = \frac{113.8 \text{ J/mol}\cdot{}^\circ\text{C}}{119.38 \text{ g/mol}} = 0.9533 \text{ J/g}\cdot{}^\circ\text{C}$$

PE 6.13 q = mass × specific heat × ΔT

q = –3.0 kJ; the sign is negative since heat is lost, m = 200 g, and specific heat = 4.2 J/g•°C. ΔT is

ΔT = q/mass × specific heat = –3000 J/200 g × 4.2 J/g•°C = –3.6°C

$\Delta T = T_{final} - T_{initial}$

$T_{final} = \Delta T + T_{initial} = -3.6°C + 80.0°C = 76.4°C$

PE 6.14 The amount of heat given off during the combustion is

$$q = \text{heat capacity calorimeter} \times \Delta T = \frac{5200 \text{ J}}{1^\circ\text{C}} \times 2.93^\circ\text{C} = 1.52 \times 10^4 \text{ J} = 15.2 \text{ kJ}$$

Hence, 0.555 g mayonnaise gives off 15.2 kJ on combustion. The heat given off in kilocalories per gram is

$$\frac{15.2 \text{ kJ}}{0.555 \text{ g}} \times \frac{1 \text{ kcal}}{4.184 \text{ kJ}} = 6.55 \text{ kcal/g}$$

A dietary Calorie is equal to one kilocalorie; hence, the caloric content of 1 g of mayonnaise is 6.55 Calories. The number of dietary Calories in a 10.0-g helping of mayonnaise is therefore 10.0 g × 6.55 Calories/1 g = 65.5 Calories.

PE 6.15 If the reaction doesn't involve gases or there is no change in the number of moles of gas, then the constant volume heat of reaction equals (approximately) the enthalpy change.

(a) $S(s) + O_2(g) \rightarrow SO_2(g)$. Equal number of gas moles before and after reaction; therefore, there will be no difference.

(b) $CaO(s) + H_2O(l) \rightarrow Ca(OH)_2(s)$. The reaction doesn't involve gases; therefore, there will be no difference.

(c) $ZnO(s) + H_2(g) \rightarrow Zn(s) + H_2O(l)$. The number of moles of gas decreases; therefore, there will be a difference.

Answer: (a) no; (b) no; (c) yes.

PE 6.16 $NH_3(aq) + HCl(aq) \rightarrow NH_4Cl(aq) \quad \Delta H^\circ = -52.0$ kJ

The number of moles of NH_3 that reacts is 0.100 L × 0.50 mol NH_3/L = 0.050 mol NH_3. The heat given off by the reaction of 0.050 mol NH_3 is 0.050 mol NH_3 × 52.0 kJ/1 mol NH_3 = 2.6 kJ. The mass of the solution is 400 mL × 1.00 g/mL = 400 g. The temperature change is

$$\Delta T = \frac{q}{\text{mass} \times \text{specific heat}} = \frac{2600 \text{ J}}{400 \text{ g} \times 4.18 \text{ J/g}\cdot{}^\circ\text{C}} = 1.6^\circ\text{C}$$

PE 6.17 $N_2(g) + 1/2\ O_2(g) \rightarrow N_2O(g) \quad \Delta H^\circ = +82.05$ kJ

$\Delta n = n_{products} - n_{reactants} = 1 \text{ mol} - 1\frac{1}{2} \text{ mol} = -0.5 \text{ mol}$

$\Delta H^\circ = \Delta E^\circ + RT\Delta n$

$$\Delta E° = \Delta H° - RT\Delta n = 82.05\ kJ - 8.314 \times 10^{-3}\ kJ/mol\bullet K \times 298\ K \times (-0.500\ mol)$$
$$= 82.05\ kJ + 1.24\ kJ = +83.29\ kJ$$

Solutions To Final Exercises

6.1 (a) The first law of thermodynamics states that energy can be converted from one form into another, but it cannot be created or destroyed.

(b) $\Delta E = q + w$, where q is the heat flowing into the system and w is the work done on the system. q is positive when heat flows into the system and negative when heat flows out. w is positive when work is done on the system and negative when work is done by the system.

6.3 Yes. The energy of the system will increase whenever the work on the system by the surroundings is greater than the heat loss. Examples:

1. Hitting a piece of metal (the system) with a hammer. The metal will heat up and start losing heat, but its overall energy will be increasing.
2. Compressing a gas, like when you pump up a tire. The gas heats up and starts losing heat to its surroundings, but the system still has an overall increase in energy.
3. Charging a battery. The battery heats up and starts losing heat, but the system has an overall increase in chemical energy.
4. Winding up a spring. Heat is lost through friction, but the overall energy of the spring increases.
5. A nonwork example is the process of eating and digestion. The process of eating and digestion produces heat that is eventually lost to the surroundings, but the organism's overall energy increases.

6.5 Heat flows into the kernels of popcorn, therefore q is positive. The popcorn kernels expand and push back the atmosphere when they explode, therefore the kernels are doing work on the surroundings and w is negative.

6.7 A state function is any variable of the system whose magnitude or value only depends on the state of the system and not on how the system arrived at that state or came to be in that state.

Other state functions besides the internal energy (E) are: temperature (T), pressure (P), volume (V), enthalpy (H), density (d), and number of moles (n). Keep in mind that any variable that is some combination of state functions will also be a state function.

6.9 (a) $q = +10.6\ J$, $w = -5.3\ J$ and $\Delta E = +10.6\ J - 5.3\ J = +5.3\ J$

(b) $q = -25.3\ J$, $w = -12.0\ J$ and $\Delta E = -25.3\ J - 12.0\ J = -37.3\ J$

(c) $q = +52.9\ J$, $w = +387.0\ J$ and $\Delta E = +52.9\ J + 387.0\ J = +439.9\ J$

(d) $q = -187.5\ J$, $w = +10.0\ J$ and $\Delta E = -187.5\ J + 10.0\ J = -177.5\ J$

6.11 The battery is the system; the rest of the universe is the surroundings.

(a) For the battery: $q_{batt} = -2\ J$, $w_{batt} = -15\ J$ and $\Delta E_{batt} = -2\ J - 15\ J = -17\ J$.

(b) For the surroundings: $q_{surr} = +2\ J$, $w_{surr} = +15\ J$ and $\Delta E_{surr} = +2\ J + 15\ J = +17\ J$.

6.13 A constant pressure reaction is one that occurs under constant external pressure. Reactions that take place in aqueous solution, like heats of ionic reaction, aqueous acid-base neutralizations, and heats of solution, as well as changes of state, are most conveniently carried out at constant pressure. Some examples are:

$NaOH(aq) + HNO_3(aq) \rightarrow NaNO_3(aq) + H_2O(l)$

$Pb(NO_3)_2(aq) + K_2CrO_4(aq) \rightarrow PbCrO_4(s) + 2KNO_3(aq)$

$H_2O(s) \xrightarrow{\Delta} H_2O(l) \xrightarrow{\Delta} H_2O(g)$

6.15 (a) Pressure-volume work is the work associated with the volume change of a system.

(b) Both the pressure and volume change of the system would be measured and the values obtained substituted into the equation $w = -P\Delta V$.

(c) 1. Heating a gas or liquid in a container with a moveable piston.
2. Compressing a gas with a piston.
3. Reactions, like combustion reactions, decomposition reactions, etc., performed under constant pressure conditions, in which gases are either produced or consumed.

6.17 (a) The heat of reaction equals the enthalpy change for a reaction at constant temperature and pressure; that is, when the system is returned, after reaction, to the temperature and pressure it had before the reaction.

(b) Any reaction performed under these conditions of constant temperature and pressure is acceptable. Several examples are:

1. $Ca(OH)_2(s) + H_2SO_4(aq) \rightarrow CaSO_4(s) + 2H_2O(l)$
2. $2Al(s) + Fe_2O_3(s) \rightarrow 2Fe(s) + Al_2O_3(s)$
3. $MgCO_3(s) \xrightarrow{\Delta} MgO(s) + CO_2(g)$

6.19 (a) Yes. The quantity of heat evolved would be smaller, because some of the heat would be consumed in vaporizing the liquid. ΔH would be a smaller negative value.

(b) Since each state of a substance has a different enthalpy value, the value of the heat of reaction, ΔH, will be different depending on which state the substance is in. The value for the heat of reaction with a substance in one state does not apply to the same reaction with the substance in a different state.

6.21 When heat is absorbed, q is positive and ΔH is positive. When heat is given off, q is negative and ΔH is negative. Endothermic reactions have positive ΔH. Exothermic reactions have negative ΔH.

(a) Heat must be absorbed to vaporize the bromine. Therefore, ΔH is positive and the change is endothermic.
(b) Heat must be absorbed to melt the ice. Therefore, ΔH is positive and the change is endothermic.
(c) The reaction produces heat; heat is given off. Therefore, q is negative, ΔH is negative, and the reaction is exothermic.

6.23 A thermochemical equation is a chemical equation that includes the enthalpy change. For standard enthalpies of combustion to apply, all reactants and all products must be in their standard states.

(a) $C_2H_2(g) + 5/2\ O_2(g) \rightarrow 2CO_2(g) + H_2O(l)$ $\Delta H^\circ = -1299.6$ kJ
(b) $CH_3OH(l) + 3/2\ O_2(g) \rightarrow CO_2(g) + 2H_2O(l)$ $\Delta H^\circ = -726.5$ kJ

6.25 A standard enthalpy of formation (ΔH_f°) is the standard enthalpy change accompanying the formation of 1 mol of a substance from its elements, each in its most stable (lowest energy) form.

(a) $2Hg(l) + Cl_2(g) \rightarrow Hg_2Cl_2(s)$ $\Delta H_f^\circ = -265.2$ kJ

(b) $Ca(s) + O_2(g) + H_2(g) \rightarrow Ca(OH)_2(s)$ $\Delta H_f^\circ = -986.1$ kJ

(c) $Hg(l) \rightarrow Hg(g)$ $\Delta H_f^\circ = +61.3$ kJ

(d) $S(s) + \frac{3}{2}O_2(g) \rightarrow SO_3(g)$ $\Delta H_f^\circ = -395.7$ kJ

6.27 The reactants in a thermochemical formation equation must be neutral elements in their most stable forms.

(a) No. $Ca^{2+}(aq)$ and $CO_3^{2-}(aq)$ are not neutral elements, they are aqueous ions.
(b) Yes. All the reactants are neutral elements in their most stable form.
(c) No. Ozone (O_3) is not the most stable form of oxygen.

6.29 Reversing an equation changes the sign of its enthalpy change, and multiplying an equation by a factor multiplies the enthalpy change by the same factor.

$CO(g) + 1/2\ O_2(g) \rightarrow CO_2(g)$ $\Delta H° = -283.0$ kJ

(a) $2CO(g) + O_2(g) \rightarrow 2CO_2(g)$ $\Delta H° = -566.0$ kJ
Original equation multiplied by a factor of two.

(b) $CO_2(g) \rightarrow CO(g) + 1/2\ O_2(g)$ $\Delta H° = +283.0$ kJ
Original equation reversed.

6.31 $CaCO_3(s) \rightarrow CaO(s) + CO_2(g)$ $\Delta H° = +178.3$ kJ

The steps are: g $CaCO_3 \rightarrow$ mol $CaCO_3 \rightarrow$ kJ.

$$750\text{ g } CaCO_3 \times \frac{1\text{ mol } CaCO_3}{100.1\text{ g } CaCO_3} \times \frac{178.3\text{ kJ}}{1\text{ mol } CaCO_3} = 1.34 \times 10^3\text{ kJ}$$

6.33 $CS_2(l) + 3O_2(g) \rightarrow CO_2(g) + 2SO_2(g)$ $\Delta H° = -1076.9$ kJ

(a)

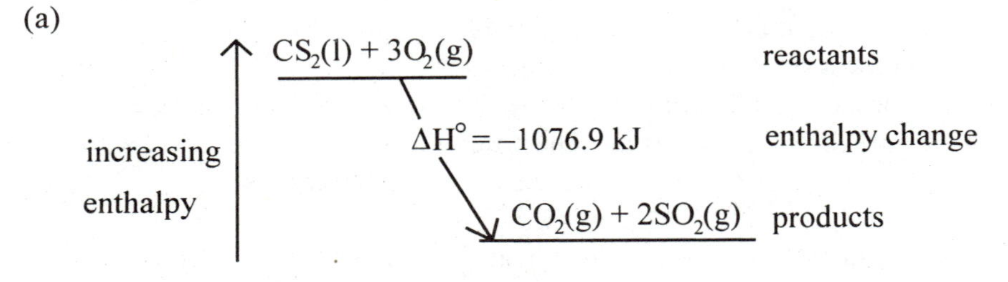

(b) The steps are: g $CS_2 \rightarrow$ mol $CS_2 \rightarrow$ kJ.

$$25.0\text{ g } CS_2 \times \frac{1\text{ mol } CS_2}{76.14\text{ g } CS_2} \times \frac{1076.9\text{ kJ}}{1\text{ mol } CS_2} = 354\text{ kJ}$$

6.35 $2N_2O(g) \rightarrow 2N_2(g) + O_2(g)$ $\Delta H° = -164.1$ kJ

The steps are: g $N_2O \rightarrow$ mol $N_2O \rightarrow$ kJ.

$$25.5\text{ g } N_2O \times \frac{1\text{ mol } N_2O}{44.01\text{ g } N_2O} \times \frac{164.1\text{ kJ}}{2\text{ mol } N_2O} = 47.5\text{ kJ}$$

6.37 $CaO(s) + H_2O(l) \rightarrow Ca(OH)_2(s)$ $\Delta H° = -65.2$ kJ

The steps are: kg CaO $\rightarrow$ g CaO $\rightarrow$ mol CaO $\rightarrow$ kJ.

$$1.00\text{ kg CaO} \times \frac{10^3\text{ g}}{1\text{ kg}} \times \frac{1\text{ mol CaO}}{56.08\text{ g CaO}} \times \frac{65.2\text{ kJ}}{1\text{ mol CaO}} = 1.16 \times 10^3\text{ kJ}$$

6.39 (a) Hess's law: The total enthalpy change for a reaction is the same whether the reaction occurs in one or several steps.

(b) Yes. Hess's law is based on the fact that enthalpy is a state function. Since the internal energy is also a state function, its change is also independent of the path and should follow Hess's law.

6.41 $F_2(g) \rightarrow 2F(g)$ $\Delta H° = ?$

Reversing and multiplying the original equations as needed, we get the following steps that add up to the desired equation.

$H_2(g) + F_2(g) \rightarrow 2HF(g)$ $\Delta H° = 2 \times -271$ kJ
$2H(g) \rightarrow H_2(g)$ $\Delta H° = -436$ kJ
$2HF(g) \rightarrow 2H(g) + 2F(g)$ $\Delta H° = 2 \times 568$ kJ

The enthalpy change is obtained by adding the enthalpy changes of the three steps:

$\Delta H° = -2 \times 271 \text{ kJ} - 436 \text{ kJ} + 2 \times 568 \text{ kJ} = +158 \text{ kJ}$

6.43 $C_2H_2(g) + 2H_2(g) \rightarrow C_2H_6(g)$ $\Delta H° = ?$
Reversing and multiplying the original equations as needed, we get the following steps that add up to the desired equation.

$C_2H_2(g) + 5/2\ O_2(g) \rightarrow 2CO_2(g) + H_2O(l)$ $\Delta H° = -1300$ kJ
$2H_2(g) + O_2(g) \rightarrow 2H_2O(l)$ $\Delta H° = 2 \times -286$ kJ
$2CO_2(g) + 3H_2O(l) \rightarrow 7/2\ O_2(g) + C_2H_6(g)$ $\Delta H° = +1560$ kJ

The enthalpy change is obtained by adding the enthalpy changes of the three steps:
$\Delta H° = -1300 \text{ kJ} - 572 \text{ kJ} + 1560 \text{ kJ} = -312 \text{ kJ}$

6.45 The standard enthalpy change for a reaction is given by the following equation.

$$\Delta H° = \sum_{\text{all products}} n_p \times (\Delta H_f^\circ)_p - \sum_{\text{all reactants}} n_r \times (\Delta H_f^\circ)_r,$$

where n_p is the number of moles of a given product and $(\Delta H_f^\circ)_p$ is its standard enthalpy of formation; n_r is the number of moles of a given reactant and $(\Delta H_f^\circ)_r$ is its standard enthalpy of formation.

(a) $CO(g) + 1/2\ O_2(g) \rightarrow CO_2(g)$; $\Delta H° = 1 \times -393.5 \text{ kJ} - (1 \times -110.5 \text{ kJ}) - (1/2 \times 0 \text{ kJ}) = -283.0 \text{ kJ}$; exothermic

(b) $N_2(g) + 3H_2(g) \rightarrow 2NH_3(g)$; $\Delta H° = 2 \times -46.1 \text{ kJ} - (1 \times 0 \text{ kJ}) - (3 \times 0 \text{ kJ}) = -92.2 \text{ kJ}$; exothermic

(c) $2O_3(g) \rightarrow 3O_2(g)$; $\Delta H° = 3 \times 0 \text{ kJ} - (2 \times 142.7 \text{ kJ}) = -285.4 \text{ kJ}$; exothermic

(d) $2KClO_3(s) \rightarrow 2KCl(s) + 3O_2(g)$; $\Delta H° = 2 \times -436.7 \text{ kJ} + (3 \times 0 \text{ kJ}) - (2 \times -397.7 \text{ kJ}) = -78.0 \text{ kJ}$; exothermic

(e) $CO(g) + 2H_2(g) \rightarrow CH_3OH(l)$; $\Delta H° = 1 \times -238.7 \text{ kJ} - (1 \times -110.5 \text{ kJ}) - (2 \times 0 \text{ kJ}) = -128.2 \text{ kJ}$; exothermic

6.47 $4NH_3(g) + 5O_2(g) \rightarrow 4NO(g) + 6H_2O(g)$ $\Delta H° = -905.4$ kJ
ΔH_f°: 4×-46.1 kJ 0 kJ ? 6×-241.8 kJ

$\Delta H° = 4 \times \Delta H_f^\circ \text{ NO} + 6 \times -241.8 \text{ kJ} - (4 \times -46.1 \text{ kJ}) - 0 \text{ kJ} = -905.4 \text{ kJ}$

$$\Delta H_f^\circ \text{ NO} = \frac{-905.4 \text{ kJ} + 6(241.8 \text{ kJ}) - 4(46.1 \text{ kJ})}{4} = +90.25 \text{ kJ}$$

6.49 (a) $CH_4(g) + 2O_2(g) \rightarrow CO_2(g) + 2H_2O(l)$; $\Delta H° = 2 \times -285.8 \text{ kJ} + (1 \times -393.5 \text{ kJ}) - (1 \times -74.8 \text{ kJ}) - (2 \times 0 \text{ kJ}) = -890.3 \text{ kJ}$; value in Table 6.2 is $\Delta H° = -890.4$ kJ

(b) $C_3H_8(g) + 5O_2(g) \rightarrow 3CO_2(g) + 4H_2O(l)$; $\Delta H° = 4 \times -285.8 \text{ kJ} + (3 \times -393.5 \text{ kJ}) - (1 \times -104.9 \text{ kJ}) - (5 \times 0 \text{ kJ}) = -2218.8 \text{ kJ}$; value in Table 6.2 is $\Delta H° = -2218.8$ kJ

(c) $C_8H_{18}(l) + 25/2\ O_2(g) \rightarrow 8CO_2(g) + 9H_2O(l)$; $\Delta H° = 9 \times -285.8 \text{ kJ} + (8 \times -393.5 \text{ kJ}) - (1 \times -269.8 \text{ kJ}) - (25/2 \times 0 \text{ kJ}) = -5450.4 \text{ kJ}$; value in Table 6.2 is $\Delta H° = -5450.5$ kJ

6.51 The heat capacity of an object is the amount of heat required to raise the temperature of the object by 1°C. The molar heat capacity is the amount of heat required to raise the temperature of 1 mol of a substance by 1°C. The specific heat is the amount of heat required to raise the temperature of 1 gram of a substance by 1°C.

6.53 The molar heat capacity is found by multiplying the specific heat by the molar mass in grams.

(a) The molar mass of ethanol (C_2H_5OH) is 46.07 g/mol.

$$\text{molar heat capacity} = \text{specific heat} \times \text{molar mass} = 2.419 \text{ J/g}\cdot°\text{C} \times 46.07 \text{ g/mol} = 111.4 \text{ J/mol}\cdot°\text{C}$$

(b) The molar mass of sodium (Na) is 22.99 g/mol.

$$\text{molar heat capacity} = 1.228 \text{ J/g}\cdot°\text{C} \times 22.99 \text{ g/mol} = 28.23 \text{ J/mol}\cdot°\text{C}$$

6.55 (a) $q = \text{mass} \times \text{specific heat} \times \Delta T = 15.0 \text{ g} \times 0.902 \text{ J/g}\cdot°\text{C} \times (30°\text{C} - 20°\text{C}) = 135 \text{ J}$

(b) $q = \text{mass} \times \text{specific heat} \times \Delta T = 1.50 \text{ mol Cu} \times \dfrac{63.55 \text{ g}}{1 \text{ mol Cu}} \times \dfrac{0.385 \text{ J}}{\text{g}\cdot°\text{C}} \times (30°\text{C} - 20°\text{C}) = 367 \text{ J}$

6.57 (a) $\text{specific heat} = \dfrac{q}{\text{mass} \times \Delta T} = \dfrac{4.20 \times 10^3 \text{ J}}{100 \text{ g} \times 5.00°\text{C}} = 8.40 \text{ J/g}\cdot°\text{C}$

(b) $\text{heat capacity} = \text{mass} \times \text{specific heat} = q/\Delta T = 4.20 \times 10^3 \text{ J}/5.00°\text{C} = 8.40 \times 10^2 \text{ J/°C}$

6.59 $\Delta T = q/\text{mass} \times \text{specific heat}$

For copper: $\Delta T = 100 \text{ J}/1.00 \text{ g} \times 0.385 \text{ J/g}\cdot°\text{C} = 260°\text{C}$
For gold: $\Delta T = 100 \text{ J}/1.00 \text{ g} \times 0.129 \text{ J/g}\cdot°\text{C} = 775°\text{C}$

Gold has a lower specific heat than copper and, therefore, a greater temperature change. Gold will have the higher final temperature.

6.61 The heat lost by the iron is equal to the heat gained by the water. The final temperature is the same for both the iron and the water. The specific heats of iron and water are in Table 6.6.

Iron: $q_{iron} = 13.0 \text{ g} \times 0.449 \text{ J/g}\cdot°\text{C} \times (T_{final} - 95°\text{C})$

Water: $q_{water} = 80.0 \text{ g} \times 4.18 \text{ J/g}\cdot°\text{C} \times (T_{final} - 15.0°\text{C})$

$q_{iron} = -q_{water}$

$13.0 \text{ g} \times 0.449 \text{ J/g}\cdot°\text{C} \times (T_{final} - 95°\text{C}) = -80.0 \text{ g} \times 4.18 \text{ J/g}\cdot°\text{C} \times (T_{final} - 15.0°\text{C})$

$5.84 \text{ J/°C} \times T_{final} - 555 \text{ J} = -334 \text{ J/°C} \times T_{final} + 5016 \text{ J}$

$340 \text{ J/°C} \times T_{final} = 5571 \text{ J}$

$T_{final} = 5571 \text{ J}/340 \text{ J/°C} = 16.4°\text{C}$

6.63 $q = 500 \text{ W} \times \dfrac{1 \text{ J/s}}{1 \text{ W}} \times 60.0 \text{ s} = 3.00 \times 10^4 \text{ J}$; mass water = 250 g

$$\Delta T = \frac{q}{\text{mass water} \times \text{specific heat}} = \frac{3.00 \times 10^4 \text{ J}}{250 \text{ g} \times 4.18 \text{ J/g}\cdot°\text{C}} = 28.7°\text{C}$$

6.65 See description and Figure 6.13 in the textbook..

6.67 (a) When the reaction is carried out at constant volume, no work can be exchanged with the surroundings. In a constant pressure reaction, work can be exchanged with the surroundings. Therefore, the two differ by this amount of work.

(b) They will be identical (approximately) when the reaction involves only solids and liquids or when the total number of moles of gas doesn't change in the reaction.

(c) They will differ when the total number of moles of gas changes during the reaction.

6.69

$$\frac{31.6 \text{ kJ}}{3.00 \text{ g urea}} \times \frac{60.06 \text{ g urea}}{1 \text{ mol}} = 633 \text{ kJ/mol}$$

The molar enthalpy of combustion of urea is –633 kJ/mol.

6.71 The constant volume heat of combustion is the heat given off when 1 mol of the substance is burned at constant volume. 29.6 kJ were given off by the combustion of 1.50 g of morphine. The combustion of 1 mol, 303.36 g, would give off

$$\frac{29.6 \text{ kJ}}{1.50 \text{ g morphine}} \times \frac{303.36 \text{ g morphine}}{1 \text{ mol}} = 5.99 \times 10^3 \text{ kJ/mol}$$

The combustion is exothermic. Hence, the constant volume heat of combustion of morphine is -5.99×10^3 kJ/mol.

6.73 The combustion causes the temperature of the calorimeter to increase by 25.75°C – 24.33°C = 1.42°C. The amount of heat given off during the combustion is 1.42°C × 11.40 kJ/°C = 16.2 kJ. 16.2 kJ were given off by the combustion of 0.5258 g of acetone. The combustion of 1 mol, 58.08 g, would give off

$$\frac{16.2 \text{ kJ}}{0.5258 \text{ g}} \times \frac{58.08 \text{ g}}{1 \text{ mol}} = 1790 \text{ kJ/mol}$$

The combustion is exothermic. Hence, the constant volume heat of combustion is –1790 kJ/mol.

6.75 (a) The heat capacity of the calorimeter is the amount of heat that would produce a one-degree rise in the temperature of the calorimeter, or 15.0 kJ/2.00°C = 7.50 kJ/°C.

(b) The combustion causes the temperature of the calorimeter to increase by 1.54°C. The amount of heat given off during the combustion of 0.300 g of camphor is 1.54°C × 7.50 kJ/°C = 11.6 kJ. The combustion of 1 mol, 152.2 g, would give off

$$\frac{11.6 \text{ kJ}}{0.300 \text{ g}} \times \frac{152.2 \text{ g}}{1 \text{ mol}} = 5.89 \times 10^3 \text{ kJ/mol}$$

The constant volume heat of combustion of camphor is -5.89×10^3 kJ/mol.

6.77 (a) C (graphite) → C (diamond). No. There are no gases produced or consumed.

(b) C (graphite) + $O_2(g) \rightarrow CO_2(g)$. No. The number of moles of gas consumed equals the number of moles of gas produced; Δn gas = 0.

(c) 2C (graphite) + $O_2(g) \rightarrow 2CO(g)$. Yes. For each mole of oxygen gas consumed 2 mol of carbon monoxide gas is produced; Δn gas does not equal zero.

6.79

$$C_{10}H_{16}O(s) + 27/2\ O_2(g) \rightarrow 10CO_2(g) + 8H_2O(l)$$

$\Delta H = \Delta E + RT\Delta n$. $\Delta E = -5.89 \times 10^3$ kJ (See Exercise 6.75), T = 298 K, and for the gaseous reactants and products $\Delta n = n_{products} - n_{reactants}$ = 10 mol – 13.5 mol = –3.5 mol. Substituting into the equation gives

$\Delta H = -5.89 \times 10^3$ kJ + $(8.314 \times 10^{-3}$ kJ/mol•K) (298 K) (–3.5 mol) = -5.89×10^3 kJ – 8.67 kJ = -5.90×10^3 kJ.

Answer: $\Delta H°$ for the combustion of camphor at 25°C is approximately -5.90×10^3 kJ/mol. $\Delta H°$ for the combustion of 1.00 mol of camphor at 25°C is 1.00 mol × -5.90×10^3 kJ/mol = -5.90×10^3 kJ.

6.81 $CaCO_3(s) \rightarrow CaO(s) + CO_2(g)$; $\Delta H° = \Delta E° + RT\Delta n$; $\Delta E° = \Delta H° - RT\Delta n$.
$\Delta H° = +178.3$ kJ, T = 298 K, and Δn = 1 mol – 0 mol = 1 mol. Substituting into the equation gives
$\Delta E° = +178.3$ kJ – $(8.314 \times 10^{-3}$ kJ/mol•K) (298 K) (1 mol) = +178.3 kJ – 2.48 kJ = + 175.8 kJ.

Answer: $\Delta E°$ = +175.8 kJ/mol for the decomposition of $CaCO_3$ at 25°C. $\Delta E°$ for the decomposition of 1.00 mol of $CaCO_3$ at 25°C is 1.00 mol × 175.8 kJ/mol = +176 kJ.

6.83 $NH_3(g) + HCl(g) \rightarrow NH_4Cl(s)$ $\Delta H° = -176.0$ kJ

(a) $w = -P\Delta V = -RT\Delta n$. For the gaseous reactants and products $\Delta n = n_{products} - n_{reactants}$ = 0 mol – 2 mol = –2 mol. The PV work done on the system is w = $-(8.314 \times 10^{-3}$ kJ/mol•K) (298 K) (–2 mol) = +4.96 kJ. The surroundings do 4.96 kJ of work on the system when 1.00 mol of ammonia reacts with hydrogen chloride at 25°C and 1 atm.

(b) $\Delta E° = \Delta H° - RT\Delta n = -176.0$ kJ + 4.96 kJ = –171.0 kJ

6.85 A plant absorbs light energy from its surroundings; this is an integral part of the process of photosynthesis. It also performs work in extracting water and minerals from its surroundings, and it evolves heat into the surroundings.

During life the plant produces complex carbohydrates and oxygen from carbon dioxide and water. In doing this the plant converts light energy into chemical bond potential energy (potential energy stored in the chemical bonds). During decay the complex carbohydrates formed undergo oxidation back into carbon dioxide and water. This decay reaction converts the chemical bond potential energy into heat energy. Throughout the plant's life cycle, energy is being converted from one form into another. However, no energy is being created or destroyed and, therefore, the plant's cycle is consistent with the first law of thermodynamics.

6.87 The internal energy of the system comprises the total chemical bond potential energy, the total internal potential energy of other forms (intermolecular forces, electrical and magnetic interactions with the surroundings, etc.), and the total internal motional energy of the system's substances or components.

(a) When the temperature changes, the motional energy component of the internal energy changes the most.
(b) Changes of state, like melting, disrupts or radically changes the internal arrangement of the particles composing the substance. The rearranged particles have a totally different intermolecular force relationship than before and, therefore, the intermolecular potential energy component of the internal energy is primarily affected. It should be pointed out that the nature of the intermolecular forces does not change, only their strengths and the number of each type of interaction. The motional energy component, on the other hand, does not change during a phase change.
(c) Chemical changes have the greatest effect on the chemical bond potential energy of the system. Therefore, the chemical bond potential energy component of the internal energy changes for the most part.

6.89 C (graphite) + $O_2(g) \rightarrow CO_2(g)$ $\Delta H_f^\circ = -393.5$ kJ

$$x \text{ g C} \times \frac{1 \text{ mol C}}{12.01 \text{ g C}} \times \frac{393.5 \text{ kJ}}{1 \text{ mol C}} = 5.000 \times 10^5 \text{ kJ}$$

$$x \text{ g} = \frac{12.01 \text{ g} \times 5.000 \times 10^5}{393.5} = 1.526 \times 10^4 \text{ g} = 15.26 \text{ kg}$$

Answer: 15.26 kg of coke.

6.91 Twenty-five percent efficient means that only one-quarter of the heat produced is actually used in melting the ice. The steps in the calculation are: 1.00 mol $H_2 \rightarrow$ kJ $\rightarrow$ effective kJ $\rightarrow$ mol ice $\rightarrow$ g ice.

$$1.00 \text{ mol } H_2 \times \frac{285.8 \text{ kJ}}{1 \text{ mol } H_2} \times 0.25 \times \frac{1.00 \text{ mol ice}}{6.02 \text{ kJ}} \times \frac{18.02 \text{ g ice}}{1 \text{ mol ice}} = 2.1 \times 10^2 \text{ g ice}$$

6.93 The amount of heat given off by the combustion of 5.0 g of fat is
q = mass × specific heat × ΔT = 1.00×10^3 g × 4.18 J/g•°C × (65°C – 20°C) = 1.9×10^5 J or 1.9×10^2 kJ.
The number of kilocalories supplied by the oxidation of 1.00 g of fat is

$$1.00 \text{ g fat} \times \frac{1.9 \times 10^2 \text{ kJ}}{5.0 \text{ g fat}} \times \frac{1 \text{ kcal}}{4.184 \text{ kJ}} = 9.1 \text{ kcal} = 9.1 \text{ Calories}$$

6.95 (a) w = force × distance = 120 lb × 5000 ft = 6.00×10^5 ft•lb

$$w = 6.00 \times 10^5 \text{ ft•lb} \times \frac{0.3048 \text{ m}}{1 \text{ ft}} \times \frac{1 \text{ kg}}{2.205 \text{ lb}} \times \frac{9.80665 \text{ m}}{1 \text{ s}^2} \times \frac{1 \text{ cal}}{4.184 \text{ J}} = 1.94 \times 10^5 \text{ calories} = 194 \text{ kcal}$$

Remember that pounds (lbs) is the English unit of force and kilograms (kg) the metric unit of mass. Therefore, in order to convert pounds into the metric unit of force newtons (N), you have to multiply the kilogram equivalent of the pounds by the gravitational constant g = 9.80665 m/s^2. The conversion factor 0.239 cal/0.738 ft•lb can be found in some tables. Alternatively, this conversion factor could be used, if you happened to know it.

(b) $$194 \text{ kcal work} \times \frac{100 \text{ kcal}}{25 \text{ kcal work}} = 7.8 \times 10^2 \text{ kcal}$$

6.97 For the man to lose the 30 lb in 20 weeks, he must take in 30 × 3500 = 105,000 kcal less than required for normal maintenance over that period. On a per day basis, he would have to take in 105,000/140 = 750 kcal less than required for normal maintenance. A complication arises in the fact that the amount of calories required for normal maintenance per day decreases as the person loses weight. In other words, as the person begins to lose weight, he needs less calories to maintain his lower weight and his daily caloric intake would likewise be lowered.

To maintain the present weight of 180 lb he would have to take in 15 × 180 = 2700 kcal per day. But to maintain the final weight of 150 lb he only needs 2250 kcal per day. Therefore, on the first day of the diet the person should have a caloric intake of no more than 2700 – 750 = 1950 kcal per day, while on the last day of the diet the person should have a caloric intake of no more than 2250 – 750 = 1500 kcal per day. The average of the two extreme daily intakes (1950 + 1500)/2 = 1725 kcal per day provides an estimate of the daily caloric intake he would need in order to lose the 30 lb in 20 weeks. Therefore, in the beginning of the diet he would be losing more weight per day than at the end of the diet, but overall he should lose approximately 30 lb.

Answer: 1725 kcal/day.

6.99 (a)

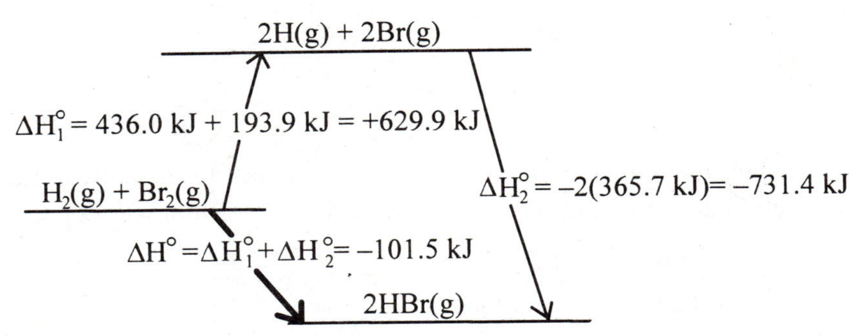

(b)

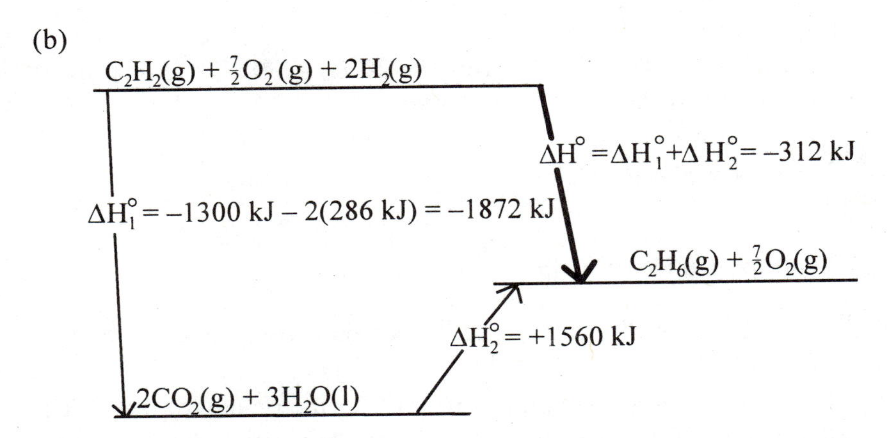

6.101 (a) The heat lost by the copper is equal to the heat gained by the water. The final temperature is the same for both the copper and the water. The specific heats are in Table 6.6.

Copper: $q_{copper} = 100\text{ g} \times 0.385\text{ J/g}\cdot{}^\circ\text{C} \times (T_{final} - 75^\circ\text{C})$

Water: $q_{water} = 500\text{ g} \times 4.18\text{ J/g}\cdot{}^\circ\text{C} \times (T_{final} - 15^\circ\text{C})$

$q_{copper} = -q_{water}$

$100\text{ g} \times 0.385\text{ J/g}\cdot{}^\circ\text{C} \times (T_{final} - 75^\circ\text{C}) = -500\text{ g} \times 4.18\text{ J/g}\cdot{}^\circ\text{C} \times (T_{final} - 15^\circ\text{C})$

$38.5\text{ J/}^\circ\text{C} \times T_{final} - 2887.5\text{ J} = -2090.0\text{ J/}^\circ\text{C} \times T_{final} + 31350\text{ J}$

$2128.5\text{ J/}^\circ\text{C} \times T_{final} = 34237.5\text{ J}$

$$T_{final} = \frac{34237.5\text{ J}}{2128.5\text{ J/}^\circ\text{C}} = 16^\circ\text{C}$$

(b) $q_{copper} = 100\text{ g} \times 0.385\text{ J/g}\cdot{}^\circ\text{C} \times (16^\circ\text{C} - 75^\circ\text{C}) = -2.3 \times 10^3\text{ J}$

(c) $q_{water} = +2.3 \times 10^3\text{ J}$

6.103 $C_2N_2(g) + O_2(g) \rightarrow 2CO(g) + N_2(g)$

(a) The combustion causes the temperature of the calorimeter to increase by 25.60°C – 24.35°C = 1.25°C. The amount of heat given off during the combustion is 1.25°C × 5.375 kJ/°C = 6.72 kJ. 6.72 kJ were given off by the combustion of 0.6596 g of cyanogen. The combustion of 1 mol, 52.04 g, would give off

$$\frac{6.72\text{ kJ}}{0.6596\text{ g}} \times \frac{52.04\text{ g}}{1\text{ mol}} = 530\text{ kJ}$$

The combustion is exothermic. Hence, the constant volume heat of combustion is $\Delta E° = -530$ kJ/mol. $\Delta E°$ for the combustion of 1.00 mol is –5.30 kJ.

$\Delta H° = \Delta E° + RT\Delta n = -530 \text{ kJ} + (8.314 \times 10^{-3} \text{ kJ/mol•K})(298 \text{ K})(1 \text{ mol}) = -530 \text{ kJ} + 2.48 \text{ kJ} = -528 \text{ kJ}$

$\Delta H°$ for the combustion of 1.00 mol is –528 kJ.

(b) $C_2N_2(g) + O_2(g) \rightarrow 2CO(g) + N_2(g) \quad \Delta H° = -528$ kJ

$\Delta H_f°$: ? 0 kJ (2 × –110.5 kJ) 0 kJ

$\Delta H° = (2 \times -110.5 \text{ kJ}) + 0 \text{ kJ} - \Delta H_f° \, C_2N_2 - 0 \text{ kJ} = -528 \text{ kJ}$

$\Delta H_f° \, C_2N_2 = 528 \text{ kJ} - 2(110.5 \text{ kJ}) = +307 \text{ kJ}$

6.105 w is defined as the work done on the system by the surroundings. Therefore, for an expansion w must be negative.

$w = -\text{force} \times \text{distance} = -f \times d$,

where f is a constant force on the system and d is the displacement of the piston.

$P = f/A$, $f = P \times A$, $d = \Delta h = h_2 - h_1$, and $V = A \times h$,

where P is the constant pressure on the system, A is the surface area of the piston, h is the height of the piston, and V is the volume of the system. Substituting and rearranging, we get

$w = -f \times d = -P \times A \times (h_2 - h_1) = -P \times (Ah_2 - Ah_1) = -P \times (V_2 - V_1) = -P\Delta V$

6.107 It would evolve more heat at constant pressure. Work is done on the system by the surroundings in the constant pressure reaction. Therefore the system would contain more energy in the constant pressure reaction, and would have to evolve more heat in order to bring it back to the original starting temperature.

6.109 Warming it at constant pressure. Gases expand when heated (at constant pressure). Therefore, the gas does work on the surroundings (and in so doing loses energy to the surroundings) when heated under constant pressure and, consequently, requires more heat in order to bring it up to the final temperature.

6.111 (a) $w = -P\Delta V = -P \times (V_{final} - V_{initial})$; $w = -1.25 \text{ atm} \times (8.50 \text{ L} - 3.75 \text{ L}) = -5.94 \text{ L•atm}$

(b) $w = -5.94 \text{ L•atm} \times 101.3 \text{ J/1 L•atm} = -602 \text{ J}$. w is the PV work done on the gas. Since the system does work on the surroundings, w is negative.

6.113 The volume change for 1.00 mol is

$$\Delta V = V_{final} - V_{initial} = 1.00 \text{ mol} \times \frac{18.02 \text{ g}}{1 \text{ mol}} \times \left(\frac{1 \text{ cm}^3}{0.9998 \text{ g}} - \frac{1 \text{ cm}^3}{0.9168 \text{ g}} \right) = -1.6 \text{ cm}^3 \times 1 \text{ L}/10^3 \text{ cm}^3$$

$$= -1.6 \times 10^{-3} \text{ L}$$

$w = -P\Delta V = -1.00 \text{ atm} \times (-1.6 \times 10^{-3} \text{ L}) = 1.6 \times 10^{-3} \text{ L•atm} \times 101.3 \text{ J/L•atm} = +0.16 \text{ J}$

Work is done on the system (the system's volume decreases); therefore, w is positive.

6.115 (a)

		Molar Heat Capacity
Al: 0.902 J/g•°C × 26.98 g/mol	=	24.3 J/°C•mol
Cu: 0.385 J/g•°C × 63.546 g/mol	=	24.5 J/°C•mol
Au: 0.129 J/g•°C × 196.97 g/mol	=	25.4 J/°C•mol
Fe: 0.449 J/g•°C × 55.847 g/mol	=	25.1 J/°C•mol
Pb: 0.128 J/g•°C × 207.2 g/mol	=	26.5 J/°C•mol
Hg: 0.140 J/g•°C × 200.59 g/mol	=	28.1 J/°C•mol
Na: 1.228 J/g•°C × 22.99 g/mol	=	28.2 J/°C•mol
Ta: 0.140 J/g•°C × 180.95 g/mol	=	25.3 J/°C•mol

(b) The reciprocals of the specific heats, in units of g•°C/J, are:

Al	Cu	Au	Fe	Pb	Hg	Na	Ta
1.11	2.60	7.75	2.23	7.81	7.14	0.814	7.14

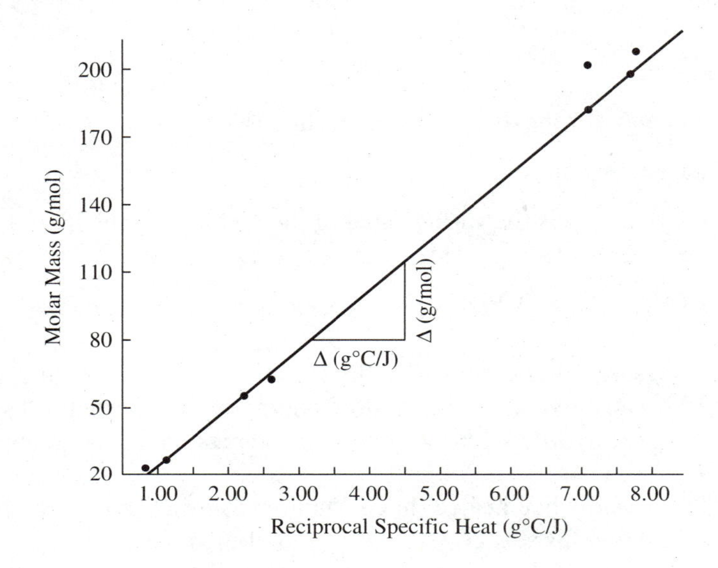

The graph shows the straight line obtained by plotting molar mass versus reciprocal specific heat. A right triangle has been drawn with part of the straight line as the hypotenuse. The legs of the triangle are Δ(g/mol) = 113.1 g/mol – 80.0 g/mol = 33.1 g/mol and Δ(g•°C/J) = 4.50 g•°C/J – 3.20 g•°C/J = 1.30 g•°C/J. The slope of the line is

$$\text{slope} = \frac{\Delta(\text{g/mol})}{\Delta(\text{g}\bullet{}^\circ\text{C/J})} = \frac{33.1\ \text{g/mol}}{1.30\ \text{g}\bullet{}^\circ\text{C/J}} = 25.5\ \text{J/}^\circ\text{C}\bullet\text{mol}$$

(c) $$\mathcal{M} = \frac{\text{molar heat capacity}}{\text{specific heat}} = \frac{25.5\ \text{J/}^\circ\text{C}\bullet\text{mol}}{0.278\ \text{J/g}\bullet{}^\circ\text{C}} = 91.7\ \text{g/mol}$$

The calculated value of 91.7 g/mol is a fairly decent approximation of the accepted value of 91.2 g/mol.

CHAPTER 7

QUANTUM THEORY AND THE HYDROGEN ATOM

Solutions To Practice Exercises

PE 7.1 $\lambda\nu = c$; $\lambda = \dfrac{c}{\nu} = \dfrac{2.998 \times 10^8 \text{ m/s}}{1.560 \times 10^6\text{/s}} = 192.2 \text{ m}$

PE 7.2 $E = h\nu$ and $\nu \times \lambda = c$. So $\nu = c/\lambda = E/h$ and

$$\lambda = \frac{c \times h}{E} = \frac{2.9979 \times 10^8 \text{ m/s} \times 6.626 \times 10^{-34} \text{ J}\bullet\text{s}}{1.50 \times 10^{-18} \text{ J}} = 1.32 \times 10^{-7} \text{ m} = 132 \text{ nm}$$

PE 7.3 $KE = h\nu - W$ and $\nu_0 = W/h$.

$$\nu = c/\lambda = \frac{2.998 \times 10^8 \text{ m/s}}{500 \times 10^{-9} \text{ m}} = 6.00 \times 10^{14} \text{ s}^{-1}$$

$$W = h\nu - KE = 6.626 \times 10^{-34} \text{ J}\bullet\text{s} \times 6.00 \times 10^{14} \text{ s}^{-1} - 1.08 \times 10^{-19} \text{ J}$$

$$= 3.98 \times 10^{-19} \text{ J} - 1.08 \times 10^{-19} \text{ J} = 2.90 \times 10^{-19} \text{ J}$$

$$\nu_0 = 2.90 \times 10^{-19} \text{ J}/6.626 \times 10^{-34} \text{ J}\bullet\text{s} = 4.38 \times 10^{14} \text{ s}^{-1}$$

PE 7.4 $n^2 = \dfrac{-2.179 \times 10^{-18} \text{ J}}{E_n} = \dfrac{-2.179 \times 10^{-18} \text{ J}}{-8.716 \times 10^{-20} \text{ J}} = 25$; $n = 5$

Answer: The fourth excited state.

PE 7.5 (a) Absorbed. The energy of the electron increases; therefore, a photon is absorbed.
(b) Emitted. The energy of the electron decreases; therefore, a photon is emitted.

PE 7.6 The energy of the electron after the transition is equal to the ground-state energy plus the energy of the photon. The energy of the photon is

$$E = h\nu = \frac{hc}{\lambda} = \frac{6.6261 \times 10^{-34} \text{ J}\bullet\text{s} \times 2.9979 \times 10^8 \text{ m/s}}{97.25 \times 10^{-9} \text{ m}} = 2.043 \times 10^{-18} \text{ J}$$

$E_n = -2.179 \times 10^{-18}$ J/n^2. The energy of the ground state, n = 1, is
$E_1 = -2.179 \times 10^{-18}$ J. The energy of the electron after the transition is
$E_{electron} = E_1 + E_{photon} = -2.179 \times 10^{-18}\text{ J} + 2.043 \times 10^{-18}\text{ J} = -1.36 \times 10^{-19}\text{ J}$

The energy level the electron is in is

$$n^2 = \frac{-2.179 \times 10^{-18}\text{ J}}{E_n} = \frac{-2.179 \times 10^{-18}\text{ J}}{-1.36 \times 10^{-19}\text{ J}} = 16.0;\ n = 4$$

Answer: The electron has an energy of -1.36×10^{-19} J and is in the fourth energy level (third excited state).

PE 7.7 The first Balmer line is caused by the electron transition from the n = 3 level to the n = 2 level.

$$E_{photon} = -2.179 \times 10^{-18}\text{ J}\left(\frac{1}{n_H{}^2} - \frac{1}{n_L{}^2}\right) = -2.179 \times 10^{-18}\text{ J}\left(\frac{1}{3^2} - \frac{1}{2^2}\right) = 3.026 \times 10^{-19}\text{ J}$$

$$\lambda = \frac{hc}{E_{photon}} = \frac{6.626 \times 10^{-34}\text{ J•s} \times 2.998 \times 10^8\text{ m/s}}{3.026 \times 10^{-19}\text{ J}} = 6.565 \times 10^{-7}\text{ m} = 656.5\text{ nm}$$

The first Balmer line thus appears as an orange-red line (Figure 7.3) in the emission spectrum of hydrogen.

PE 7.8 100 miles/h = 44.7 m/s

$$\lambda = \frac{h}{mv} = \frac{6.626 \times 10^{-34}\text{ J•s}}{0.140\text{ kg} \times 44.7\text{ m/s}} = 1.06 \times 10^{-34}\text{ m}$$

The wavelength of the baseball is approximately $1/10^{29}$ that of the electron. The wavelength of the baseball is extremely small compared to that of the electron.

PE 7.9 For an f subshell, l = 3 and 2l + 1 = 7. Any f subshell will always contain seven orbitals regardless of its n value. These correspond to the seven possible m_l values: –3, –2, –1, 0, 1, 2, 3.

PE 7.10 (a) If n is 1, the only allowed l value is 0. Hence, the n = 1 shell contains only a single s orbital. There are no 1p orbitals.

(b) If n is 2, the allowed l values are 0 and 1 and the n = 2 shell contains s and p orbitals. Hence, there are 2p orbitals.

PE 7.11 There are five orbitals in a d subshell, l = 2 and 2l + 1 = 5. Two spin states are possible for each of these orbitals. The total number of states for the electron is 5 × 2 = 10.

Solutions To Final Exercises

7.1 Higher frequency radiation is more energetic than lower frequency radiation. From Figure 7.3 the spectral regions in order of increasing frequency (low energy → high energy) are: radio, microwave, infrared, visible, ultraviolet, x-rays, gamma rays.

7.3 Consult Figure 7.3. (a) The gamma ray region. (b) The radio region. (c) The gamma ray region.

7.5 The infrared region.
From Figure 7.3 we see that the infrared waves are less energetic than visible light waves.

7.7 (a) An emission spectrum is a spectrum obtained from a glowing source; that is, the substance under examination is emitting the electromagnetic waves being examined with the spectroscope. A continuous emission spectrum contains a range of wavelengths, while a line spectrum contains only a number of specific wavelengths.

(b) See Figure 7.5.

7.9 The signal was in the form of electromagnetic radiation and, therefore, moved at the speed of light (c).

$$\text{time} = \frac{\text{distance}}{\text{speed}} = \frac{1.69 \times 10^{12}\ \text{miles} \times 1609\ \text{m/mile}}{3.00 \times 10^{8}\ \text{m/s} \times 3600\ \text{s/h}} = 2.52 \times 10^{3}\ \text{h}$$

7.11 (a) $\nu = \frac{c}{\lambda} = \frac{3.00 \times 10^{8}\ \text{m/s}}{650 \times 10^{-9}\ \text{m}} = 4.62 \times 10^{14}\ \text{s}^{-1}$; the visible region.

(b) $\lambda = \frac{c}{\nu} = \frac{3.00 \times 10^{8}\ \text{m/s}}{6.00 \times 10^{16}/\text{s}} = 5.00 \times 10^{-9}\ \text{m} = 5.00\ \text{nm}$

In the region where the ultraviolet and x-ray regions overlap.

7.13 (a) $\nu = \frac{c}{\lambda} = \frac{2.9979 \times 10^{8}\ \text{m/s}}{589.0 \times 10^{-9}\ \text{m}} = 5.090 \times 10^{14}\ \text{Hz}$

$\nu = \frac{2.9979 \times 10^{8}\ \text{m/s}}{589.6 \times 10^{-9}\ \text{m}} = 5.085 \times 10^{14}\ \text{Hz}$

(b) The visible region.

7.15 The radiation spectra from hot glowing objects (oftentimes referred to as blackbody radiation), the photoelectric effect, and line spectra are presently explainable only if we assume that radiation consists of particle-like quanta called photons. Also, the fact that a chemical reaction (e.g., sunburn) that is brought about by high-frequency radiation may not be induced by radiation of lower frequency, no matter what the intensity, suggests that the energy comes in "packets" that are greater for the high-frequency than for the low-frequency radiation. According to wave theory, on the other hand, the low-frequency radiation could be made to deliver more energy, and thus induce the reaction, by simply increasing the intensity.

7.17 $E = h\nu = hc/\lambda$

(a) As the wavelength increases, the energy of the photon decreases.

(b) As the frequency increases, the energy of the photon increases.

7.19 $E = h\nu = hc/\lambda$ and $\lambda\nu = c$.

Range In Which Wave Is Found	Frequency, ν	Wavelength, λ	Energy, E
Ultraviolet	1×10^{16} Hz	30 nm	7×10^{-18} J
Radio	1×10^{6} Hz	300 m	7×10^{-28} J
Visible (green)	6×10^{14} Hz	500 nm	4×10^{-19} J

7.21 (a) $\nu = \frac{c}{\lambda} = \frac{2.9979 \times 10^{8}\ \text{m/s}}{589.6 \times 10^{-9}\ \text{m}} = 5.085 \times 10^{14}\ \text{Hz}$

$E = h\nu = 6.6261 \times 10^{-34}\ \text{J}\cdot\text{s} \times 5.085 \times 10^{14}/\text{s} = 3.369 \times 10^{-19}\ \text{J}$

(b) From Figure 7.3 we see that radiation with wavelengths in the region of 600 nm lie in the orange-yellow portion of the visible spectrum. Therefore, objects will appear to be orange-yellow when they reflect this light.

7.23 (a) $$\nu = \frac{c}{\lambda} = \frac{3.00 \times 10^{8}\ \text{m/s}}{460 \times 10^{-9}\ \text{m}} = 6.52 \times 10^{14}\ \text{Hz}$$

$$\nu = \frac{c}{\lambda} = \frac{3.00 \times 10^{8}\ \text{m/s}}{610 \times 10^{-9}\ \text{m}} = 4.92 \times 10^{14}\ \text{Hz}$$

(b) $E = h\nu = 6.63 \times 10^{-34}\ \text{J}\bullet\text{s} \times 6.52 \times 10^{14}/\text{s} = 4.32 \times 10^{-19}\ \text{J}$

$E = h\nu = 6.63 \times 10^{-34}\ \text{J}\bullet\text{s} \times 4.92 \times 10^{14}/\text{s} = 3.26 \times 10^{-19}\ \text{J}$

(c) The energy of 1.00 mol of photons is:

$$460\ \text{nm:}\ 1.00\ \text{mol} \times \frac{4.32 \times 10^{-19}\ \text{J}}{1\ \text{photon}} \times \frac{6.022 \times 10^{23}\ \text{photons}}{1\ \text{mol}} = 2.60 \times 10^{5}\ \text{J} = 260\ \text{kJ}$$

$$610\ \text{nm:}\ 1.00\ \text{mol} \times \frac{3.26 \times 10^{-19}\ \text{J}}{1\ \text{photon}} \times \frac{6.022 \times 10^{23}\ \text{photons}}{1\ \text{mol}} = 1.96 \times 10^{5}\ \text{J} = 196\ \text{kJ}$$

7.25 (a) The energy of one photon is

$$E = \frac{hc}{\lambda} = \frac{6.6261 \times 10^{-34}\ \text{J}\bullet\text{s} \times 2.9979 \times 10^{8}\ \text{m/s}}{285.2 \times 10^{-9}\ \text{m}} = 6.965 \times 10^{-19}\ \text{J}$$

The number of photons required to provide 1.00 J of energy is

$$1.00\ \text{J} \times \frac{1\ \text{photon}}{6.965 \times 10^{-19}\ \text{J}} = 1.44 \times 10^{18}\ \text{photons}$$

(b) The energy of one photon is

$$E = \frac{6.6261 \times 10^{-34}\ \text{J}\bullet\text{s} \times 2.9979 \times 10^{8}\ \text{m/s}}{501.5 \times 10^{-9}\ \text{m}} = 3.961 \times 10^{-19}\ \text{J}$$

The number of photons required to provide 1.00 J of energy is

$$1.00\ \text{J} \times \frac{1\ \text{photon}}{3.961 \times 10^{-19}\ \text{J}} = 2.52 \times 10^{18}\ \text{photons}$$

(c) The energy of one photon is

$$E = \frac{hc}{\lambda} = \frac{6.63 \times 10^{-34}\ \text{J}\bullet\text{s} \times 3.00 \times 10^{8}\ \text{m/s}}{45 \times 10^{-6}\ \text{m}} = 4.4 \times 10^{-21}\ \text{J}$$

The number of photons required to provide 1.00 J of energy is

$$1.00\ \text{J} \times \frac{1\ \text{photon}}{4.4 \times 10^{-21}\ \text{J}} = 2.3 \times 10^{20}\ \text{photons}$$

7.27 The photoelectric effect is the light-induced emission of electrons from a metal surface.

The photoelectric effect showed that not only was light emitted as quanta, but that light also consisted of particle-like quanta (photons). Therefore, it extended the concept of quantization of energy, the quantum theory, to radiation itself. The argument for quantization of radiation based on the photoelectric effect goes as follows:

Based on the wave theory one would expect the release of some electrons at every frequency of radiation employed, since according to wave theory the energy of the radiation beam depends on the intensity and, therefore, a sufficiently intense beam of any frequency should supply enough energy to cause the ejection of electrons. However, no electrons are ejected until a threshold frequency is reached, no matter how intense the beam employed. This implies that the radiation energy is contained in tiny bundles or packets, called photons, and, therefore, supports the concept of the quantization of electromagnetic radiation.

Also, based on the wave theory one would expect the kinetic energy of the released electrons to depend on the intensity of the radiation. The more intense the radiation, the greater the kinetic energy of the ejected electrons. However, it is found that the kinetic energy of an ejected electron is independent of the intensity and just depends on the frequency of the radiation employed. This also implies that the radiation energy is quantized.

7.29 For the light to eject electrons, its frequency must be greater than the threshold frequency of the metal. The threshold frequency required to dislodge an electron from the potassium metal's surface is

$$\nu_0 = \frac{W}{h} = \frac{3.69 \times 10^{-19}\ \text{J}}{6.63 \times 10^{-34}\ \text{J}\bullet\text{s}} = 5.57 \times 10^{14}\ \text{Hz}$$

The frequency of radiation with a wavelength of 600 nm is

$$\nu = \frac{c}{\lambda} = \frac{3.00 \times 10^{8}\ \text{m/s}}{600 \times 10^{-9}\ \text{m}} = 5.00 \times 10^{14}\ \text{Hz}$$

Since the frequency of the light beam is less than the threshold frequency, no photoelectrons will be produced.

7.31

$$\text{KE} = h\nu - W = \frac{hc}{\lambda} - W = \frac{6.626 \times 10^{-34}\ \text{J}\bullet\text{s} \times 2.998 \times 10^{8}\ \text{m/s}}{400 \times 10^{-9}\ \text{m}} - 3.69 \times 10^{-19}\ \text{J}$$

$$= 4.97 \times 10^{-19}\ \text{J} - 3.69 \times 10^{-19}\ \text{J} = 1.28 \times 10^{-19}\ \text{J}$$

7.33 (a) Transitions from the excited states (n = 2, 3, 4, . . .) down to the ground state (n = 1) give rise to the Lyman series. The frequencies of the photons emitted lie in the ultraviolet region of the spectrum.

(b) Transitions from the second excited state on up (n = 3, 4, 5, . . .) down to the first excited state (n = 2) give rise to the Balmer series. The frequencies of the photons emitted lie in the visible region of the spectrum.

7.35 $E_n = -2.179 \times 10^{-18}\ \text{J/n}^2$, where n can only take on positive integer values (n = 1, 2, 3, . . .).

(a) The fact that n must be a positive integer limits the possible values of the energy. In other words, the energy is no longer continuous in the sense that it can take on any value. Any quantity that is not continuous, but is limited to certain values or is composed of distinct units, is considered to be quantized. Since it is the integral limitation on the number n that causes the discontinuity or quantization of the hydrogen energies, the number n is called or referred to as a "quantum number."

(b) The lowest allowed energy is -2.179×10^{-18} J.

(c) Since E_n is negative, the greater the magnitude the lower the energy. Since n is in the denominator, the lower the value of n, the greater the magnitude. The lowest possible value for n is one. Therefore, the lowest allowed energy is associated with n = 1.

7.37 To say the hydrogen atom is in the ground state means that the hydrogen electron is in its lowest energy, most strongly bound (most stable), state.

The value of n corresponding to the ground state is n = 1.

7.39 One photon is emitted. According to Bohr, one photon of energy is given off or absorbed each time an electron moves from one energy level to another. Since the electron is going directly from the n = 4 level to the n = 1 level, only one photon is emitted.

7.41
$E_n = -2.179 \times 10^{-18}$ J/n^2
$E_1 = -2.179 \times 10^{-18}$ J
$E_2 = -2.179 \times 10^{-18}$ J/4 $= -5.448 \times 10^{-19}$ J
$E_3 = -2.179 \times 10^{-18}$ J/9 $= -2.421 \times 10^{-19}$ J
$E_4 = -2.179 \times 10^{-18}$ J/16 $= -1.362 \times 10^{-19}$ J
$E_5 = -2.179 \times 10^{-18}$ J/25 $= -8.716 \times 10^{-20}$ J
$E_6 = -2.179 \times 10^{-18}$ J/36 $= -6.053 \times 10^{-20}$ J

See Figure 7.16 in the textbook.

7.43
$r_n = 52.9 \text{ pm} \times n^2$
$r_1 = a_0 = 52.9 \text{ pm} \times (1)^2 = 52.9 \text{ pm}$
$r_2 = 52.9 \text{ pm} \times (2)^2 = 212 \text{ pm}$
$r_3 = 52.9 \text{ pm} \times (3)^2 = 476 \text{ pm}$
$r_4 = 52.9 \text{ pm} \times (4)^2 = 846 \text{ pm}$
$r_5 = 52.9 \text{ pm} \times (5)^2 = 1.32 \times 10^3 \text{ pm}$
$r_6 = 52.9 \text{ pm} \times (6)^2 = 1.90 \times 10^3 \text{ pm}$

Draw a graph similar to that in Figure 7.15 in the textbook using these r values, with the nucleus centered at the origin (0, 0).

7.45
$E_n = -2.179 \times 10^{-18}$ J/n^2
(a) $E_4 = -2.179 \times 10^{-18}$ J/16 $= -1.362 \times 10^{-19}$ J
(b) $E_3 = -2.179 \times 10^{-18}$ J/9 $= -2.421 \times 10^{-19}$ J

7.47
(a) The electron has greater (more) energy in the n = 3 level than in the n = 2 level; therefore, a photon must be absorbed when the electron makes a transition from level 2 to level 3.

(b) $$E_{\text{photon}} = -2.179 \times 10^{-18}\text{ J}\left(\frac{1}{n_H^2} - \frac{1}{n_L^2}\right) = -2.179 \times 10^{-18}\text{ J}\left(\frac{1}{9} - \frac{1}{4}\right) = 3.026 \times 10^{-19}\text{ J}$$

$$\nu = \frac{E}{h} = \frac{3.026 \times 10^{-19}\text{ J}}{6.6261 \times 10^{-34}\text{ J}\bullet\text{s}} = 4.567 \times 10^{14}\text{ Hz}$$

$$\lambda = \frac{hc}{E} = \frac{6.6261 \times 10^{-34}\text{ J}\bullet\text{s} \times 2.9979 \times 10^{8}\text{ m/s}}{3.026 \times 10^{-19}\text{ J}} = 656.5\text{ nm}$$

(c) The visible region of the spectrum.

7.49
$$\Delta E = E_2 - E_1 = -2.179 \times 10^{-18}\text{ J}\left(\frac{1}{4} - \frac{1}{1}\right) = 1.634 \times 10^{-18}\text{ J}$$

The frequency of the photon emitted when an electron drops from the second energy level to the first (ground state) is

$\nu = 1.634 \times 10^{-18}$ J/6.6261×10^{-34} J•s $= 2.466 \times 10^{15}$ Hz

From Figure 7.3 we see that this photon falls in the ultraviolet region of the spectrum. The Lyman series arises when electrons in excited states (n = 2, 3, 4, . . .) emit photons while returning to the ground state (n = 1). The transition n = 2 → n = 1 gives rise to the first spectral line in the Lyman series and it lies in the ultraviolet region, as just shown. Since the first excited state (n = 2) is closest in energy to the ground state, all of the other Lyman series transitions will emit photons of greater energy or higher frequency than the n = 2 → n = 1 transition. Therefore, all the rest of the Lyman lines will also be in the ultraviolet region.

7.51 (a) The energy of the electron before emission is equal to the electron's final energy plus the energy of the photon. The energy of the photon is

$$E = \frac{hc}{\lambda} = \frac{6.6261 \times 10^{-34}\ \text{J}\bullet\text{s} \times 2.9979 \times 10^{8}\ \text{m/s}}{3.740 \times 10^{-6}\ \text{m}} = 5.311 \times 10^{-20}\ \text{J}$$

$$E_{electron} = -8.716 \times 10^{-20}\ \text{J} + 5.311 \times 10^{-20}\ \text{J} = -3.405 \times 10^{-20}\ \text{J}$$

The energy level the electron was in is

$$n^2 = \frac{-2.179 \times 10^{-18}\ \text{J}}{E_n} = \frac{-2.179 \times 10^{-18}\ \text{J}}{-3.405 \times 10^{-20}\ \text{J}} = 64.0;\ n = 8$$

The electron had an energy of -3.405×10^{-20} J and was in the eighth energy level (seventh excited state).

(c) The wavelength of the emitted photon lies in the infrared region of the spectrum. Therefore, you should look in the Pfund spectral series for the emission line.

7.53 (a) The de Broglie relationship is $\lambda = h/mv$, where m and v are the particle's mass and speed, respectively, and h is Planck's constant.

(b) The particle's wavelength decreases with increasing speed. The particle's wavelength decreases with increasing particle mass.

7.55 Reflecting electrons, protons, and other subatomic and atomic size particles off crystal surfaces produces interference patterns from which the wavelengths associated with these particles can be calculated. Wavelengths are a property of waves. Also, passing a beam of electrons or protons through thin metal foils produces an interference pattern. Interference patterns can only be produced by waves.

7.57 The masses of the electron and proton are 9.10939×10^{-31} kg and 1.67262×10^{-27} kg, respectively. The associated wavelengths are

(a) electron: $\lambda = \frac{h}{mv} = \frac{6.626 \times 10^{-34}\ \text{J}\bullet\text{s}}{9.109 \times 10^{-31}\ \text{kg} \times 50\ \text{m/s}} = 1.5 \times 10^{-5}$ m

(b) proton: $\lambda = \frac{6.626 \times 10^{-34}\ \text{J}\bullet\text{s}}{1.673 \times 10^{-27}\ \text{kg} \times 50\ \text{m/s}} = 7.9 \times 10^{-9}$ m

(c) 0.40-kg piece: $\lambda = \frac{6.626 \times 10^{-34}\ \text{J}\bullet\text{s}}{0.40\ \text{kg} \times 50\ \text{m/s}} = 3.3 \times 10^{-35}$ m

The wavelength of the proton is considerably smaller than that of the electron, and the wavelength of the 0.40-kg piece is virtually nonexistent when compared to that of the electron and proton.

7.59 The uncertainty principle states that it is impossible to make simultaneous and exact measurements of both the position and momentum of a particle. According to classical physics, to predict the future behavior of a particle we have to make at least one simultaneous and exact measurement of both the position and momentum of the particle. Since the uncertainty principle tells us that such a measurement is impossible for very small particles, there is no way that we can obtain the necessary information needed to predict the future behavior of these particles and, therefore, we cannot obtain or predict the future behavior for very small particles.

7.61 Since particles (electrons, protons, etc.) have very small masses, the photons will alter both the position and momentum of the particle.

A uranium atom is considerably more massive than a subatomic particle and, therefore, the effect of the photon on the uranium atom should be less.

7.63 In the Bohr theory the electron has a definite trajectory or precise path. Its position and momentum are known exactly and each possible trajectory has a definite energy associated with it. However, the uncertainty principle showed that an electron cannot have a precise path and a definite energy cannot be associated with any particular precise path. Consequently, the Bohr theory had to be abandoned.

7.65 (a) Bohr's quantized energy levels are retained in wave mechanics. Wave mechanics also retains the Bohr hypothesis that the absorption and emission of energy by one atom is accompanied by the transition of an electron from one energy level to another. His derived energy formula for hydrogen is also retained, since it gives the correct spectrum results and, therefore, must be correct.

(b) Bohr's model of the electron as a definite particle going around the nucleus in a precise path or orbit is abandoned. Also, any consequences directly related to this classical orbit model have been abandoned (see comment).

Comment: The Bohr orbits, since they are classical in nature, all have nonzero angular momentum values associated with them. Therefore, the Bohr model predicts that the hydrogen atom ground state has a definite angular momentum. Wave mechanics, on the other hand, predicts correctly a zero angular momentum for the hydrogen atom ground state (1s orbital).

Also, the Bohr model with its definite orbits doesn't allow the electron to get any closer than 52.9 pm to the nucleus. However, there is experimental evidence that indicates the s orbital electrons in an atom spend a fraction of their time inside the nucleus. This is explainable in wave mechanics by the fact that the s orbital electron density or electron cloud is spread out in space, including the space in and around the nucleus. Therefore, in wave mechanics there is a slight probability of finding the electron in the nucleus. These are just two interesting differences between the Bohr model and wave mechanics.

7.67 (a) Three quantum numbers are needed to specify an orbital.

(b)

quantum number	symbol
principal	n
azimuthal	l
magnetic	m_l

(c) The orbital quantum numbers are not independent. The restrictions on the quantum numbers values (or allowed values) are:
1. $n = 1, 2, 3, \ldots$, any positive integer (0 not allowed).
2. For a given value of n, l has integral values ranging from 0 to $n - 1$; $l = 0, 1, 2, \ldots, (n - 1)$.
3. For a given l value, m_l may have positive or negative integral values ranging from $-l$ to $+l$, including zero; $m_l = -l, -(l - 1), \ldots, (l - 1), +l$ or $m_l = 0, \pm 1, \pm 2, \ldots, \pm l$.

7.69 (a) All orbitals in a shell have the same principal quantum number (n).
(b) All orbitals in a subshell have the same principal (n) and azimuthal (l) quantum numbers.

7.71 The quantum numbers $n = 3$ and $l = 2$ are shared by all the orbitals in the 3d subshell.

The quantum numbers $n = 4$ and $l = 3$ are shared by all the orbitals in the 4f subshell.

7.73 Within a given shell, all orbitals with the same l value constitute a subshell. The number of orbitals in a subshell is 2l + 1; l is the azimuthal quantum number. Each orbital is specified by a different combination of the three orbital quantum numbers. No two orbitals have the same combination of three quantum numbers.

(a) All p orbitals have l = 1. Therefore, a p subshell has 2(1) + 1 = 3 orbitals.
(b) In a subshell with l = 2 there would be 2(2) + 1 = 5 orbitals.
(c) In the lowest energy shell the only allowed quantum numbers are n = 1, l = 0, and $m_l = 0$. Only one combination of the three quantum numbers exists and, therefore, there is only one orbital.
(d) The number of orbitals in the seventh shell is equal to the number of different combinations of the three orbital quantum numbers possible. In the seventh shell the allowed quantum numbers and numbers of orbitals are

n = 7
l = 0, 1, 2, 3, 4, 5, 6

l = 0:	$m_l = 0$	→	1 orbital
l = 1:	$m_l = -1, 0, 1$	→	3 orbitals
l = 2:	$m_l = -2, -1, 0, 1, 2$	→	5 orbitals
l = 3:	$m_l = -3, -2, -1, 0, 1, 2, 3$	→	7 orbitals
l = 4:	$m_l = -4, -3, -2, -1, 0, 1, 2, 3, 4$	→	9 orbitals
l = 5:	$m_l = -5, -4, -3, -2, -1, 0, 1, 2, 3, 4, 5$	→	11 orbitals
l = 6:	$m_l = -6, -5, \ldots, 5, 6$	→	13 orbitals
		Total =	49 orbitals

Answer: The seventh shell has 49 orbitals.

Comment: Thus, in general, for principle level n, there are n^2 possible orbitals in n subshells.

7.75 (a) n = 2, l = 1, $m_l = 0$; allowed.
(b) n = 2, l = 2, $m_l = 2$; not allowed. For any given n, the maximum l value is n – 1.
(c) n = 2, l = 1, $m_l = -1$; allowed.
(d) n = 2, l = 1, $m_l = -2$; not allowed. For any given l, the highest and lowest m_l values are ±l.

7.77 (a) 2s; possible orbital. (b) 3p; possible orbital. (c) 2d; impossible orbital. If n is 2, the allowed l values are 0 and 1 and, therefore, the n = 2 shell contains no d (l = 2) orbitals. (d) 2f; impossible orbital. See (c); the n = 2 shell contains no f (l = 3) orbitals.

7.79 There are n^2 orbitals in a shell and 2l + 1 orbitals in a subshell.
(a) n = 3 shell; $n^2 = 3^2 = 9$ orbitals.
(b) 3d subshell; for a d subshell, l = 2 and 2l + 1 = 5 orbitals.
(c) 4d subshell; for a d subshell, l = 2 and 2l + 1 = 5 orbitals.
(d) 5g subshell; for a g subshell, l = 4 and 2l + 1 = 9 orbitals.

7.81 (a) Spin explains the fact that the electron behaves like a tiny magnet.
(b) The spin quantum number, m_s. The allowed values of the spin quantum number are +1/2 and –1/2.

7.83

n	l	m_l	m_s
1	0	0	+1/2
1	0	0	–1/2

7.85 A beam of silver atoms issuing from a hole in an oven where silver is being vaporized, is first collimated and then passed through an inhomogeneous magnetic field and on to a detector plate. The original beam of silver atoms split into two, showing that two spin states exist for the electron.

7.87 The shapes of the orbitals are the surfaces or contours of equal electron density. Usually the shapes are chosen such that the probability of finding the electron residing in that orbital within the representative surface is 90% or better.

(a) The surfaces of all s orbitals are spheres. Therefore, the shape of all s orbitals is considered to be spherical.

(b) The surfaces of all p orbitals are dumbbell-shaped. Therefore, the shape of all p orbitals is considered to be like dumbbells.

7.89 See Figure 7.32 in the textbook.

7.91 None of them represents the path of the electron. Because of the Heisenberg uncertainty principle it is impossible to specify a path for the electron. The best we can do is talk about or specify regions in space where there is a high probability of finding the electron. If you view the electron as being in a particle form in the atom, then the electron spends most of its time in these high-probability regions. If you view the electrons as being spread out in space as a three-dimensional electron cloud about the nucleus, then the regions of high probability are the regions in space that contain a relatively large amount or proportion of the overall or total electron cloud. (Remember, these are regions in space, not points. The small volumes around individual points may contain relatively high probabilities, while the region of space they're in contains only a small fraction of the overall electron cloud.)

7.93 The electron density is a measure of the probability of finding the electron in a small volume of space centered about a given point.

For an s electron the electron density is a maximum at the nucleus.

For a 2p electron, refer to Figure 7.31 in the textbook, the position of the greatest electron density corresponds to the distance from the nucleus of the maximum in the Ψ^2_{2p} versus r plot.

7.95 A nodal plane is a plane in space where the electron density equals zero. See Figure 7.32 for appropriate balloon diagrams.

(a) The p_x and d_{xy} orbitals have nodal planes; s orbitals don't have nodal planes.

(b) 1. The y-z plane is a nodal plane for the p_x orbital.
2. The x-z and y-z planes are nodal planes for the d_{xy} orbital.

7.97 The most probable distance corresponds to the maximum radial electron density (maximum spherical shell probability). The highest peak in a plot corresponds to the most probable electron distance.

(a) Rough estimates of the most probable distances are: 1s, 0.05 nm; 2s, 0.28 nm; 2p, 0.2 nm; 3s, 0.7 nm; 3p, 0.62 nm; and 3d, 0.49 nm.

(b) The graphs show that the most probable distance increases with increasing n.

(c) From the graphs we see that within a shell, the most probable distance decreases with increasing l.

7.99 The gaseous atoms in the fluorescent tube are under bombardment by the electrons traveling from the cathode to the anode. These electrons strip electrons from the atoms, forming ions. When these ions recombine with free electrons, the neutral atoms formed are in unstable excited states. Light is emitted by these excited atoms as they return to their stable ground states.

7.101 The energy of the absorbed ultraviolet radiation is converted into heat energy.

7.103 The energy of one photon having a wavelength of 510 nm is

$$E_{photon} = h\nu = \frac{hc}{\lambda} = \frac{6.626 \times 10^{-34}\ \text{J}\bullet\text{s} \times 2.998 \times 10^{8}\ \text{m/s}}{510 \times 10^{-9}\ \text{m}} = 3.90 \times 10^{-19}\ \text{J}$$

The number of 510-nm photons in 3.15×10^{-17} J is

3.15×10^{-17} J × 1 photon/3.90×10^{-19} J = 81 photons

7.105

$$q = 250 \text{ mL} \times \frac{1.00 \text{ g}}{1 \text{ mL}} \times \frac{4.18 \text{ J}}{\text{g•°C}} \times (95°\text{C} - 25°\text{C}) = 7.3 \times 10^4 \text{ J}$$

$$E_{photon} = \frac{hc}{\lambda} = \frac{6.626 \times 10^{-34} \text{ J•s} \times 2.998 \times 10^8 \text{ m/s}}{0.12 \text{ m}} = 1.7 \times 10^{-24} \text{ J}$$

The number of photons required is

7.3×10^4 J × 1 photon/1.7×10^{-24} J = 4.3×10^{28} photons

7.107

(a) One mole of Br_2 molecules absorbed 193.9 kJ; therefore, the energy absorbed per Br_2 molecule is

$$\frac{193.9 \text{ kJ}}{1 \text{ mol } Br_2} \times \frac{1 \text{ mol } Br_2}{6.022 \times 10^{23} \text{ } Br_2 \text{ molecules}} = 3.220 \times 10^{-19} \text{ J/}Br_2 \text{ molecule}$$

Answer: The minimum energy of the photon is $E_{photon} = 3.220 \times 10^{-19}$ J.

(b) The maximum wavelength of the photon is

$$\lambda = \frac{hc}{E} = \frac{6.6261 \times 10^{-34} \text{ J•s} \times 2.9979 \times 10^8 \text{ m/s}}{3.220 \times 10^{-19} \text{ J}} = 6.169 \times 10^{-7} \text{ m or } 616.9 \text{ nm}$$

From Figure 7.3 we see that the visible region of the spectrum provides light of this wavelength.

7.109

The quantum theory of radiation postulates that radiation is made up of tiny bundles of energy called photons. Each photon acts like a tiny particle with an energy equal to $h\nu$, where ν is the frequency of the radiation. All photons of a given ν are identical. Therefore, according to the quantum theory of radiation, increasing the intensity of the radiation just means that there are more photons present in the beam.

Photoelectrons are ejected after the absorption of a single photon. Therefore, the maximum kinetic energy a photoelectron can achieve depends on the amount of energy carried by the photon. Since all the photons have the same energy, the maximum kinetic energy of the photoelectron cannot increase, no matter how intense the light becomes. More photoelectrons are ejected, because there are more photons available to interact with the metal surface electrons.

7.111

Let $k = 2.179 \times 10^{-18}$ J. Then

$$E_{photon} = h\nu \text{ and } E_{photon} = -k\left(\frac{1}{n_H^2} - \frac{1}{n_L^2}\right).$$

Equating these two expressions we get: $h\nu = -k\left(\frac{1}{n_H^2} - \frac{1}{n_L^2}\right)$.

Since $\nu = c/\lambda$: $\frac{hc}{\lambda} = -k\left(\frac{1}{n_H^2} - \frac{1}{n_L^2}\right)$ or $\frac{1}{\lambda} = -R_H\left(\frac{1}{n_H^2} - \frac{1}{n_L^2}\right)$,

where $R_H = k/hc = 2.179 \times 10^{-18}$ J/6.6261×10^{-34} J•s × 2.9979×10^8 m/s = 1.097×10^7 m^{-1} is the Rydberg constant for hydrogen.

7.113 (a) An electron just removed from the atom has zero energy. Therefore, the amount of energy required to remove an electron from any level in the atom is the amount needed to make its energy zero. The energies of the electrons in the n = 2 and n = 3 levels are

$n = 2$: $E = -2.179 \times 10^{-18}\ J/4 = -5.448 \times 10^{-19}\ J$

$n = 3$: $E = -2.179 \times 10^{-18}\ J/9 = -2.421 \times 10^{-19}\ J$

Therefore, it would take 5.448×10^{-19} J to remove the electron from the n = 2 level of the first hydrogen atom and 2.421×10^{-19} J to remove the electron from the n = 3 level of the second hydrogen atom.

(b) The one in the n = 2 level in the sense that it requires more energy to remove it.

(c) The second hydrogen atom is more easily ionized in the sense that it requires less energy to remove its electron.

7.115

$$\frac{1}{\lambda} = -R_H\left(\frac{1}{n_H^2} - \frac{1}{n_L^2}\right) = -1.097 \times 10^7\ m^{-1}\left(\frac{1}{n_H^2} - \frac{1}{n_L^2}\right)$$

For the first Humphreys line we get ($n = 7 \rightarrow n = 6$)

$$\frac{1}{\lambda} = -1.097 \times 10^7\ m^{-1}\left(\frac{1}{49} - \frac{1}{36}\right) = 8.084 \times 10^4\ m^{-1}$$

$\lambda = 1/8.084 \times 10^4\ m^{-1} = 1.237 \times 10^{-5}\ m$

7.117 No. The energy of the photon emitted when the electron jumps or moves from the higher level to the lower level is

$$E_{photon} = E_H - E_L = -9.68 \times 10^{-19}\ J + 21.8 \times 10^{-19}\ J = 1.21 \times 10^{-18}\ J$$

The wavelength associated with a photon of this energy is

$$\lambda = \frac{hc}{E} = \frac{6.63 \times 10^{-34}\ J{\bullet}s \times 3.00 \times 10^8\ m/s}{1.21 \times 10^{-18}\ J} = 164\ nm$$

From Figure 7.3 we see that radiation of this wavelength falls in the ultraviolet region of the spectrum and, therefore, a green line cannot be due to a transition between these levels.

7.119

$$v = \frac{h}{m\lambda} = \frac{6.626 \times 10^{-34}\ J{\bullet}s}{9.109 \times 10^{-31}\ kg \times 1.50 \times 10^{-4}\ m} = 4.85\ m/s$$

7.121 (a) The phrase *standing wave* is a term applied to a wave which does not destructively interfere with itself. An important property of standing waves is that they contain either a whole number of wavelengths or a whole number of half-wavelengths.

The phrase *stationary state* refers to a state in which the wave associated with that state does not destructively interfere with itself, i.e., a standing wave. Such states have the characteristics of stability, i.e., their properties (energy, number and position of nodes, frequency, etc.) are independent of time (constant or fixed).

(b) Classical electromagnetic theory predicts that the system of an electron revolving around a proton is not stable; i.e., it cannot exist for any length of time. However, quantum mechanics or matter wave theory predicts stable states for this system. The allowed states of hydrogen are often referred to as stationary states, because the matter wave associated with the electron is considered to be a standing wave.

7.123 (a) 5p → 3s. The smaller the photon energy the greater the wavelength.

(b) 5p → 2s. Transitions from higher excited states (n = 3, 4, 5, . . .) down to the first excited state (n = 2) gives rise to the Balmer series, the spectral lines of which lie in the visible region of the electromagnetic spectrum.

(c) 5p → 1s. The Lyman series arises when electrons in excited states (n = 2, 3, 4, . . .) emit photons while returning to the ground state (n = 1).

7.125 The radial electron density is the total amount of electron density contained within a thin spherical shell. At the nucleus the volume of the spherical shell is zero and, therefore, the radial electron density is zero, even though the electron density is at a maximum. As you go away from the nucleus, the shell volume continually increases while the electron density (ψ^2) continually decreases. At first the increasing volume factor wins out over the decreasing electron density factor and the radial electron density increases with increasing r. However, a point is reached where the decreasing electron density factor takes over and the radial electron density starts decreasing, rapidly approaching zero with increasing r after this maximum point. The maximum in the radial electron density corresponds to the most probable electron distance, i.e., the distance from the nucleus where you are most likely to find the electron.

7.127 radial electron density $= 4\left(\dfrac{1}{a_0}\right)^3 r^2\, e^{-2r/a_0} = k\, r^2\, e^{-2r/a_0}$

To simplify things, one can scale the units of r in multiples of a_0, i.e., r = 0, $a_0/4$, $a_0/2$, $3a_0/4$, a_0, $3a_0/2$, $2a_0$, . . . , ∞, and the units of the radial electron density in terms of the constant $K = ka_0^2$.

$r = 0$: r.e.d. = 0

$r = a_0/4$: r.e.d. $= (ka_0^2/16)e^{-1/2} = 0.0379ka_0^2 = 0.0379K$

$r = a_0/2$: r.e.d. $= (ka_0^2/4)e^{-1} = 0.0920K$

$r = 3a_0/4$: r.e.d. $= (9ka_0^2/16)e^{-3/2} = 0.126K$

$r = a_0$: r.e.d. $= ka_0^2e^{-2} = 0.135K$

$r = 3a_0/2$: r.e.d. $= (9ka_0^2/4)e^{-3} = 0.112K$

$r = 2a_0$: r.e.d. $= 4ka_0^2e^{-4} = 0.0733K$

$r = 5a_0/2$: r.e.d. $= (25ka_0^2/4)e^{-5} = 0.0421K$

$r = \infty$: r.e.d. = 0 (The e^{-2r/a_0} factor goes to zero faster than the kr^2 factor goes to infinity.)

Plot these points on a piece of graph paper. See Figure 7.34 in the textbook for a similar plot.

Maximum radial electron density is at $r = a_0 = 52.9$ pm.

Minimum radial electron density is at r = 0 and r = ∞.

CHAPTER 8

MANY-ELECTRON ATOMS AND THE PERIODIC TABLE

Solutions To Practice Exercises

PE 8.1 The energy of an electron in a given orbital decreases with increasing atomic number. Within a shell the energy of an electron in an orbital increases with increasing l.

(a) The 1s electron in carbon would have a lower energy, since carbon's atomic number (Z = 6) is greater than hydrogen's (Z = 1).

(b) The 3s electron in chlorine would have a lower energy, since l = 0 for an s orbital and 1 for a p orbital.

PE 8.2 The (n + l) values for the orbitals in question are: 1s, (1 + 0) = 1; 2s, (2 + 0) = 2; 2p, (2 + 1) = 3; 3s, (3 + 0) = 3; 3p, (3 + 1) = 4; 3d, (3 + 2) = 5; 4s, (4 + 0) = 4; and 4p, (4 + 1) = 5. Hence, the filling order for these orbitals based on the "n + l rule" is: 1s2s2p3s3p4s3d4p.

This orbital filling order is identical to the given orbital filling order.

PE 8.3 See Table 8.1.

PE 8.4 (a) V (Z = 23): $[Ar]4s^2 3d^3$ or $[Ar]3d^3 4s^2$

(b) ↑ ↑ ↑ _ _ (3d) ⇅ (4s)

PE 8.5 (a) Iodine, a Group 7A element, has seven valence electrons.

(b) Magnesium, a Group 2A element, has two valence electrons.

(c) He ($1s^2$) has two valence electrons.

PE 8.6 Lead is element 82; lead is in the sixth period, Group 4A. The preceding noble gas is xenon in the fifth period, therefore the lead configuration starts with [Xe]. Traveling from xenon to lead takes us 28 spaces across the sixth period, adding 2 electrons in the s block, 14 electrons in the f block, 10 electrons in the d block, and finally two p electrons. The configuration of lead is $[Xe]6s^2 4f^{14} 5d^{10} 6p^2$.

PE 8.7 When an atom forms a positive ion, its electrons are removed in order of decreasing principal quantum number. Within a shell electrons are removed in order of decreasing l value.

(a) The ground-state configuration of the barium atom is Ba (Z = 56): [Xe]$6s^2$. Barium will lose two 6s electrons in forming Ba^{2+}. The electron configuration of Ba^{2+} is [Xe].
(b) The configuration of the lead atom is Pb (Z = 82): [Xe]$4f^{14}5d^{10}6s^2 6p^2$. Lead will lose two 6p electrons in forming Pb^{2+}. The electron configuration of Pb^{2+} is [Xe]$4f^{14}5d^{10}6s^2$.

PE 8.8 The atomic number of iodine is 53, and the neutral atom has 53 electrons. The iodide ion has one additional electron, or 54 in all. Its electron configuration is identical to that of xenon, [Xe].

PE 8.9 (a) Ca (Z = 20): [Ar]$4s^2$. Diamagnetic – calcium has no unpaired electrons.
(b) Na (Z = 1): [Ne]$3s^1$. Paramagnetic – sodium has one unpaired electron in the 3s orbital.
(c) S (Z = 16): [Ne]$3s^2 3p^4$. Paramagnetic – sulfur has 2 unpaired electrons in the 3p orbitals.

PE 8.10 In general, atomic radii increase from top to bottom within a group. Also, in general, atomic radii decrease from left to right across a period.
(a) I lies just below Br in Group 7A; therefore I has the larger atomic radius.
(b) Si (Group 4A) lies to the left of P (Group 5A) in the third period. Si therefore has a larger atomic radius.

PE 8.11 Monatomic positive ions (cations) are smaller than the parent atoms. Monatomic negative ions (anions) are larger than the parent atoms.
(a) Ba^{2+} is a positive ion; therefore Ba will have the larger radius.
(b) S^{2-} is a negative ion; therefore S^{2-} will have the larger radius.

PE 8.12 Ionization energies tend to decrease going down a group. Also, ionization energies tend to increase from left to right across a period.
(a) Br lies just below Cl in Group 7A; therefore Cl should have a higher first ionization energy.
(b) Mg (Group 2A) lies just to the right of Na (Group 1A) in the third period; therefore Mg should have a higher first ionization energy.

Solutions To Final Exercises

8.1 (a) The orbitals of many-electron atoms can be described in terms of the same quantum numbers, n, l, and m_l, and the same shapes, s, p, d, and f, as hydrogen orbitals.
(b) The orbitals of many-electron atoms differ from those of hydrogen in their electron energies and in the most probable distances of the electron from the nucleus. Also, the energies in a given shell increase with l in many-electron atoms.

8.3 In many-electron atoms, the energy of an electron in a given shell increases with the l value of its orbital; therefore, 3s < 3p < 3d.

8.5 The energy of an electron in a given orbital decreases with increasing atomic number.
(a) An electron in the 2p orbital of barium would have a lower energy, since barium's atomic number (Z = 56) is greater than strontium's (Z = 38).
(b) The electron is more tightly bound to the atom in the 2p orbital of barium. This is because the barium nucleus has a greater charge and, therefore, exerts a greater attractive force. Moreover, the energy of the barium electron is lower, so that more energy would be required to remove it.

8.7 (a) The nuclear attraction for an electron is a Coulomb's law attraction, i.e., two charges of opposite sign attracting one another. Electrons, on the other hand, exhibit Coulomb's law repulsions. When electrons exist in underlying orbitals between the nucleus and an electron, the electron-electron repulsions counteract the nuclear-electron attraction, leading to a decrease in the nuclear attraction for the electron. This decrease in nuclear attraction for an electron due to the presence of other electrons in underlying orbitals is called the shielding effect.

(b) The form of the radial electron densities depends on both the n and l quantum numbers. As l increases, the electron's ability to penetrate underlying orbitals and spend some time nearer the nucleus decreases, since less radial electron density exists close to the nucleus as l increases. The overall effect of this, is that different orbitals within the same shell are not all shielded to the same extent. The amount of shielding increases with the l value, and since the better the shielding (less nuclear attraction) the higher the energy, the electron energies increase with increasing l within a shell.

8.9 The electron will be more tightly bound in the 3s orbital. Electrons in s orbitals are not shielded as effectively as electrons in p orbitals and, therefore, s electrons experience a greater effective nuclear charge. Since the actual positive charge experienced by the 3s electron is greater than that experienced by the 3p electron, it will be more tightly bound.

8.11 (1) Only two electrons can be in the same orbital, and these electrons must have opposite spins.
(2) No two electrons in an atom can have the same values for all four quantum numbers.

8.13 The number of electron states in a shell is $2n^2$, where n is the principal quantum number of the shell.
(a) The second shell (n = 2) can hold a maximum of $2 \times (2)^2 = 8$ electrons.
(b) The fourth shell (n = 4) can hold a maximum of $2 \times (4)^2 = 32$ electrons.
(c) The nth shell can hold a maximum of $2n^2$ electrons.

8.15 (a) A full 2s subshell contains 2 electrons. The four quantum numbers for each electron are 2 0 0 $+\frac{1}{2}$ and 2 0 0 $-\frac{1}{2}$.
(b) A full 2p subshell contains 6 electrons. The four quantum numbers for each electron are 2 1 −1 $+\frac{1}{2}$; 2 1 −1 $-\frac{1}{2}$; 2 1 0 $+\frac{1}{2}$; 2 1 0 $-\frac{1}{2}$; 2 1 +1 $+\frac{1}{2}$; and 2 1 +1 $-\frac{1}{2}$.

8.17 A ground-state electron configuration is the electron configuration the atom has when it's in its lowest energy state.

O (Z = 8): $1s^22s^22p^4$; $[He]2s^22p^4$
H (Z = 1): $1s^1$

8.19 The orbital filling order for the first 36 elements is 1s 2s 2p 3s 3p 4s 3d 4p.

8.21 The aufbau procedure is a procedure used to obtain the ground-state electron configuration of an atom by adding electrons into successive subshells according to an experimentally established subshell filling order. When filling the orbitals in a given subshell the Pauli exclusion principle and Hund's rule must be obeyed.

The subshell or orbital filling order for the first 36 elements is 1s 2s 2p 3s 3p 4s 3d 4p. Carbon has six electrons. According to the aufbau procedure and Pauli exclusion principle, the first two electrons enter the 1s orbital, the second two electrons enter the 2s orbital, and the last two electrons enter the 2p orbitals. The ground-state electron configuration of the carbon atom according to the aufbau procedure is therefore $1s^2\ 2s^2\ 2p^2$.

8.23 (a) Be (Z = 4): $1s^22s^2$; $[He]2s^2$
(b) P (Z = 15): $1s^22s^22p^63s^23p^3$; $[Ne]3s^23p^3$
(c) Ca (Z = 20): $1s^22s^22p^63s^23p^64s^2$; $[Ar]4s^2$
(d) V (Z = 23): $1s^22s^22p^63s^23p^63d^34s^2$; $[Ar]3d^34s^2$

(e) Se (Z = 34): $1s^2 2s^2 2p^6 3s^2 3p^6 3d^{10} 4s^2 4p^4$; $[Ar]3d^{10} 4s^2 4p^4$
(f) F (Z = 9): $1s^2 2s^2 2p^5$; $[He]2s^2 2p^5$

The configurations agree with those in Table 8.1.

8.25 (a) Be (Z = 4): $\underset{1s}{\uparrow\downarrow}\ \underset{2s}{\uparrow\downarrow}$; no unpaired electrons.

(b) P (Z = 15): $\underset{1s}{\uparrow\downarrow}\ \underset{2s}{\uparrow\downarrow}\ \underset{2p}{\uparrow\downarrow\ \uparrow\downarrow\ \uparrow\downarrow}\ \underset{3s}{\uparrow\downarrow}\ \underset{3p}{\uparrow\ \uparrow\ \uparrow}$; three unpaired electrons.

(c) Ca (Z = 20): $\underset{1s}{\uparrow\downarrow}\ \underset{2s}{\uparrow\downarrow}\ \underset{2p}{\uparrow\downarrow\ \uparrow\downarrow\ \uparrow\downarrow}\ \underset{3s}{\uparrow\downarrow}\ \underset{3p}{\uparrow\downarrow\ \uparrow\downarrow\ \uparrow\downarrow}\ \underset{3d}{\_\ \_\ \_\ \_\ \_}\ \underset{4s}{\uparrow\downarrow}$; no unpaired electrons.

(d) V (Z = 23): $[Ar]\ \underset{3d}{\uparrow\ \uparrow\ \uparrow\ \_\ \_}\ \underset{4s}{\uparrow\downarrow}$; three unpaired electrons.

(e) Se (Z = 34): $[Ar]\ \underset{3d}{\uparrow\downarrow\ \uparrow\downarrow\ \uparrow\downarrow\ \uparrow\downarrow\ \uparrow\downarrow}\ \underset{4s}{\uparrow\downarrow}\ \underset{4p}{\uparrow\downarrow\ \uparrow\ \uparrow}$; two unpaired electrons.

(f) F (Z = 9): $\underset{1s}{\uparrow\downarrow}\ \underset{2s}{\uparrow\downarrow}\ \underset{2p}{\uparrow\downarrow\ \uparrow\downarrow\ \uparrow}$; one unpaired electron.

8.27 The orbital filling order for the first 36 elements based on the "n + l rule" is: 1s 2s 2p 3s 3p 4s 3d 4p. The electron configurations for chromium and copper based on the "n + l rule" are

Cr (Z = 24): $1s^2 2s^2 2p^6 3s^2 3p^6 3d^4 4s^2$; $[Ar]3d^4 4s^2$
Cu (Z = 29): $1s^2 2s^2 2p^6 3s^2 3p^6 3d^9 4s^2$; $[Ar]3d^9 4s^2$

These configurations do not agree with those in Table 8.1.

8.29 Periodic means that the same thing happens or returns at generally regular intervals of some variable.
(a) Yes. The same day reappears every eighth day, i.e., in seven day intervals.
(b) No. Your age always keeps increasing; you never have the same age twice.
(c) No. Atomic numbers just increase; no two elements have the same atomic number.
(d) Yes. Similar valence shell electron configurations reappear at certain, but not totally regular, intervals as the atomic number increases.

8.31

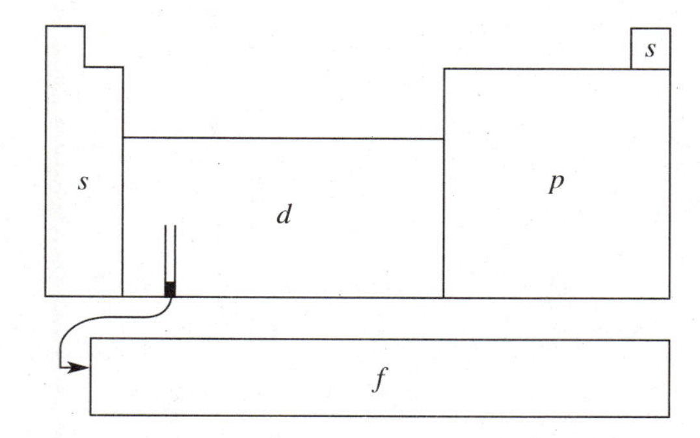

See Figure 8.4 in the textbook for further details.

(a) s block: two columns (plus He); d block: 10 columns; p block: six columns (excluding He)
(b) fourth row
(c) fourteen spaces

8.33
(a) Second period: 2s and 2p.
(b) Third period: 3s and 3p.
(c) Fourth period: 4s, 3d, and 4p. (Note that Cr and Cu are exceptions to this filling pattern.)

8.35
(a) Representative elements: s and p blocks
(b) Transition elements: d and f blocks
(c) Inner transition elements: f block
(d) Metals: s, p, d, and f blocks
(e) Nonmetals: p block, plus H and He from the s block
(f) Rare earth elements (lanthanides): f block

8.37
(a) Be, [He] (b) Fe, [Ar] (c) Sb, [Kr] (d) Al, [Ne] (e) U, [Rn] (f) Ra, [Rn]

8.39
From the periodic table:
(a) Be (Z = 4): $[He]2s^2$
(b) Fe (Z = 26): $[Ar]3d^6 4s^2$
(c) Sb (Z = 51): $[Kr]4d^{10}5s^2 5p^3$
(d) Al (Z = 13): $[Ne]3s^2 3p^1$
(e) U (Z = 92): $[Rn]5f^3 6d^1 7s^2$
(f) Ra (Z = 88): $[Rn]7s^2$

None of these elements show irregularities.

8.41
From the periodic table:
(a) Cr (Z = 24): $[Ar]3d^4 4s^2$
(b) Sm (Z = 62): $[Xe]4f^5 5d^1 6s^2$
(c) Mo (Z = 42): $[Kr]4d^4 5s^2$
(d) I (Z = 53): $[Kr]4d^{10}5s^2 5p^5$
(e) Hg (Z = 80): $[Xe]4f^{14}5d^{10}6s^2$
(f) Ba (Z = 56): $[Xe]6s^2$

Cr, Sm, and Mo show irregularities.

8.43
(b) ↑↓ ↑ ↑ ↑ ↑ (3d)
(c) ↑ ↑ ↑ (5p)
(d) ↑ _ _ (3p)
(e) ↑ ↑ ↑ _ _ _ _ (5f) ↑ _ _ _ _ (6d)

8.45
Sc (Z = 21): ↑ _ _ _ _ (3d) ↑↓ (4s)
Ti (Z = 22): ↑ ↑ _ _ _ (3d) ↑↓ (4s)
V (Z = 23): ↑ ↑ ↑ _ _ (3d) ↑↓ (4s)
Cr (Z = 24): ↑ ↑ ↑ ↑ ↑ (3d) ↑ (4s)
Mn (Z = 25): ↑ ↑ ↑ ↑ ↑ (3d) ↑↓ (4s)
Fe (Z = 26): ↑↓ ↑ ↑ ↑ ↑ (3d) ↑↓ (4s)
Co (Z = 27): ↑↓ ↑↓ ↑ ↑ ↑ (3d) ↑↓ (4s)

Ni (Z = 28): ⇅ ⇅ ⇅ ↑ ↑ (3d) ⇅ (4s)

Cu (Z = 29): ⇅ ⇅ ⇅ ⇅ ⇅ (3d) ↑ (4s)

Zn (Z = 30): ⇅ ⇅ ⇅ ⇅ ⇅ (3d) ⇅ (4s)

Note that chromium and copper are exceptions to the usual filling rule; they show irregularities.

8.47 For the representative elements, the number of valence electrons is equal to the number of electrons in the outermost occupied electron shell. Ground-state configurations can be obtained from Table 8.1 or the periodic table. Remember that the periodic table always gives the correct ground-state electron configuration for the representative elements.

(a) Sr (Z = 38): [Kr]$5s^2$, two
(b) In (Z = 49): [Kr]$4d^{10}5s^25p^1$, three
(c) Cs (Z = 55): [Xe]$6s^1$, one
(d) Si (Z = 14): [Ne]$3s^23p^2$, four

Comment: For the representative elements, elements in the s and p blocks of the periodic table, the number of valence electrons is numerically equal to the group number of the group the element is in. (Helium is the only exception.) Therefore, for representative elements you can just consult the periodic table to determine the number of valence electrons in the atom. Check these answers with the answers you get from the periodic table. The results should be the same.

8.49 For the representative elements the valence shell is always the outermost occupied shell and the periodic table always gives the correct electron configuration.

(a) $3s^1$ (b) $4s^2$ (c) $5s^25p^4$ (d) $4s^24p^5$

8.51
(a) Halogens: ns^2np^5
(b) Alkali metals: ns^1
(c) Group 2A: ns^2
(d) Alkaline earth metals: ns^2
(e) Group 4A: ns^2np^2
(f) Group 2B: $ns^2(n-1)d^{10}$

8.53
(a) Zn (Z = 30): [Ar]$3d^{10}4s^2$. Zinc will lose two 4s electrons in forming Zn^{2+}. Zn^{2+}: [Ar]$3d^{10}$.
(b) Co (Z = 27): [Ar]$3d^74s^2$. Cobalt will lose two 4s electrons and one 3d electron in forming Co^{3+}. Co^{3+}: [Ar]$3d^6$.
(c) Al (Z = 13): [Ne]$3s^23p^1$. Aluminum will lose one 3p electron and two 3s electrons in forming Al^{3+}. Al^{3+}: [Ne].
(d) Ni (Z = 28): [Ar]$3d^84s^2$. Nickel will lose two 4s electrons in forming Ni^{2+}. Ni^{2+}: [Ar]$3d^8$.
(e) V (Z = 23): [Ar]$3d^34s^2$. Vanadium will lose two 4s electrons in forming V^{2+}. V^{2+}: [Ar]$3d^3$.
(f) F (Z = 9): [He]$2s^22p^5$. Fluorine gains one electron and, therefore, F^- has the same configuration as the atom with Z = 10. F^-: [Ne].
(g) S (Z = 16): [Ne]$3s^23p^4$. Sulfur gains two electrons and, therefore, S^{2-} has the same configuration as the atom with Z = 18. S^{2-}: [Ar].
(h) Ag (Z = 47): [Kr]$4d^{10}5s^1$. Silver will lose one 5s electron in forming Ag^+. Ag^+: [Kr]$4d^{10}$.

8.55 Paramagnetic substances have unpaired electrons. Unpaired electrons in atoms or ions gives the overall atom or ion a magnetic field. This magnetic field can interact with external magnetic fields. Therefore, paramagnetic substances are drawn into external magnetic fields because its constituent particles are magnetic.

Paired electrons, two electrons in the same orbital, have opposed spins and their magnetic fields cancel. Atoms and ions whose electrons are all paired, i.e., atoms and ions containing no singly filled orbitals, have no net magnetic field. Such atoms and ions are called diamagnetic, and since they have no net magnetic field they cannot interact with an external magnetic field. Therefore, diamagnetic substances are not drawn into external magnetic fields because its constituent particles are not magnetic.

8.57 Because of electrostatic repulsive forces, atoms can only approach each other up to a certain distance. To bring the atoms closer together would require the application of a tremendous amount of force. Therefore, under normal conditions the atoms appear to have an outer spherical surface and a definite radius, which as we see is electrostatic in nature.

8.59 From Figure 8.9 we see that Fr (Z = 87), the bottom or last element in Group 1A, has the largest atomic radius and H (Z = 1), the top or first element in Group 1A, has the smallest atomic radius.

8.61 Consult Figures 8.11 and 8.12. Helium has the greatest first ionization energy. Francium has the smallest first ionization energy. Note that cesium has the smallest tabulated value, but francium should be below cesium if a value were available.

8.63 (a) Ionization energies are always positive because the removal of a negative electron from the positive field of the nucleus requires the input of energy.
(b) The second ionization energy is greater than the first because the loss of one electron leaves a positive charge that makes subsequent electrons more difficult to remove.

8.65 Be ($1s^2 2s^2$) lies just to the right of Li ($1s^2 2s^1$) in period 2. Because of beryllium's greater nuclear charge, it requires more energy to remove a 2s electron from Be than from Li. Therefore, Be's first ionization energy should be greater than Li's.

After the first ionization the configurations of the ions are: Be^+ $1s^2 2s^1$ and Li^+ $1s^2$. The second ionization removes a 1s electron from Li^+ and a 2s electron from Be^+. Since a 2s electron is much further from the nucleus than a 1s electron, it requires substantially more energy to remove the 1s electron from Li^+ than the 2s electron from Be^+. Therefore, Li's second ionization energy should be substantially greater than Be's.

8.67 The electron affinity is the enthalpy change accompanying the addition of an electron to a gaseous ground-state atom or ion. In general, the more energy released, the greater the tendency for the process to occur. Therefore, the value of the electron affinity should be a measure of the tendency of an atom to gain an electron.

8.69 The second electron affinity of an atom is always positive (endothermic) because energy is required to overcome the electrostatic repulsion between the negative ion and the incoming electron.

8.71 H, Li, B, C, N, O, and F have paramagnetic atoms. He, Be, and Ne have diamagnetic atoms.

8.73 (a) N (Z = 7): $[He]2s^2 2p^3$. Nitrogen has three unpaired electrons in the 2p orbitals.
(b) Zn (Z = 20): $[Ar]3d^{10}4s^2$. Zinc has no unpaired electrons.
(c) Fe (Z = 26): $[Ar]3d^6 4s^2$. Iron has four unpaired electrons in the 3d orbitals.
(d) Ag (Z = 47): $[Kr]4d^{10}5s^1$. Silver has one unpaired electron in the 5s orbital.

8.75 The ions with unpaired electrons are paramagnetic. Co^{3+}, Ni^{2+}, and V^{2+} are paramagnetic ions.

8.77 The degree of paramagnetism increases with the number of unpaired electrons. The configurations of Cu, Cu^+, and Cu^{2+} are: Cu (Z = 29): $[Ar]3d^{10}4s^1$; Cu^+: $[Ar]3d^{10}$; and Cu^{2+}: $[Ar]3d^9$. Cu and Cu^{2+} both have one unpaired electron. Cu^+ has no unpaired electrons. Therefore, both Cu and Cu^{2+} should be the most paramagnetic. Cu^+ should be the least paramagnetic, since it is actually diamagnetic.

8.79 In general, atomic radii increase from top to bottom within a group. Also, in general, atomic radii decrease from left to right across a period.

(a) S (Group 6A) lies to the left of Cl (Group 7A) in the third period. S therefore has a larger atomic radius.
(b) Ba lies below Ca in Group 2A; therefore Ba has the larger atomic radius.
(c) Si (Group 4A) lies to the left of S (Group 6A) in the third period. Si therefore has a larger atomic radius.
(d) Na (Group 1A) lies to the left of Mg (Group 2A) in the third period. Na therefore has a larger atomic radius.
(e) C (Group 4A) lies to the left of N (Group 5A) in the second period; therefore C has the larger atomic radius.
(f) I lies below Br in Group 7A; therefore I has the larger atomic radius.

8.81 (a) I lies just below Br in Group 7A; therefore I is larger than Br. Br lies just below Cl in Group 7A; therefore Br is larger than Cl. Cl is the smallest and I the largest: $Cl < Br < I$.
(b) C (Group 4A) lies to the left of F (Group 7A) in the second period. C is therefore larger than F. Si lies just below C in Group 4A; therefore Si is larger than C. F is the smallest and Si the largest: $F < C < Si$.
(c) Ca lies just below Mg in Group 2A; therefore Ca is larger than Mg. K (Group 1A) lies to the left of Ca (Group 2A) in the fourth period. K is therefore larger than Ca. Mg is the smallest and K is the largest: $Mg < Ca < K$.

8.83 Monatomic positive ions (cations) are smaller than the parent atoms. Monatomic negative ions (anions) are larger than the parent atoms.

(a) K^+ is a positive ion; therefore K should have the larger radius.
(b) Cl^- is a negative ion; therefore Cl^- should have the larger radius.
(c) Fe^{3+} is more positive than Fe^{2+}; therefore Fe^{2+} should have the larger radius.
(d) P^{3-} is a negative ion; therefore P^{3-} should have the larger radius.
(e) $S > O$; therefore S^{2-} should have the larger radius.
(f) $I > Cl$; therefore I^- should have the larger radius.

8.85 Ionization energies tend to decrease going down a group. Also, ionization energies tend to increase from left to right across a period. (Keep in mind that as you go across a period exceptions occur in Groups 3A and 6A.)

(a) Ca (Group 2A) lies just to the right of K (Group 1A) in the fourth period; therefore Ca should have the greater first ionization energy.
(b) F (Group 7A) lies to the right of C (Group 4A) in the second period; therefore F should have the greater first ionization energy.
(c) Se lies just below S in Group 6A; therefore S should have a higher first ionization energy.
(d) Ba lies below Mg in Group 2A; therefore Mg should have the higher first ionization energy.
(e) P (Group 5A) lies to the right of Al (Group 3A) in the third period; therefore P should have the greater first ionization energy.
(f) F (Group 7A) lies to the right of N (Group 5A) in the second period; therefore F should have the greater first ionization energy.

8.87 The second ionization energy is always greater than the first ionization energy, and the third ionization energy is always greater than the second ionization energy; i.e., third ionization energy > second ionization energy > first ionization energy. Monoatomic anions (negative ions) always have a lower ionization energy than their neutral parent atom.

(a) K^+ is a positive ion; therefore K^+ should have the greater first ionization energy.
(b) Cl^- is a negative ion; therefore Cl should have the greater first ionization energy.
(c) Fe^{3+} is more positive than Fe^{2+}; therefore Fe^{3+} should have the greater first ionization energy.
(d) P^{3-} is a negative ion; therefore P should have the greater first ionization energy.
(e) S^{2-} has a larger radius than O^{2-}; therefore O^{2-} should have the greater first ionization energy.
(f) I^- has a larger radius than Cl^-; therefore Cl^- should have the greater first ionization energy.

8.89 The ionization energy trends are the reverse of the atomic radius trends. (Keep in mind that as you go across a period exceptions occur in Groups 3A and 6A.)
(a) I < Br < Cl (b) Si < C < F (c) K < Ca < Mg

8.91 Refer to Figures 8.11 and 8.15 in the textbook. There are 22 representative elements that are metals, if you include the metalloids Ge and Sb among the metals. Otherwise, there are 20 representative elements that are metals.

There are 22 representative elements that are nonmetals, if you include the metalloids B, Si, As, and Te among the nonmetals. Otherwise, there are 18 representative elements that are nonmetals.

8.93 There are 87 elements that are metals, if you include the metalloids Ge and Sb among the metals. Otherwise, there are 85 elements that are metals.

There are 22 elements that are nonmetals, if you include the metalloids B, Si, As, and Te among the nonmetals. Otherwise, there are 18 elements that are nonmetals.

8.95 A metal can be distinguished from a nonmetal by its luster, electrical and thermal conductivity, malleability, and ductility. Also, see Table 8.4 in the textbook.

8.97 Let N stand for the nucleus.

Electrostatic attractions: $N \leftrightarrow 1$; $N \leftrightarrow 2$; and $N \leftrightarrow 3$
Electrostatic repulsions: $1 \leftrightarrow 2$; $1 \leftrightarrow 3$; and $2 \leftrightarrow 3$

8.99 (a) The configurations based on the periodic table are:

La (Z = 57): [Xe]$5d^1 6s^2$; Ce (Z = 58): [Xe]$4f^1 5d^1 6s^2$; Pr (Z =59): [Xe]$4f^2 5d^1 6s^2$; Nd (Z = 60): [Xe]$4f^3 5d^1 6s^2$; Pm (Z = 61): [Xe]$4f^4 5d^1 6s^2$; Sm (Z = 62): [Xe]$4f^5 5d^1 6s^2$; Eu (Z = 63): [Xe]$4f^6 5d^1 6s^2$; Gd (Z = 64): [Xe]$4f^7 5d^1 6s^2$; Tb (Z = 65): [Xe]$4f^8 5d^1 6s^2$; Dy (Z = 66): [Xe]$4f^9 5d^1 6s^2$; Ho (Z = 67): [Xe]$4f^{10} 5d^1 6s^2$; Er (Z = 68): [Xe]$4f^{11} 5d^1 6s^2$; Tm (Z = 69): [Xe]$4f^{12} 5d^1 6s^2$; Yb (Z = 70): [Xe]$4f^{13} 5d^1 6s^2$; Lu (Z = 71): [Xe]$4f^{14} 5d^1 6s^2$; and Hf (Z = 72): [Xe]$4f^{14} 5d^2 6s^2$.

(b) According to the "n + l rule" the 4f should fill before the 5d, since (n + l) = 4 + 3 = 7 for 4f and (n + l) = 5 + 2 = 7 for 5d and when two orbitals have the same (n + l) value, the orbital with the lower n value fills first. The configurations based on the "n + l rule" are:

La (Z = 57): [Xe]$4f^1 6s^2$; Ce (Z = 58): [Xe]$4f^2 6s^2$; Pr (Z = 59): [Xe]$4f^3 6s^2$; Nd (Z = 60): [Xe]$4f^4 6s^2$; Pm (Z = 61): [Xe]$4f^5 6s^2$; Sm (Z = 62): [Xe]$4f^6 6s^2$; Eu (Z = 63): [Xe]$4f^7 6s^2$; Gd (Z = 64): [Xe]$4f^8 6s^2$; Tb (Z = 65): [Xe]$4f^9 6s^2$; Dy (Z = 66): [Xe]$4f^{10} 6s^2$; Ho (Z = 67): [Xe]$4f^{11} 6s^2$; Er (Z = 68): [Xe]$4f^{12} 6s^2$; Tm (Z = 69): [Xe]$4f^{13} 6s^2$; Yb (Z = 70): [Xe]$4f^{14} 6s^2$; Lu (Z = 71): [Xe]$4f^{14} 5d^1 6s^2$; and Hf (Z = 72): [Xe]$4f^{14} 5d^2 6s^2$.

Comparison of the configurations based on the periodic table with those in Table 8.1 shows that the configurations for Pr, Nd, Pm, Sm, Eu, Tb, Dy, Ho, Er, Tm, and Yb differ from those given in Table 8.1. Eleven configurations differ.

Comparison of the configurations based on the "n + l rule" with those in Table 8.1 shows that the configurations for La, Ce, and Gd differ from those given in Table 8.1. Three configurations differ.

The "n + l rule" gives fewer wrong configurations and, therefore, gives better results for these elements.

8.101 (a) From Figure 8.11 we see that the energy needed to ionize a mole of gaseous sodium atoms is 495.8 kJ/mol. The energy needed to ionize one gaseous sodium atom is

$$\frac{495.8 \times 10^3\ \text{J}}{1\ \text{mol}} \times \frac{1\ \text{mol}}{6.022 \times 10^{23}\ \text{atoms}} = 8.233 \times 10^{-19}\ \text{J/atom}$$

The wavelength of a photon with this energy is

$$\lambda = \frac{hc}{E} = \frac{6.6261 \times 10^{-34}\ \text{J}\bullet\text{s} \times 2.9979 \times 10^8\ \text{m/s}}{8.233 \times 10^{-19}\ \text{J}} = 2.413 \times 10^{-7}\ \text{m or } 241.3\ \text{nm}$$

Answer: The longest wavelength of radiation that will ionize gaseous sodium atoms is 241.3 nm.

(b) From Figure 7.3 we see that radiation with this wavelength falls in the ultraviolet region of the spectrum.

8.103 (a) The lanthanide contraction is the name given to the decrease in size of the lanthanide elements as you go across the lanthanide series from left to right.

(b) As you go across the lanthanide series, part of the sixth period of the periodic table, the nuclear charge increases while you continue to fill an inner 4f subshell. The increase in nuclear charge, without starting a new electron shell, i.e., while filling an already existing inner shell, leads to an increased nuclear attractive force, because there is a greater effective nuclear charge, that brings the outer electrons closer to the nucleus.

8.105 The electron affinity is the energy change accompanying the formation of the negative ion which results when an electron is added to a gaseous neutral atom. When this process is exothermic, the electron affinity is negative, and, in general, the larger the negative number the more stable the negative ion that is formed. A positive number indicates an unstable negative ion is formed. The noble gases have the very stable valence shell configuration ns^2np^6, the so called noble-gas configuration. Changing this configuration by adding an additional electron will not lead to a more stable particle; i.e., the ion formed will be less stable than the original parent atom. Consequently, the noble gases have positive electron affinites.

8.107 Most of the physical and chemical properties of metals are associated with loosely held valence electrons. Atoms of metals have lower ionization energies than those of nonmetals, and they have less tendency to accept electrons.

The increase in metallic character from top to bottom within a group can be explained in terms of ionization energies and atomic sizes. Ionization energies decrease and atomic sizes increase in going from top to bottom within a group; therefore we would expect the "metallic character" of the elements to increase in going from top to bottom within a group, which agrees with the observed trend.

The decrease in metallic character from left to right across a period can be explained in terms of ionization energies, atomic sizes, and electron affinities. Ionization energies increase and atomic sizes and electron affinities decrease in going from left to right across a period. Therefore, we would expect the "metallic character" of the elements to decrease in going from left to right across a period, which agrees with the observed trend.

8.109 (a) K (Group 1A) should have a considerably larger atomic size than Br (Group 7A), since it's at the beginning of the fourth period, while Br is next to the end of the fourth period.

(b) K should have a considerably smaller first ionization energy than Br.

(c) Br should have a much lower (greater negative number) electron affinity than K.

(d) K should exhibit almost exclusively a metallic character, while Br should exhibit almost exclusively a nonmetallic character.

(e) K should have a much greater tendency to give up electrons than Br.

(f) Br should have a much greater tendency to accept electrons than K.

8.111 Rearranging gives

$$Z^* = \sqrt{\frac{-n^2E_n}{2.179 \times 10^{-18}\,J}}$$

The first ionization energy of Li ([He]$2s^1$) is 520 kJ/mol and corresponds to the removal of the 2s electron. The energy required to remove the 2s electron from a single Li atom is

$$\frac{5.20 \times 10^5\,J}{1\,mol} \times \frac{1\,mol}{6.022 \times 10^{23}\,atoms} = 8.64 \times 10^{-19}\ J/atom$$

The energy of the Li 2s level is -8.64×10^{-19} J.

$$2s:\ Z^* = \sqrt{\frac{4(8.64 \times 10^{-19}\,J)}{2.179 \times 10^{-18}\,J}} = 1.26 \text{ and } \sigma = Z - Z^* = 3 - 1.26 = 1.74$$

The second ionization energy of Li is 7298 kJ/mol and corresponds to the removal of a 1s electron. The energy required to remove the 1s electron from a single Li atom is

$$\frac{7.298 \times 10^6\,J}{1\,mol} \times \frac{1\,mol}{6.022 \times 10^{23}\,atoms} = 1.212 \times 10^{-17}\ J/atom$$

The energy of the Li 1s level is -1.212×10^{-17} J.

$$1s:\ Z^* = \sqrt{\frac{(1)(1.212 \times 10^{-17}\,J)}{2.179 \times 10^{-18}\,J}} = 2.358 \text{ and } \sigma = Z - Z^* = 3 - 2.358 = 0.642$$

The lithium 2s valence electron has two inner electrons between it and the nuclear charge. Their electron repulsions toward the 2s electron diminishes the nuclear attraction for this electron; i.e., they shield the 2s electron from some of the nuclear charge. If each inner electron could cancel the effects of one nuclear proton (one nuclear positive charge), it would be completely effective in shielding, and the effective nuclear charge would be 3 – 2 = 1. The fact that Z^* is 1.26 shows that the shielding is not totally effective. This result makes sense in light of the fact that s electrons tend to spend some time near the nucleus, a phenomenon often referred to as the penetration effect, which is the ability of outer electrons to penetrate inner shells and spend some time nearer the nucleus.

A lithium 1s electron basically only has the other 1s electron to shield it from the nucleus. Outer electrons do not effectively shield because they are further out from the nucleus. The calculated Z^* of 2.358 for the 1s electron, reduced only somewhat from the actual charge of 3 by the electron repulsions between the two 1s electrons, shows this to be the case. Keep in mind that both 1s electrons have a Z^* value of 2.358.

SURVEY OF THE ELEMENTS 1
GROUPS 1A, 2A, AND 8A

Solutions To Practice Exercises

PE S1.1 $H^- \ 1s^2$

PE S1.2 (a) $2K(s) + 2H_2O(l) \rightarrow H_2(g) + 2KOH(aq)$
(b) $2Na(s) + Br_2(l) \rightarrow 2NaBr(s)$
(c) $2Li(s) + H_2(g) \rightarrow 2LiH(s)$

PE S1.3 (a) $Ca(s) + 2H_2O(l) \rightarrow Ca(OH)_2(aq) + H_2(g)$
(b) $Ba(s) + Cl_2(g) \rightarrow BaCl_2(s)$
(c) $Sr(s) + H_2(g) \xrightarrow{\text{heat}} SrH_2(s)$
(d) $3Ca(s) + N_2(g) \xrightarrow{\text{heat}} Ca_3N_2(s)$

PE S1.4 (a) He has the lowest boiling point at –268.9°C. Therefore, air must be cooled down to –268.9°C in order to liquefy all of its components.
(b) From first to last: He, Ne, N_2, Ar, O_2, CH_4, Kr, Xe, CO_2, and Rn.

Solutions To Final Exercises

S1.1 (a) hydrogen atom, H (b) hydrogen molecule, H_2 (c) hydrogen ion, H^+ (d) hydride ion, H^-

S1.3 (a) Hydrogen ion, H^+, and hydride ion, H^-.
(b) H^-. H^- is larger because of electron repulsion between electrons in the same orbital.
(c) H^+. The hydrogen ion is stable in aqueous solution and is generated by many substances, collectively referred to as acids, while the hydride ion is unstable in aqueous solution; the main solution in our environment.

S1.5 The earth's atmosphere contains so little hydrogen because hydrogen is light enough to escape the earth's gravity.

Yes. Jupiter is considerably more massive than the earth and, therefore, the escape velocity required to overcome Jupiter's gravitational field is considerably higher than earth's. Consequently, fewer hydrogen atoms on Jupiter possess a velocity equal to or greater than the escape velocity. Also, the temperature on Jupiter is much colder than the earth's, since it is further from the sun, and, consequently, the average kinetic energy of the molecules in its atmosphere is lower.

S1.7 (a) Synthesis gas is a mixture of carbon monoxide (CO) and hydrogen (H_2).
(b) Water gas is an equimolar mixture of carbon monoxide and hydrogen. Yes.
(c) Synthesis gas can be prepared from small-molecule hydrocarbons, like methane, and carbon, in the form of coke or coal, by reacting them with steam at high temperature.
(d) Synthesis gas is used as the starting material for the synthesis of various organic compounds and as an industrial fuel.

S1.9 Hydrogen is used in metallurgy and chemical synthesis. For instance, in the production of ammonia (NH_3) and in the preparation of metals from their oxides by reduction. Hydrogen is also used as a fuel. For instance, hydrogen gas is used in cutting and welding torches, and liquid hydrogen is used as a rocket fuel.

The reasons why hydrogen is especially suitable for these purposes are:
Rocket fuel: light weight, high energy per gram content, clean, and efficient.
Welding: produces very high temperatures and nonpolluting (clean burning).
Metallurgy: very good reducing agent and produces a very pure product.
Chemical synthesis: hydrogen is relatively cheap and reactive; it forms strong covalent bonds with all nonmetals except the noble gases. So, it is usually easier and more economical to use a synthesis route employing hydrogen to form or obtain the desired product than some other route that doesn't directly use hydrogen.

S1.11 (a) $C(s) + H_2O(g) \xrightarrow{\Delta} CO(g) + H_2(g)$
(b) $CH_4(g) + H_2O(g) \xrightarrow{Ni,\ \Delta} CO(g) + 3H_2(g)$
(c) $CH_4(g) + H_2O(g) \xrightarrow{Ni,\ \Delta} CO(g) + 3H_2(g)$
$CO(g) + H_2O(g) \xrightarrow{\text{iron oxide},\ \Delta} CO_2(g) + H_2(g)$

S1.13 (a) $3H_2(g) + N_2(g) \rightarrow 2NH_3(g)$; ammonia.
(b) $H_2(g) + Cl_2(g) \rightarrow 2HCl(g)$; hydrogen chloride.
(c) $2H_2(g) + O_2(g) \rightarrow 2H_2O(l)$; water.

S1.15 (a) $N_2(g) + 3H_2(g) \xrightarrow[\text{catalyst}]{\text{500 atm, 450°C}} 2NH_3(g)$
(b) $CO(g) + 2H_2(g) \xrightarrow[\text{catalyst}]{\text{heat}} CH_3OH(g)$
(c) $H_2(g) + Cl_2(g) \rightarrow 2HCl(g)$

S1.17 (a) Na (Z = 11): [Ne]$3s^1$ and Mg (Z = 12): [Ne]$3s^2$
(b) Na > Mg
(c) Mg > Na
(d) Na > Mg
(e) Na^+ and Mg^{2+}

S1.19 (a) K (Z = 19): [Ar]$4s^1$ and Cs (Z = 55): [Xe]$6s^1$
(b) Cs > K
(c) K > Cs
(d) $Cs^+ > K^+$

S1.21 The following compares neighboring (adjacent) Group 1A and 2A metals (elements).
(a) The alkali metals are more reactive.
(b) The alkaline earth metals are harder.
(c) The alkaline earth metals are denser.
(d) The alkali metals have lower melting points.

The atoms of the alkali metals have lower ionization energies than those of the alkaline earth metals. This accounts for their being more reactive. The atoms of the alkaline earth metals are heavier and smaller than those of the alkali metals. This accounts for their greater density. The atoms of the alkaline earth metals have a greater nuclear charge and smaller size than those of the alkali metals. Therefore, they can pack in closer and be held together more tightly. This accounts for their greater hardness. Also, because the atoms of the alkaline earth metals are packed in closer and held together more tightly, the amount of energy needed to separate their atoms is greater than that required to separate the atoms of the alkali metals. This accounts for their higher melting points. In summary, the greater hardness, density, and melting points of the Group 2A metals are physical properties caused by their smaller sizes and greater interatomic attractive forces in the solid state.

S1.23 The crust is composed of sodium carbonate. When exposed to the air, sodium unites with the oxygen and moisture in the air to form sodium hydroxide. The sodium hydroxide then reacts with the carbon dioxide present in air forming sodium carbonate.

Metallic sodium is so soft that it can easily be cut with a knife. Cutting away the tarnish crust reveals sodium's silvery-white metallic luster.

S1.25 (a) Na; sodium chloride (b) Mg; $MgCO_3$, $MgCl_2$, $MgSO_4$, and dolomite (c) Be; beryl (d) Ca; limestone, gypsum, fluorite, and dolomite.

S1.27 (a) Sodium is used as a liquid heat exchanger in some nuclear reactors and for making sodium vapor lamps for outdoor lighting. Sodium is used to remove traces of moisture from nonaqueous solvents and as a reducing agent. It is also employed as a reagent in the manufacture of various organic compounds, one of which is the now banned gasoline additive tetraethyl lead.
(b) Beryllium is used to make lightweight structural alloys. It is also used for windows in x-ray tubes.
(c) Magnesium is used to make lightweight structural alloys. It is also used in flares and flashbulbs, and as an igniter for thermite.
(d) Cesium is used in various photoelectric devices.

S1.29 (a) Group 1A: Li, Na, K, Rb, and Cs ions
Group 2A: Ca, Sr, and Ba ions
(b) The energy of the flame produces higher energy electron configurations in the ions. The flame color is produced when these excited ions give off visible photons in the process of returning to their ground-state configuration.
(c) Sodium nitrate, yellow; strontium nitrate, scarlet; calcium nitrate, brick red; potassium nitrate, violet; barium nitrate, pale green.

S1.31 Sodium carbonate, Na_2CO_3; used in detergents and making glass.
Sodium hydrogen carbonate, $NaHCO_3$; used in antacids and baking.
Sodium hydroxide, NaOH; used in the manufacture of plastics, paper, petroleum, and soap.
Lithium carbonate, Li_2CO_3; used to control manic depression.
Sodium chloride, NaCl; essential mineral in the diet and raw material for the manufacture of many other compounds.

S1.33 (a) caustic soda, NaOH
(b) soda ash, Na_2CO_3
(c) baking soda, $NaHCO_3$
(d) washing soda, Na_2CO_3
(e) bicarbonate of soda, $NaHCO_3$
(f) lye, NaOH and KOH

S1.35 (a) $2KCl(l) \xrightarrow{\text{electrolysis}} 2K(l) + Cl_2(g)$
(b) $CaCl_2(l) \xrightarrow{\text{electrolysis}} Ca(s) + Cl_2(g)$

S1.37 (a) $2Li(s) + H_2(g) \rightarrow 2LiH(s)$, lithium hydride, heat
$2Na(s) + H_2(g) \rightarrow 2NaH(s)$, sodium hydride, heat
$2K(s) + H_2(g) \rightarrow 2KH(s)$, potassium hydride, heat
$2Rb(s) + H_2(g) \rightarrow 2RbH(s)$, rubidium hydride, heat
$2Cs(s) + H_2(g) \rightarrow 2CsH(s)$, cesium hydride, heat

(b) $6Li(s) + N_2(g) \rightarrow 2Li_3N(s)$, lithium nitride, heat

(c) $4Li(s) + O_2(g) \rightarrow 2Li_2O(s)$, lithium oxide
$2Na(s) + O_2(g) \rightarrow Na_2O_2(s)$, sodium peroxide
$2K(s) + O_2(g) \rightarrow K_2O_2(s)$, potassium peroxide
$K(s) + O_2(g) \rightarrow KO_2(s)$, potassium superoxide
$Rb(s) + O_2(g) \rightarrow RbO_2(s)$, rubidium superoxide
$Cs(s) + O_2(g) \rightarrow CsO_2(s)$, cesium superoxide

(d) $2Li(s) + Cl_2(g) \rightarrow 2LiCl(s)$, lithium chloride
$2Na(s) + Cl_2(g) \rightarrow 2NaCl(s)$, sodium chloride
$2K(s) + Cl_2(g) \rightarrow 2KCl(s)$, potassium chloride
$2Rb(s) + Cl_2(g) \rightarrow 2RbCl(s)$, rubidium chloride
$2Cs(s) + Cl_2(g) \rightarrow 2CsCl(s)$, cesium chloride

(e) $2Li(s) + 2H_2O(l) \rightarrow 2LiOH(aq) + H_2(g)$, lithium hydroxide and hydrogen
$2Na(s) + 2H_2O(l) \rightarrow 2NaOH(aq) + H_2(g)$, sodium hydroxide and hydrogen
$2K(s) + 2H_2O(l) \rightarrow 2KOH(aq) + H_2(g)$, potassium hydroxide and hydrogen
$2Rb(s) + 2H_2O(l) \rightarrow 2RbOH(aq) + H_2(g)$, rubidium hydroxide and hydrogen
$2Cs(s) + 2H_2O(l) \rightarrow 2CsOH(aq) + H_2(g)$, cesium hydroxide and hydrogen

S1.39 (a) $2K(s) + 2H_2O(l) \rightarrow 2KOH(aq) + H_2(g)$. The products are potassium hydroxide and hydrogen.
(b) $CaO(s) + H_2O(l) \rightarrow Ca(OH)_2(aq)$. The product is calcium hydroxide.
(c) $NaH(s) + H_2O(l) \rightarrow NaOH(aq) + H_2(g)$. The products are sodium hydroxide and hydrogen.
(d) Nitrides form ammonia gas when decomposed by water.
$Li_3N(s) + 3H_2O(l) \rightarrow NH_3(g) + 3LiOH(aq)$. The products are ammonia gas and lithium hydroxide.

S1.41 (a) $Ba(s) + Br_2(l) \rightarrow BaBr_2(s)$; barium bromide
(b) $2K(s) + O_2(g) \rightarrow K_2O_2(s)$; potassium peroxide
$K(s) + O_2(g) \rightarrow KO_2(s)$; potassium superoxide
(c) $6Li(s) + N_2(g) \xrightarrow{\text{heat}} 2Li_3N(s)$; lithium nitride
(d) $Sr(s) + H_2(g) \xrightarrow{\text{heat}} SrH_2(s)$; strontium hydride

S1.43 (a) $2KCl(aq) + 2H_2O(l) \xrightarrow{\text{electrolysis}} 2KOH(aq) + H_2(g) + Cl_2(g)$
(b) $2NaHCO_3(s) \xrightarrow{\text{heat}} Na_2CO_3(s) + H_2O(g) + CO_2(g)$
(c) $Na_2CO_3(s) + 2HCl(aq) \rightarrow 2NaCl(aq) + H_2O(l) + CO_2(g)$
(d) $2NaHCO_3(s) + H_2SO_4(aq) \rightarrow Na_2SO_4(aq) + 2H_2O(l) + 2CO_2(g)$

S1.45 The noble gases remained undiscovered for so long because they are chemically inert, colorless, odorless, tasteless, liquefy only at very low temperatures, and scarce. In other words, they have no properties that affect our senses and, therefore, required the development of the appropriate and necessary scientific instruments for their discovery.

S1.47 Helium, argon, and radon.

S1.49 Helium is used to provide a protective (non-reactive) atmosphere for welding and other high-temperature work. It is especially suitable for this because it is inert.

It is used for filling balloons and blimps, because it has a low density and is not flammable.

It is used in breathing mixtures for divers, because helium is less soluble in blood than nitrogen, and, therefore, reduces the chances of getting the "bends."

S1.51
(a) Neon is used in advertising signs and beacons for airports and lighthouses.
(b) Krypton is used in electric light bulbs.
(c) Argon is used in electric light bulbs and advertising signs.
(d) Radon is used in the treatment of malignant tumors and earthquake prediction.

S1.53
(a) He, $1s^2$; Ne, $2s^22p^6$; Ar, $3s^23p^6$; Kr, $4s^24p^6$; Xe, $5s^25p^6$; and Rn, $6s^26p^6$.
(b) Helium has the highest ionization energy. Because of this helium is chemically inert and acts like the Group 8A elements, which are also, with a few minor exceptions, chemically inert. On the other hand, the Group 2A elements have relatively low ionization energies and, consequently, are very chemically active. Therefore, even though He has the same similar outer electron configuration as the Group 2A elements, it is placed in Group 8A because its chemical properties are dissimilar to those of the Group 2A elements and similar to those of the Group 8A elements.

S1.55 Hydrogen (Group 1A, $1s^1$) has the same similar outer electron configuration as the alkali metals, but its physical and chemical properties are not at all similar to theirs. Hydrogen is a nonmetal that forms mainly covalent compounds, while the alkali metals are highly metallic in nature and form mainly ionic compounds. Therefore, it can be argued that hydrogen doesn't really belong in Group 1A. Some chemists prefer to put hydrogen in both Groups 1A and 7A of the periodic table, since (molecular) hydrogen is a single bond diatomic molecule just like the halogens, and achieves the stable noble gas configuration, that of helium, by either sharing an electron in a covalent bond or gaining an electron to form the anion (hydride ion, H^-), just like the halogens do. However, the halogens have much lower electron affinites than hydrogen. Therefore, they form many ionic compounds and their anions are very stable (nonreactive), while hydrogen only forms ionic compounds with the more active metals and its anion is highly reactive (unstable). Therefore, hydrogen doesn't totally belong in Group 7A either. The other groups are ruled out due to basic differences in electron configuration.

S1.57 The heat evolved per mole of fuel combusted is its fuel value per mole.

(a) $2H_2(g) + O_2(g) \rightarrow 2H_2O(l)$

$\Delta H° = 2 \times -285.8 \text{ kJ} - (2 \times 0 \text{ kJ}) - (1 \times 0 \text{ kJ}) = -571.6 \text{ kJ}$

fuel value of hydrogen per mole = +571.6 kJ/2 mol = +285.8 kJ/mol

(b) $CH_4(g) + 2O_2(g) \rightarrow CO_2(g) + 2H_2O(l)$

$\Delta H° = 1 \times -393.5 \text{ kJ} + 2 \times -285.8 \text{ kJ} - (1 \times -74.8 \text{ kJ}) - (2 \times 0 \text{ kJ}) = -890.3 \text{ kJ}$

fuel value of methane per mole = +890.3 kJ/mol

(c) Water gas is composed of equimolar amounts of carbon dioxide and hydrogen. The fuel value of hydrogen per mole is known; the fuel value of carbon dioxide per mole has to be calculated.

$2CO(g) + O_2(g) \rightarrow 2CO_2(g)$

$\Delta H° = 2 \times -393.5 \text{ kJ} - (2 \times -110.5 \text{ kJ}) = -566.0 \text{ kJ}$

fuel value of carbon monoxide per mole = +566.0 kJ/2 mol = +283.0 kJ/mol

The fuel value of water gas per mole = 0.5 mol H_2 × 285.8 kJ/mol H_2 + 0.5 mol CO × 283.0 kJ/mol CO = +284.4 kJ/mol

S1.59 (a) The reaction for the production of water gas from graphite and liquid water is $C(s) + H_2O(l) \rightarrow CO(g) + H_2(g)$

$\Delta H° = 1 \times -110.5 \text{ kJ} + 1 \times 0 \text{ kJ} - (1 \times 0 \text{ kJ}) - (1 \times -285.8 \text{ kJ}) = +175.3 \text{ kJ}$

The amount of heat absorbed during the conversion of 1.00 kg of carbon to water gas is (1.00 kg C → mol C → kJ absorbed)

$$1.00 \text{ kg C} \times \frac{10^3 \text{ g}}{1 \text{ kg}} \times \frac{1 \text{ mol C}}{12.01 \text{ g C}} \times \frac{175.3 \text{ kJ}}{1 \text{ mol C}} = 1.46 \times 10^4 \text{ kJ}$$

(b) $C(s) + O_2(g) \rightarrow CO_2(g)$

$\Delta H° = 1 \times -393.5 \text{ kJ} - (1 \times 0 \text{ kJ}) - (1 \times 0 \text{ kJ}) = -393.5 \text{ kJ}$

The heat energy released by burning 1.00 kg of carbon directly is (1.00 kg C → mol C → kJ)

$$1.00 \text{ kg C} \times \frac{10^3 \text{ g}}{1 \text{ kg}} \times \frac{1 \text{ mol C}}{12.01 \text{ g C}} \times \frac{393.5 \text{ kJ}}{1 \text{ mol C}} = 3.28 \times 10^4 \text{ kJ}$$

The net amount of heat energy released by converting 1.00 kg of carbon to water gas (fuel value = +284.4 kJ/mol, see S1.57) and then burning the gas is (1.00 kg C → mol C → mol water gas → released kJ water gas combustion → net kJ released)

net heat released = heat released during water gas combustion – heat consumed in water gas production

$$= 1.00 \times 10^3 \text{ g C} \times \frac{1 \text{ mol C}}{12.01 \text{ g C}} \times \frac{2 \text{ mol water gas}}{1 \text{ mol C}} \times \frac{284.4 \text{ kJ}}{1 \text{ mol water gas}} - 1.46 \times 10^4 \text{ kJ}$$

$$= 47{,}400 \text{ kJ} - 14{,}600 \text{ kJ} = 3.28 \times 10^4 \text{ kJ}$$

The net water gas process releases 32,800 kJ – 32,800 kJ = 0 kJ more heat energy than the direct burning of carbon.

(c) The combustion of water gas is clean, efficient, and nonpolluting. The combustion of coke is dirty, inefficient, and polluting. Therefore, water gas can be used in areas of high-population density while coke shouldn't be; the use of coke is oftentimes banned from such areas. Also, since water gas is a gas it can be centrally stored and dispensed to many locations through a system of pipes. Therefore, water gas also has handling and dispensing advantages over coke.

Note: The result in part (b) could have also been obtained by the use of Hess's law:

$C(s) + H_2O(l) \rightarrow CO(g) + H_2(g)$	$\Delta H° = +175.3 \text{ kJ}$
$CO(g) + \frac{1}{2}O_2(g) \rightarrow CO_2(g)$	$\Delta H° = -283.0 \text{ kJ}$
$H_2(g) + \frac{1}{2}O_2(g) \rightarrow H_2O(l)$	$\Delta H° = -285.8 \text{ kJ}$
$C(s) + O_2(g) \rightarrow CO_2(g)$	$\Delta H° = -393.5 \text{ kJ}$

This shows that the same amount of heat has to be evolved either way. Notice that we have to use liquid water in the water gas equation in order to account for the energy (amount of heat) used to create the steam employed in the actual reaction.

S1.61 In order for an ionic compound to dissolve, the cations and anions making up the ionic compound must be separated from one another and surrounded by the solvent molecules. In general, the stronger the forces holding the cations and anions together, the less soluble the substance. Since the alkaline earth metal cations carry a greater charge (2+ versus 1+) and have a smaller radius than the alkali metal cations, their cation-anion attractions are much greater than those in the alkali metal compounds and, therefore, they are less soluble.

S1.63 Potassium (fourth period) has considerably larger atoms than lithium (second period) and sodium (third period). As the atomic size increases, the attractive forces between the atoms in the solid metal decreases. Therefore, the potassium atoms are held together less tightly than those of lithium and sodium and, consequently, they can be more easily and rapidly removed during reaction with the water molecules, and this is why potassium has a greater reaction rate.

Also, since potassium has a lower ionization energy than lithium and sodium, more heat energy should be released in its reaction, raising the temperature of its reactants faster than theirs, leading to potassium's reaction rate increasing faster than lithium's and sodium's. In general, the reaction rate increases as the temperature increases.

S1.65 In He, Ne, and Ar the electrons are not sufficiently removed from the nucleus to allow them to interact with fluorine. He, Ne, and Ar have much smaller atomic radii and larger ionization energies than Kr and Xe, which makes their valence energy levels too low relative to fluorine's for there to be an effective interaction between them.

S1.67 (a) From the ideal gas law we have $d = \mathcal{M}P/RT$. At 1.00 atm and 25°C the approximate densities are

$$H_2\text{: } d = \frac{2.016 \text{ g/mol} \times 1.00 \text{ atm}}{0.0821 \text{ L•atm/mol•K} \times 298 \text{ K}} = 0.0824 \text{ g/L}$$

$$\text{He: } d = \frac{4.003 \text{ g/mol} \times 1.00 \text{ atm}}{0.0821 \text{ L•atm/mol•K} \times 298 \text{ K}} = 0.164 \text{ g/L}$$

$$\text{Ar: } d = \frac{39.948 \text{ g/mol} \times 1.00 \text{ atm}}{0.0821 \text{ L•atm/mol•K} \times 298 \text{ K}} = 1.63 \text{ g/L}$$

$$\text{Kr: } d = \frac{83.80 \text{ g/mol} \times 1.00 \text{ atm}}{0.0821 \text{ L•atm/mol•K} \times 298 \text{ K}} = 3.43 \text{ g/L}$$

(b) Balloons filled with Ar and Kr will sink to the ground, since both gases have greater densities than air.
(c) Decrease with increasing altitude, since argon is denser and, therefore, should displace nitrogen at the lower altitudes. Consequently, the ratio of argon to nitrogen will be greatest at ground level and will steadily decrease with increasing altitude. The same holds for krypton.

S1.69 $CaH_2(s) + 2H_2O(l) \rightarrow Ca(OH)_2(aq) + 2H_2(g)$
100.0 g mol = ?

The number of moles of H_2 produced in the decomposition is (g CaH_2 → mol CaH_2 → mol H_2)

$$100.0 \text{ g } CaH_2 \times \frac{1 \text{ mol } CaH_2}{42.09 \text{ g } CaH_2} \times \frac{2 \text{ mol } H_2}{1 \text{ mol } CaH_2} = 4.752 \text{ mol } H_2$$

$H_2(g) + 1/2\ O_2(g) \rightarrow H_2O(l) \quad \Delta H° = -285.8 \text{ kJ}$

The heat released by burning 4.752 moles of hydrogen is

$$4.752 \text{ mol } H_2 \times \frac{285.8 \text{ kJ}}{1 \text{ mol } H_2} = 1.358 \times 10^3 \text{ kJ}$$

The mass of water than can be heated from 25°C to 100°C by 1.358×10^3 kJ is

mass = q/specific heat × ΔT = 1.358×10^6 J/4.2 J/g•°C × (100°C – 25°C) = 4.3×10^3 g.

The density of water is approximately 1.0 g/mL. Therefore, the number of liters of water that could be heated from 25°C to 100°C is

$$4.3 \times 10^3 \text{ g water} \times \frac{1 \text{ mL}}{1.0 \text{ g water}} \times \frac{1 \text{ L}}{10^3 \text{ mL}} = 4.3 \text{ L}$$

Answer: 4.3 L of water.

S1.71 The equations for the burning of Mg in air are:

$2Mg(s) + O_2(g) \rightarrow 2MgO(s)$ — x moles, x moles

$3Mg(s) + N_2(g) \rightarrow Mg_3N_2(s)$ — $3y$ moles, y moles

Let x and y equal the number of moles of MgO and Mg_3N_2 formed, respectively. The molar masses of MgO and Mg_3N_2 are 40.30 g/mol and 100.9 g/mol, respectively. Then we have the following pair of simultaneous equations to solve

$x + 3y = 1.000$, $x = 1.000 - 3y$ and $x \times 40.30 + y \times 100.9 = 38.20$

Substituting the value of x into the last equation we get

$$\begin{aligned} 40.30\,(1.000 - 3y) + 100.9y &= 38.20 \\ 20.0y &= 2.10 \\ y &= 0.105 \text{ mol } Mg_3N_2 \\ x &= 1.000 - 3y = 1.000 - 3(0.105) = 0.685 \text{ mol MgO} \end{aligned}$$

The fraction of the product that was MgO is

$$\text{fraction MgO} = \frac{\text{mass MgO}}{\text{mass product}} = \frac{0.685 \text{ mol MgO} \times 40.30 \text{ g/mol MgO}}{38.20 \text{ g}} = 0.723$$

CHAPTER 9

THE CHEMICAL BOND

Solutions To Practice Exercises

PE 9.1 (a) :Ṗ· (b) :B̈r· (c) ·Ṡi·

PE 9.2 (a) Mg^{2+} (b) I^- (c) S^{2-}

PE 9.3 (a) Li_2O: Both Li^+, [He], and O^{2-}, [Ne], do.
(b) $PbCl_2$: Cl^-, [Ar], does, but Pb^{2+}, $[Xe]4f^{14}5d^{10}6s^2$, doesn't.
(c) $ZnBr_2$: Br^-, [Kr], does, but Zn^{2+}, $[Ar]3d^{10}$, doesn't.

PE 9.4 (a) The cations Mg^{2+} and Ca^{2+} have the same charge and the anions F^- and Cl^- have the same charge. If the lattice energies differ, it will be because of size differences. In agreement with periodic trends, the sizes are $Mg^{2+} < Ca^{2+}$ and $F^- < Cl^-$. Therefore, MgF_2 should have a higher lattice energy, since smaller ions generally have higher lattice energies.
(b) F and O are adjacent to each other in the second period. They both form anions with the electron configuration [Ne]. Therefore, we can expect the sizes of F^- and O^{2-} to be similar. Consequently, if the lattice energies differ, it will be mainly because of charge differences. O^{2-} carries a larger charge than F^-; therefore, MgO should have a higher lattice energy.
(c) Al lies just to the right of Mg in the third period; therefore Al^{3+} should be smaller than Mg^{2+}. Also, Al^{3+} carries a charge of 3+ while Mg^{2+} carries a charge of 2+. Consequently, AlF_3 should have a higher lattice energy.

PE 9.5 The enthalpy of sublimation is the enthalpy change accompanying the reaction

$Li(s) \rightarrow Li(g) \qquad \Delta H_{sublimation} = ?$

The following steps add up to the desired equation.

(1) $Li(s) + 1/2\ F_2(g)$	$\rightarrow LiF(s)$	ΔH_1	$= -612.1$ kJ
(2) $LiF(s)$	$\rightarrow Li^+(g) + F^-(g)$	ΔH_2	$= +1036$ kJ
(3) $Li^+(g) + e^-$	$\rightarrow Li(g)$	ΔH_3	$= -520.3$ kJ
(4) $F^-(g)$	$\rightarrow F(g) + e^-$	ΔH_4	$= +322$ kJ
(5) $F(g)$	$\rightarrow 1/2\ F_2(g)$	ΔH_5	$= -78.5$ kJ

The enthalpy of sublimation of lithium metal is obtained by adding the enthalpy changes of these steps.

$$\Delta H_{sublimation} = \Delta H_1 + \Delta H_2 + \Delta H_3 + \Delta H_4 + \Delta H_5$$
$$= -612.1 \text{ kJ} + 1036 \text{ kJ} - 520.3 \text{ kJ} + 322 \text{ kJ} - 78.5 \text{ kJ} = +147 \text{ kJ}$$

PE 9.6 (a) HCl and (b) HBr.
Since Cl is smaller than Br, HCl should have a higher bond energy and HBr the longer bond length. The predictions agree with the data in Table 9.3.

PE 9.7 $2H_2(g) + N{\equiv}N(g) \rightarrow N_2H_4(g) \quad \Delta H° = ?$

The enthalpy change is the sum of the enthalpy changes for the following steps:

		$\Delta H°$
1. Breaking 2 mol of H–H bonds:	2 × +436 kJ	= +872 kJ
2. Breaking 1 mol of N≡N bonds:		+945 kJ
3. Forming 4 mol of N–H bonds:	4 × –391 kJ	= –1564 kJ
4. Forming 1 mol of N–N bonds:		–170 kJ
	$\Delta H°$ = Sum =	+83 kJ

The enthalpy change is +83 kJ per mole of hydrazine.

PE 9.8 CH_4 has 4 + 4(1) = 8 valence electrons or 4 electron pairs.

```
   H
   |
H–C–H
   |
   H
```

PE 9.9 SO_2 has 6 + 2(6) = 18 valence electrons or 9 electron pairs.

:Ö – S̈ = Ö:

PE 9.10 HNC has 1 + 5 + 4 = 10 valence electrons or 5 electron pairs.

H–N≡C:

PE 9.11 (a) H_2CO_3 has 2(1) + 4 + 3(6) = 24 valence electrons or 12 electron pairs.

```
      :O:
       ||
H–Ö–C–Ö–H
```

(b) HCO_3^- has 1 + 4 + 3(6) + 1 = 24 valence electrons or 12 electron pairs.

```
[       :O:      ]-
        ||
[ H-Ö-C-Ö: ]
```

PE 9.12 (a) $:\ddot{O}=C=\ddot{O}:$

C: formal charge = 4 – 0 – 1/2(8) = 0
O: formal charge = 6 – 4 – 1/2(4) = 0

(b)

```
        :Ö:
         ||
  H-Ö-P-Ö-H
         |
        :O:
         |
         H
```

H: formal charge = 1 – 0 – 1/2(2) = 0
P: formal charge = 5 – 0 – 1/2(8) = +1
O in OH: formal charge = 6 – 4 – 1/2(4) = 0
other O: formal charge = 6 – 6 – 1/2(2) = –1

(c)

```
[    H    ]+
     |
[ H-N-H ]
     |
[    H    ]
```

H: formal charge = 1 – 0 – 1/2(2) = 0
N: formal charge = 5 – 0 – 1/2(8) = +1

(d) $[:\ddot{O}-H]^-$

O: formal charge = 6 – 6 – 1/2(2) = –1
H: formal charge = 1 – 0 – 1/2(2) = 0

PE 9.13 CO_3^{2-} has 4 + 3(6) + 2 = 24 valence electrons or 12 electron pairs.

```
[    Ö:    ]2-          [    :Ö:    ]2-          [    :Ö:    ]2-
     ||          <-->        |           <-->         |
[ :Ö-C-Ö: ]             [ :Ö=C-Ö: ]              [ :Ö-C=Ö: ]
```

These structures differ only in the position of the double bond and, therefore, should contribute equally to the hybrid. Consequently, the carbon-oxygen bonds should be equal and intermediate in length between single and double bonds.

PE 9.14 (a) N_2O has 2(5) + 6 = 16 valence electrons or 8 electron pairs. The resonance structures with their formal charges are

```
(-1)  (+1)              (+1) (-1)             (-2)  (+1) (+1)
 N=N=O       <-->     :N≡N-O:       <-->      :N-N≡O:
   I                     II                      III
```

(b) Structure III is the least stable of the three structures since it has more formal charge on its atoms than structures I and II, and has positive charges on adjacent atoms as well. Therefore, structure III contributes least to the hybrid.

PE 9.15 (a) BrF_3 has 7 + 3(7) = 28 valence electrons or 14 electron pairs.

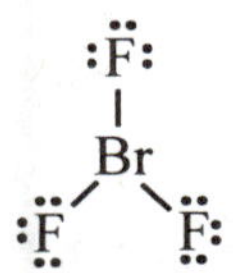

(b) IF_7 has 7 + 7(7) = 56 valence electrons or 28 electron pairs.

:F: :F: :F: :F: :F: :F: :F: bonded to central I

PE 9.16 (a) $HClO_3$ has 1 + 7 + 3(6) = 26 valence electrons or 13 electron pairs.

H–O–Cl–O: with :O: below Cl (formal charges: Cl +2, terminal O −1, −1)

H: formal charge = 1 – 0 – 1/2(2) = 0
Cl: formal charge = 7 – 2 – 1/2(6) = +2
O in OH: formal charge = 6 – 4 – 1/2(4) = 0
other O: formal charge = 6 – 6 – 1/2(2) = –1

(b) A Lewis structure without formal charge is

H–O–Cl=O with =O: below Cl

The second structure makes the larger contribution to the resonance hybrid, because it eliminates formal charge and provides stronger bonding between the chlorine and oxygen atoms.

PE 9.17 XeF_2 has 8 + 2(7) = 22 valence electrons or 11 electron pairs.

:F–Xe–F:

PE 9.18 (a) $BeCl_2$ has 2 + 2(7) = 16 valence electrons or 8 electron pairs.

:Cl–Be–Cl:

For $BeCl_2$ to follow the octet rule, Be would have to have a negative formal charge and each Cl a positive formal charge, which diminishes the importance of such a contributing structure to the hybrid. Consequently, $BeCl_2$ doesn't follow the octet rule.

(b) BCl_3 has 3 + 3(7) = 24 valence electrons or 12 electron pairs.

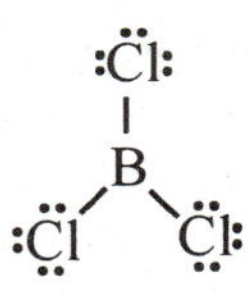

BCl_3 doesn't follow the octet rule for the same reasons that $BeCl_2$ doesn't.

(c) $SnCl_2$ has 4 + 2(7) = 18 valence electrons or 9 electron pairs.

:Cl–Sn–Cl:

$SnCl_2$ doesn't follow the octet rule for the same reasons that $BeCl_2$ doesn't.

PE 9.19 ClO_2 has 7 + 2(6) = 19 valence electrons or 9 electron pairs plus one odd electron.

O=Cl=O ⟷ O–Cl=O· ⟷ ·O–Cl=O

PE 9.20 (a) For BCl_3 the electronegativity difference, 3.2 – 2.0 = 1.2, is less than 1.7 and, therefore, each B–Cl bond is polar covalent.

(b) For RbCl the electronegativity difference, 3.2 – 0.82 = 2.4, exceeds 1.7 and, therefore, RbCl is predominantly ionic: Rb^+Cl^-.

Solutions To Final Exercises

9.1 (a) 1. An ionic bond involves the transfer of electrons between different atoms. A metallic bond involves no electron transfers and the atoms are all the same.
2. In an ionic bond the positive and negative charge centers (ions) resulting from the electron transfers are at fixed positions in space. In a metallic bond, the atoms or positive ions are fixed in space, but the valence electrons are free to travel throughout the metal.
(b) An ionic bond involves the transfer of electrons between different atoms. One atom loses electrons while another atom gains electrons. In a covalent bond, one or more pairs of electrons are shared between two atoms. The atoms can be the same or different and both atoms can be thought of as gaining electrons. Also, ionic bonds form between metals and nonmetals generally, while covalent bonds generally form between two nonmetals. However, keep in mind that many of the metals do participate in covalent bonding with nonmetals.

9.3 Ammonium nitrate, NH_4NO_3, is composed of the polyatomic ions NH_4^+ and NO_3^-. The covalent bonds are the nitrogen to hydrogen bonds in NH_4^+ and the nitrogen to oxygen bonds in NO_3^-. The ionic bond is between the NH_4^+ and NO_3^- ions.

9.5 (a) ·Ċ̣· (b) :Ȧ̤s· (c) :Ṅ· (d) K· (e) ·Ȧl· (f) ·Ö̤· (g) ·Mg· (h) :C̤̈l·

9.7 The energy required to remove four electrons from an atom is greater than that available under normal conditions. The same holds for placing four extra electrons on the same atom.

9.9 (a) Sodium has the configuration $[He]2s^22p^63s^1$. The 3s electron is loosely held and, therefore, easily removed. The second electron would have to come from the 2p orbitals. The 2p electrons are tightly held since they are in a shell closer to the nucleus and the atom is now carrying a positive charge. Therefore, under normal conditions it requires too much energy to remove a second electron from sodium and, consequently, sodium does not normally form a Na^{2+} ion. The univalent ion, Na^+, also has a noble gas configuration, [Ne], which is quite stable.
(b) Cl^- has the stable noble gas configuration [Ar]. To form Cl^{2-} you would have to place an electron in a 4s orbital, i.e., in an s orbital one shell further from the nucleus. Such an electron would be loosely held, making Cl^{2-} highly reactive and unstable. Therefore, chlorine does not form a Cl^{2-} ion.

9.11 The octet rule states that when atoms react to form bonds they lose, gain, or share enough electrons to achieve the configuration of a noble gas.

The valence shell configuration of all the noble gases, except He, is ns^2np^6. It takes eight electrons, an octet, to achieve this configuration; hence, the name octet rule.

9.13 (a) The lattice energy increases with increasing ionic charge.
(b) The lattice energy decreases with increasing ionic radii.

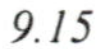

9.15

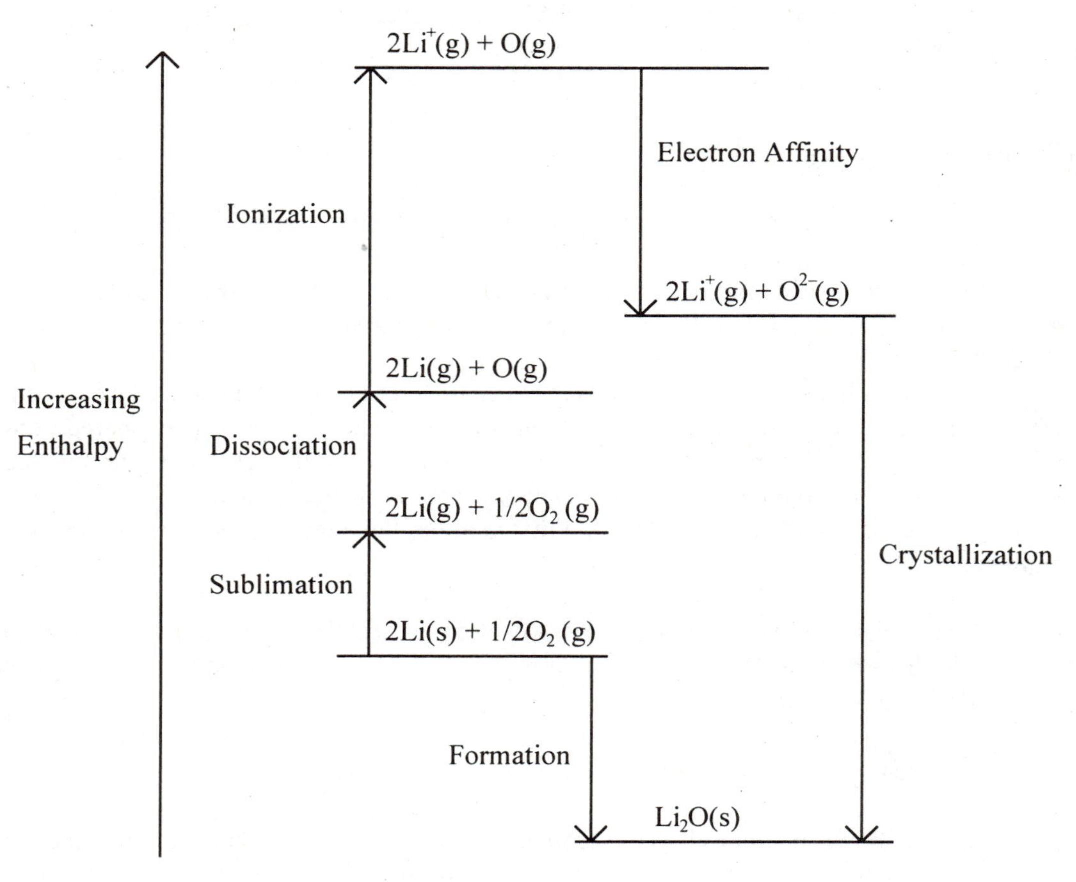

9.17 (a) Se^{2-} (b) Rb^{+} (c) F^{-} (d) O^{2-} (e) P^{3-} (f) Ba^{2+}

9.19 The alkali (Group 1A) and alkaline earth (Group 2A) cations are isoelectronic with the noble gas atom preceding its period. Since Kr ends period 4, either Rb^{+} or Sr^{2+} from period 5 elements Rb and Sr will be acceptable. In general, a negative monatomic ion is isoelectric with the noble gas atom at the end of its period. Since Kr ends period 4, either Se^{2-} or Br^{-} from period 4 elements Se and Br will be acceptable. (As doesn't form an As^{3-} ion).

Answer: RbBr, $SrBr_2$, Rb_2Se, or SrSe.

9.21 (a) Sn^{2+} (b) Fe^{2+}; Fe^{3+}

9.23 (a) $Be(NO_3)_2$. Since the charges are the same, any differences in lattice energies will be because of size differences. Mg (third period) lies just below Be (second period) in Group 2A. Therefore, $Be^{2+} < Mg^{2+}$ and, consequently, $Be(NO_3)_2$ should have the higher lattice energy.

(b) $Ca(OH)_2$. The sizes of the ions involved are approximately the same and any differences in lattice energy will be because of charge differences. Ca^{2+} has a greater charge than Na^{+} and, consequently, $Ca(OH)_2$ should have the higher lattice energy.

9.25 The Born-Haber Cycle for $CaCl_2$ is

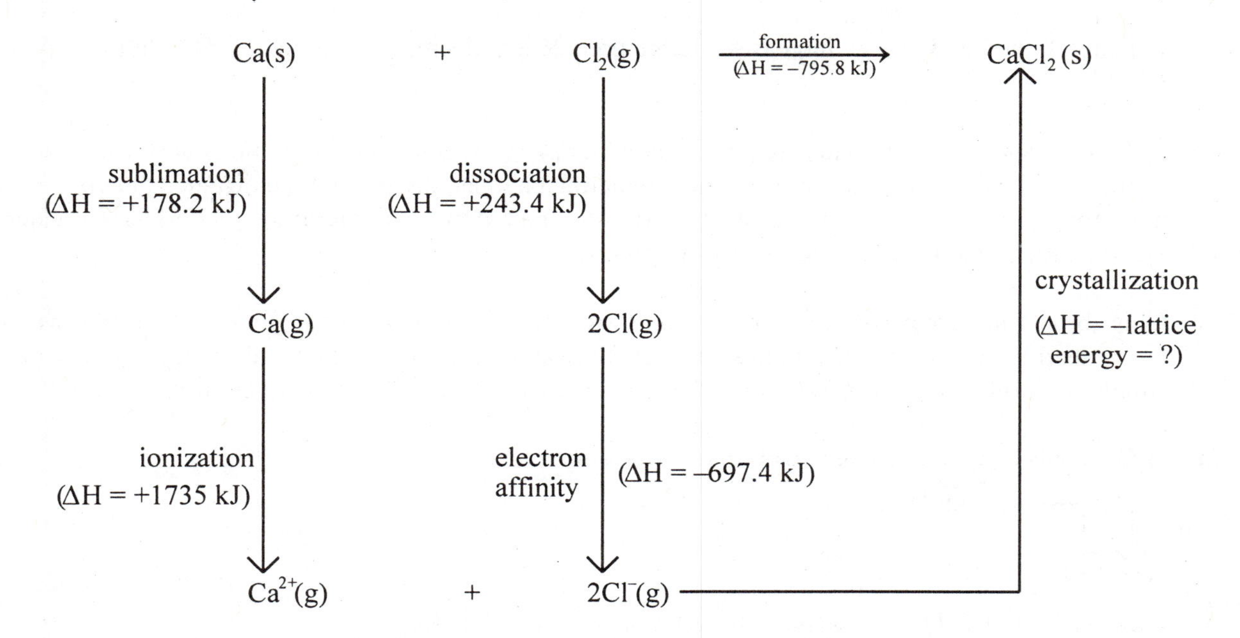

The lattice energy is the enthalpy change accompanying the reaction

$CaCl_2(s) \rightarrow Ca^{2+}(g) + 2Cl^-(g)$ lattice energy = ?

The following steps add up to the desired equation.

$$
\begin{array}{llll}
(1) & Ca(s) & \rightarrow Ca(g) & \Delta H_1 = +178 \text{ kJ} \\
(2) & Ca(g) & \rightarrow Ca^{2+}(g) + 2e^- & \Delta H_2 = +1735 \text{ kJ} \\
(3) & Cl_2(g) & \rightarrow 2Cl(g) & \Delta H_3 = +243 \text{ kJ} \\
(4) & 2Cl(g) + 2e^- & \rightarrow 2Cl^-(g) & \Delta H_4 = -697 \text{ kJ} \\
(5) & CaCl_2(s) & \rightarrow Ca(s) + Cl_2(g) & \Delta H_5 = +796 \text{ kJ}
\end{array}
$$

The lattice energy is obtained by adding the enthalpy changes of these steps:

$$
\begin{aligned}
\text{lattice energy} &= \Delta H_1 + \Delta H_2 + \Delta H_3 + \Delta H_4 + \Delta H_5 \\
&= 178 \text{ kJ} + 1735 \text{ kJ} + 243 \text{ kJ} - 697 \text{ kJ} + 796 \text{ kJ} = +2255 \text{ kJ}
\end{aligned}
$$

9.27 A Lewis structure is a representation of a molecule, polyatomic ion, or ionic compound that shows all the valence electrons. Shared electron pairs are shown as dashes between the bonded atoms and unshared electrons are shown as dots.

9.29 The bond length is the equilibrium distance between the centers of two vibrating bonded atoms, i.e., the distance at which the attractive and repulsive forces causing the atomic vibration are in balance and the net force on the atoms is zero.

The atoms in a covalent bond are constantly in motion, stretching and compressing about the bond length. The frequency of this vibration depends on the masses of the bonded atoms and the characteristics of the bond. The atoms act like they are attached to a spring, with a spring constant determined mainly or basically by the bond strength; hence, the comparison of a covalent bond to a spring.

9.31 The bond energy is the average energy required to break a chemical bond in a gaseous molecule.

(a) In general, the bond energy increases and the bond length decreases with decreasing size of the bonded atoms.

(b) In general, the bond energy increases and the bond length decreases with increasing multiplicity of the bond.

9.33 Both ionic and covalent bonds are electrostatic in nature. Electrostatic attractions vary with the reciprocal of the square of the distance between the interacting charges; the smaller the distance the greater the attractive force. The small atoms can get closer to other atoms and, therefore, tend to have greater electrostatic attractions leading to stronger bonds.

9.35 Cl is above I in Group 7A; therefore Cl atoms are smaller than I atoms. In general, bonds between small atoms tend to be stronger than bonds between large atoms; therefore, the bond energy in Cl_2, with its two smaller chlorine atoms, is greater than the bond energy in I_2, with its two larger atoms.

9.37 (a) H_2O has 6 + 2(1) = 8 valence electrons or 4 electron pairs.

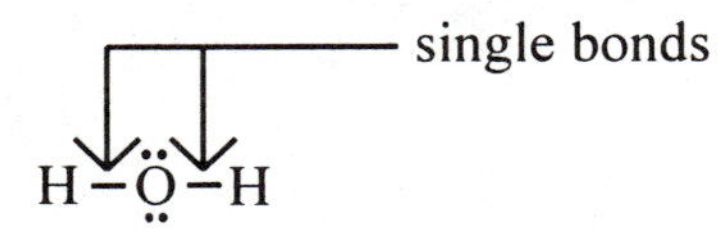

(b) C_2H_2 has 2(4) + 2(1) = 10 valence electrons or 5 electron pairs.

H–C≡C–H

triple bond

(c) NH_3 has 5 + 3(1) = 8 valence electrons or 4 electron pairs.

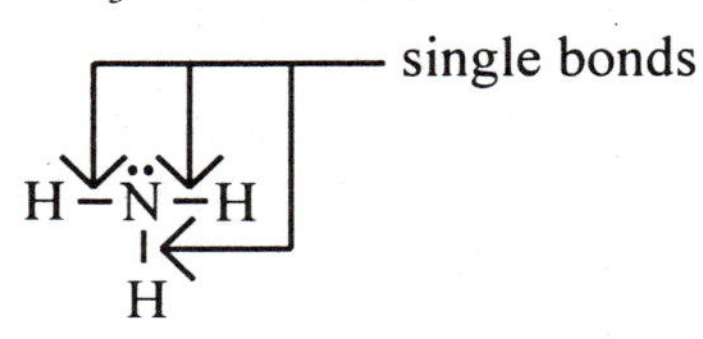

9.39 Stronger bond: NH_3, since nitrogen is smaller than phosphorus.
Shorter bond: NH_3, since nitrogen is smaller than phosphorus.

9.41 Since S is smaller than Se, the H–S bond should have the higher bond energy and the H–Se bond the longer bond length.
(a) H–S (b) H–Se

9.43 $2HCl(g) + F_2(g) \rightarrow 2HF(g) + Cl_2(g) \quad \Delta H° = ?$

The enthalpy change is the sum of the enthalpy changes for the following steps:

		$\Delta H°$
(1) Breaking 2 mol of H–Cl bonds:	2 × +432.0 kJ =	+864.0 kJ
(2) Breaking 1 mol of F–F bonds:	+158.0 kJ =	+158.0 kJ
(3) Forming 2 mol of H–F bonds:	2 × –568.1 kJ =	–1136 kJ
(4) Forming 1 mol of Cl–Cl bonds:	–243.4 kJ =	–243.4 kJ
	$\Delta H°$ = Sum =	–357 kJ

The enthalpy change for the reaction is $\Delta H° = -357$ kJ.

9.45 $1/2H_2(g) + 1/2F_2(g) \rightarrow HF(g) \quad \Delta H°_f = -268.6$ kJ

The F–F bond energy is the enthalpy change accompanying the reaction

$F_2(g) \rightarrow 2F(g) \quad \Delta H° = ?$

The sum of the enthalpy changes of the following steps gives the required answer.

		$\Delta H°$
(1) $H_2(g) + F_2(g) \rightarrow 2HF(g)$	2×-268.6 kJ =	-537.2 kJ
(2) $2H–F(g) \rightarrow 2H(g) + 2F(g)$	$2 \times +568.1$ kJ =	$+1136$ kJ
(3) $2H(g) \rightarrow H_2(g)$		-435.9 kJ
	$\Delta H°$ = Sum =	$+163$ kJ

The F–F bond energy is 163 kJ per mole.

Comment: An alternate approach to solving this problem goes as follows:

$1/2H_2(g) + 1/2F_2(g) \rightarrow HF(g) \quad \Delta H_f^\circ = -268.6 \text{ kJ}$

If we treat the thermochemical equation as an energy equality equation, it can be rearranged and rewritten as follows (after first multiplying the equation by 2):

$$\begin{aligned} D(F\text{–}F) &= 2 \times D(H\text{–}F) - D(H\text{–}H) - 2 \times 268.6 \text{ kJ} \\ &= 2 \times +568.1 \text{ kJ} - 435.9 \text{ kJ} - 2 \times 268.6 \text{ kJ} = +163 \text{ kJ} \end{aligned}$$

Since this is an exothermic reaction, the products are more stable (have larger negative energies) than the reactants and, therefore, the $\Delta H°$ reaction energy of -268.6 kJ has to be added to the product dissociation energy side in order for it to be equal to the reactant dissociation energy side, since it is more stable by that amount of energy. For endothermic reactions, the $\Delta H°$ reaction energy has to be subtracted from the reactant dissociation energy side.

9.47 $CH_4(g) + Cl_2(g) \rightarrow CH_3Cl(g) + HCl(g) \quad \Delta H° = ?$

The enthalpy change is the sum of the enthalpy changes for the following steps:

	$\Delta H°$
(1) Breaking 1 mol of C–H bonds:	+413 kJ
(2) Breaking 1 mol of Cl–Cl bonds:	+243 kJ
(3) Forming 1 mol of C–Cl bonds:	–330 kJ
(4) Forming 1 mol of H–Cl bonds:	–432 kJ
$\Delta H°$ = Sum =	–106 kJ

The enthalpy change is –106 kJ per mole of methyl chloride.

9.49 Hydrogen, $1s^1$, only has one orbital in its valence shell and, therefore, can only form one covalent bond. In order to be a central atom, an atom has to be able to form at least two covalent bonds.

9.51 The sum of the formal charges in a neutral molecule equals zero.
The sum of the formal charges in an ion equals the charge on the ion.

9.53 (a) H_2S has 6 + 2(1) = 8 valence electrons or 4 electron pairs.

H–S̤̈–H

(b) NI_3 has 5 + 3(7) = 26 valence electrons or 13 electron pairs.

:Ï̤–N̈–Ï̤:
|
:I̤:

(c) SiH_4 has 4 + 4(1) = 8 valence electrons or 4 electron pairs.

```
     H
     |
  H—Si—H
     |
     H
```

(d) C_2H_6 has 2(4) + 6(1) = 14 valence electrons or 7 electron pairs.

```
     H  H
     |  |
  H—C—C—H
     |  |
     H  H
```

(e) H_2CO has 2(1) + 4 + 6 = 12 valence electrons or 6 electron pairs.

```
     H
     |   ..
  H—C=O
         ..
```

9.55 (a) H_3O^+ has 3(1) + 6 – 1 = 8 valence electrons or 4 electron pairs.

```
 [     ..     ]+
 [  H—O—H  ]
 [     |      ]
 [     H      ]
```

(b) OH^- has 6 + 1 + 1 = 8 valence electrons or 4 electron pairs.

```
 [  ..    ]-
 [ :O—H  ]
 [  ..    ]
```

(c) OCl^- has 6 + 7 + 1 = 14 valence electrons or 7 electron pairs.

```
 [  ..  ..  ]-
 [ :O—Cl: ]
 [  ..  ..  ]
```

(d) NH_4^+ has 5 + 4(1) – 1 = 8 valence electrons or 4 electron pairs.

```
 [     H      ]+
 [     |      ]
 [  H—N—H  ]
 [     |      ]
 [     H      ]
```

9.57 Refer to structures in Exercises 9.53 and 9.54.

(a) 1. H_2S; none. 2. NI_3; none. 3. SiH_4; none. 4. C_2H_6; none. 5. H_2CO; none.

(b) 1. N_2; none. 2. N_2H_4; none. 3. H_2O_2; none. 4. C_2N_2; none. 5. S_8; none.

9.59 (a) Certain molecules and ions cannot be accurately depicted by a single Lewis structure. Such molecules and ions are depicted by two or more Lewis structures, which, when employed or taken together, give a better description than any single Lewis structure. The separate Lewis structures are called resonance structures or contributing structures, and the actual molecule or ion is referred to as a resonance hybrid. The actual molecule or ion, the resonance hybrid, has properties that are some combination of those of the contributing structures.

(b) Individual contributing structures do not represent the actual molecule. The properties of the individual contributing structures are different from those of the actual molecule or ion. For instance, bond length predictions based on individual contributing structures are incorrect.

9.61 A delocalized bond is a chemical bond that bonds three or more atoms together as compared to a localized bond, which only bonds two atoms together. The bonding electrons in a delocalized bond spread out over several atoms, creating a bond that binds them all together.

A delocalized bond cannot be accurately represented by a single Lewis (resonance) structure. Delocalized bonds are represented by a composite of two or more resonance structures. (Resonance structures are the different Lewis structures, each with the same arrangement of atoms and the same number of electron pairs, that can be drawn for a given molecule or ion).

9.63 (a) NO_2^- has 5 + 2(6) + 1 = 18 valence electrons or 9 electron pairs.

[:Ö–N̈=Ö]$^-$ ⟷ [Ö=N̈–Ö:]$^-$

(b) N_2O_4 has 2(5) + 4(6) = 34 valence electrons or 17 electron pairs.

:Ö: :Ö: | | Ö=N—N=Ö ⟷ :O: :O: ‖ ‖ :Ö–N—N–Ö: ⟷ :O: :Ö: ‖ | :Ö–N—N=Ö ⟷ :Ö: :O: | ‖ Ö=N—N–Ö:

(c) HN_3 has 1 + 3(5) = 16 valence electrons or 8 electron pairs.

H–N̈=N=N̈ ⟷ H–N̈–N≡N: ⟷ H–N≡N–N̈:

(d) SO_2 has 6 + 2(6) = 18 valence electrons or 9 electron pairs.

:Ö–S̈=Ö ⟷ Ö=S̈–Ö:

Comment: A number of the less stable resonance structures have not been included. In (d), octet expansion is not explored.

9.65 (a) [(–1):Ö–N̈=Ö]$^-$ ⟷ [Ö=N̈–Ö:(–1)]$^-$

I II

Both contribute equally.

(b) (–1):Ö: :Ö:(–1) | | Ö=N(+1)—N(+1)=Ö ⟷ :O: :O: ‖ ‖ (–1):Ö–N(+1)—N(+1)–Ö:(–1) ⟷ :O: :Ö:(–1) ‖ | (–1):Ö–N(+1)—N(+1)=Ö ⟷ (–1):Ö: :O: | ‖ Ö=N(+1)—N(+1)–Ö:(–1)

I II III IV

All contribute equally.

(c) H–N̈=N(+1)=N̈(–1) ⟷ H–N̈(–1)–N(+1)≡N: ⟷ H–N(+1)≡N(+1)–N̈:(–2)

I II III

Structures I and II make greater contributions to the hybrid.

(d) :Ö(–1)–S̈(+1)=Ö ⟷ Ö=S̈(+1)–Ö:(–1)

I II

Both contribute equally.

9.67

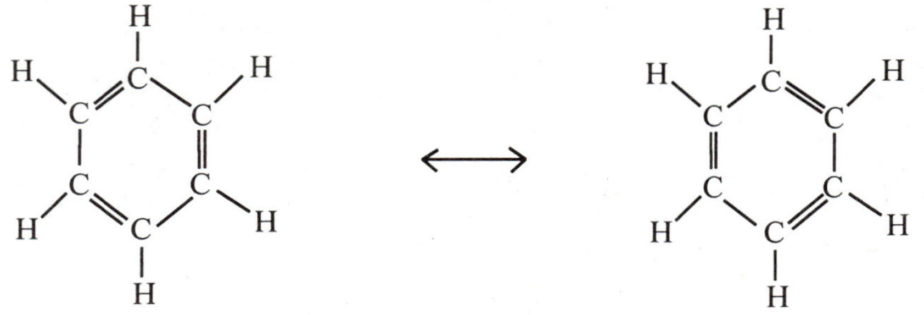

$\longleftrightarrow$

9.69 No. The best Lewis structure is the one that most closely resembles the actual molecule in stability and placement of the bonding electrons. Usually, this is the Lewis structure with the least amount of formal charge. Often, octet expansion, where possible, will reduce formal charge and lead to a Lewis structure that better represents the actual molecule, even though the octet rule is being broken or not followed.

9.71 (a) "Incomplete octet" means fewer than four electron pairs around a bonded atom.
(b) Beryllium, boron, and aluminum.

9.73 (a) SF_4 has 6 + 4(7) = 34 valence electrons or 17 electron pairs.

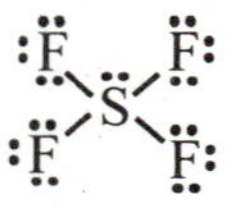

(b) SbF_6^- has 5 + 6(7) + 1 = 48 valence electrons or 24 electron pairs.

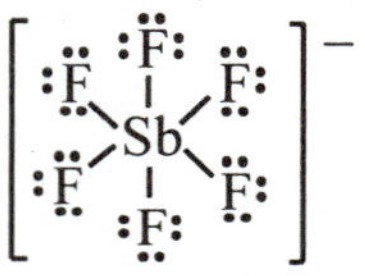

9.75 (a) SO_3 has 6 + 3(6) = 24 valence electrons or 12 electron pairs.

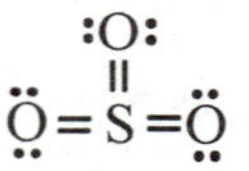

(b) $HBrO_4$ has 1 + 7 + 4(6) = 32 valence electrons or 16 electron pairs.

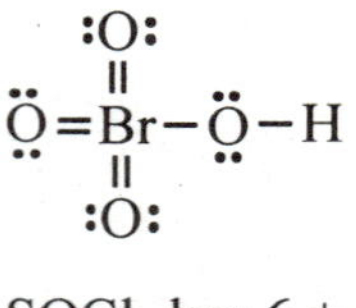

(c) $SOCl_2$ has 6 + 6 + 2(7) = 26 valence electrons or 13 electron pairs.

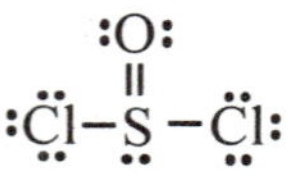

(d) SO_2Cl_2 has 6 + 2(6) + 2(7) = 32 valence electrons or 16 electron pairs.

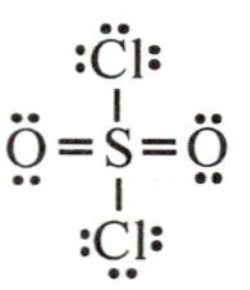

9.77 (a) SO_2 has 6 + 2(6) = 18 valence electrons or 9 electron pairs.

:Ö–S̈=Ö: ⟷ :Ö=S̈–Ö:

O_3 has 3(6) = 18 valence electrons or 9 electron pairs.

:Ö–Ö=Ö: ⟷ :Ö=Ö–Ö:

Both are triatomic molecules composed of Group 6A elements. They are both resonance hybrids and have the same number of valence electrons (18). They differ in the atom occupying the middle position.

(b) :Ö=S̈=Ö:

No. Because oxygen has no d orbitals in its valence shell.

9.79 (a) $AlCl_3$ has 3 + 3(7) = 24 valence electrons or 12 electron pairs.

```
   :C̈l:
    |
:C̈l–Al–C̈l:
```

(b) Li_2 has 2(1) = 2 valence electrons or 1 electron pair.

Li–Li

(c) $TlCl_2^+$ has 3 + 2(7) – 1 = 16 valence electrons or 8 electron pairs.

[:C̈l–Tl–C̈l:]$^+$

9.81 (a) CH_3 has 4 + 3(1) = 7 valence electrons or 3 electron pairs plus one odd electron.

```
   H
   |
H–C–H
   ·
```

(b) ClO_4 has 7 + 4(6) = 31 valence electrons or 15 electron pairs plus one odd electron.

```
     :O:
     ||
 Ö=Cl–Ö·
     ||
     :O:
```

9.83 (a) A polar covalent bond is a covalent bond in which there is not equal sharing of the electron density by the two bonded atoms. The bonding electrons spend more time near one atom than near the other. (Polar covalent bonds are not perfectly symmetrical; they have positive and negative charge centers and some ionic character.)

A nonpolar covalent bond is a covalent bond in which the electron density is equally shared by the two bonded atoms. The bonding electrons spend an equal amount of time near each atom. (Nonpolar covalent bonds are perfectly symmetrical; they have no positive and negative charge centers and zero ionic character.)

(b) In a polar covalent bond, while the two bonding electrons are not being shared equally, each atom still has a share of both bonding electrons. In an ionic bond, on the other hand, one atom has accepted complete control of the transferred electron(s), while the other atom has completely given up the transferred electron(s). In a pure ionic bond there is no sharing taking place.

9.85 The dipole moment, μ, is a molecular constant which gives a quantitative measure of the distribution or separation of electrical charges in a molecule. The dipole moment of a dipole is equal to the magnitude of the separated charges multiplied by the distance between their centers. Dipole moments are vectors and the dipole moment of a molecule is the resultant of the bond dipole moments. The unit used for measuring molecular dipole moments is the debye (D): 1 D = 3.33×10^{-30} C•m. The SI unit for dipole moment is the coulomb meter (C•m), which is a much larger unit.

Dipole moments are measured by placing the molecules between two charged plates, essentially a parallel plate capacitor. Polar molecules, molecules that have a permanent separation of their positive and negative centers of charge, tend to orient themselves in respect to the external field. This alignment allows the plates to hold a greater charge, and the dipole moment can be calculated from the magnitude of this effect.

9.87 The "ionic character" in a covalent bond refers to the degree of polarity of the bond. Nonpolar covalent bonds have no dipole moments, no permanent separation of their positive and negative charge centers, and, therefore, can be considered to have zero ionic character. Pure ionic bonds, on the other hand, have the maximum amount of charge separation and, therefore, can be considered to have 100% or total ionic character. Polar covalent bonds have varying degrees of polarity (charge separation) and fall somewhere in between these two extremes in bonding. They can be considered to have varying degrees of ionic character depending on the polarity of the bond; the greater the polarity, the greater the ionic character of the bond.

H–F (1.82 D), H–Cl (1.08 D), and H–O–H (1.87 D) are examples of covalent bonds with considerable ionic character.

9.89 The electronegativity is a measure of the relative tendency of an atom to attract bonding or shared electrons to itself when it is covalently bonded with another atom.

The electronegativity is a measure of the desire for electrons or electron density by an atom involved in a bonding situation where another atom is also competing for the same shared electrons, while the electron affinity is a measure of the desire a single gaseous atom has for acquiring an additional electron.

9.91 From Figure 9.14 we see that (a) fluorine has the highest electronegativity, (b) oxygen has the second highest, and (c) francium has the lowest.

9.93
(a) Cl is just above Br in Group 7A; therefore, Cl should be more electronegative.
(b) Si (Group 4A, period 3) is below and to the left of O (Group 6A, period 2); therefore, O should be more electronegative.
(c) N (Group 5A) lies to the left of O (Group 6A) in the second period; therefore, O should be more electronegative.
(d) Ga (Group 3A, period 4) is below and to the left of S (Group 6A, period 3); therefore, S should be more electronegative.

9.95
(a) O is to the right of N in the second period; therefore, O will be more electronegative and the bond polarity is

$^{\delta+}$N–O$^{\delta-}$

(b) Cl is just above Br in Group 7A; therefore, Cl will be more electronegative and the bond polarity is

$^{\delta+}$Br–Cl$^{\delta-}$

(c) S (Group 6A, third period) on the right side of the periodic table should be more electronegative than H (Group 1A, first period) on the left side of the periodic table. Therefore, the bond polarity is

$^{\delta-}$S–H$^{\delta+}$

9.97
(a) For $BeCl_2$ the electronegativity difference, 3.2 – 1.6 = 1.6, is less than 1.7 and, therefore, $BeCl_2$ is predominantly covalent.
(b) For BF_3 the electronegativity difference, 4.0 – 2.0 = 2.0, exceeds 1.7 and, therefore, BF_3 is predominantly ionic.

(c) For KF the electronegativity difference, 4.0 – 0.82 = 3.18, exceeds 1.7 and, therefore, KF is predominantly ionic.

Note: BF_3 is covalent, contrary to this prediction. It is a pungent, colorless gas (b.p. –99.9°C).

9.99 (a) HBr has 1 + 7 = 8 valence electrons or 4 electron pairs.

$$\overset{\delta+}{\text{H}}-\overset{\delta-}{\ddot{\underset{\cdot\cdot}{\text{Br}}}}\text{:}$$

(b) H_3BO_3 has 3(1) + 3(6) = 24 valence electrons or 12 electron pairs.

```
        H
        |
       :O:
        |
  H−Ö−B−Ö−H
```

(c) NH_3 has 5 + 3(1) = 8 valence electrons or 4 electron pairs.

```
    H
    |
  H−N−H
    ..
```

9.101 MgO. In general, the greater the lattice energy the higher the melting point of the substance. The lattice energy depends on the size and charge of the ions composing the ionic compound. Smaller ions and larger charges, i.e., greater charge density, lead to greater lattice energies. Mg^{2+} is the same for both substances; therefore any difference in melting point between MgO and $MgCl_2$ just depends on the size and charge differences between O^{2-} and Cl^-. Since O is in period 2 and Cl is in period 3, O^{2-} should be smaller than Cl^-, $O^{2-} < Cl^-$. Also, O^{2-} carries a larger charge than Cl^-. Therefore, MgO should have a higher lattice energy and, hence, a higher melting point than $MgCl_2$. The actual melting points are: MgO, 2852°C; $MgCl_2$, 714°C. MgO is actually used as a refractory in making firebricks, crucibles, and furnace linings.

9.103 $1/2H_2(g) + 1/2Br_2(l) \rightarrow HBr(g) \quad \Delta H_f^\circ = ?$

The sum of the enthalpy changes of the following steps gives the required answer:

			ΔH°
1.	$H(g) + Br(g) \rightarrow HBr(g)$		–366.2 kJ
2.	$1/2\ Br_2(l) \rightarrow 1/2\ Br_2(g)$	1/2 × +30.9 kJ =	+15.4 kJ
3.	$1/2\ H_2(g) \rightarrow H(g)$	1/2 × +435.9 kJ =	+218.0 kJ
4.	$1/2\ Br_2(g) \rightarrow Br(g)$	1/2 × +192.9 kJ =	+96.4 kJ
		ΔH_f° = Sum =	–36.4 kJ

The standard enthalpy of formation of HBr is –36.4 kJ per mole.

9.105 $C(s) + 2H_2(g) \rightarrow CH_4(g) \quad \Delta H_f^\circ = -74.8\ kJ$

The C–H bond energy is one-fourth the enthalpy change accompanying the reaction

$CH_4(g) \rightarrow C(g) + 4H(g)$ or,

$1/4CH_4(g) \rightarrow 1/4C(g) + H(g) \quad \Delta H^\circ = ?$

The sum of the enthalpy changes of the following steps gives the required answer.

			ΔH°
(1)	1/4 $CH_4(g) \rightarrow$ 1/4 C(s) + 1/2 $H_2(g)$	1/4 × +74.8 kJ =	+18.7 kJ
(2)	1/4 C(s) $\rightarrow$ 1/4 C(g)	1/4 × +716.7 kJ =	+179.2 kJ
(3)	1/2 $H_2(g) \rightarrow$ H(g)	1/2 × +435.9 kJ =	+218.0 kJ
		ΔH° = Sum =	+415.9 kJ

The C–H bond energy is +415.9 kJ per mole.

9.107

(Resonance structures of chlorobenzene: two Kekulé structures with alternating C=C double bonds, joined by ⟷)

9.109 (a) BF_3 has 3 + 3(7) = 24 valence electrons or 12 electron pairs.

B: formal charge = 3 – 0 – 3 = 0

NH_3 has 3 + 5 = 8 valence electrons or 4 electron pairs.

N: formal charge = 5 – 2 – 3 = 0

F_3BNH_3 has 3(7) + 3 + 5 + 3(1) = 32 valence electrons or 16 electron pairs.

N: formal charge = 5 – 0 – 4 = +1
B: formal charge = 3 – 0 – 4 = –1

(b) Yes. Boron's formal charge goes from 0 to –1 and nitrogen's goes from 0 to +1.

9.111 (a) CN has 4 + 5 = 9 valence electrons or 4 electron pairs plus one odd electron.

·C≡N: ⟷ :C≐N: ⟷ $\overset{(-1)}{:C}\equiv\overset{(+1)}{N}\cdot$

(b) NH has 5 + 1 = 6 valence electrons.

·N̈–H

(c) N_2^+ has 2(5) – 1 = 9 valence electrons.

$$[:N\equiv N\cdot]^+ \longleftrightarrow [:N\dot{=}N:]^+$$

(d) CO^+ has 4 + 6 – 1 = 9 valence electrons.

$$[\cdot C\equiv O:]^+ \longleftrightarrow [:C\dot{=}O:]^+$$

(e) CO_2^+ has 4 + 2(6) – 1 = 15 valence electrons

$$[:\ddot{O}=C=\ddot{O}\cdot]^+$$

9.113 (a) O_3 has 3(6) = 18 valence electrons or 9 electron pairs. The resonance structures for ozone are

$$:\ddot{O}=\overset{+1}{\ddot{O}}-\overset{-1}{\ddot{\underset{..}{O}}}: \longleftrightarrow :\overset{-1}{\ddot{\underset{..}{O}}}-\overset{+1}{\ddot{O}}=\ddot{O}:$$

The formal charges on the oxygen atoms suggest that the center oxygen tends to be positive with respect to the end oxygens, and, consequently, a bond dipole moment exists.

(b) Since the resonance structures show that the two O–O bonds are equivalent, the ozone molecule cannot be linear because then it would have a zero dipole moment. Since the dipole moment is not zero, the molecule must be bent:

$$:\overset{..}{O}\overset{\ddot{O}}{=\!\!\!/\ \ \diagdown}\ddot{\underset{..}{O}}: \longleftrightarrow :\ddot{\underset{..}{O}}\overset{\ddot{O}}{\diagup\ \ \diagdown\!\!\diagdown}\underset{..}{O}:$$

9.115 The electronegativity is a measure of the relative tendency of an atom to attract bonding or shared electrons to itself when it is covalently bonded with another atom. If the electronegativity difference between the two bonded atoms is too large, the bonding electrons will go over to the atom with the greater electronegativity, and the bonding will become ionic bonding. The electron transfer between the two atoms making up the bond, let's call them A and B, would lead to either A^-B^+ or A^+B^-, depending on which atom was the more electronegative. Now, if the electronegativities of A and B were identical, the two ionic pairs would be of equal energies. The energy of A^-B^+ depends on the ionization energy of B and the electron affinity of A, while the energy of A^+B^- depends on the ionization energy of A and the electron affinity of B. The ionization energy and electron affinity are the energies accompanying the following processes:

$$IE_A:\ A \rightarrow A^+ + e^-;\quad EA_A:\ A + e^- \rightarrow A^-$$

The IE is always positive and the EA is usually, but not always, negative.

If the original energy of A plus the original energy of B is designated as E, then the energy of A^-B^+ is E + $IE_B + EA_A$, and the energy of A^+B^- is $E + IE_A + EA_B$. If the energies of the two ionic structures are equal, then $E + IE_B + EA_A = E + IE_A + EA_B$ or $IE_A - EA_A = IE_B - EA_B$ is the condition that specifies equal A and B electronegativities. If A is more electronegative than B, then the energy of A^-B^+ is less than the energy of A^+B^-, $IE_B + EA_A < IE_A + EA_B$, and $IE_A - EA_A > IE_B - EA_B$. Therefore, the electronegativity of an atom should be proportional or related to the quantity (IE – EA), and that's why Mulliken defined the electronegativity of an atom to be $X = k \times (IE - EA)$. The negative sign in the Mulliken formula is due to the way the electron affinity is defined. If the EA is defined as the energy accompanying the process $A^- \rightarrow A + e^-$, then the sign would be positive.

CHAPTER 10

MOLECULAR GEOMETRY AND CHEMICAL BONDING THEORY

Solutions To Practice Exercises

PE 10.1 OF_2 has 6 + 2(7) = 20 valence electrons or 10 electron pairs. The Lewis structure for OF_2 is

:F̤̈–Ö̤–F̤̈:

Since both O–F bonds in OF_2 are equivalent, OF_2 can have a dipole moment only if the molecule is bent.

F–O–F (bent)

PE 10.2 (a) The Lewis structure for CO_2 is

:Ö=C=Ö:

The two double bonds each contribute one VSEPR pair, there are no single bonds or lone pairs; hence, there are two VSEPR pairs.

(b) The Lewis structure for NH_3 is

H–N̈–H
  |
  H

The three single bonds each contribute one VSEPR pair, there is one lone pair, and there are no double bonds; hence, there are four VSEPR pairs.

PE 10.3 (a) NO_3^- has 5 + 3(6) + 1 = 24 valence electrons and is trigonal planar.

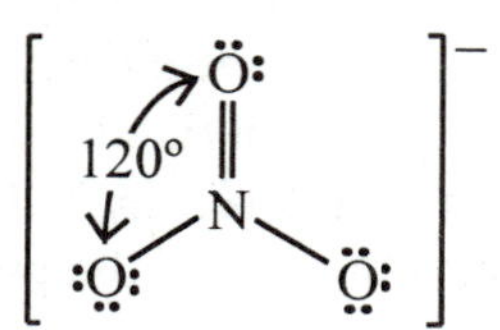

(a resonance hybrid)

(b) CO_3^{2-} has 4 + 3(6) + 2 = 24 valence electrons and is trigonal planar.

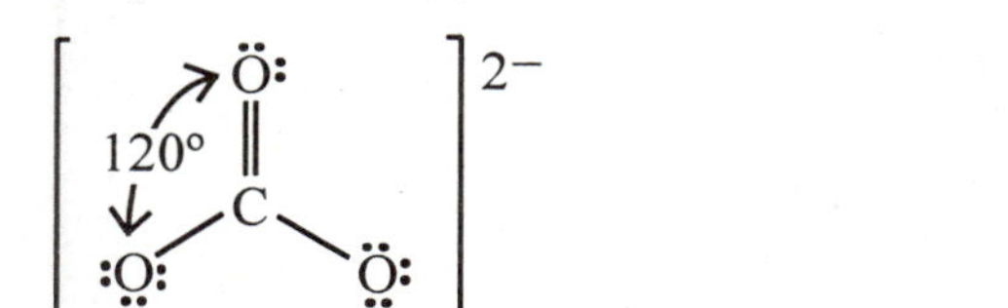

(a resonance hybrid)

PE 10.4 The Lewis structure for ozone (9 electron pairs) is

:Ö–Ö=Ö:

The central oxygen has three VSEPR pairs, which adopt a trigonal planar arrangement. The molecule is bent (V-shaped).

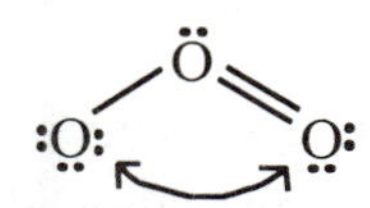

around 120°

PE 10.5 NH_4^+, SO_4^{2-}, and SiF_4 are tetrahedral.

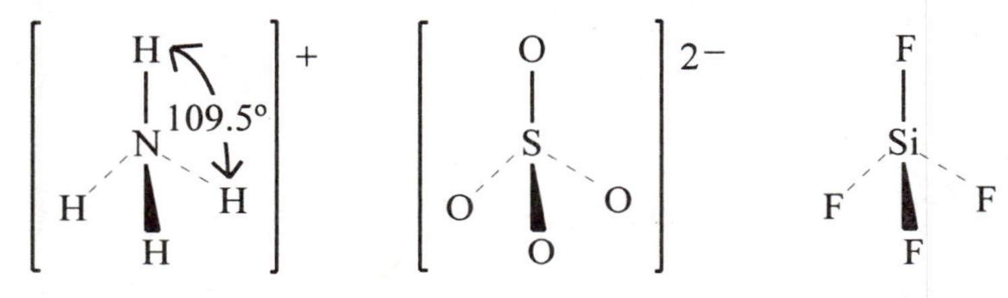

PE 10.6 (a) The phosphine molecule, PH_3, has three bonds and one lone pair tetrahedrally arranged around the central phosphorus atom.

trigonal pyramidal

(b) The hydrogen sulfide molecule has two bonds and two lone pairs tetrahedrally arranged around the sulfur atom.

V-shaped or bent

PE 10.7 These molecules all have five VSEPR pairs:

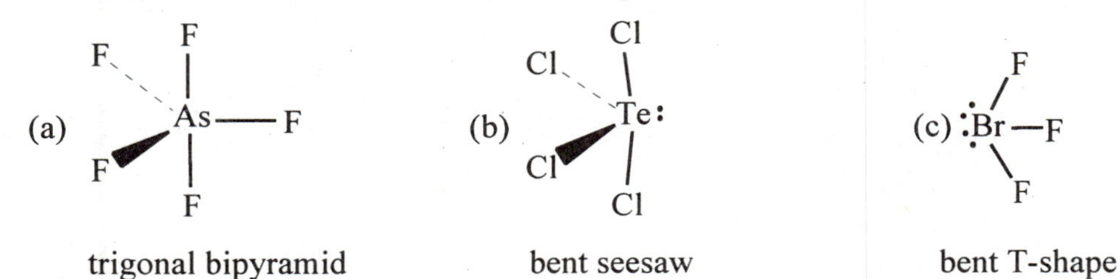

trigonal bipyramid   bent seesaw   bent T-shape

PE 10.8 These molecules have six VSEPR pairs.

F—Br—F (with F, F, F below; lone pair on Br)

slightly bent
square pyramid

Cl—I—Cl (Cl above and below; 90°)

square planar

PE 10.9 NO_2^- has 5 + 2(6) + 1 = 18 valence electrons or 9 electron pairs. The Lewis structure is

$[:\ddot{\underset{..}{O}}-\ddot{N}=\ddot{O}:]^-$ (a resonance hybrid)

The central N has three VSEPR pairs, which adapt a trigonal planar arrangement. Since one of the pairs is a lone pair, the molecule is bent (V-shaped). Compression caused by the lone pair should make the bond angle slightly less than 120°.

PE 10.10 NF_3 has four VSEPR pairs and a trigonal pyramidal shape. Since nitrogen is less electronegative than fluorine, the molecule will be polar with the nitrogen end positive.

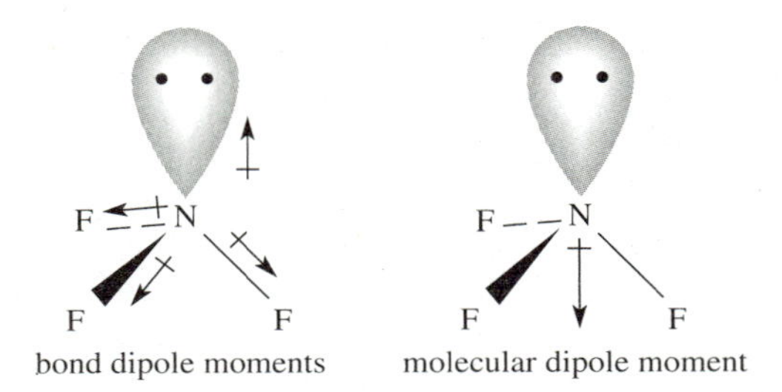

Comment: Lone pairs also have dipole moments and can have a significant affect on the resultant molecular dipole moment. You should be aware of this, but you can ignore their effects for now.

PE 10.11 (a) polar (b) nonpolar (c) nonpolar (d) polar (e) nonpolar

PE 10.12 (a) and (b) are the same molecule; (c) is a different isomer.

PE 10.13 No. The carbon atom at the right of the double bond does not have two different substituents.

PE 10.14

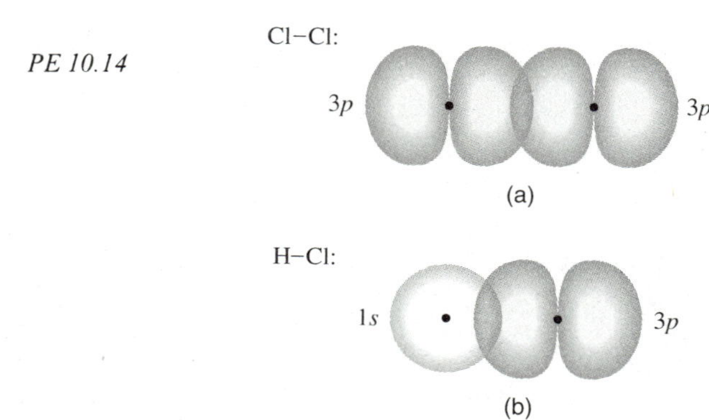

PE 10.15 The central Be atom is sp hybridized.

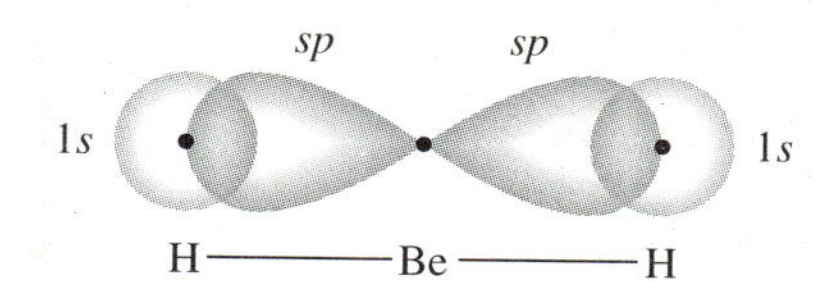

Note: Small sp lobes have been omitted for clarity.

PE 10.16 The central Al atom is sp^2 hybridized.

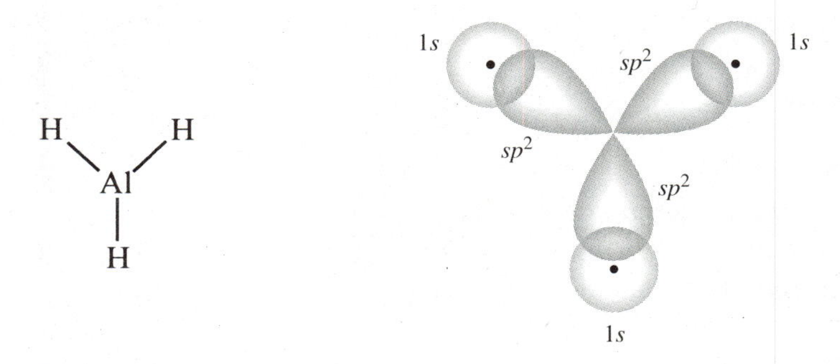

Note: Small sp^2 lobes have been omitted for clarity.

PE 10.17 Si ($3s^23p^2$) mixes the 3s and 3p orbitals to form the required sp^3 hybrid orbitals.

PE 10.18

```
  H  H
  |  |
H-N--C-H
  ••  |
     H
```

N and C, each with four VSEPR pairs, are sp^3 hybridized. Consequently, the bond angles (H–N–H, H–N–C, N–C–H, and H–C–H) will all be approximately 109.5°. (Although it is likely that the H–N–H and H–N–C bond angles will be somewhat less than 109.5° due to the nonbonding pair on N.)

PE 10.19 (a) The ground-state configuration for H_2^+ is $(\sigma_{1s})^1$. It has one bonding electron and no antibonding electrons. The bond order is (1 – 0)/2 = 1/2.

(b) The ground-state configuration for H_2^- is $(\sigma_{1s})^2\,(\sigma^*_{1s})^1$. It has two bonding electrons and one antibonding electron. The bond order is (2 – 1)/2 = 1/2.

(c) The ground-state configuration for He_2 is $(\sigma_{1s})^2\,(\sigma^*_{1s})^2$. It has two bonding electrons and two antibonding electrons. The bond order is (2 – 2)/2 = 0.

PE 10.20 (a) Be_2 has eight electrons; its molecular orbital configuration is $(\sigma_{1s})^2\,(\sigma^*_{1s})^2\,(\sigma_{2s})^2\,(\sigma^*_{2s})^2$.

(b) Be_2 has four bonding electrons and four antibonding electrons. The bond order is (4 – 4)/2 = 0. Be_2 is unstable with respect to its atoms, since the bond order is zero. Be_2 shouldn't exist.

PE 10.21 (a) N_2 has 14 electrons; its molecular orbital configuration is
$(\sigma_{1s})^2\,(\sigma^*_{1s})^2\,(\sigma_{2s})^2\,(\sigma^*_{2s})^2\,(\pi_{2px})^2\,(\pi_{2py})^2\,(\sigma_{2pz})^2$.

(b) No. N_2 has no unpaired electrons and, therefore, is not paramagnetic.

PE 10.22 First we will determine their molecular orbital configurations and bond orders.

O_2^{2-} has 18 electrons; its molecular orbital configuration is

$(\sigma_{1s})^2(\sigma_{1s}^*)^2\,(\sigma_{2s})^2\,(\sigma_{2s}^*)^2\,(\pi_{2px})^2\,(\pi_{2py})^2\,(\sigma_{2pz})^2\,(\pi_{2px}^*)^2\,(\pi_{2py}^*)^2$.

O_2^{2-} has 10 bonding electrons and 8 antibonding electrons. The bond order is (10 – 8)/2 = 1.
O_2^- has 17 electrons; its molecular orbital configuration is

$(\sigma_{1s})^2\,(\sigma_{1s}^*)^2\,(\sigma_{2s})^2\,(\sigma_{2s}^*)^2\,(\pi_{2px})^2\,(\pi_{2py})^2\,(\sigma_{2pz})^2\,(\pi_{2px}^*)^2\,(\pi_{2py}^*)^1$.

O_2^- has 10 bonding electrons and 7 antibonding electrons. The bond order is (10 – 7)/2 = 3/2.

(a) O_2^{2-} is not paramagnetic because it has no unpaired electrons. O_2^- is paramagnetic because it has an unpaired electron.
(b) O_2^- should have a higher bond energy.
(c) O_2^- should have a shorter bond length.

PE 10.23 (a) HeH^+ has 2 electrons; its molecular orbital configuration is $(\sigma_{1s})^2$.
(b) Yes. HeH^+ has two bonding electrons and no antibonding electrons. The bond order is (2 – 0)/2 = 1. Since the bond order is greater than zero, HeH^+ should be stable.

Solutions To Final Exercises

10.1 A bond angle is the equilibrium angle between the lines that join two atoms to a third atom.

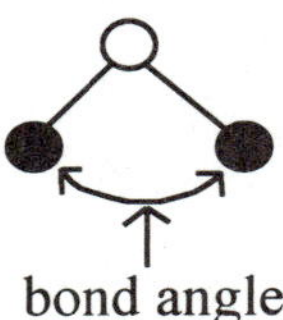

bond angle

10.3 (a) A bond dipole moment is the dipole moment associated with an individual bond. A molecular dipole moment is the net dipole moment associated with the whole molecule. The net dipole moment is obtained by vector addition of the individual bond dipole moments.
(b) They are identical for polar diatomic molecules and for polyatomic molecules where all but one of the bond dipole moments cancel each other out.

10.5 (a) Bent. If the molecule was linear, the two S–H bond dipole moments would cancel each other out, since they are pointed in opposite directions on the same line and have equal magnitudes. Therefore, if the molecule was linear, the molecular dipole moment would be zero and not 0.97 D. Since the dipole moment is not zero, the molecule must be bent.

(b)

S
H H

direction of individual bond dipole moments

S
H H

direction of net (molecular) dipole moment

10.7 It tells us that BF_3 must be flat with its bonds pointed to the corners of an equilateral triangle. BF_3 has 3 + 3(7) = 24 valence electrons or 12 electron pairs. The Lewis structure is

:F:
|
:F–B–F:

From the Lewis structure we see that all three B–F bonds are equivalent. In this situation BF_3 can have a zero dipole moment only if the molecule is flat and the bonds point to the corners of an equilateral triangle.

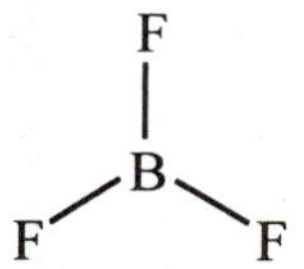

F
|
B
F   F

10.9 VSEPR stands for Valence Shell Electron Pair Repulsion. The basic assumption of the theory is that valence electron pairs tend to stay as far apart as possible.

10.11 A lone pair is just under the influence of one nucleus and, therefore, the electron density of a lone pair is concentrated on one nucleus. A bonding pair, on the other had, is under the influence of two nuclei and, therefore, the electron density of a bond pair is less concentrated since it is drawn out between two nuclei. The greater charge density of the lone pair repels the lower charge density of adjacent bond pairs and, in so doing, pushes the bond pairs closer together.

10.13 (a) linear;  CO_2 (b) bent;  H_2O

(c) trigonal planar; 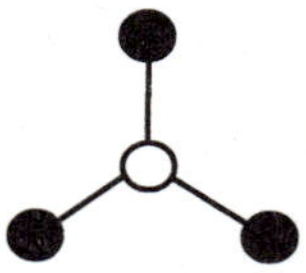 SO_3 (d) tetrahedral; 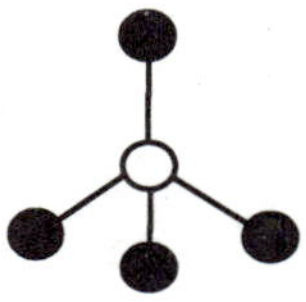 CCl_4

(e) square planar; 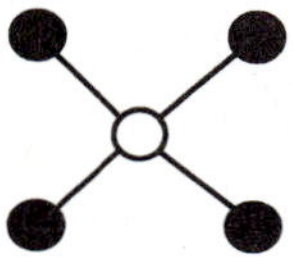 IF_4^- (f) trigonal pyramidal;  PF_3

10.15

	Lewis Structure	VSEPR Pairs	Lone Pairs	Shape	Bond Angles
(a)	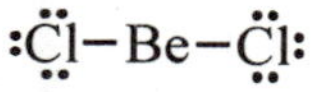 :C̈l–Be–C̈l:	2	0	linear	180°
(b)	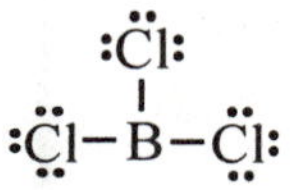 :C̈l–B(–C̈l:)–C̈l:	3	0	trigonal planar	120°
(c)	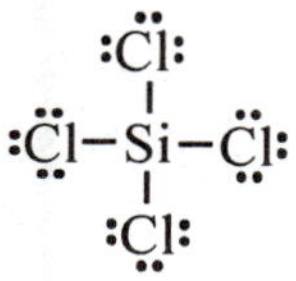 :C̈l–Si(–C̈l:)(–C̈l:)–C̈l:	4	0	tetrahedral	109.5°
(d)	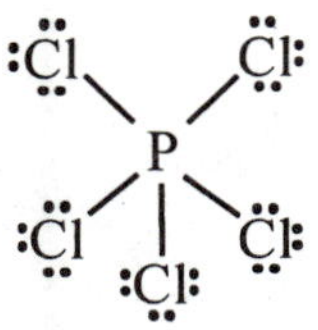 P with five :C̈l:	5	0	trigonal bipyramidal	120°, 90°, 180°
(e)	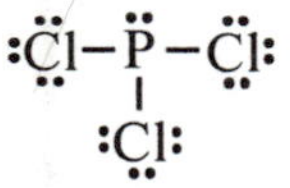 :C̈l–P̈(–C̈l:)–C̈l:	4	1	trigonal pyramidal	<109.5°

	Lewis Structure	VSEPR Pairs	Lone Pairs	Shape	Bond Angles
(f)	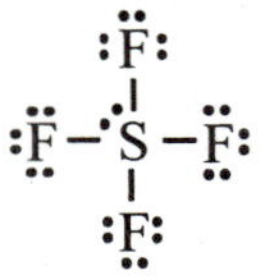 :F: / :F–S–F: / :F:	5	1	bent seesaw	120°, <90°, <180°

10.17 A molecule has a dipole moment if the individual bond dipole moments do not cancel. (e) and (f) have dipole moments.

10.19

	Lewis Structure	VSEPR Pairs	Lone Pairs	Shape	Bond Angles
(a)	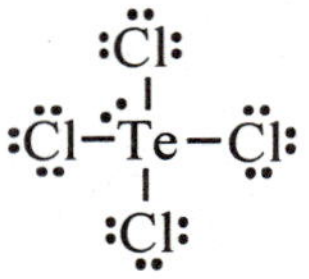 :Cl: / :Cl–Te–Cl: / :Cl:	5	1	bent seesaw	120°, <90°, <180°
(b)	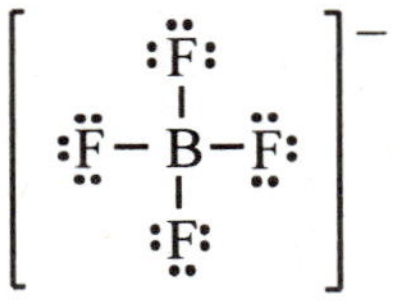 [:F: / :F–B–F: / :F:]$^-$	4	0	tetrahedral	109.5°
(c)	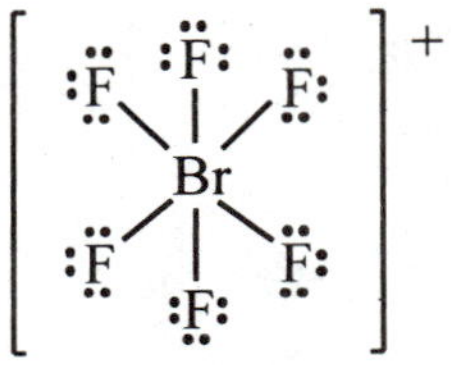 [:F, :F:, F: / Br / :F, :F:, F:]$^+$	6	0	octahedral	90°, 180°
(d)	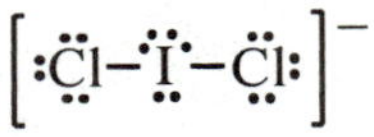 [:Cl–I–Cl:]$^-$	5	3	linear	180°
(e)	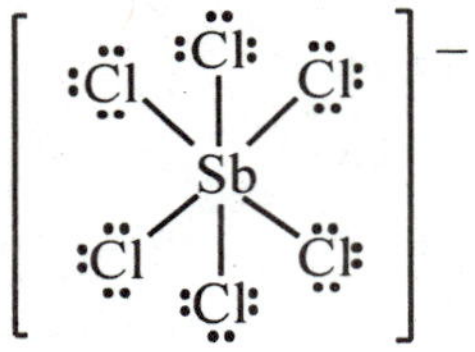 [:Cl, :Cl:, Cl: / Sb / :Cl, :Cl:, Cl:]$^-$	6	0	octahedral	90°, 180°
(f)	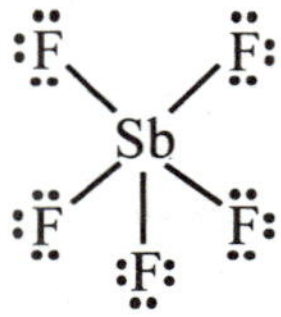 :F, F: / Sb / :F, :F:, F:	5	0	trigonal bipyramidal	120°, 90°, 180°

10.21

	Lewis Structure	VSEPR Pairs	Lone Pairs	Shape	Bond Angles
(a)	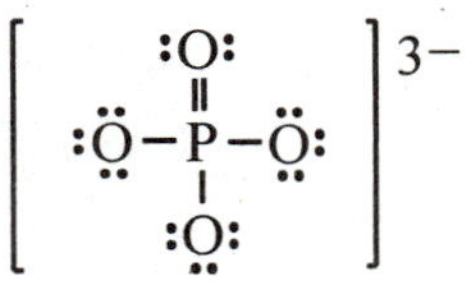 [:O: / ‖ / :O–P–O: / :O:]$^{3-}$	4	0	tetrahedral	109.5°
(b)	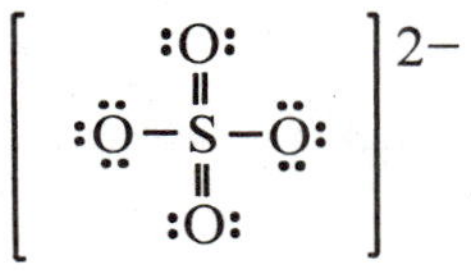 [:O: / ‖ / :O–S–O: / ‖ / :O:]$^{2-}$	4	0	tetrahedral	109.5°

	Lewis Structure	VSEPR Pairs	Lone Pairs	Shape	Bond Angles
(c)	$[\text{:}\ddot{F}\text{–}\ddot{I}(\text{–}\ddot{F}\text{:})_2\text{–}\ddot{F}\text{:}]^{+}$	5	1	bent seesaw	120°, <90°, <180°
(d)	$[\text{:}\ddot{O}\text{–}C(=\ddot{O})\text{–}\ddot{O}\text{:}]^{2-}$	3	0	trigonal planar	120°

10.23

	Lewis Structure	VSEPR Pairs	Lone Pairs	Shape	Bond Angles
(a)	$H\text{–}\ddot{O}\text{–}H$	4	2	bent	<109.5°
	$[H\text{–}\ddot{O}(\text{–}H)\text{–}H]^{+}$	4	1	trigonal pyramidal	<109.5°
(b)	$[\text{:}\ddot{O}=N=\ddot{O}\text{:}]^{+}$	2	0	linear	180°
	$[\text{:}\ddot{O}=\ddot{N}\text{–}\ddot{O}\text{:}]^{-}$	3	1	bent	<120°
(c)	$\ddot{O}=S(=\ddot{O}\text{:})=\ddot{O}\text{:}$	3	0	trigonal planar	120°
	$[\ddot{O}=\ddot{S}(\text{–}\ddot{O}\text{:})\text{–}\ddot{O}\text{:}]^{2-}$	4	1	trigonal pyramidal	<109.5°

10.25

1. The carbon-carbon bond is stable. This factor enables carbon atoms to form long chains containing hundreds of atoms.
2. Carbon has four valence electrons and, therefore, can form four covalent bonds in compounds. This factor enables carbon atoms to form an unending array of branched chains of varying lengths and complexity and rings (single, multiple, and substituted) of different sizes.
3. The carbon atom is small in size. This factor enables the unhybridized valence p orbitals of the carbon atom to participate in π bonding, forming both double, C=C, and triple, C≡C, bonds with other carbon atoms, thus allowing for whole new categories of compounds containing these unsaturated bonds. In addition, carbon atoms bond with most of the elements in the periodic table.

10.27 Alkanes are hydrocarbons containing only carbon-carbon single bonds. An alkene is a hydrocarbon that contains a carbon-carbon double bond. And an alkyne is a hydrocarbon that contains a carbon-carbon triple bond.

10.29 Isomers either have different bonding patterns (structural isomers) or different relative positions of the bonds, atoms, or groups of atoms (geometrical and optical isomers) in their molecules, which leads them to have different properties. Different conformations are not isomers, because the bonding patterns are the same and the relative positions of the bonds are the same in each conformer. In conformers, the relative positions of the atoms are only temporarily different, because the molecules are continually twisting, bending, and rotating their parts. Consequently, individual conformations only exist for a very short time, and on the average all the conformers or molecules have identical properties.

10.31 The bonds are arranged or oriented tetrahedrally around each of the carbon atoms. All the bond angles have the tetrahedral value of 109.5°.

10.33 Because of the lack of free rotation around a double bond, and the trigonal planar arrangement of the sigma bonds attached to each of the carbon atoms participating in the double bond, the doubly bonded carbon atoms and the atoms directly bonded to them, six in total, all lie in a plane. Therefore, the effect a double bond has on the geometry of a hydrocarbon molecule is to introduce a plane or flat, rigid region into the molecule.

10.35 (a)

```
                                 H
                                 |
                               H-C-H
    H H H H                   H  |  H
    | | | |                   |  |  |
  H-C-C-C-C-H               H-C--C--C-H
    | | | |                   |  |  |
    H H H H                   H  H  H

    n-butane                  isobutane
```

(b)

```
                                                       H
                                                       |
                                                     H-C-H
    H H H H            H H H H                         |  H
    | | | |            | | | |                         |  |
  H-C=C-C-C-H        H-C-C=C-C-H                   H-C=C--C-H
        | |            |     |                       | |  |
        H H            H     H                       H H  H

    1-butene           2-butene                    isobutene
```

10.37 The carbon skeletons are:

```
1.  C-C-C-C-C-C

        C
        |
2.  C-C-C-C-C

      C C
      | |
3.  C-C-C-C

        C
        |
4.  C-C-C-C
        |
        C

          C
          |
5.  C-C-C-C-C
```

10.39 (a)

```
H3C       CH2-CH3          H3C       H
   \     /                    \     /
    C = C                      C = C
   /     \                    /     \
  H       H                  H       CH2-CH3

     cis                        trans
```

(b) Do not exist.

10.41 Refer to structural formulas in Section 10.3 and Table 10.1 in the textbook.

10.43 A sigma bond is a shared pair of electrons, the electron density of which is cylindrically symmetric about the line of atomic centers (bond axis). Sigma bonds form when two atomic orbitals, each on a different atom and each with cylindrical symmetry about the line joining the nuclei, overlap to build up electron density along the axis between the two nuclei.

Refer to Figure 10.17 in the textbook.

10.45 Hybrid orbitals account for the bond angles; i.e., they give a good approximation of the observed bond angles.

10.47 See Figure 10.19 in the textbook.

10.49 (a) sp: 180°; linear (b) sp^2: 120°; trigonal planar (c) sp^3: 109.5°; tetrahedral

10.51 A pi bond is a shared pair of electrons, the electron density of which lies above and below the line connecting the atomic centers (bond axis). Pi bonds are not cylindrically symmetric like sigma bonds, but do have or possess planar symmetry. Pi bonds form when two parallel p orbitals, each on a different atom and each perpendicular to the line joining the nuclei, overlap to build up electron density above and below the axis between the two nuclei.

Refer to Figure 10.26 in the textbook.

10.53 $\ddot{N}\equiv\ddot{N}$

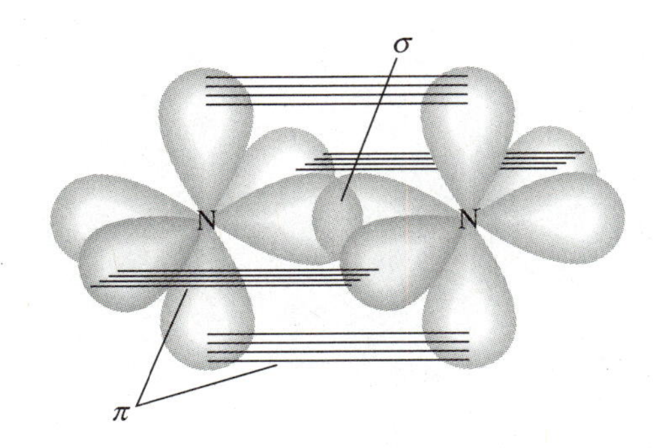

10.55 The number of hybrid orbitals used by a central atom is equal to the number of VSEPR electron pairs around it.

(a)

Central Atom	VSEPR pairs	Type of Hybridization
(a) Be	2	sp
(b) B	3	sp^2
(c) Si	4	sp^3
(d) P	5	sp^3d
(e) P	4	sp^3
(f) S	5	sp^3d

(b)

Central Atom	VSEPR pairs	Type of Hybridization
(a) S	4	sp^3
(b) C	2	sp
(c) I	6	sp^3d^2
(d) O	4	sp^3
(e) Sn	3	sp^2
(f) N	2	sp

10.57 (a) N has four VSEPR electron pairs; these will be tetrahedrally oriented and the hybridization is sp^3.

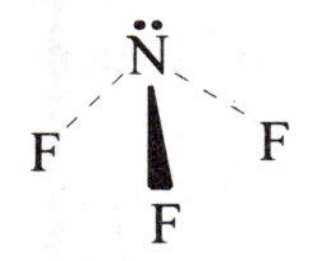

trigonal pyramidal

(b) C has four VSEPR electron pairs; these will be tetrahedrally oriented and the hybridization is sp^3.

tetrahedral

(c) C has two VSEPR electron pairs; these will be linearly oriented and the hybridization is sp.

H−C≡N: linear

10.59 (a) NF_3: CCl_4:

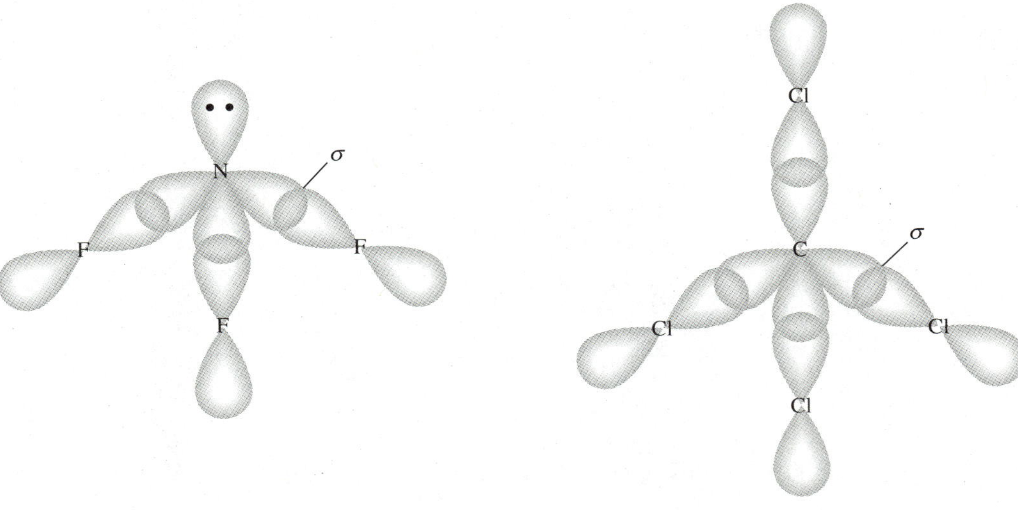

HCN:

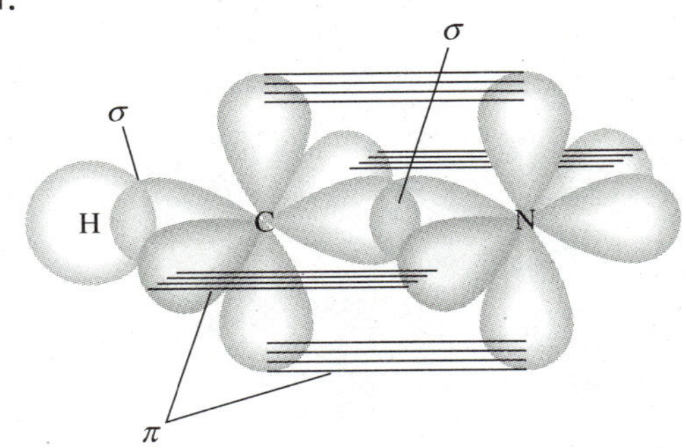

(b) N has one lone pair in an sp^3 orbital.

10.61 S in period 3 has 3d orbitals available for bonding purposes in its valence shell. Therefore, the octet limit of four covalent bonds does not apply to S and S forms many compounds, like SF_6, that have more than four covalent bonds.

O in period 2, on the other hand, has no available d orbitals in its valence shell. It can only form bonds with the s and p orbitals and is therefore limited to an octet or four covalent bonds. Consequently, it cannot form molecules like OF_6 that have more than four covalent bonds.

10.63 (a) ethane (C_2H_6):

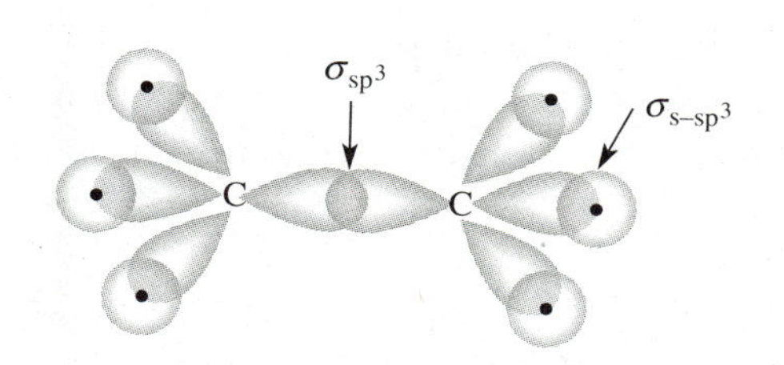

ethene (C_2H_4):

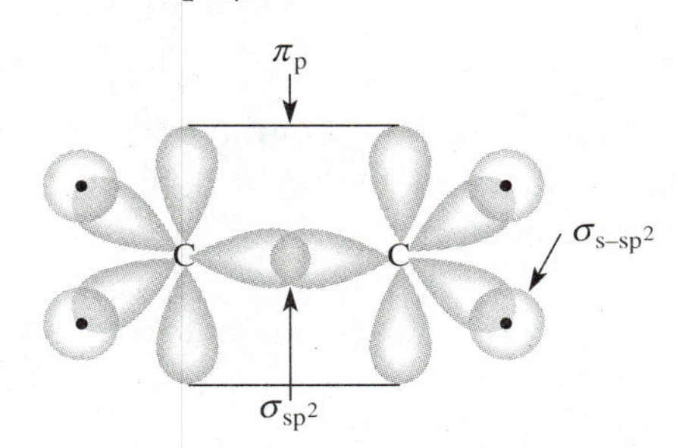

acetylene (C_2H_2):

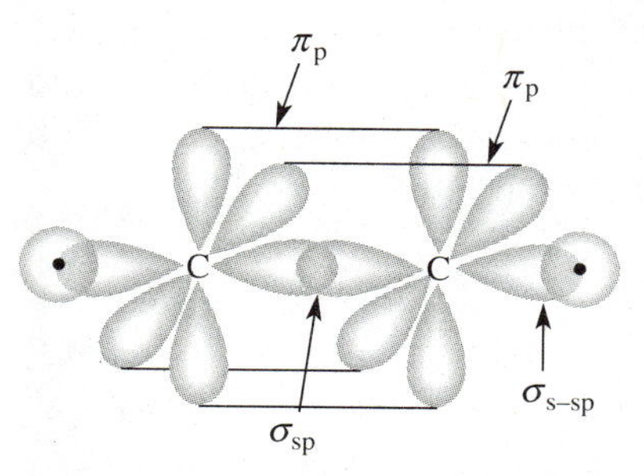

(b) Acetylene with its triple bond will have the greatest carbon-carbon bond energy and shortest carbon-carbon bond length.

10.65 (a)

```
   ┌─ sp3 ─┐
   ↓       ↓ ,H
   :Ö──────Ö:
  /
 H
```

Both H–O–O angles are < 109.5°.

(b)

```
   H   H
   |   |   ..
 H-C — C - O - H
   |   |   ..
   H ↖ H ↖   ↖
        \___\___\______ sp3
```

All angles are approximately 109.5°. The C–O–H angle is slightly smaller than the others.

(c)

```
        sp3
   H  ↙ ↓ ↘  H
   |         |
 H-C — N — C-H
   |   |   |
   H   H   H
```

All angles are approximately 109.5°. The C–N–C and C–N–H angles are slightly smaller than the others.

(d)

```
               H      sp3
               |  ↙
 H   1     2  3C - H
  \       /     \
   C ═ C         H
  /     \
 H  ↑↑   H
    sp2
```

C^1: H–C^1–H angle is < 120°; both H–C^1–C^2 angles are > 120°.

C^2: C^1–C^2–C^3 angle is > 120°; C^1–C^2–H angle is > 120°; C^3–C^2–H angle is < 120°.

C^3: All H–C^3–H angles are 109.5°; all H–C^3–C^2 angles are 109.5°.

(e)

```
      ↙────↙──── sp3
   ..    ..
 H-N  —  N-H
   |     |
   H     H
```

All angles are < 109.5°.

10.67 See Figures 10.33 and 10.36 in the textbook.

10.69 (a) Sigma bonding orbitals have a lower energy than sigma antibonding orbitals.
(b) Sigma bonding orbitals have a buildup of electron density between the nuclei and along the line of atomic centers, while sigma antibonding orbitals have a buildup of electron density beyond the nuclei and along the line of atomic centers.

10.71 A sigma bond is a chemical bond or shared pair of electrons. A sigma orbital, on the other hand, is one of the allowed electron states in a molecule. Sigma orbitals can be either bonding or antibonding in nature, while a sigma bond is only bonding in nature.

10.73 Bonding electrons spend most of their time between the nuclei; the electrostatic attraction between the bonding electrons and the nuclei keep the molecule together. Antibonding electrons spend most of their time beyond the equilibrium distance of the nuclei; the electrostatic attraction between the antibonding electrons and the nuclei pulls the nuclei further apart, thus destabilizing the molecule.

10.75 See Figure 10.38 in the textbook.

10.77

1. (a) B_2 has ten electrons; its molecular orbital configuration is $(\sigma_{1s})^2 (\sigma^*_{1s})^2 (\sigma_{2s})^2 (\sigma^*_{2s})^2 (\pi_{2px})^1 (\pi_{2py})^1$. B_2 has six bonding electrons and four antibonding electrons. The bond order is (6 – 4)/2 = 1.
(b) Yes. B_2 will be stable, since it has a bond order of 1.
(c) Yes. B_2 is paramagnetic, since it has two unpaired electrons.

2. (a) C_2 has 12 electrons; its molecular orbital configuration is $(\sigma_{1s})^2 (\sigma^*_{1s})^2 (\sigma_{2s})^2 (\sigma^*_{2s})^2 (\pi_{2px})^2 (\pi_{2py})^2$. C_2 has eight bonding electrons and four antibonding electrons. The bond order is (8 – 4)/2 = 2.
(b) Yes. C_2 will be stable, since it has a bond order of 2.
(c) No. C_2 has no unpaired electrons and, therefore, is not paramagnetic.

3. (a) F_2 has 18 electrons; its molecular orbital configuration is $(\sigma_{1s})^2 (\sigma^*_{1s})^2 (\sigma_{2s})^2 (\sigma^*_{2s})^2 (\pi_{2px})^2 (\pi_{2py})^2 (\sigma_{2pz})^2 (\pi^*_{2px})^2 (\pi^*_{2py})^2$. F_2 has ten bonding electrons and eight antibonding electrons. The bond order is (10 – 8)/2 = 1.
(b) Yes. F_2 will be stable, since it has a bond order of 1.
(c) No. F_2 has no unpaired electrons and, therefore, is not paramagnetic.

4. (a) Ne_2 has 20 electrons; its molecular orbital configuration is $(\sigma_{1s})^2 (\sigma^*_{1s})^2 (\sigma_{2s})^2 (\sigma^*_{2s})^2 (\pi_{2px})^2 (\pi_{2py})^2 (\sigma_{2pz})^2 (\pi^*_{2px})^2 (\pi^*_{2py})^2 (\sigma^*_{2pz})^2$. Ne_2 has ten bonding electrons and ten antibonding electrons. The bond order is (10 – 10)/2 = 0.
(b) No. Ne_2 will be unstable, since it has a bond order of zero.
(c) No. If Ne_2 existed, it would not be paramagnetic, since it would have no unpaired electrons.

10.79 (a)

Molecule	Molecular Orbital Configuration	Bond Order
O_2	$(\sigma_{2s})^2 (\sigma^*_{2s})^2 (\pi_{2px})^2 (\pi_{2py})^2 (\sigma_{2pz})^2 (\pi^*_{2px})^1 (\pi^*_{2py})^1$	2
O_2^{2-}	$(\sigma_{2s})^2 (\sigma^*_{2s})^2 (\pi_{2px})^2 (\pi_{2py})^2 (\sigma_{2pz})^2 (\pi^*_{2px})^2 (\pi^*_{2py})^2$	1
O_2^-	$(\sigma_{2s})^2 (\sigma^*_{2s})^2 (\pi_{2px})^2 (\pi_{2py})^2 (\sigma_{2pz})^2 (\pi^*_{2px})^2 (\pi^*_{2py})^1$	1 1/2
O_2^+	$(\sigma_{2s})^2 (\sigma^*_{2s})^2 (\pi_{2px})^2 (\pi_{2py})^2 (\sigma_{2pz})^2 (\pi^*_{2px})^1$	2 1/2

(b) Most stable: O_2^+; least stable: O_2^{2-}. O_2^+ has the highest bond order, and O_2^{2-} has the lowest bond order.
(c) O_2, O_2^-, and O_2^+ are paramagnetic.

10.81 (a) 1. N_2^+ has 13 electrons; its molecular orbital configuration is $(\sigma_{1s})^2 (\sigma_{1s}^*)^2 (\sigma_{2s})^2 (\sigma_{2s}^*)^2 (\pi_{2px})^2 (\pi_{2py})^2 (\sigma_{2pz})^1$.

2. N_2^+ has nine bonding electrons and four antibonding electrons. The bond order is (9 – 4)/2 = 2 1/2.

(b) Yes. B_2 is a free radical.

10.83 (a) HeH; yes. HeH has three electrons; its molecular orbital configuration is $(\sigma_{1s})^2 (\sigma_{1s}^*)^1$. The bond order is (2 – 1)/2 = 1/2. Since the bond order is greater than zero, HeH should be stable.

(b) HeH^-; no. HeH^- has four electrons; its molecular orbital configuration is $(\sigma_{1s})^2 (\sigma_{1s}^*)^2$. The bond order is (2 – 2)/2 = 0. HeH^- should be unstable, since it has a bond order of zero.

10.85 (a) 1. CN^- has 14 electrons; its ground-state configuration is $(\sigma_{1s})^2 (\sigma_{1s}^*)^2 (\sigma_{2s})^2 (\sigma_{2s}^*)^2 (\pi_{2px})^2 (\pi_{2py})^2 (\sigma_{2pz})^2$.

2. CN^- has ten bonding electrons and four antibonding electrons. The bond order is (10 – 4)/2 = 3.

(b) 1. BN has twelve electrons; its ground-state configuration is $(\sigma_{1s})^2 (\sigma_{1s}^*)^2 (\sigma_{2s})^2 (\sigma_{2s}^*)^2 (\pi_{2px})^2 (\pi_{2py})^2$.

2. BN has eight bonding electrons and four antibonding electrons. The bond order is (8 – 4)/2 = 2.

10.87 Valence bond theory assumes that each pair of bonding electrons spends most of its time located between two atoms, while molecular orbital theory assumes that the bonding electrons in a molecule are distributed over bonding and antibonding molecular orbitals, which can be spread over several atoms or the entire molecule.

The valence bond theory provides the simplest explanation of Lewis structures with localized bonds. This is because sigma bonds are easy to visualize.

The molecular orbital theory best accounts for resonance structures with delocalized bonds. This is because molecular orbitals may encompass several atoms or the entire molecule, thus allowing or providing for the delocalization of the electrons.

10.89 (a) Each carbon atom in ethene (C_2H_4) is sp^2 hybridized. Therefore, all the atoms lie in the same plane and the sigma bond framework with all atoms lying in the plane of the paper is

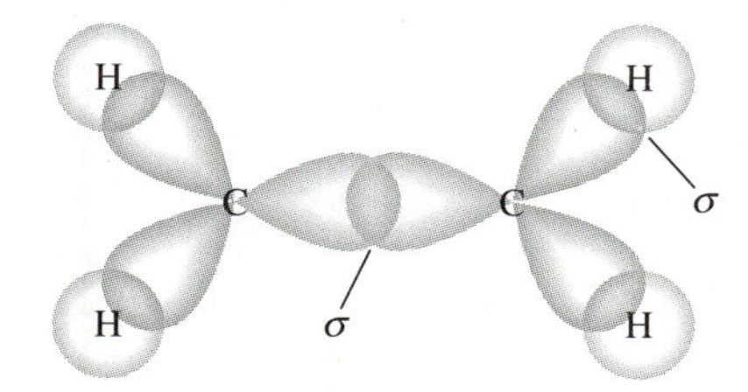

The remaining unhybridized p orbitals, one on each carbon atom, are perpendicular to the molecular plane and overlap to form one bonding (π) and one antibonding (π*) pi molecular orbital.

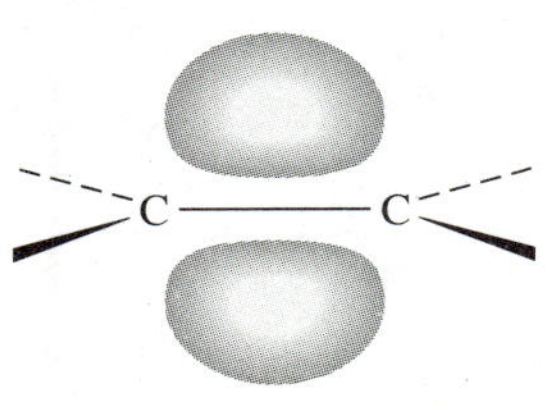

π molecular orbital

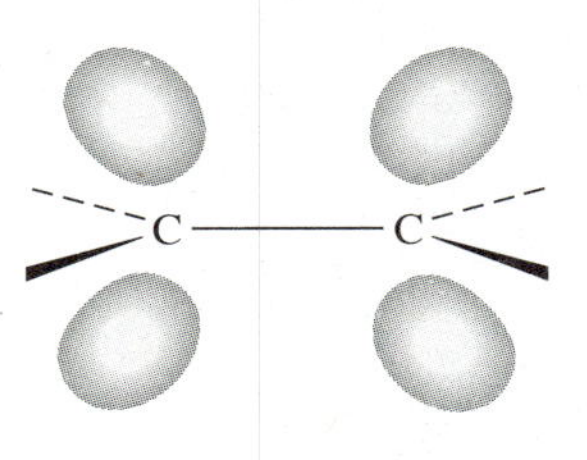

π* molecular orbital

(b) The Lewis structure for formaldehyde is

$$\mathrm{H_2C{=}\ddot{O}\!:}$$

The central carbon atom has three VSEPR electron pairs and, therefore, uses sp^2 hybrid orbitals to form the sigma bonding skeleton or framework. The molecule is therefore planar with a trigonal planar geometry around the carbon atom. The sigma bond framework with all atoms lying in the plane of the paper is

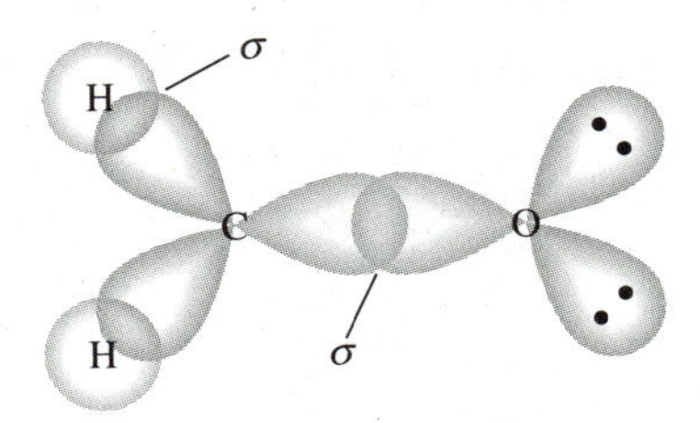

(Oxygen can be portrayed as either unhybridized or sp^2 hybridized, depending on your viewpoint.)

The remaining unhybridized 2p orbital on carbon is perpendicular to the molecular plane and overlaps with a similarly situated 2p orbital on oxygen forming bonding and antibonding pi molecular orbitals.

π molecular orbital $\qquad$ π^* molecular orbital

10.91 (a) All four of 1,3-butadiene's carbon atoms are sp^2 hybridized.

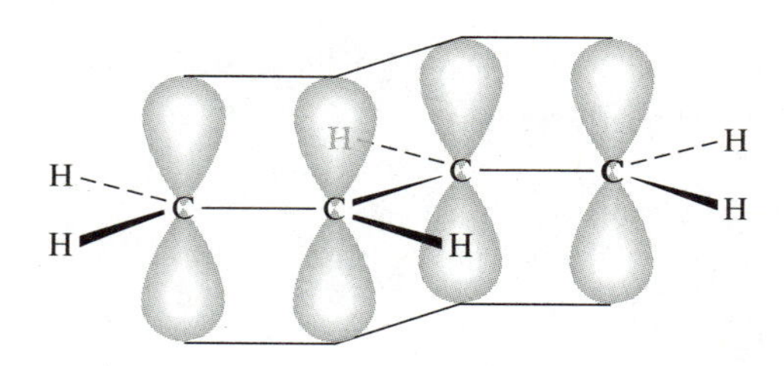

(b) Four.

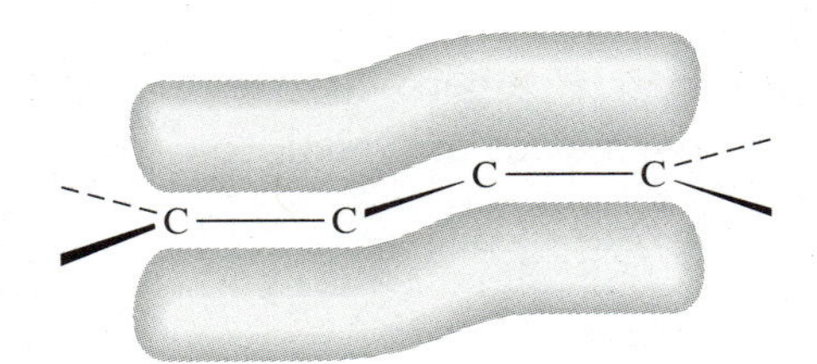

(c) 1,3-Butadiene has 6(1) + 4(4) = 22 valence electrons. The sigma-bonded skeleton accounts for 18 of the 22 valence electrons. Therefore, the remaining 4 electrons must be in the delocalized pi system. Since there are four pi molecular orbitals in the delocalized pi system and each pi orbital can only hold 2 electrons, the two lowest energy pi orbitals must be doubly occupied. The two higher energy ones are empty.

10.93 (a) O_3 with 3(6) = 18 valence electrons has the contributing Lewis structures

:Ö–Ö=Ö ⟷ Ö=Ö–Ö:

The central oxygen atom has three VSEPR electron pairs and, therefore, uses sp^2 hybrid orbitals to form the σ bonding skeleton or framework. The remaining 2p orbital not used in the sp^2 hybrid set is perpendicular to the molecular plane of the molecule and overlaps with similarly situated 2p orbitals on the other oxygen atoms forming delocalized pi molecular orbitals. (Four of the valence electrons are in the sigma bonds, ten more are lone pairs in nonbonding orbitals, and four are in the pi molecular orbitals.)

(b) CO_3^{2-} with 4 + 3(6) + 2 = 24 valence electrons has the following contributing Lewis structures:

[:Ö–C(=Ö:)–Ö:]$^{2-}$ ⟷ [:Ö=C(–Ö:)–Ö:]$^{2-}$ ⟷ [:Ö–C(–Ö:)=Ö:]$^{2-}$

The CO_3^{2-} ion is triangular planar, and each O–C–O bond angle is 120°. The central C atom uses sp^2 hybrid orbitals to form the σ bonding skeleton or framework. The remaining carbon 2p orbital is perpendicular to the plane of the ion and overlaps with similarly situated 2p orbitals on each of the oxygen atoms forming delocalized pi molecular orbitals. (Six of the valence electrons are in the sigma bonds, 12 more are lone pairs in nonbonding orbitals, and six are in the pi molecular orbitals.)

10.95 In orthodichlorobenzene the two C–Cl bond dipole moments reinforce each other and, therefore, the molecule has a net dipole moment. In paradichlorobenzene the two C–Cl bond dipole moments oppose each other and cancel each other, giving the molecule a zero net dipole moment. Consequently, a dipole moment measurement could distinguish between the two isomers.

10.97

1. Lone pairs on the central atom cause bond angles to be somewhat smaller (compressed) than predicted, because lone pairs are more repulsive than bonding pairs, since they possess a greater charge density.
2. Double bonds on the central atom cause the bond angles containing the double bond to be somewhat larger (expanded) than predicted, and bond angles not containing the double bond to be somewhat smaller (compressed) than predicted. Double bonds contain more electrons than single bonds, and their greater repulsive force widens the bond angle. A double bond has a similar effect on bond angles (not containing the double bond) as a lone electron pair.
3. Slight distortions of the bond angles occur if the bonded atoms are not all the same. For instance, if some of the bonded atoms are much larger than others, or if some of the bonded atoms have a much greater electronegativity than others, then we can expect the actual bond angles to be somewhat different from the ideal VSEPR bond angles.

Nonpolar molecules should have bond angles that are the same as the ideal angles. Some examples are: methane (CH_4), carbon tetrachloride (CCl_4), sulfur trioxide (SO_3), carbon dioxide (CO_2), phosphorus pentachloride (PCl_5), and sulfur hexafluoride (SF_6).

10.99 (a) Yes. The VSEPR predicted structures are the ones that minimize the repulsions between the electron pairs. This IF_7 pentagonal bipyramidal structure is one of the predicted structures that minimizes the repulsions for seven electron pairs.

(b) The five equatorial F atoms lie in a plane and point to the corners of a regular pentagon. Since the five equal F–I–F angles must add up to 360°, each F–I–F angle is 360°/5 = 72°.

10.101 (a)

```
   H   H H   H
 H  \ /   \ / H
    C-----C
   /       \
  C         C
   \       /
 H  C-----C   H
   / \   / \
  H   H H   H
```

(b) Puckered. Each C atom in the ring is sp^3 hybridized. Therefore, the carbon atoms cannot all lie in the same plane, but must zigzag up and down.

10.103 In a free atom there is no evidence for an electron occupying anything having the properties of a hybrid orbital. However, experimentally determined bond angles and molecular shapes strongly indicate that something akin to hybrid orbitals exists in covalently bonded substances. Or, to put it another way, the experimental data are best explained in terms of hybrid orbitals.

10.105 (a) Since the nitrogen and central carbon (bonded to oxygen) atoms are sp^2 hybridized, they both have 2p orbitals perpendicular to the plane of the peptide linkage. These two p orbitals and a similarly situated 2p orbital on oxygen overlap forming delocalized pi molecular orbitals. There is no rotation around the central C–N bond, since rotation would destroy the delocalized pi molecular orbital structure. This may be represented by the resonance structures

```
  \      ..  C            \    (+1) C
   C — N                   C = N
  //     \                /      \
:O:       H            :O:        H
                      (-1)
```

(b) Since both central C (bonded to O) and central N are sp^2 hybridized, all bond angles should be around 120°.

10.107 (a) S_8 has 8(6) = 48 valence electrons or 24 electron pairs.

```
     S — S — S
    /         \
 :S:           :S:
    \         /
     S — S — S
```

(b) Since the S atoms are sp^3 hybridized, S_8 cannot be planar. The S–S–S tetrahedral bond angles requires that S_8 take on the shape of a puckered ring having a crownlike structure.

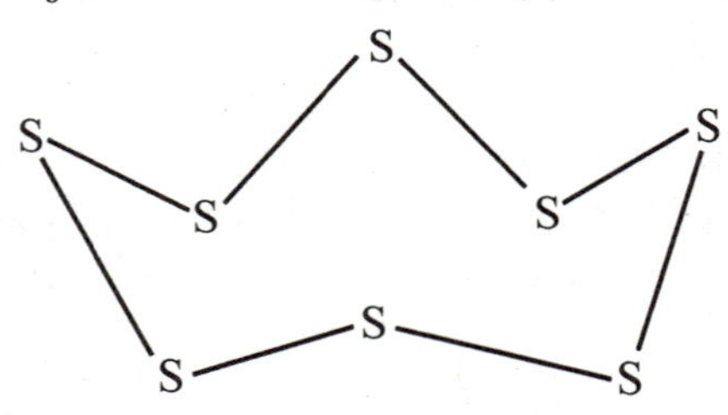

10.109 (a) H_2. The orbital occupancy diagram for H_2 is

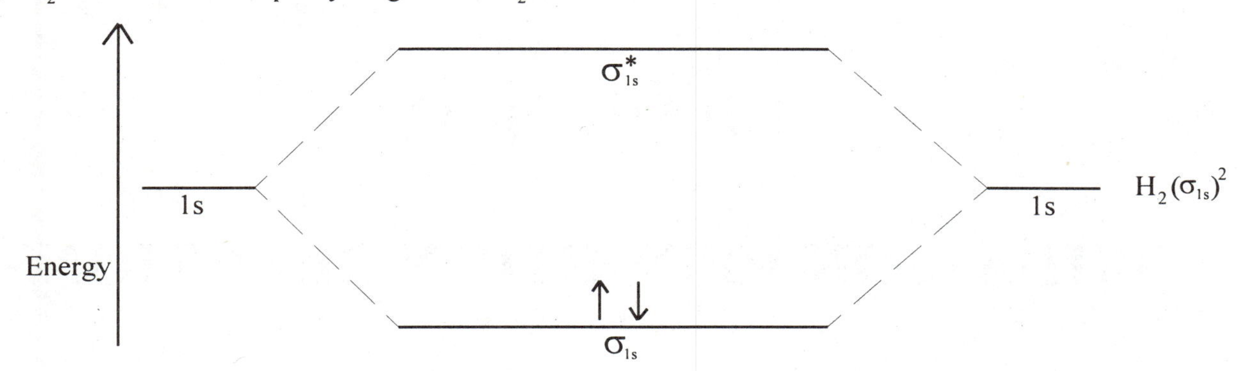

From the diagram we see that in the ground state both of the H_2 electrons are lower in energy than the ground-state hydrogen atom electron. Therefore, H_2 should have a greater first ionization energy, since it would require more energy to remove one of its ground-state electrons.

(b) O. The orbital occupancy diagram for O_2 is given in the textbook (Figure 10.39). Referring to Figure 10.39 we see that in the ground state O_2 has two electrons that are higher in energy than any of those in the ground state of the oxygen atom (O). Therefore, O should have the greater first ionization energy.

10.111 (a)

H   1   2   3   H
  C=C=C
H   H

The central carbon atom (C^2) has two VSEPR pairs; these will be linearly oriented and the hybridization is sp. The two remaining unhybridized 2p orbitals on C^2 are situated perpendicular to one another. C^1 has three VSEPR pairs and is sp^2 hybridized. It uses its three sp^2 orbitals to form two sigma bonds with hydrogen and a sigma bond with C^2. Its remaining 2p orbital forms a pi bond with one of the remaining 2p orbitals on C^2. Likewise, C^3 is sp^2 hybridized and forms two sigma bonds with hydrogen and a sigma bond with C^2. Its remaining 2p orbital forms a pi bond with the other remaining 2p orbital of C^2. Since the 2p orbitals on C^2 are situated perpendicular to one another, the two pi bonds are also situated perpendicular to one another and, therefore, do not coalesce into one extended delocalized system, but remain as separate pi electron systems or bonds. Therefore, the allene molecule has two distinct double bonds, instead of one delocalized pi electron system.

(b)

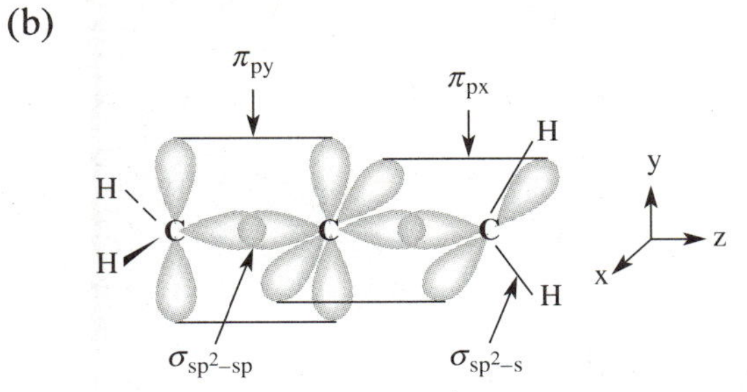

Both C^1 and C^3 have trigonal planar geometries. However, since the pi bond formed between C^1 and C^2 is situated perpendicular to the pi bond formed between C^2 and C^3, these two trigonal planar geometries must also be situated perpendicular to one another. Consequently, the two sets of hydrogen atoms must lie in different, perpendicular planes.

CHAPTER 11

OXIDATION-REDUCTION REACTIONS

Solutions To Practice Exercises

PE 11.1 $Zn(s) + 2H^+(ag) \rightarrow Zn^{2+}(aq) + H_2(g)$

(a) $Zn(s)$ is oxidized to $Zn^{2+}(aq)$ and $H^+(aq)$ is reduced to $H_2(g)$. The half-reactions are

$Zn(s) \rightarrow Zn^{2+}(aq) + 2e^-$ (oxidation)

$2H^+(aq) + 2e^- \rightarrow H_2(g)$ (reduction)

(b) $H^+(aq)$ accepts electrons and is the oxidizing agent. $Zn(s)$ gives up electrons and is the reducing agent.

PE 11.2 (a) The oxidizing agent is reduced. (b) The reducing agent is oxidized.

PE 11.3 x_O is –2 (Rule 4) and x_{Cl} is –1 (Rule 5). x_C is calculated as follows:

$x_C + x_O + 2x_{Cl} = 0$ (Rule 2)

$x_C - 2 + 2(-1) = 0$

and $x_C = +4$

PE 11.4 (a) The sum of the oxidation numbers is –1 (Rule 2) and x_O is –2 (Rule 4). Hence,

$x_N + 3x_O = -1$

$x_N + 3(-2) = -1$

and $x_N = +5$

(b) The sum of the oxidation numbers is –1 (Rule 2) and x_O is –2 (Rule 4). Hence,

$x_N + 2x_O = -1$

$x_N + 2(-2) = -1$

and $x_N = +3$

(c) The sum of the oxidation numbers is –1 (Rule 2) and x_N is –3 (Rule 6). Hence,

$x_C + x_N = -1$

$x_C - 3 = -1$

and $x_C = +2$

PE 11.5 $x_{Na} = +1$ (Rule 3) and $x_O = -2$ (Rule 4).

$2x_{Na} + 4x_S + 6x_O = 0$

$2(+1) + 4x_S + 6(-2) = 0$

and $x_S = +10/4$ or $+2.5$

PE 11.6 In each compound $x_H = +1$ (Rule 3) and $x_O = -2$ (Rule 4).

$HBrO_3$: $x_H + x_{Br} + 3x_O = 0$

$+1 + x_{Br} + 3(-2) = 0$

and $x_{Br} = +5$

HBr: $x_{Br} = -1$

$HBrO_4$: $x_H + x_{Br} + 4x_O = 0$

$+1 + x_{Br} + 4(-2) = 0$

and $x_{Br} = +7$

$HBrO_2$: $x_H + x_{Br} + 2x_O = 0$

$+1 + x_{Br} + 2(-2) = 0$

and $x_{Br} = +3$

The order of increasing oxidation number of bromine is

$HBr < HBrO_2 < HBrO_3 < HBrO_4$

PE 11.7 P goes from +3 in PCl_3 to +5 in PCl_5. The increase in oxidation number signifies that P is oxidized. It is also correct to say that PCl_3 is oxidized.

PE 11.8 $2FeCl_3(aq) + 3H_2S(aq) \rightarrow 2FeS(s) + S(s) + 6HCl(aq)$

Fe goes from +3 in $FeCl_3$ on the reactant side to +2 in FeS on the product side. Therefore, $FeCl_3$ is reduced. S goes from –2 in H_2S on the reactant side to –2 in FeS and 0 in S on the product side. Therefore, some H_2S is oxidized.

PE 11.9 The oxidation number of S in SO_3 is +6. From Figure 11.1 we see that +6 is the highest oxidation state of sulfur. So, sulfur trioxide cannot be oxidized, but it can be reduced. Therefore, sulfur trioxide can be used as an oxidizing agent, but not as a reducing agent.

The oxidation number of S in SO_2 is +4. From Figure 11.1 we see that +4 lies in between the highest and lowest oxidation states of sulfur. So, sulfur dioxide can be either oxidized or reduced. Therefore, sulfur dioxide can be used as an oxidizing agent and a reducing agent.

PE 11.10 Step 2: $IO_3^-(aq) \rightarrow I_2(s)$ (unbalanced)

Step 3:

(a) The oxidation state of iodine changes from +5 in IO_3^- to 0 in I_2. The iodine atoms are balanced by putting a 2 in from of the IO_3^- ion:

$2IO_3^-(aq) \rightarrow I_2(s)$

(b) Each iodine atom gains five electrons; hence, ten electrons are added to the reactant side:

$2IO_3^-(aq) + 10e^- \rightarrow I_2(s)$

(c) The solution is acidic, so the charge is balanced by adding twelve H^+ ions to the reactant side:

$2IO_3^-(aq) + 12H^+(aq) + 10e^- \rightarrow I_2(s)$

(d) The hydrogen and oxygen atoms are balanced by adding six H_2O molecules to the product side:

$2IO_3^-(aq) + 12H^+(aq) + 10e^- \rightarrow I_2(s) + 6H_2O(l)$ (balanced)

PE 11.11 (a) 1. $Ag(s) + H^+(aq) + NO_3^-(aq) \rightarrow Ag^+(aq) + NO(g)$

2. $Ag(s) \rightarrow Ag^+(aq)$ (oxidation)
$NO_3^-(aq) \rightarrow NO(g)$ (reduction)

3. $Ag(s) \rightarrow Ag^+(aq) + e^-$ (balanced oxidation)
$NO_3^-(aq) + 4H^+(aq) + 3e^- \rightarrow NO(g) + 2H_2O(l)$ (balanced reduction)

4. $3Ag(s) \rightarrow 3Ag^+(aq) + 3e^-$
$NO_3^-(aq) + 4H^+(aq) + 3e^- \rightarrow NO(g) + 2H_2O(l)$

5. $3Ag(s) + 4H^+(aq) + NO_3^-(aq) \rightarrow 3Ag^+(aq) + NO(g) + 2H_2O(l)$

(b) 6. $3Ag(s) + 4HNO_3(aq) \rightarrow 3AgNO_3(aq) + NO(g) + 2H_2O(l)$

PE 11.12 Step 2: $S_2O_4^{2-}(aq) \rightarrow SO_3^{2-}(aq)$ (unbalanced)
Step 3:

(a) The oxidation state of sulfur changes from +3 in $S_2O_4^{2-}$ to +4 in SO_3^{2-}. The sulfur atoms are balanced by putting a 2 in front of the SO_3^{2-} ion:

$S_2O_4^{2-}(aq) \rightarrow 2SO_3^{2-}(aq)$

(b) Each sulfur atom loses one electron; hence, two electrons are added to the product side:

$S_2O_4^{2-}(aq) \rightarrow 2SO_3^{2-}(aq) + 2e^-$

(c) The solution is basic, so the charge is balanced by adding four OH^- ions to the reactant side:

$S_2O_4^{2-}(aq) + 4OH^-(aq) \rightarrow 2SO_3^{2-}(aq) + 2e^-$

(d) The hydrogen and oxygen atoms are balanced by adding two H_2O molecules to the product side:

$S_2O_4^{2-}(aq) + 4OH^-(aq) \rightarrow 2SO_3^{2-}(aq) + 2H_2O(l) + 2e^-$ (balanced)

PE 11.13 1. $CrO_4^{2-}(aq) + I^-(aq) \rightarrow Cr^{3+}(aq) + IO_3^-(aq)$

2. $I^-(aq) \rightarrow IO_3^-(aq)$ (oxidation)
$CrO_4^{2-}(aq) \rightarrow Cr^{3+}(aq)$ (reduction)

3. $I^-(aq) + 6OH^-(aq) \rightarrow IO_3^-(aq) + 3H_2O(l) + 6e^-$ (balanced oxidation)
$CrO_4^{2-}(aq) + 4H_2O(l) + 3e^- \rightarrow Cr^{3+}(aq) + 8OH^-(aq)$ (balanced reduction)

4. $I^-(aq) + 6OH^-(aq) \rightarrow IO_3^-(aq) + 3H_2O(l) + 6e^-$
$2CrO_4^{2-}(aq) + 8H_2O(l) + 6e^- \rightarrow 2Cr^{3+}(aq) + 16OH^-(aq)$

5. $2CrO_4^{2-}(aq) + I^-(aq) + 5H_2O(l) \rightarrow 2Cr^{3+}(aq) + IO_3^-(aq) + 10OH^-(aq)$

PE 11.14 1. $MnO_4^{2-}(aq) \rightarrow MnO_4^-(aq) + MnO_2(s)$

2. $MnO_4^{2-}(aq) \rightarrow MnO_4^-(aq)$ (oxidation)
$MnO_4^{2-}(aq) \rightarrow MnO_2(s)$ (reduction)

3. $MnO_4^{2-}(aq) \rightarrow MnO_4^-(aq) + e^-$ (balanced oxidation)
$MnO_4^{2-}(aq) + 4H^+(aq) + 2e^- \rightarrow MnO_2(s) + 2H_2O(l)$ (balanced reduction)

4. $2MnO_4^{2-}(aq) \rightarrow 2MnO_4^-(aq) + 2e^-$
$MnO_4^{2-}(aq) + 4H^+(aq) + 2e^- \rightarrow MnO_2(s) + 2H_2O(l)$

5. $3MnO_4^{2-}(aq) + 4H^+(aq) \rightarrow 2MnO_4^-(aq) + MnO_2(s) + 2H_2O(l)$

PE 11.15 (a) 1. $Zn(s) + H_2O(l) \rightarrow Zn(OH)_4^{2-}(aq) + H_2(g)$
2. Zn increases from 0 in Zn to +2 in $Zn(OH)_4^{2-}$. H decreases from +1 in H_2O to zero in H_2.
3. Zn atoms are oxidized from 0 to +2; H atoms are reduced from +1 to 0. Because the total change in oxidation number is zero, two H atoms are reduced for every Zn atom oxidized. Hence, there must be 2 mol of H_2O, since each mole of H_2O supplies just 1 mol of H atoms for reduction, for each mole of Zn on the reactant side of the equation, and 1 mol of H_2 for each mole of $Zn(OH)_4^{2-}$ on the product side:

$Zn(s) + 2H_2O(l) \rightarrow Zn(OH)_4^{2-}(aq) + H_2(g)$

4. By inspection the H and O atoms are balanced by adding two OH^- to the reactant side.

$Zn(s) + 2OH^-(aq) + 2H_2O(l) \rightarrow Zn(OH)_4^-(aq) + H_2(g)$

(b) 1. $Cl_2(g) \rightarrow Cl^-(aq) + ClO_3^-(aq)$

2. Cl goes from zero in Cl_2 to −1 in Cl^-. Cl goes from zero in Cl_2 to +5 in ClO_3^-.

3. Cl atoms are oxidized from 0 to +5; Cl atoms are reduced from 0 to −1. Because the total change in oxidation number is zero, five Cl atoms are reduced for every Cl atom oxidized. Hence, there must be 5 mol of Cl^- for every mole of ClO_3^- on the product side:

$3Cl_2(g) \rightarrow 5Cl^-(aq) + ClO_3^-(aq)$

4. Balance the oxygen in the usual way for a basic solution.

$3Cl_2(g) + 6OH^-(aq) \rightarrow 5Cl^-(aq) + ClO_3^-(aq) + 3H_2O(l)$

PE 11.16 The balanced net ionic equation is

$5Fe^{2+}(aq) + MnO_4^-(aq) + 8H^+(aq) \rightarrow 5Fe^{3+}(aq) + Mn^{2+}(aq) + 4H_2O(l)$

(a) For every mole of MnO_4^- that reacts, 5 mol of Fe^{2+} reacts. Therefore, the mole ratio of MnO_4^- to Fe^{2+} is 1:5.

(b) The problem is solved in the usual way: $V \times M$ (of $KMnO_4$) $\rightarrow$ mol $KMnO_4$ $\rightarrow$ mol $FeSO_4$.

$$0.1000 \text{ L } KMnO_4 \text{ soln} \times \frac{0.02118 \text{ mol } KMnO_4}{1 \text{ L } KMnO_4 \text{ soln}} \times \frac{5 \text{ mol } FeSO_4}{1 \text{ mol } KMnO_4} = 0.01059 \text{ mol } FeSO_4$$

Solutions To Final Exercises

11.1 Oxidation is the increase in oxidation number of an atom or the removal of electrons from one atom or reactant, which are then transferred to another atom or reactant; the atom or reactant accepting the electrons being reduced in the process.

When a substance combines with elemental or atomic oxygen (O_2 or O; $x_O = 0$), that substance gives up or transfers electrons to oxygen. Therefore, the substance is oxidized and oxygen is reduced to the –2 oxidation state.

11.3 (a) When an atom or substance loses an electron, the electron is almost immediately captured by another atom or substance, because bare electrons are too reactive to remain uncombined for any length of time. This process could also be viewed as the simultaneous relinquishing of an electron by one atom or substance and capturing of said electron by another atom or substance. Therefore, oxidation and reduction always occur together, because when one atom or substance gives up an electron, another atom or substance immediately accepts that electron.

(b) Oxygen (O_2)

11.5 A half-reaction is a reaction in which either oxidation only takes place or reduction only takes place. It is the oxidation half or the reduction half of an oxidation-reduction reaction.

The electron always appears in a half-reaction equation but never appears in a net equation.

11.7 (a) $Mg(s) + 2HCl(aq) \rightarrow MgCl_2(aq) + H_2(g)$

1. $Mg(s) \rightarrow Mg^{2+}(aq) + 2e^-$ (oxidation); $2H^+(aq) + 2e^- \rightarrow H_2(g)$ (reduction)
2. $Mg(s) + 2H^+(aq) \rightarrow Mg^{2+}(aq) + H_2(g)$ (net ionic)

(b) $Al(s) + 3AgNO_3(aq) \rightarrow Al(NO_3)_3(aq) + 3Ag(s)$

1. $Al(s) \rightarrow Al^{3+}(aq) + 3e^-$ (oxidation); $Ag^+(aq) + e^- \rightarrow Ag(s)$ (reduction)
2. $Al(s) + 3Ag^+(aq) \rightarrow Al^{3+}(aq) + 3Ag(s)$ (net ionic)

(c) $Zn(s) + Br_2(aq) \rightarrow ZnBr_2(aq)$

1. $Zn(s) \rightarrow Zn^{2+}(aq) + 2e^-$ (oxidation); $Br_2(aq) + 2e^- \rightarrow 2Br^-(aq)$ (reduction)
2. $Zn(s) + Br_2(aq) \rightarrow Zn^{2+}(aq) + 2Br^-(aq)$

11.9 (a)

1. $H^+(aq)$ is the oxidizing agent; Mg(s) is the reducing agent.
2. Mg(s) is oxidized and $H^+(aq)$ is reduced.

(b)

1. $Ag^+(aq)$ is the oxidizing agent; Al(s) is the reducing agent.
2. Al(s) is oxidized and $Ag^+(aq)$ is reduced.

(c)

1. $Br_2(aq)$ is the oxidizing agent; Zn(s) is the reducing agent.
2. Zn(s) is oxidized and $Br_2(aq)$ is reduced.

11.11 $N_2(g) + 3H_2(g) \rightarrow 2NH_3(g)$

Hydrogen ($0 \rightarrow +1$) is oxidized and nitrogen ($0 \rightarrow -3$) is reduced.

11.13 (a) The group number of a main group element is equal to the number of valence electrons that element has. For a main group element to have an oxidation number greater than its group number, electrons from the next lower shell would have to be used in bonding and this is excluded on energy grounds.

(b) From Figure 11.1 we see, excluding the noble gases (Group 8A), that oxygen, fluorine, polonium, and astatine do not exhibit the maximum oxidation number of their group.

11.15 In order for a substance to be or act as an oxidizing agent, it has to contain atoms that are capable of being reduced. This means that it has to contain an atom in an oxidation state above its minimum oxidation state, since reduction is a decrease in oxidation number. Therefore, by examining the oxidation numbers of the atoms in a substance, you can determine if it's a potential oxidizing agent or not. Substances that contain multiple oxidation state atoms in their maximum oxidation state, or one of the higher oxidation states, are capable or have the potential of being either strong or good oxidizing agents. Some of the nonmetallic elements are also good oxidizing agents, since some of them have a strong tendency to exist in a negative oxidation state.

11.17 Because they contain elements that have oxidation numbers that lie in between the maximum and minimum oxidation numbers exhibited by these elements. In the case of H_2O_2 it is oxygen with $x_O = -1$. In SO_2 it is sulfur with $x_S = +4$.

11.19 O has an oxidation number of –2 in each compound.
N_2O_4: $x_N = +4$; N_2O: $x_N = +1$; NO_2: $x_N = +4$; NO: $x_N = +2$; N_2O_3: $x_N = +3$; N_2O_5: $x_N = +5$.
The order of increasing oxidation number of N is: $N_2O < NO < N_2O_3 < NO_2 = N_2O_4 < N_2O_5$.

11.21
(a) $XeOF_2$. $x_O = -2$ (Rule 4) and $x_F = -1$ (Rule 4). $x_{Xe} = +4$ (Rule 2).
(b) H_3BO_3. $x_H = +1$ (Rule 3) and $x_O = -2$ (Rule 4). $x_B = +3$ (Rule 2).
(c) KIF_4. $x_K = +1$ (Rule 3) and $x_F = -1$ (Rule 4). $x_I = +3$ (Rule 2).
(d) BrF_3. $x_F = -1$ (Rule 4) and $x_{Br} = +3$ (Rule 2).
(e) $KSbF_6$. $x_K = +1$ (Rule 3) and $x_F = -1$ (Rule 4). $x_{Sb} = +5$ (Rule 2).
(f) $H_4I_2O_9$. $x_H = +1$ (Rule 3) and $x_O = -2$ (Rule 4). $x_I = +7$ (Rule 2).
(g) $Na_2Sn(OH)_6$. $x_{Na} = +1$ (Rule 3), $x_H = +1$ (Rule 3), and $x_O = -2$ (Rule 4). $x_{Sn} = +4$ (Rule 2).
(h) $KCrO_3Cl$. $x_K = +1$ (Rule 3), $x_O = -2$ (Rule 4), and $x_{Cl} = -1$ (Rule 5). $x_{Cr} = +6$ (Rule 2).
(i) $(NH_4)_2Cr_2O_7$ is $2NH_4^+ + Cr_2O_7^{2-}$. $x_H = +1$ (Rule 3), $x_N = -3$ (Rule 2), $x_O = -2$ (Rule 4), and $x_{Cr} = +6$ (Rule 2).
(j) $(CH_3)_2S$. $x_H = +1$ (Rule 3) and $x_S = -2$ (Rule 6). $x_C = -2$ (Rule 2).

11.23
(a) H–O–O–H. $x_H = +1$ (Rule 3) and $x_O = -1$ (Rule 2).
(b) BaO_2. $x_{Ba} = +2$ (Rule 3) and $x_O = -1$ (Rule 2).
(c) O_2^-. $x_O = -1/2$ (Rule 2).
(d) OF_2. $x_F = -1$ (Rule 4) and $x_O = +2$ (Rule 2).
(e) O_2F_2. $x_F = -1$ (Rule 4) and $x_O = +1$ (Rule 2).

11.25
(a) CO_2. $x_O = -2$ (Rule 4) and $x_C = +4$ (Rule 2).
(b) CH_4. $x_H = +1$ (Rule 3) and $x_C = -4$ (Rule 2).
(c) CH_3OH. $x_H = +1$ (Rule 3), $x_O = -2$ (Rule 4), and $x_C = -2$ (Rule 2).
(d) CCl_4. $x_{Cl} = -1$ (Rule 5) and $x_C = +4$ (Rule 2).
(e) $HCOO^-$. $x_H = +1$ (Rule 3), $x_O = -2$ (Rule 4), and $x_C = +2$ (Rule 2).
(f) C_{60}. $x_C = 0$ (Rule 1).

11.27
(a) No. $x_N = +5$ in N_2O_5 on the reactant side and +5 in HNO_3 on the product side.
(b) Yes. $x_N = +4$ in NO_2 on the reactant side and +5 in HNO_3 and +2 in NO on the product side. NO_2 is both the oxidizing and reducing agent; i.e., some of it gets oxidized and some of it gets reduced.
(c) Yes. $x_S = +4$ in SO_2 on the reactant side and +6 in SO_3 on the product side. O_2 is the oxidizing agent and SO_2 the reducing agent.
(d) Yes. $x_S = -2$ in K_2S on the reactant side and zero in S on the product side. $K_2Cr_2O_7$ is the oxidizing agent and K_2S is the reducing agent.

11.29
(a) The maximum oxidation number of a main group element is equal to its group number. Therefore, arsenic in Group 5A has a maximum oxidation number of +5.
(b) The most negative oxidation state for a nonmetal from groups 4A through 7A is the group number minus eight. Therefore, arsenic's lowest oxidation state is $5 - 8 = -3$.

11.31 A substance can be oxidized if it contains a readily oxidizable atom in an oxidation state that is below its (chemical) maximum oxidation state.

(a) No. $x_N = +5$ and the maximum oxidation state of nitrogen is +5. Hydrogen also has its maximum oxidation state ($x_H = +1$) and oxygen ($x_O = -2$) is not readily oxidizable.
(b) Yes. $x_{Se} = -2$ and this is below its maximum oxidation state of +6.
(c) Yes. $x_{Mn} = +4$ and this is below its maximum oxidation state of +7.
(d) Yes. $x_{Co} = +2$ and this is below its maximum oxidation state of +3.
(e) Yes. $x_I = -1$ and therefore HI can be oxidized.
(f) Yes. $x_P = +3$ and this is below its maximum oxidation state of +5.
(g) No. $x_{Cr} = +6$ and the maximum oxidation state of chromium is +6.
(h) No. $x_{Cl} = +7$ and the maximum oxidation state of chlorine is +7.

11.33 If a substance has an atom whose oxidation state lies in between the highest and lowest oxidation states of that atom, then it can act as both an oxidizing agent and a reducing agent. Otherwise, it has to be one or the other.
(a) HNO_3, CrO_4^{2-}, and $HClO_4$ (b) H_2Se, Co^{2+}, and HI (c) MnO_2 and H_3PO_3

11.35 Refer to Section 11.3 in the textbook for the steps.

11.37 The balanced half-reaction equations and net ionic equations are:

(a) $HSO_3^-(aq) + H_2O(l) \rightarrow HSO_4^-(aq) + 2H^+(aq) + 2e^-$ (oxidation)
$MnO_4^-(aq) + 8H^+(aq) + 5e^- \rightarrow Mn^{2+}(aq) + 4H_2O(l)$ (reduction)
$2MnO_4^-(aq) + 6H^+(aq) + 5HSO_3^-(aq) \rightarrow 2Mn^{2+}(aq) + 5HSO_4^-(aq) + 3H_2O(l)$

(b) $CuS(s) \rightarrow Cu^{2+}(aq) + S(s) + 2e^-$ (oxidation)
$NO_3^-(aq) + 4H^+(aq) + 3e^- \rightarrow NO(g) + 2H_2O(l)$ (reduction)
$3CuS(s) + 2NO_3^-(aq) + 8H^+(aq) \rightarrow 3Cu^{2+}(aq) + 3S(s) + 2NO(g) + 4H_2O(l)$

(c) $H_2S(aq) \rightarrow S(s) + 2H^+(aq) + 2e^-$ (oxidation)
$Fe^{3+}(aq) + H_2S(aq) + e^- \rightarrow FeS(s) + 2H^+(aq)$ (reduction)
$2Fe^{3+}(aq) + 3H_2S(aq) \rightarrow 2FeS(s) + S(s) + 6H^+(aq)$

(d) $H_2S(aq) \rightarrow S(s) + 2H^+(aq) + 2e^-$ (oxidation)
$NO_3^-(aq) + 4H^+(aq) + 3e^- \rightarrow NO(g) + 2H_2O(l)$ (reduction)
$2H^+(aq) + 2NO_3^-(aq) + 3H_2S(aq) \rightarrow 3S(s) + 2NO(g) + 4H_2O(l)$

(e) $Pb(s) + SO_4^{2-}(aq) \rightarrow PbSO_4(s) + 2e^-$ (oxidation)
$PbO_2(s) + SO_4^{2-}(aq) + 4H^+(aq) + 2e^- \rightarrow PbSO_4(s) + 2H_2O(l)$ (reduction)
$Pb(s) + PbO_2(s) + 4H^+(aq) + 2SO_4^{2-}(aq) \rightarrow 2PbSO_4(s) + 2H_2O(l)$

(f) $MnO_2(s) + 2H_2O(l) \rightarrow MnO_4^-(aq) + 4H^+(aq) + 3e^-$ (oxidation)
$PbO_2(s) + 4H^+(aq) + 2e^- \rightarrow Pb^{2+}(aq) + 2H_2O(l)$ (reduction)
$2MnO_2(s) + 3PbO_2(s) + 4H^+(aq) \rightarrow 2MnO_4^-(aq) + 3Pb^{2+}(aq) + 2H_2O(l)$

(g) $I_2(s) + 6H_2O(l) \rightarrow 2IO_3^-(aq) + 12H^+(aq) + 10e^-$ (oxidation)
$ClO^-(aq) + 2H^+(aq) + 2e^- \rightarrow Cl^-(aq) + H_2O(l)$ (reduction)
$5ClO^-(aq) + I_2(s) + H_2O(l) \rightarrow 5Cl^-(aq) + 2IO_3^-(aq) + 2H^+(aq)$

(h) $SO_2(g) + 2H_2O(l) \rightarrow SO_4^{2-}(aq) + 4H^+(aq) + 2e^-$ (oxidation)
$Br_2(aq) + 2e^- \rightarrow 2Br^-(aq)$ (reduction)
$Br_2(aq) + SO_2(g) + 2H_2O(l) \rightarrow 2Br^-(aq) + SO_4^{2-}(aq) + 4H^+(aq)$

11.39 The balanced half-reaction equations and net ionic equations are:

(a) $Al(s) + 4OH^-(aq) \rightarrow Al(OH)_4^-(aq) + 3e^-$ (oxidation)
$2H_2O(l) + 2e^- \rightarrow H_2(g) + 2OH^-(aq)$ (reduction)
$2Al(s) + 2OH^-(aq) + 6H_2O(l) \rightarrow 2Al(OH)_4^-(aq) + 3H_2(g)$

(b) $S^{2-}(aq) \rightarrow S(s) + 2e^-$ (oxidation)
$MnO_4^-(aq) + 2H_2O(l) + 3e^- \rightarrow MnO_2(s) + 4OH^-(aq)$ (reduction)
$2MnO_4^-(aq) + 3S^{2-}(aq) + 4H_2O(l) \rightarrow 2MnO_2(s) + 3S(s) + 8OH^-(aq)$

(c) $Zn(s) + 4OH^-(aq) \rightarrow Zn(OH)_4^{2-}(aq) + 2e^-$ (oxidation)
$NO_3^-(aq) + 6H_2O(l) + 8e^- \rightarrow NH_3(aq) + 9OH^-(aq)$ (reduction)
$4Zn(s) + 7OH^-(aq) + 6H_2O(l) + NO_3^-(aq) \rightarrow 4Zn(OH)_4^{2-}(aq) + NH_3(aq)$

(d) $Pb(OH)_3^-(aq) + OH^-(aq) \rightarrow PbO_2(s) + 2H_2O(l) + 2e^-$ (oxidation)
$OCl^-(aq) + H_2O(l) + 2e^- \rightarrow Cl^-(aq) + 2OH^-(aq)$ (reduction)
$Pb(OH)_3^-(aq) + OCl^-(aq) \rightarrow PbO_2(s) + Cl^-(aq) + OH^-(aq) + H_2O(l)$

11.41 (a) $H_2O_2(aq) + 2H^+(aq) + 2e^- \rightarrow 2H_2O(l)$
(b) $H_2O_2(aq) \rightarrow O_2(g) + 2H^+(aq) + 2e^-$

11.43 (a) $H_2O_2(aq) \rightarrow O_2(g) + 2H^+(aq) + 2e^-$ (oxidation)
$Cr_2O_7^{2-}(aq) + 14H^+(aq) + 6e^- \rightarrow 2Cr^{3+}(aq) + 7H_2O(l)$ (reduction)
$Cr_2O_7^{2-}(aq) + 3H_2O_2(aq) + 8H^+(aq) \rightarrow 2Cr^{3+}(aq) + 3O_2(g) + 7H_2O(l)$

(b) $H_2S(aq) \rightarrow S(s) + 2H^+(aq) + 2e^-$ (oxidation)
$H_2O_2(aq) + 2H^+(aq) + 2e^- \rightarrow 2H_2O(l)$ (reduction)
$H_2S(aq) + H_2O_2(aq) \rightarrow S(s) + 2H_2O(l)$

(c) $H_2O_2(aq) \rightarrow O_2(g) + 2H^+(aq) + 2e^-$ (oxidation)
$MnO_4^-(aq) + 8H^+(aq) + 5e^- \rightarrow Mn^{2+}(aq) + 4H_2O(l)$ (reduction)
$2MnO_4^-(aq) + 5H_2O_2(aq) + 6H^+(aq) \rightarrow 2Mn^{2+}(aq) + 5O_2(g) + 8H_2O(l)$

(d) $Sn^{2+}(aq) + 6Cl^-(aq) \rightarrow SnCl_6^{2-}(aq) + 2e^-$ (oxidation)
$H_2O_2(aq) + 2H^+(aq) + 2e^- \rightarrow 2H_2O(l)$ (reduction)
$Sn^{2+}(aq) + H_2O_2(aq) + 2H^+(aq) + 6Cl^-(aq) \rightarrow SnCl_6^{2-}(aq) + 2H_2O(l)$

(e) $HAsO_2(aq) + 2H_2O(l) \rightarrow H_3AsO_4(aq) + 2H^+(aq) + 2e^-$ (oxidation)
$H_2O_2(aq) + 2H^+(aq) + 2e^- \rightarrow 2H_2O(l)$ reduction
$HAsO_2(aq) + H_2O_2(aq) \rightarrow H_3AsO_4(aq)$

11.45 The balanced half-reaction equations and net ionic equations are:
(a) $NO_2(g) + H_2O(l) \rightarrow NO_3^-(aq) + 2H^+(aq) + e^-$ (oxidation)
$NO_2(g) + 2H^+(aq) + 2e^- \rightarrow NO(g) + H_2O(l)$ (reduction)
$3NO_2(g) + H_2O(l) \rightarrow 2NO_3^-(aq) + NO(g) + 2H^+(aq)$

(b) $ClO_2(g) + 2OH^-(aq) \rightarrow ClO_3^-(aq) + H_2O(l) + e^-$ (oxidation)
$ClO_2(g) + e^- \rightarrow ClO_2^-(aq)$ (reduction)
$2ClO_2(g) + 2OH^-(aq) \rightarrow ClO_2^-(aq) + ClO_3^-(aq) + H_2O(l)$

11.47 (a) 1. $HAsO_3^{2-}(aq) + BrO_3^-(aq) \rightarrow Br^-(aq) + H_3AsO_4(aq)$
2. $HAsO_3^{2-}(aq) \rightarrow H_3AsO_4(aq)$ (oxidation)
$BrO_3^-(aq) \rightarrow Br^-(aq)$ (reduction)
3. $HAsO_3^{2-}(aq) + H_2O(l) \rightarrow H_3AsO_4(aq) + 2e^-$ (balanced oxidation)
$BrO_3^-(aq) + 6H^+(aq) + 6e^- \rightarrow Br^-(aq) + 3H_2O(l)$ (balanced reduction)
4. $3HAsO_3^{2-}(aq) + 3H_2O(l) \rightarrow 3H_3AsO_4(aq) + 6e^-$
$BrO_3^-(aq) + 6H^+(aq) + 6e^- \rightarrow Br^-(aq) + 3H_2O(l)$
5. $3HAsO_3^{2-}(aq) + BrO_3^-(aq) + 6H^+(aq) \rightarrow 3H_3AsO_4(aq) + Br^-(aq)$

(b) 6. $3Na_2HAsO_3(aq) + KBrO_3(aq) + 6HCl(aq) \rightarrow 3H_3AsO_4(aq) + KBr(aq) + 6NaCl(aq)$

11.49 The balanced equations are (the oxidation number changes are given in parentheses):

(a) (+1) × 4

$$4Mn(OH)_2(s) + O_2(g) + 2H_2O(l) \rightarrow 4Mn(OH)_3(s)$$

(−2) × 2

(b) (+1) × 3

$$3Ti^{3+}(aq) + RuCl_5^{2-}(aq) + 6OH^-(aq) \rightarrow Ru(s) + 3TiO^{2+}(aq) + 5Cl^-(aq) + 3H_2O(l)$$

(−3) × 1

(c) (+2) × 2

$$2H_2S(g) + SO_2(g) \rightarrow 3S(s) + 2H_2O(g)$$

(−4) × 1

(d) (+1) × 6

$$6FeSO_4(aq) + K_2Cr_2O_7(aq) + 8H_2SO_4(aq) \rightarrow 3Fe_2(SO_4)_3(aq) + Cr_2(SO_4)_3(aq) + 2KHSO_4(aq) + 7H_2O(l)$$

(−3) × 2

11.51 The balanced equation is

$$3Cu(s) + 8HNO_3(aq) \rightarrow 3Cu(NO_3)_2(aq) + 2NO(g) + 4H_2O(l)$$

The number of moles of NO produced is

$$n = \frac{PV}{RT} = \frac{1.00 \text{ atm} \times 0.500 \text{ L}}{0.821 \text{ L•atm/mol•K} \times 293 \text{ K}} = 0.0208 \text{ mol}$$

The remaining steps in the calculation are: 0.0208 mol NO → mol Cu → g Cu.

$$0.0208 \text{ mol NO} \times \frac{3 \text{ mol Cu}}{2 \text{ mol NO}} \times \frac{63.55 \text{ g Cu}}{1 \text{ mol Cu}} = 1.98 \text{ g Cu}$$

11.53 The balanced equation is

$$5Na_2C_2O_4(aq) + 2KMnO_4(aq) + 16HCl(aq) \rightarrow$$
$$2MnCl_2(aq) + 10CO_2(g) + 10NaCl(aq) + 2KCl(aq) + 8H_2O(l)$$

The steps in the calculation are: g $KMnO_4$ → mol $KMnO_4$ → mol $Na_2C_2O_4$ → g $Na_2C_2O_4$.

$$0.500 \text{ g } KMnO_4 \times \frac{1 \text{ mol } KMnO_4}{158.0 \text{ g } KMnO_4} \times \frac{5 \text{ mol } Na_2C_2O_4}{2 \text{ mol } KMnO_4} \times \frac{134.0 \text{ g } Na_2C_2O_4}{1 \text{ mol } Na_2C_2O_4} = 1.06 \text{ g } Na_2C_2O_4$$

11.55 The balanced net ionic equation is $I_2(s) + 2S_2O_3^{2-}(aq) \rightarrow 2I^-(aq) + S_4O_6^{2-}(aq)$

The calculation proceeds as follows: g I_2 → mol I_2 → mol $Na_2S_2O_3$ → mL $Na_2S_2O_3$ soln.

$$7.50 \text{ g } I_2 \times \frac{1 \text{ mol } I_2}{253.81 \text{ g } I_2} \times \frac{2 \text{ mol } Na_2S_2O_3}{1 \text{ mol } I_2} = 0.0591 \text{ mol } Na_2S_2O_3$$

The number of milliliters of 0.100 M $Na_2S_2O_3$ needed is

$$0.0591 \text{ mol } Na_2S_2O_3 \times \frac{1 \text{ L}}{0.100 \text{ mol } Na_2S_2O_3} = 0.591 \text{ L or } 591 \text{ mL}$$

11.57 The balanced equation is

$5Fe^{2+}(aq) + MnO_4^-(aq) + 8H^+(aq) \rightarrow 5Fe^{3+}(aq) + Mn^{2+}(aq) + 4H_2O(l)$

The steps in the calculation are: V × M (of $KMnO_4$) → mol $KMnO_4$ → mol $FeSO_4$ → g $FeSO_4$.

$$0.02500 \text{ L } KMnO_4 \text{ soln} \times \frac{0.600 \text{ mol } KMnO_4}{1 \text{ L } KMnO_4 \text{ soln}} \times \frac{5 \text{ mol } FeSO_4}{1 \text{ mol } KMnO_4} \times \frac{151.91 \text{ g } FeSO_4}{1 \text{ mol } FeSO_4} = 11.4 \text{ g } FeSO_4$$

11.59 The balanced equation is

$5Na_2C_2O_4(s) + 2KMnO_4(aq) + 16HCl(aq) \rightarrow 2MnCl_2(aq) + 10CO_2(g) + 10NaCl(aq) + 2KCl(aq) + 8H_2O(l)$

First the limiting reactant has to be determined. 1.500 g of $Na_2C_2O_4$ is equal to 1.119×10^{-2} mol $Na_2C_2O_4$, and 30.0 mL of 0.100 M $KMnO_4$ is equal to 3.00×10^{-3} mol $KMnO_4$. From the balanced equation we see that $KMnO_4$ is the limiting reagent.

The steps in the calculation are: V × M (of $KMnO_4$) → mol $KMnO_4$ → mol CO_2 → g CO_2.

$$0.0300 \text{ L } KMnO_4 \text{ soln} \times \frac{0.100 \text{ mol } KMnO_4}{1 \text{ L } KMnO_4 \text{ soln}} \times \frac{10 \text{ mol } CO_2}{2 \text{ mol } KMnO_4} \times \frac{44.01 \text{ g } CO_2}{1 \text{ mol } CO_2} = 0.660 \text{ g } CO_2$$

11.61 Yes. In a pure ionic compound the oxidation number of an element can correspond to an actual charge. For example, in NaCl sodium exists as the Na^+ ion (charge 1+) and has an oxidation number of +1, and chlorine exists as the Cl^- ion (charge 1–) and has an oxidation number of –1. Similarly, in MgO magnesium exists as the Mg^{2+} ion (charge 2+) and has an oxidation number of +2, and oxygen exists as the O^{2-} ion (charge 2–) and has an oxidation number of –2.

11.63 (a) True. As long as the oxidation sums equal zero for neutral particles and the particle charge for ions, and the oxidation number increases equals the oxidation number decreases in balanced oxidation-reduction equations, then it does not matter what rules have been used to assign oxidation numbers. In essence, the assigning of oxidation numbers is an electron bookkeeping technique that is very useful for balancing oxidation-reduction reactions, among other things.

(b) Nothing. You would get the same balanced equations and the same calculated values.

Yes. This is consistent with the answer given in part (a). As long as the oxidation number system employed meets the basic criteria mentioned in part (a), and any appropriate system that is just shifted up or down by a fixed amount still does, then the results will not vary.

11.65 (a) The oxidation number of sulfur would be +8 if none of the oxygens were bonded to each other. Since sulfur is in Group 6A, its maximum oxidation number is +6; therefore, Caro's acid must contain peroxide oxygen.

(b) One. Each oxygen in a peroxide bond has an oxidation number of –1. Therefore, for S to be +6 there could only be one O–O bond in H_2SO_5.

11.67 2. $CN^-(aq) \rightarrow CNO^-(aq)$ (oxidation)
$Cu(NH_3)_4^{2+}(aq) \rightarrow CuCN(s)$ (reduction)

3. $CN^-(aq) + 2OH^-(aq) \rightarrow CNO^-(aq) + H_2O(l) + 2e^-$ (balanced oxidation)
   $Cu(NH_3)_4^{2+}(aq) + CN^-(aq) + e^- \rightarrow CuCN(s) + 4NH_3(aq)$ (balanced reduction)
4. $CN^-(aq) + 2OH^-(aq) \rightarrow CNO^-(aq) + H_2O(l) + 2e^-$
   $2Cu(NH_3)_4^{2+}(aq) + 2CN^-(aq) + 2e^- \rightarrow 2CuCN(s) + 8NH_3(aq)$
5. $2Cu(NH_3)_4^{2+}(aq) + 3CN^-(aq) + 2OH^-(aq) \rightarrow 2CuCN(s) + CNO^-(aq) + 8NH_3(aq) + H_2O(l)$

11.69

(+2) × 1

$$K_4Fe(CN)_6(s) + K_2CO_3(s) \rightarrow 5KCN(s) + KOCN(s) + Fe(s) + CO_2(g)$$

(−2) × 1

11.71 (a) $I_2(s) + 2S_2O_3^{2-}(aq) \rightarrow 2I^-(aq) + S_4O_6^{2-}(aq)$

(b) The steps in the calculation are: $V \times M\ (Na_2S_2O_3) \rightarrow \text{mol } Na_2S_2O_3 \rightarrow \text{mol } I_2 \rightarrow \text{g } I_2$.

$$0.0500\text{ L } Na_2S_2O_3 \text{ soln} \times \frac{0.362\text{ mol } Na_2S_2O_3}{1\text{ L } Na_2S_2O_3\text{ soln}} \times \frac{1\text{ mol } I_2}{2\text{ mol } Na_2S_2O_3} \times \frac{253.81\text{ g } I_2}{1\text{ mol } I_2} = 2.30\text{ g } I_2$$

11.73 The balanced equation is $SO_2(g) + 2H_2S(g) \rightarrow 3S(s) + 2H_2O(g)$

(a) From the equation, we see that H_2S is the limiting reactant. The steps in the calculation are: mol $H_2S \rightarrow$ mol S $\rightarrow$ g S.

$$1.00\text{ mol } H_2S \times \frac{3\text{ mol S}}{2\text{ mol } H_2S} \times \frac{32.07\text{ g S}}{1\text{ mol S}} = 48.1\text{ g S}$$

(b) 0.50 mol SO_2 and 1.00 mol H_2O

11.75 The balanced equation is $4Fe^{2+}(aq) + O_2(g) + 4H^+(aq) \rightarrow 4Fe^{3+}(aq) + 2H_2O(l)$

The steps in the calculation are: g $FeCl_2 \rightarrow$ mol $FeCl_2 \rightarrow$ mol $O_2 \rightarrow$ volume O_2 at STP.

$$5.0\text{ g } FeCl_2 \times \frac{1\text{ mol } FeCl_2}{126.75\text{ g } FeCl_2} \times \frac{1\text{ mol } O_2}{4\text{ mol } FeCl_2} = 9.9 \times 10^{-3}\text{ mol } O_2$$

One mole of gas occupies approximately 22.4 L at STP. The volume of O_2, measured in milliliters at STP, is

$$9.9 \times 10^{-3}\text{ mol } O_2 \times \frac{22.4\text{ L}}{1\text{ mol } O_2} = 0.22\text{ L or } 2.2 \times 10^2\text{ mL}$$

11.77 (a)
1. $Co(NO_2)_6^{3-}(aq) + MnO_4^-(aq) \rightarrow Co^{2+}(aq) + Mn^{2+}(aq) + NO_3^-(aq)$
2. $Co(NO_2)_6^{3-}(aq) \rightarrow Co^{2+}(aq) + NO_3^-(aq)$ (oxidation)
   $MnO_4^-(aq) \rightarrow Mn^{2+}(aq)$ (reduction)
3. $Co(NO_2)_6^{3-}(aq) + 6H_2O(l) \rightarrow Co^{2+}(aq) + 6NO_3^-(aq) + 12H^+(aq) + 11e^-$ (balanced oxidation)
   $MnO_4^-(aq) + 8H^+(aq) + 5e^- \rightarrow Mn^{2+}(aq) + 4H_2O(l)$ (balanced reduction)
4. $5Co(NO_2)_6^{3-}(aq) + 30H_2O(l) \rightarrow 5Co^{2+}(aq) + 30NO_3^-(aq) + 60H^+(aq) + 55e^-$
   $11MnO_4^-(aq) + 88H^+(aq) + 55e^- \rightarrow 11Mn^{2+}(aq) + 44H_2O(l)$
5. $5Co(NO_2)_6^{3-}(aq) + 11MnO_4^-(aq) + 28H^+(aq) \rightarrow 5Co^{2+}(aq) + 11Mn^{2+}(aq) + 30NO_3^-(aq) + 14H_2O(l)$

The reduction of the cobalt is treated within the overall oxidation of the nitrogen.

(b) The steps in the calculation are: V × M ($KMnO_4$) → mol $KMnO_4$ → mol $K_2Na[Co(NO_2)_6]$ → mg K.

$$0.0555 \text{ L } KMnO_4 \text{ soln} \times \frac{0.0250 \text{ mol } KMnO_4}{1 \text{ L } KMnO_4 \text{ soln}} \times \frac{5 \text{ mol } K_2Na[Co(NO_2)_6]}{11 \text{ mol } KMnO_4} =$$

$$6.31 \times 10^{-4} \text{ mol } K_2Na[Co(NO_2)_6] \times \frac{78.196 \text{ g K}}{1 \text{ mol } K_2Na[Co(NO_2)_6]} = 0.0493 \text{ g K}$$

Answer: The precipitate contains 6.31×10^{-4} mol $K_2Na[Co(NO_2)_6]$ and 49.3 mg K.

SURVEY OF THE ELEMENTS 2
OXYGEN, NITROGEN, AND THE HALOGENS

Solutions To Practice Exercises

PE S2.1 $2I^-(aq) \rightarrow I_2(aq) + 2e^-$ (oxidation)
$O_3(g) + H_2O(l) + 2e^- \rightarrow O_2(g) + 2OH^-(aq)$ (reduction)

PE S2.2 (a) $SO_3(g) + H_2O(l) \rightarrow H_2SO_4(aq)$. Sulfuric acid.
(b) $SO_3(g) + 2NaOH(aq) \rightarrow Na_2SO_4(aq) + H_2O(l)$. Sulfate ion and water.

PE S2.3 (a) $H_2O_2(aq) + 2H^+(aq) + 2e^- \rightarrow 2H_2O(l)$ (reduction)
(b) $H_2O_2(aq) + 2e^- \rightarrow 2OH^-(aq)$ (reduction)
(c) $H_2O_2(aq) \rightarrow O_2(g) + 2H^+(aq) + 2e^-$ (oxidation)

PE S2.4 Linear. The Lewis contributing structures for N_2O (eight electron pairs) are

:N̈=N=Ö: ⟷ :N≡N–Ö:

The central nitrogen has two VSEPR pairs (both bonding), which adopt a linear arrangement. The molecule is linear.

PE S2.5 Bent. The Lewis structure for the nitrite ion (nine electron pairs) is

[:Ö–N̈=Ö:]⁻

The central nitrogen has three VSEPR pairs (two bonding and one lone), which adopt a trigonal planar arrangement. The molecule is bent.

PE S2.6 (a) $HNO_2(aq) + H^+(aq) + e^- \rightarrow NO(g) + H_2O(l)$
(b) $NO_2^-(aq) + 2OH^-(aq) \rightarrow NO_3^-(aq) + H_2O(l) + 2e^-$

PE S2.7

	Lewis Structures	VSEPR Pairs	Lone Pairs	Shape
(a)	[:Ö=N=Ö:]$^+$	2	0	linear
(b)	[:Ö–N(–:O:)=Ö:]$^-$	3	0	trigonal planar

PE S2.8 1. $H^+(aq) + NO_3^-(aq) + H_2O(l) \xrightarrow{h\nu} NO_2(g) + O_2(g)$

2. and 3. $2H_2O(l) \rightarrow O_2(g) + 4H^+(aq) + 4e^-$ (balanced oxidation)

$2H^+(aq) + NO_3^-(aq) + e^- \rightarrow NO_2(g) + H_2O(l)$ (balanced reduction)

4. $2H_2O(l) \rightarrow O_2(g) + 4H^+(aq) + 4e^-$

$8H^+(aq) + 4NO_3^-(aq) + 4e^- \rightarrow 4NO_2(g) + 4H_2O(l)$

5. $4H^+(aq) + 4NO_3^-(aq) \xrightarrow{h\nu} 4NO_2(g) + O_2(g) + 2H_2O(l)$

6. $4HNO_3(aq) \xrightarrow{h\nu} 4NO_2(g) + O_2(g) + 2H_2O(l)$

PE S2.9 (a) Carbon reduces concentrated HNO_3 to NO_2.

$C(s) + 2H_2O(l) \rightarrow CO_2(g) + 4H^+(aq) + 4e^-$ (oxidation)

$NO_3^-(aq) + 2H^+(aq) + e^- \rightarrow NO_2(g) + H_2O(l)$ (reduction)

(b) Copper, a less active metal, reduces nitric acid to NO or NO_2, depending on the acid concentration. With 6 M HNO_3 the principal reduction product is NO.

$Cu(s) \rightarrow Cu^{2+}(aq) + 2e^-$ (oxidation)

$NO_3^-(aq) + 4H^+(aq) + 3e^- \rightarrow NO(g) + 2H_2O(l)$ (reduction)

PE S2.10 $2Br^-(aq) \rightarrow Br_2(l) + 2e^-$ (oxidation)

$Cr_2O_7^{2-}(aq) + 14H^+(aq) + 6e^- \rightarrow 2Cr^{3+}(aq) + 7H_2O(l)$ (reduction)

PE S2.11

	Lewis Structures	VSEPR Pairs	Lone Pairs	Shape
(a)	IF_3: F–I(:)–F with F below I	5	2	bent T-shape
(b)	BrF_5: Br(:) with five F	6	1	bent square pyramid

PE S2.12 (a) F is –1 (Rule 4); (b) O is –2 (Rule 4) and Cl is +1 (Rule 2); (c) O is –2 (Rule 4) and Br is +4 (Rule 2); (d) O is –2 (Rule 4) and I is +7 (Rule 2).

PE S2.13 (a) $HBrO_3$: Br is +5; bromic acid
(b) HIO_4: I is +7; periodic acid
(c) NaOBr: Br is +1; sodium hypobromite
(d) $NaIO_3$: I is +5; sodium iodate

PE S2.14 (a) potassium chlorite: Cl is +3; $KClO_2$
(b) hypobromous acid: Br is +1; HOBr
(c) perbromic acid: Br is +7; $HBrO_4$
(d) sodium periodate: I is +7; $NaIO_4$

PE S2.15

	Lewis Structures	VSEPR Pairs	Lone Pairs	Shape
(a)	$[:\ddot{O}-\ddot{Cl}-\ddot{O}:]^-$	4	2	bent
(b)	$[:\ddot{O}-\ddot{Cl}(-\ddot{O}:)-\ddot{O}:]^-$	4	1	trigonal pyramid
(c)	$[:\ddot{O}-Cl(-\ddot{O}:)(-\ddot{O}:)-\ddot{O}:]^-$	4	0	tetrahedral

Solutions To Final Exercises

S2.1

(a) Elemental oxygen (O_2) constitutes 20.9% of the atmosphere by molecule count, or 20.9% of the atmosphere by volume.

(b) The lithosphere (earth's solid crust) contains 62.6% oxygen by atom count and 46.6% oxygen by mass, where it is present mainly in the form of oxides, silicates, and carbonates.

(c) The hydrosphere or natural waters cover approximately 70% of the earth's surface, and water is 88.81% oxygen by mass.

(d) Oxygen compounds are major constituents of the biosphere. In fact, the most important biosphere reaction, the conversion of carbon dioxide and water into glucose by photosynthesis, involves oxygen compounds. Also, the process of respiration uses oxygen in the slow combustion of bioorganic material to CO_2 and H_2O.

S2.3

(a) Ozone in the upper atmosphere absorbs much of the sun's ultraviolet radiations, thus protecting life on earth from overexposure to ultraviolet radiation and its concomitant afflictions.

(b) Harmful. Ozone in the lower atmosphere is a pollutant. Direct contact with ozone is deleterious, and prolonged exposure to concentrations above 0.1 ppm are considered unsafe.

S2.5

The presence of ozone in air can be detected by passing the air through a solution of potassium iodide and starch. Ozone oxidizes iodide ion to iodine and the deep blue color of the iodine-starch complex indicates the presence of ozone.

$$2I^-(aq) + O_3(g) + H_2O(l) \rightarrow 2OH^-(aq) + I_2(aq) + O_2(g)$$

colorless — blue color with starch

S2.7

(a) Na_2O, K_2O, and CaO. In general, the metallic elements.

(b) SO_2, SO_3, and P_4O_{10}. In general, the nonmetallic elements.

(c) Test an aqueous solution of the oxide with litmus paper. (Litmus is blue in alkaline solution and red in acid solution). Also, see if the oxide dissolves in basic or in acidic solution, since many oxides are insoluble in water.

S2.9

1. A 30% aqueous solution of H_2O_2 is used in water purification, in bleaching wood pulp, and in chemical synthesis. 2. A 3% solution is used as a mild antiseptic. 3. H_2O_2 is used to bleach hair. 4. In the restoration of old paintings, where hydrogen peroxide oxidizes the black lead sulfide to white lead sulfate.

S2.11 See Figure S2.3 in the textbook.

Each oxygen has four VSEPR pairs (two bonding and two lone pairs). Therefore, each oxygen is sp^3 hybridized and both H–O–O bond angles should be close to the ideal tetrahedral angle of 109.5°. However, the two lone pairs of electrons on each of the oxygen atoms compress the bond angles to about 95°, well below the ideal tetrahedral angle of 109.5°.

S2.13 (a) O_2 has 16 electrons and its molecular orbital configuration is

$$(\sigma_{1s})^2(\sigma^*_{1s})^2(\sigma_{2s})^2(\sigma^*_{2s})^2(\pi_{2px})^2(\pi_{2py})^2(\sigma_{2pz})^2(\pi^*_{2px})^1(\pi^*_{2py})^1.$$

(b) 1. The configuration shows that oxygen should be a paramagnetic diradical, and it is.
2. The configuration shows that O_2 has a bond order of 2, which roughly corresponds to a double bond in valence bond theory terminology. In actual fact, O_2 has a bond length and bond energy that is consistent with a double bond between the oxygen atoms.
3. The configurations shows that O_2 is a diradical and, since radicals are usually reactive, we would expect O_2 to be reactive. In actual fact, O_2 is the most reactive element in Group 6A.

S2.15 (a) $2H_2O(l) \xrightarrow{\text{electrolysis}} 2H_2(g) + O_2(g)$

(b) $2KClO_3(s) \xrightarrow[\text{heat}]{MnO_2(s)} 2KCl(s) + 3O_2(g)$

(c) $2H_2O_2(l) \rightarrow 2H_2O(l) + O_2(g)$

(d) $6CO_2(g) + 6H_2O(l) \xrightarrow{h\nu} C_6H_{12}O_6(aq) + 6O_2(g)$

S2.17 (a) $3Fe(s) + 2O_2(g) \rightarrow Fe_3O_4(s)$ or
$4Fe(s) + 3O_2(g) \rightarrow 2Fe_2O_3(s)$
(b) $2Zn(s) + O_2(g) \rightarrow 2ZnO(s)$
(c) $S(s) + O_2(g) \rightarrow SO_2(g)$
(d) $P_4(s) + 5O_2(g) \rightarrow P_4O_{10}(s)$
(e) $2SO_2(g) + O_2(g) \rightarrow 2SO_3(g)$
(f) $2C_2H_2(g) + 5O_2(g) \rightarrow 4CO_2(g) + 2H_2O(g)$

S2.19 (a) $Mg(OH)_2(s) \xrightarrow{\Delta} MgO(s) + H_2O(g)$
(b) $2LiOH(s) \xrightarrow{\Delta} Li_2O(s) + H_2O(g)$
(c) $Fe(OH)_2(s) \xrightarrow{\Delta} FeO(s) + H_2O(g)$
(d) $2Fe(OH)_3(s) \xrightarrow{\Delta} Fe_2O_3(s) + 3H_2O(g)$

S2.21 (a) $BaO(s) + H_2O(l) \rightarrow Ba(OH)_2(aq)$
(b) $Na_2O(s) + H_2O(l) \rightarrow 2NaOH(aq)$

S2.23 (a) $SO_2(g) + 2NaOH(aq) \rightarrow Na_2SO_3(aq) + H_2O(l)$
$SO_2(g) + Ca(OH)_2(aq) \rightarrow CaSO_3(s) + H_2O(l)$
(b) $SO_3(g) + 2NaOH(aq) \rightarrow Na_2SO_4(aq) + H_2O(l)$
$SO_3(g) + Ca(OH)_2(aq) \rightarrow CaSO_4(s) + H_2O(l)$
(c) $CO_2(g) + 2NaOH(aq) \rightarrow Na_2CO_3(aq) + H_2O(l)$
$CO_2(g) + Ca(OH)_2(aq) \rightarrow CaCO_3(s) + H_2O(l)$

S2.25 Small amounts of aqueous hydrogen peroxide can be prepared in the laboratory by treating barium peroxide with cold dilute sulfuric acid:

$$BaO_2(s) + H_2SO_4(aq) \xrightarrow{0°C} BaSO_4(s) + H_2O_2(aq)$$

S2.27 See Table S2.4 in the textbook.

S2.29 (a) $PbS(s) + 4H_2O_2(aq) \rightarrow PbSO_4(s) + 4H_2O(l)$
(b) $2Na_2O_2(s) + 2H_2O(l) \rightarrow 4NaOH(aq) + O_2(g)$
(c) $2Na_2O_2(s) + 2CO_2(g) \rightarrow 2Na_2CO_3(s) + O_2(g)$
(d) $4KO_2(s) + 2CO_2(g) \rightarrow 2K_2CO_3(s) + 3O_2(g)$

S2.31 (a) 1. Elemental nitrogen (N_2) constitutes 78% of the atmosphere by molecule count, or 78% of the atmosphere by volume.
2. The lithosphere (earth's solid crust) contains very little nitrogen. Only deposits of KNO_3 and $NaNO_3$ exist in certain arid regions.
3. Nitrogen compounds occur throughout the biosphere. For instance, nitrogen is one of the essential elements in proteins and DNA.

(b) Very few nitrogen compounds are found in the lithosphere because most inorganic nitrogen compounds are water soluble.

S2.33 (a) From liquid air by fractional distillation.
(b) See Table S2.7 in the textbook.

S2.35 The electric discharge breaks the bonds of the N_2 molecules. The resulting atoms are reactive, since they have incomplete octets.

S2.37 (a) $N_2(g) + 3H_2(g) \xrightarrow[\text{catalyst}]{400°C,\ 250\ atm} 2NH_3(g)$
(b) $6Li(s) + N_2(g) \xrightarrow{\Delta} 2Li_3N(s)$
(c) $3Mg(s) + N_2(g) \xrightarrow{\Delta} Mg_3N_2(s)$

S2.39 (a) Nitrogen (N_2) and hydrogen (H_2) react (a combination reaction) at elevated pressure and temperature in the presence of a special iron oxide catalyst to form ammonia (NH_3).

$$N_2(g) + 3H_2(g) \xrightarrow[\text{catalyst}]{400°C,\ 250\ atm} 2NH_3(g)$$

The Haber process is a commercial nitrogen fixation process.

(b) Because in order to feed the earth's ever growing population, an ever greater agricultural production is required and ammonia is the most widely used synthetic nitrogen fertilizer. Ammonia is also a basic raw material used in the manufacture of some widely used fibers, plastics, and explosives.

S2.41 See Table S2.9 in the textbook.

S2.43 Some of the possible reactions are:
(a) $NH_4NO_3(s) \xrightarrow{\Delta} N_2O(g) + 2H_2O(g)$

(b) $N_2(g) + O_2(g) \xrightarrow[\text{electric discharge}]{\Delta\ \text{or}} 2NO(g)$

(Also reactions in which HNO_3 or HNO_2 is reduced)

(c) $2NO(g) + O_2(g) \rightarrow 2NO_2(g)$ (Also reactions in which HNO_3 is reduced)

(d) $2NO_2(g) \rightleftharpoons N_2O_4(g)$

S2.45 (a) $2HCl(aq) + Ca(NO_2)_2(aq) \xrightarrow{\text{cold}} 2HNO_2(aq) + CaCl_2(aq)$
(b) $3HNO_2(aq) \rightarrow HNO_3(aq) + 2NO(g) + H_2O(l)$

S2.47 (a)

Step 1: Oxidation of ammonia to nitrogen monoxide. Ammonia is burned in excess oxygen over a platinum catalyst to form NO:

$$4NH_3(g) + 5O_2(g) \xrightarrow[900°C]{Pt} 4NO(g) + 6H_2O(g)$$

Step 2: Oxidation of nitrogen monoxide to nitrogen dioxide. Additional air is added to cool the mixture and oxidize NO to NO_2:

$$2NO(g) + O_2(g) \xrightarrow{cool} 2NO_2(g)$$

Step 3: Conversion of nitrogen dioxide to nitric acid. The NO_2 gas is bubbled into warm water where it disproportionates to nitric acid and NO. The NO is recycled to Step 2:

$$3NO_2(g) + H_2O(l) \rightarrow 2HNO_3(aq) + \underset{\text{recycled}}{NO(g)}$$

(b) The first step, since it requires high temperatures, which means there will also be high gas pressures, and a platinum catalyst.

S2.49
(a) $NO_3^-(aq) + 2H^+(aq) + e^- \rightarrow NO_2(g) + H_2O(l)$
(b) $NO_3^-(aq) + 4H^+(aq) + 3e^- \rightarrow NO(g) + 2H_2O(l)$
(c) $2NO_3^-(aq) + 10H^+(aq) + 8e^- \rightarrow N_2O(g) + 5H_2O(l)$
(d) $2NO_3^-(aq) + 12H^+(aq) + 10e^- \rightarrow N_2(g) + 6H_2O(l)$
(e) $NO_3^-(aq) + 10H^+(aq) + 8e^- \rightarrow NH_4^+(aq) + 3H_2O(l)$

S2.51 The principal reduction product for the nonmetals phosphorus, sulfur, and iodine with concentrated nitric acid is NO_2. The balanced half-reaction, net ionic, and molecular equations are:

(a) $P(s) + 4H_2O(l) \rightarrow H_3PO_4(aq) + 5H^+(aq) + 5e^-$ (oxidation)
$NO_3^-(aq) + 2H^+(aq) + e^- \rightarrow NO_2(g) + H_2O(l)$ (reduction)
$P(s) + 5NO_3^-(aq) + 5H^+(aq) \rightarrow H_3PO_4(aq) + 5NO_2(g) + H_2O(l)$
$P(s) + 5HNO_3(aq) \rightarrow H_3PO_4(aq) + 5NO_2(g) + H_2O(l)$

(b) $S(s) + 4H_2O(l) \rightarrow SO_4^{2-}(aq) + 8H^+(aq) + 6e^-$ (oxidation)
$NO_3^-(aq) + 2H^+(aq) + e^- \rightarrow NO_2(g) + H_2O(l)$ (reduction)
$S(s) + 6NO_3^-(aq) + 4H^+(aq) \rightarrow SO_4^{2-}(aq) + 6NO_2(g) + 2H_2O(l)$
$S(s) + 6HNO_3(aq) \rightarrow H_2SO_4(aq) + 6NO_2(g) + 2H_2O(l)$

(c) $I_2(s) + 6H_2O(l) \rightarrow 2IO_3^-(aq) + 12H^+(aq) + 10e^-$ (oxidation)
$NO_3^-(aq) + 2H^+(aq) + e^- \rightarrow NO_2(g) + H_2O(l)$ (reduction)
$I_2(s) + 10NO_3^-(aq) + 8H^+(aq) \rightarrow 2IO_3^-(aq) + 10NO_2(g) + 4H_2O(l)$
$I_2(s) + 10HNO_3(aq) \rightarrow 2HIO_3(aq) + 10NO_2(g) + 4H_2O(l)$

S2.53
(a) $Na_2CO_3(s) + 2HNO_3(aq) \rightarrow 2NaNO_3(aq) + CO_2(g) + H_2O(l)$
(b) $Mg(OH)_2(s) + 2HNO_3(aq) \rightarrow Mg(NO_3)_2(aq) + 2H_2O(l)$
(c) $N_2O_5(s) + H_2O(l) \rightarrow 2HNO_3(aq)$
(d) $Li_3N(s) + 3H_2O(l) \rightarrow NH_3(g) + 3LiOH(aq)$
(e) $NaCN(s) + HCl(aq) \rightarrow NaCl(aq) + HCN(g)$
(f) $2NH_3(l) + OCl^-(aq) \rightarrow N_2H_4(aq) + Cl^-(aq) + H_2O(l)$

S2.55 and S2.57(a)

(a) Nitrogen monoxide (NO) has 5 + 6 = 11 valence electrons or 5 electron pairs plus one odd electron. The contributing Lewis structures are:

:Ṅ=Ö: ⟷ (−1):N̈=Ȯ:(+1)

Since NO is a diatomic molecule, its shape is linear, N–O.

(b) Dinitrogen oxide (N_2O) has 2(5) + 6 = 16 valence electrons or 8 electron pairs. The contributing Lewis structures are:

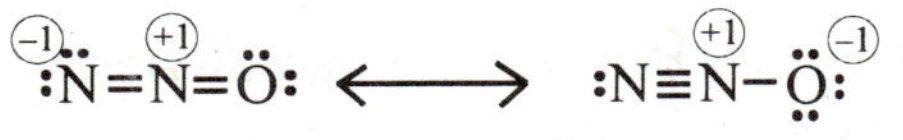

(−1):N̈=N(+1)=Ö: ⟷ :N≡N(+1)−Ö:(−1)

The central nitrogen atom has two VSEPR electron pairs and, therefore, uses sp hybrid orbitals to form the σ bonding skeleton or framework. The shape of the molecule is therefore linear; N–N–O.

(c) Nitrogen dioxide (NO_2) has 5 + 2(6) = 17 valence electrons or 8 electron pairs plus one odd electron. The contributing Lewis structures are:

(−1):Ö−Ṅ(+1)=Ö: ⟷ :Ö=Ṅ(+1)−Ö:(−1) ⟷ :Ö=N̈−Ö: ⟷ :Ö−N̈=Ö:

The central nitrogen atom has three VSEPR electron pairs and, therefore, uses sp^2 hybrid orbitals to form the σ bonding skeleton. Since one of the VSEPR pairs is a lone pair, the shape of the molecule is bent.

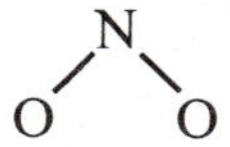

O–N–O (bent)

(d) Dinitrogen trioxide (N_2O_3) has 2(5) + 3(6) = 28 valence electrons or 14 electron pairs. The contributing Lewis structures are:

(:Ö=)(:Ö−)N−N̈=Ö: ⟷ (:Ö−)(:Ö=)N−N̈=Ö:

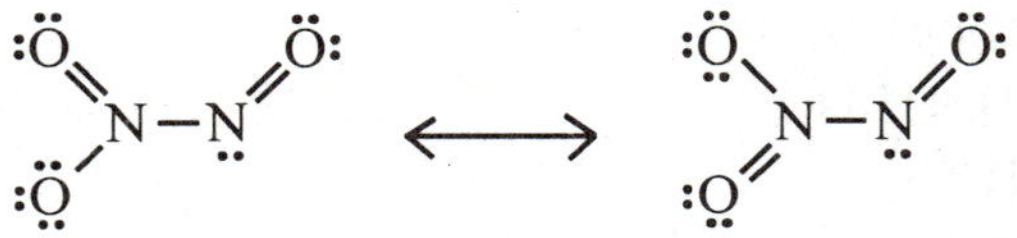

The left nitrogen atom has three bonding VSEPR electron pairs and, therefore, uses sp^2 hybrid orbitals to form its portion of the σ bonding skeleton or framework. Consequently, the N–NO_2 portion of the molecule is planar. The right nitrogen atom also has three VSEPR electron pairs and, therefore, also uses sp^2 hybrid orbitals to form its portion of the σ bonding skeleton. However, since one of the VSEPR pairs is a lone pair, the N–N–O portion of the molecule is bent. The geometric shape is (The lone right oxygen atom is not necessarily restricted to the plane of the paper or the other atoms.):

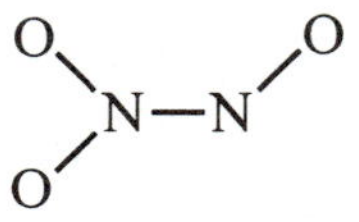

(O)(O)N—N–O

(e) Dinitrogen tetroxide (N_2O_4) has 2(5) + 4(6) = 34 valence electrons or 17 electron pairs. The contributing Lewis structures are:

Ö=N(−Ö:)—N(−Ö:)=Ö ⟷ :Ö−N(=O:)—N(=O:)−Ö: ⟷ :Ö−N(=O:)—N(−Ö:)=Ö ⟷ Ö=N(−Ö:)—N(=O:)−Ö:

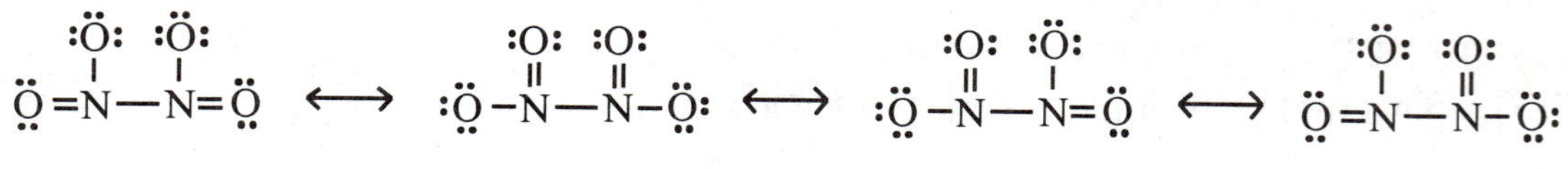

Each nitrogen atom has three bonding VSEPR electron pairs and, therefore, each uses sp^2 hybrid orbitals to form its portion of the σ bonding skeleton. Consequently, the arrangement of atoms about each nitrogen atom is trigonal planar and the overall structure of N_2O_4 is probably planar, even though the nitrogens are joined by a single bond. The geometric shape is

```
O         O
 \       /
  N --- N
 /       \
O         O
```

(f) Dinitrogen pentoxide (N_2O_5) has 2(5) + 5(6) = 40 valence electrons or 20 electron pairs. The contributing Lewis structures are:

```
   :O:    :O:                :Ö:    :Ö:
    ||     ||                  |      |
 :Ö–N–Ö–N–Ö:      ⟷        Ö=N–Ö–N=Ö
```

Each nitrogen atom has three bonding VSEPR electron pairs and, therefore, each uses sp^2 hybrid orbitals to form its portion of the σ bonding skeleton. Consequently, the arrangement of atoms about each nitrogen atom is trigonal planar. The central oxygen atom has four VSEPR electron pairs and, therefore, uses sp^3 hybrid orbitals to form its portion of the σ bonding skeleton. However, since two of the VSEPR pairs are lone pairs, the N–O–N portion of the molecule is bent. The geometric shape is

```
    O  O
    |   \
    N    N—O
   / \  /
  O   O
```

S2.56 and S2.57 (b)

(a) Nitrous acid (HNO_2) has 1 + 5 + 2(6) = 18 valence electrons or 9 electron pairs. The Lewis structure is

```
H–Ö–N̈=Ö
```

The central nitrogen atom has three VSEPR electron pairs and, therefore, uses sp^2 hybrid orbitals to form the σ bonding skeleton. However, since one of the VSEPR pairs is a lone pair, the O–N–O portion of the molecule is bent. The geometric shape is

```
H
 \
  O—N
     \
      O
```

(b) Nitric acid (HNO_3) has 1 + 5 + 3(6) + 24 valence electrons or 12 electron pairs. The contributing Lewis structures are

```
        :O:                        (-1)
         ||                        :Ö:
                                    |
  H–Ö–N–Ö:          ⟷          H–Ö–N=Ö
     (+1) (-1)                     (+1)
```

The central nitrogen atom has three bonding VSEPR electron pairs and, therefore, uses sp^2 hybrid orbitals to form the σ bonding skeleton. Consequently, the arrangement of atoms about nitrogen is trigonal planar and the HNO_3 molecule is planar. The geometric shape is

```
H      O
 \    /
  O—N
     \
      O
```

(c) Hydroxylamine (NH_2OH) has 3(1) + 5 + 6 = 14 valence electrons or 7 electron pairs. The Lewis structure is

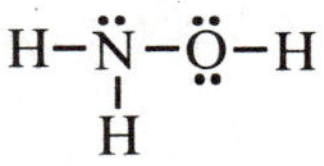

The central nitrogen atom has four VSEPR electron pairs and, therefore, uses sp^3 hybrid orbitals to form the σ bonding skeleton. However, since one of the VSEPR pairs is a lone pair, the shape of the molecule is trigonal pyramidal.

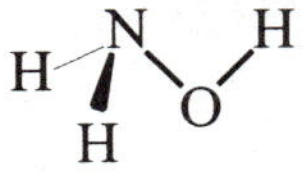

(d) Hydrazine (N_2H_4) has 2(5) + 4(1) = 14 valence electrons or 7 electron pairs. The Lewis structure is

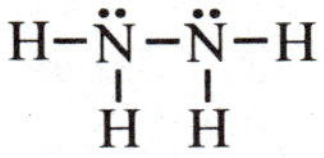

Each nitrogen atom has four VSEPR electron pairs and, therefore, each uses sp^3 hybrid orbitals to form the σ bonding skeleton. However, one of the VSEPR pairs on each nitrogen is a lone pair and, consequently, the arrangement of atoms about each of the nitrogen atoms is trigonal pyramidal. The geometric shape is

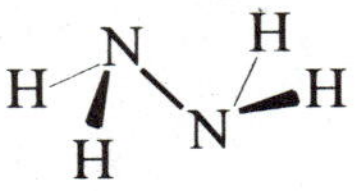

(e) Cyanide ion (CN^-) has 4 + 5 + 1 = 10 valence electrons or 5 electron pairs. The Lewis structure is

Since CN^- is a diatomic ion its shape is linear, C–N.

(f) Amide ion (NH_2^-) has 5 + 2(1) + 1 = 8 valence electrons or 4 electron pairs. The Lewis structure is

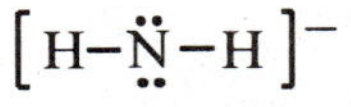

The central nitrogen atom has four VSEPR electron pairs and, therefore, uses sp^3 hybrid orbitals to form the σ bonding skeleton. However, since two of the VSEPR pairs are lone pairs, the shape of the ion is bent.

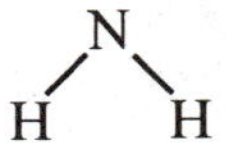

S2.59 (a) The halogens occur in nature as halide ions, X^-.
(b) Fluorspar (CaF_2), cryolite (Na_3AlF_6), fluorapatite ($Ca_5(PO_4)_3F$), sodium chloride (NaCl), sodium bromide (NaBr), sodium iodide (NaI), and sodium iodate ($NaIO_3$).
(c) NaCl.

S2.61 (a) The physical properties of the halogens are compared in Table S2.11 of the textbook.
(b) A group trend is a consistent pattern of variation in some property from top to bottom in a group. Group trends can generally be explained on the basis of atomic and molecular structure. The halogen elements are all diatomic molecules. In general, the melting points and boiling points of nonpolar molecules depend on the strength of the London dispersion forces (see Chapter 12, Section 12.3) between the molecules, which generally increase as the number of electrons per molecule increases. Therefore, we would expect the melting points and boiling points of the halogens to increase as we proceed down the group, which they do. Likewise, we would expect the physical state of the halogens to change from gas → liquid → solid with the increasing London dispersion forces as we proceed down the group, which also happens. Furthermore,

we would imagine that the colors of the elements should get darker as we proceed down the group, since each succeeding element has more electrons, which are spread over a wider range of energy levels, and which can therefore absorb more of the visible light frequencies. The more of the visible light frequencies absorbed by a substance, the darker the substance will appear, and this is also what happens.

S2.63 Fluorine is used in uranium enrichment and the production of fluorocarbon plastics, refrigerants, and water fluoridation and glass etching chemicals.

Chlorine is extensively used in the synthesis of important and widely used commercial compounds, chloroform ($CHCl_3$) and carbon tetrachloride (CCl_4) being two examples. It is also widely used as an industrial and household bleaching agent and as a bactericide in public water supplies.

S2.65
(a) HOI: I is +1; hypoiodous acid
(b) $KClO_3$: Cl is +5; potassium chlorate
(c) $HClO_4$: Cl is +7; perchloric acid
(d) HIO_3: I is +5; iodic acid
(e) $NaBrO_4$: Br is +7; sodium perbromate
(f) $Mg(OCl)_2$: Cl is +1; magnesium hypochlorite

S2.67
(a) $CaF_2(s) + H_2SO_4(18\ M) \xrightarrow{\Delta} 2HF(g) + CaSO_4(s)$

$2HF\ (in\ KF) \xrightarrow{\Delta,\ electrolysis} H_2(g) + F_2(g)$

(b) $2NaCl(aq) + 2H_2O(l) \xrightarrow{electrolysis} 2NaOH(aq) + Cl_2(g) + H_2(g)$

(c) $2Br^-(aq) + Cl_2(g) \rightarrow Br_2(l) + 2Cl^-(aq)$
in seawater

(d) $2I^-(aq) + Cl_2(g) \rightarrow I_2(s) + 2Cl^-(aq)$

S2.69
(a) $2Na(s) + F_2(g) \rightarrow 2NaF(s)$
$2Na(s) + Cl_2(g) \rightarrow 2NaCl(s)$
$2Na(s) + Br_2(l) \rightarrow 2NaBr(s)$
$2Na(s) + I_2(s) \rightarrow 2NaI(s)$
(b) $Zn(s) + F_2(g) \rightarrow ZnF_2(s)$
$Zn(s) + Cl_2(g) \rightarrow ZnCl_2(s)$
$Zn(s) + Br_2(l) \rightarrow ZnBr_2(s)$
$Zn(s) + I_2(s) \rightarrow ZnI_2(s)$
(c) $H_2(g) + F_2(g) \rightarrow 2HF(g)$
$H_2(g) + Cl_2(g) \rightarrow 2HCl(g)$
$H_2(g) + Br_2(l) \rightarrow 2HBr(g)$
$H_2(g) + I_2(s) \rightarrow 2HI(g)$
(d) $2H_2O(l) + 2F_2(g) \rightarrow 4HF(g) + O_2(g)$
$H_2O(l) + Cl_2(g) \rightleftharpoons H^+(aq) + Cl^-(aq) + HOCl(aq)$
$H_2O(l) + Br_2(l) \rightleftharpoons H^+(aq) + Br^-(aq) + HOBr(aq)$
$H_2O(l) + I_2(s) \rightleftharpoons H^+(aq) + I^-(aq) + HOI(aq)$

S2.71 (a) Yes. $Cl_2(g) + 2NaI(aq) \rightarrow 2NaCl(aq) + I_2(s)$ (b) No.

S2.73
(a) $CaF_2(s) + H_2SO_4(18\ M) \rightarrow CaSO_4(s) + 2HF(g)$
(b) $NaCl(s) + H_2SO_4(18\ M) \rightarrow NaHSO_4(s) + HCl(g)$
(c) $NaBr(s) + H_3PO_4(15\ M) \rightarrow HBr(g) + NaH_2PO_4(aq)$
(d) $NaI(s) + H_3PO_4(15\ M) \rightarrow HI(g) + NaH_2PO_4(aq)$

S2.75
(a) $Mg(s) + 2HF(g) \rightarrow MgF_2(s) + H_2(g)$
(b) $HF(aq) \rightleftharpoons H^+(aq) + F^-(aq)$
(c) $SiO_2(s) + 4HF(g\ or\ aq) \rightarrow SiF_4(g) + 2H_2O(g)$

S2.77
(a) HOBr(+1); $HBrO_3$(+5); $HBrO_4$(+7)
(b) HOBr; $HBrO_3$; $HBrO_4$
(c) $HBrO_4$; $HBrO_3$; HOBr

S2.79 (a) $I_2(s) + 2KOH(aq) \xrightarrow{\text{cool}} KI(aq) + KOI(aq) + H_2O(l)$

(b) $2HOCl(aq) \rightarrow 2HCl(aq) + O_2(g)$

(c) $2KClO_3(s) \xrightarrow[\text{heat}]{MnO_2(s)} 2KCl(s) + 3O_2(g)$

S2.81

	Lewis Structure	VSEPR Pairs	Lone Pairs	Shape
(a)	H–Ö–Cl(–:Ö:)–Ö: (Cl bonded to three O, one lone pair on Cl)	4	1	trigonal pyramid
(b)	$[:\ddot{O}-\ddot{I}-\ddot{O}:]^-$	4	2	bent
(c)	$H-\ddot{O}-\ddot{I}:$	4	2	bent
(d)	$[:\ddot{I}-\ddot{I}-\ddot{I}:]^-$	5	3	linear
(e)	$:\ddot{O}-\dot{Cl}-\ddot{O}:$	4	2	bent
(f)	$[ClO_4]^-$: Cl bonded to four :Ö:	4	0	tetrahedral

Refer to sketches of similar shaped species in Chapter 10 of the textbook.

S2.83 (a) Blackish crystalline solid. See solution to Exercise S2.61.

(b) 1. Since At is a nonmetal like the other halogens, it should react with most metals forming metal astatides, e.g., NaAt.

2. At should be slightly soluble in water and should react as follows:

$$At_2(s) + H_2O(l) \rightleftharpoons H^+(aq) + At^-(aq) + HOAt(aq)$$

(c) No. The oxidizing strength of the halogens decreases as you proceed down the group. Since At is below I, it would be unable to oxidize I^- to I_2.

S2.85 (a) Fluorine cannot attack fluorspar, CaF_2, because the calcium is already in its maximum oxidation state of +2. Therefore, there is nothing for the fluorine to oxidize.

(b) The fluorine initially attacks or reacts with the specially formulated steel producing a thin adherent coating of metal fluoride. This adherent metal fluoride coating, safe from fluorine attack since it is fully oxidized, protects the remaining unoxidized metal from further attack by not allowing the fluorine to come in contact with it.

S2.87 (a) Laughing gas; N_2O, dinitrogen oxide. Also known as nitrous oxide.
(b) Tincture of iodine; alcohol solution of iodine.
(c) Aqua regia; a mixture of one part concentrated nitric acid to three parts concentrated hydrochloric acid.
(d) Saltpeter; KNO_3, potassium nitrate.
(e) Chile saltpeter; $NaNO_3$, sodium nitrate.
(f) Fluorspar; CaF_2, calcium fluoride.
(g) Teflon; polytetrafluoroethylene, a fluorocarbon polymer.
(h) Freon-12, CCl_2F_2, dichlorodifluoromethane.
(i) Bleaching powder; $Ca(OCl)_2$, calcium hypochlorite.

S2.89 (a) A bromine concentration of 4000 ppm by mass means that there is 4000 g of Br_2 in one million (1.00×10^6) grams of brine. The volume of brine that will yield 1.00 kg of bromine is

$$1.00 \times 10^3 \text{ g } Br_2 \times \frac{1.00 \times 10^6 \text{ g brine}}{4000 \text{ g } Br_2} \times \frac{1.0 \text{ mL brine}}{1.1 \text{ g brine}} = 230 \times 10^3 \text{ mL or } 2.3 \times 10^2 \text{ L brine}$$

(b) The equation for the reaction is $Cl_2(g) + 2Br^-(aq) \rightarrow Br_2(l) + 2Cl^-(aq)$

The steps in the calculation are: g $Br_2 \rightarrow$ mol $Br_2 \rightarrow$ mol $Cl_2 \rightarrow$ volume Cl_2.

$$1.00 \times 10^3 \text{ g } Br_2 \times \frac{1 \text{ mol } Br_2}{159.81 \text{ g } Br_2} \times \frac{1 \text{ mol } Cl_2}{1 \text{ mol } Br_2} = 6.26 \text{ mol } Cl_2$$

The volume of Cl_2 measured at STP needed to produce 1.00 kg Br_2 is

$$6.26 \text{ mol } Cl_2 \times \frac{22.4 \text{ L } Cl_2}{1 \text{ mol } Cl_2} = 140 \text{ L } Cl_2$$

S2.91 A fluoride ion concentration of 1.0 ppm by mass means that there is 1.0 g of F^- in one million (1.0×10^6) grams of water.

$$\frac{100 \text{ gal water}}{1 \text{ day} \times 1 \text{ person}} \times 1.5 \times 10^6 \text{ persons} \times \frac{3.785 \text{ L}}{1 \text{ gal}} \times \frac{10^3 \text{ mL}}{1 \text{ L}} \times \frac{1.00 \text{ g water}}{1 \text{ mL water}} \times \frac{1.0 \text{ g } F^-}{1.0 \times 10^6 \text{ g water}} \times \frac{41.99 \text{ g NaF}}{19.00 \text{ g } F^-}$$

$$= 1.3 \times 10^6 \text{ g/day or } 1.3 \times 10^3 \text{ kg NaF/day}$$

S2.93

$$\text{molarity} = \frac{\text{moles of solute}}{\text{liters of solution}}$$

(a) $$\text{molarity} = \frac{9.5 \times 10^2 \text{ g soln}}{1 \text{ L soln}} \times \frac{25 \text{ g } NH_3}{100 \text{ g soln}} \times \frac{1 \text{ mol } NH_3}{17.03 \text{ g } NH_3} = 14 \text{ mol } NH_3\text{/L} = 14 \text{ M } NH_3$$

(b) $$\text{molarity} = \frac{1.0 \times 10^3 \text{ g soln}}{1 \text{ L soln}} \times \frac{3.0 \text{ g } H_2O_2}{100 \text{ g soln}} \times \frac{1 \text{ mol } H_2O_2}{34.01 \text{ g } H_2O_2} = 0.88 \text{ mol } H_2O_2\text{/L} = 0.88 \text{ M } H_2O_2$$

(c) $$\text{molarity} = \frac{1.0 \times 10^3 \text{ g soln}}{1 \text{ L soln}} \times \frac{0.90 \text{ g NaCl}}{100 \text{ g soln}} \times \frac{1 \text{ mol NaCl}}{58.44 \text{ g NaCl}} = 0.15 \text{ mol NaCl/L} = 0.15 \text{ M NaCl}$$

S2.95 (a) $2H_2O_2(aq) \rightarrow 2H_2O(l) + O_2(g)$

The steps in the calculation are: g $H_2O_2 \rightarrow$ mol $H_2O_2 \rightarrow$ mol $O_2 \rightarrow$ volume O_2.

$$1.00 \times 10^3 \text{ mL soln} \times \frac{1.00 \text{ g soln}}{1 \text{ mL soln}} \times \frac{6.0 \text{ g } H_2O_2}{100 \text{ g soln}} \times \frac{1 \text{ mol } H_2O_2}{34.01 \text{ g } H_2O_2} \times \frac{1 \text{ mol } O_2}{2 \text{ mol } H_2O_2} = 0.88 \text{ mol } O_2$$

The volume of O_2 produced at 25°C and 1.00 atm is

$$V = \frac{nRT}{P} = \frac{0.88 \text{ mol} \times 0.0821 \text{ L} \bullet \text{atm/mol} \bullet \text{K} \times 298 \text{ K}}{1.00 \text{ atm}} = 22 \text{ L}$$

(b) "20-volume H_2O_2" means that 20 mL of O_2 is produced from 1 mL of peroxide solution or 20 L of O_2 from 1 L of peroxide solution. One liter of a 6.0% peroxide solution produces 22 L of O_2 (to two significant figures). Therefore, the older designation is fairly accurate.

S2.97 (a) $N_2H_4(l) + 2H_2O_2(l) \rightarrow N_2(g) + 4H_2O(g)$

(b) $\Delta H° = 1 \times 0\ kJ + 4 \times -241.8\ kJ - (1 \times +50.63\ kJ) - (2 \times -187.78\ kJ) = -642.3\ kJ$

Answer: The heat released per mole of hydrazine is 642.3 kJ/mol.

S2.99

	$2NH_4ClO_4(s)$	$\rightarrow$	$N_2(g)$	+	$Cl_2(g)$	+	$2O_2(g)$	+	$4H_2O(g)$
ΔH_f°:	$2 \times -295.3\ kJ$		$1 \times 0\ kJ$		$1 \times 0\ kJ$		$2 \times 0\ kJ$		$4 \times -241.8\ kJ$

$\Delta H° = 1 \times 0\ kJ + (1 \times 0\ kJ) + (2 \times 0\ kJ) + (4 \times -241.8\ kJ) - (2 \times -295.3\ kJ) = -376.6\ kJ$

The heat released when 1.5 million pounds of ammonium perchlorate decomposes is

$$1.5 \times 10^6\ lb\ NH_4ClO_4 \times \frac{453.6\ g}{1\ lb} \times \frac{1\ mol\ NH_4ClO_4}{117.49\ g\ NH_4ClO_4} \times \frac{376.6\ kJ}{2\ mol\ NH_4ClO_4} = 1.1 \times 10^9\ kJ$$

CHAPTER 12

LIQUIDS, SOLIDS, AND INTERMOLECULAR FORCES

Solutions To Practice Exercises

PE 12.1 The molar heat of vaporization of water is +40.67 kJ/mol (see Table 12.3). Therefore, the molar heat of condensation is –40.67 kJ/mol and

$$q = 10.0\text{ g} \times \frac{1\text{ mol}}{18.02\text{ g}} \times \frac{-40.67\text{ kJ}}{1\text{ mol}} = -22.6\text{ kJ}$$

q is negative because heat is evolved.

Answer: 22.6 kJ of heat is evolved in the condensation.

PE 12.2 First we must find the value of the constant B for ethanol. We know that the vapor pressure P of ethanol is 1.00 atm at 78.5°C = 351.6 K. R = 8.314 J/mol•K and the heat of vaporization is +38.6 kJ/mol.

$$B = \ln P + \Delta H^\circ_{vap}/RT = \ln 1.00\text{ atm} + \frac{38.6 \times 10^3\text{ J/mol}}{8.314\text{ J/mol}\bullet\text{K} \times 351.6\text{ K}} = 13.2$$

Now we can use the value of B to find the vapor pressure P that corresponds to T = 298 K.

$$\ln P = \frac{-38.6 \times 10^3\text{ J/mol}}{8.314\text{ J/mol}\bullet\text{K} \times 298\text{ K}} + 13.2 = -15.6 + 13.2 = -2.4$$

P = 0.091 atm

Answer: The pressure has to be reduced to 0.091 atm.

Alternatively, Equation 12.3 can be used. Substituting P_2 = 1.00 atm, T_2 = 351.6 K, T_1 = 298 K, ΔH°_{vap} = +38.6 × 10^3 J/mol, and R = 8.314 J/mol•K into Equation 12.3 gives

$$\ln\left(\frac{P_1}{1.00\text{ atm}}\right) = \frac{38.6 \times 10^3\text{ J/mol}}{8.314\text{ J/mol}\bullet\text{K}}\left(\frac{1}{351.6\text{ K}} - \frac{1}{298\text{ K}}\right)$$

$\ln P_1 = -2.4$ and P_1 = 0.091 atm

PE 12.3 (a) liquid (b) liquid

PE 12.4 (a) Liquid to solid to vapor. Decreasing the pressure with no change in temperature is equivalent to moving vertically downward in the phase diagram. At 1.1 atm and 0°C, water is liquid. The phase diagram (Figure 12.10) shows that the liquid turns to solid and then to vapor as the pressure decreases at 0°C.

(b) Liquid to vapor. Heating under a constant pressure of 1.1 atm is represented by moving from left to right along a horizontal line at 1.1 atm in the phase diagram (Figure 12.10). The phase diagram shows that the liquid water warms until it reaches the vaporization curve; further heating vaporizes the liquid.

PE 12.5 Increases. For CO_2 the fusion curve slants upward to the right.
Increases. For CO_2 the sublimation curve slants upward to the right.

PE 12.6 (a) No. The pressure must be at least 5.2 atm for liquid CO_2 to exist.
(b) Yes. Liquid H_2O is stable above 0.0060 atm for certain temperatures within this temperature range.

PE 12.7 (a) H_2S (b) H_2S.
H_2S has a considerably greater dipole moment than PH_3. The greater dipole-dipole attractions in H_2S should cause it to have a higher boiling point and greater heat of vaporization.

PE 12.8 (a) No. The size of P prevents the formation of a strong hydrogen bond.
(b) No. The C–H bonds are not strongly polar.
(c) Yes. The N–H bonds are strongly polar and the hydrogen atom in the N–H bond will be attracted to a nitrogen atom in a nearby CH_3NH_2 molecule and will form a strong hydrogen bond.

PE 12.9 (a) I_2. I_2 has more electrons and is larger than Br_2.
(b) SiH_4. C is in the second period and Si is in the third; hence, SiH_4 is larger and has more electrons than CH_4.
(c) CH_3CH_3. CH_3CH_3 is larger than CH_4.

PE 12.10 A body-centered cubic lattice contains one-eighth of each corner atom and the entire center atom:

8 corner atoms × 1/8 = 1 atom
1 center atom × 1 = 1 atom
Sum = 2 atoms per cell

PE 12.11 Equation 12.5 applies to face-centered unit cells.

$$r = \frac{\text{edge}}{2\sqrt{2}} = \frac{408\text{ pm}}{2\sqrt{2}} = 144\text{ pm}$$

PE 12.12 We can calculate the molar mass from the number of atoms per unit cell, the mass of the unit cell, and Avogadro's number. Our steps will be (1) find the number of atoms in the unit cell, (2) find the mass of this number of atoms from the unit cell length and density data, (3) use this number of atoms and mass to find the mass of 6.022×10^{23} atoms, and (4) use the calculated molar mass to identify the element.

Step 1: The unit cell is a face-centered cube and therefore contains four atoms (Table 12.8).

Step 2: The volume of any cubic cell is equal to its edge cubed.

$V = (\text{edge})^3 = (404 \times 10^{-10}\text{ cm})^3 = 6.59 \times 10^{-23}\text{ cm}^3$

The density of the unit cell is the same as the density of the metal. The mass m of the unit cell is the product of its volume V and density d.

$m = V \times d = 6.59 \times 10^{-23}\text{ cm}^3 \times 2.70\text{ g/cm}^3 = 1.78 \times 10^{-22}\text{ g}$

Step 3: The molar mass is equal to the mass of 6.022×10^{23} atoms. The mass of a single atom is equal to the unit cell mass divided by the number of atoms per unit cell. Hence, the molar mass is

$$\frac{1.78 \times 10^{-22}\text{ g}}{4\text{ atoms}} \times \frac{6.022 \times 10^{23}\text{ atoms}}{1\text{ mol}} = 26.8\text{ g/mol}$$

Step 4: The table of atomic weights shows that the element closest to this molar mass is aluminum.

PE 12.13 (a) Table 12.8 shows that each unit cell in a body-centered cubic lattice contains the equivalent of two atoms. Hence, the percent of occupied space is

$$\%\text{ occupied space} = \frac{\text{volume of two atoms}}{\text{unit cell volume}} \times 100\%$$

volume of two atoms $= 2 \times 4\pi r^3/3$

body-centered unit cell volume $= (\text{edge})^3 = \left(4r/\sqrt{3}\right)^3 = 64r^3/3\sqrt{3}$

Substituting these volumes into the percent expression gives

$$\%\text{ occupied space} = \frac{2 \times 4\pi r^3/3}{64r^3/3\sqrt{3}} \times 100\% = \frac{\sqrt{3}\pi}{8} \times 100\% = 68.02\%$$

(b) Table 12.8 shows that each unit cell in a face-centered cubic lattice contains the equivalent of four atoms. Hence, the percent of occupied space is

$$\%\text{ occupied space} = \frac{\text{volume of four atoms}}{\text{unit cell volume}} \times 100\%$$

volume of four atoms $= 4 \times 4\pi r^3/3$

face-centered unit cell volume $= \left(2\sqrt{2}r\right)^3 = 16\sqrt{2}r^3$

Substituting these volumes into the percent expression gives

$$\%\text{ occupied space} = \frac{4 \times 4\pi r^3/3}{16\sqrt{2}r^3} \times 100\% = \frac{\pi}{3\sqrt{2}} \times 100\% = 74.05\%$$

PE 12.14 (a) For chloride ions:

$8\text{ corner } Cl^- \times 1/8 = 1\ Cl^-$

$6\text{ face } Cl^- \times \underline{1/2} = \underline{3\ Cl^-}$

$\text{Sum} = 4\ Cl^-$ per cell

(b) For sodium ions:

12 edge Na^+ × 1/4 = 3 Na^+
1 center Na^+ × 1 = 1 Na^+
Sum = 4 Na^+ per cell

PE 12.15

$$\sin\theta = \frac{n\lambda}{2d} = \frac{2 \times 0.05869 \text{ nm}}{2 \times 0.282 \text{ nm}} = 0.208$$

$\theta = 12.0°$

Solutions To Final Exercises

12.1 (a) gas < liquid < solid (b) solid < liquid < gas

12.3 (a) Molecular substances are held together in the liquid and solid state by intermolecular forces of attraction. The strength of these forces is highly dependent on the distance between the interacting molecules. High pressures force the molecules closer together, thus increasing the strength of the intermolecular forces and favoring the formation of the liquid and solid states.
(b) Low temperatures slow down the molecular speeds and, consequently, decrease the amount of random molecular motion. At these lower speeds and less chaotic molecular motions, the intermolecular attractive forces are strong enough to form groups or clusters of molecules, thus leading to the formation of the liquid and solid states.

12.5 (a) A dynamic equilibrium is a state in which no net change occurs in the system because opposing processes are taking place at the same rate.
(b)
1. A saturated solution in contact with excess solute.
2. A solid in contact with liquid at any point on the fusion curve.
3. A liquid in contact with vapor at any point on the vaporization curve.
4. The ionization of a weak electrolyte.

12.7 The vapor pressure of a liquid increases with increasing temperature.

12.9 Less. The pressure on the top of Mt. Everest is lower than that at sea level. The boiling point of a liquid is the temperature at which its vapor pressure equals the pressure on the liquid. Since the pressure is less, the boiling point will be lower. Remember that the vapor pressure of a liquid increases with increasing temperature.

12.11 Steam contains 40.67 kJ/mol ($= \Delta H^\circ_{vap}$) more energy at 100°C than water at 100°C. Therefore, a given mass of steam at 100°C can transfer more heat energy to your skin as it condenses and cools than an equivalent mass of water at 100°C and consequently, causes a more severe burn.

12.13 During a phase change such as melting or vaporization, the added energy is used to overcome the attractive forces holding the particles together. Therefore, the average potential energy of the particles increases, while the average kinetic energy of the particles remains the same, during the phase change. The temperature depends on the average kinetic energy of the particles. The temperature remains constant during the phase change, since the average kinetic energy doesn't change during the phase change.

12.15

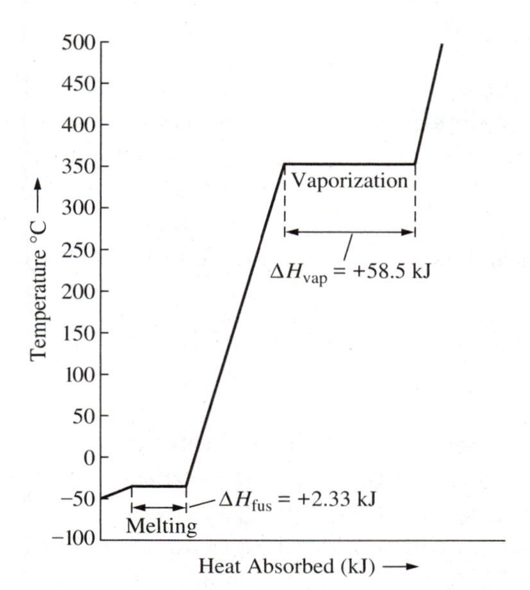

Heating curve for 1 mole of mercury, not drawn to scale

12.17 (a) $1.00 \times 10^3 \text{ g} \times \dfrac{1 \text{ mol}}{18.02 \text{ g}} \times \dfrac{6.01 \text{ kJ}}{1 \text{ mol}} = 334 \text{ kJ}$

(b) $1.00 \times 10^3 \text{ g} \times \dfrac{1 \text{ mol}}{18.02 \text{ g}} \times \dfrac{40.67 \text{ kJ}}{1 \text{ mol}} = 2.26 \times 10^3 \text{ kJ}$

12.19 (a) In going from ice at 0°C to steam at 100°C, we have to (1) melt the ice at 0°C, (2) bring the liquid water at 0°C up to 100°C, and (3) vaporize the liquid at 100°C. The amount of kilojoules required to carry out each step is:

Step 1: $1.00 \text{ mol} \times \dfrac{6.01 \text{ kJ}}{1 \text{ mol}} = 6.01 \text{ kJ}$

Step 2: $q = \text{mass} \times \text{specific heat} \times \Delta T$

$= 18.02 \text{ g} \times 4.18 \text{ J/g}\bullet°\text{C} \times (100°\text{C} - 0°\text{C}) = 7530 \text{ J or } 7.53 \text{ kJ}$

Step 3: $1.00 \text{ mol} \times \dfrac{40.67 \text{ kJ}}{1 \text{ mol}} = 40.7 \text{ kJ}$

The total kilojoules required is the sum of the three steps:

6.01 kJ + 7.53 kJ + 40.7 kJ = 54.2 kJ

(b) $54.2 \text{ kJ} \times \dfrac{1 \text{ kcal}}{4.184 \text{ kJ}} = 13.0 \text{ kcal}$

12.21 The value of B is 13.1 (see Example 12.2 in textbook). Substituting P = 0.28 atm, B = 13.1, and ΔH°_{vap} = +40.7 kJ/mol into Equation 12.2 gives

$$\ln 0.28 = \frac{-40.7 \times 10^3 \text{ J/mol}}{8.314 \text{ J/mol}\bullet\text{K} \times T} + 13.1$$

Solving for T gives T = 341 K = 68°C

12.23 (a) gas (b) supercritical fluid

12.25 Increase. Since solid benzene is more dense than liquid benzene, a vertical line drawn in the direction of increasing pressure from any point on the fusion curve should take liquid benzene into solid benzene. This would mean that the fusion curve slants upward to the right and, consequently, the melting point increases with increasing pressure.

12.27 A triple point on a phase diagram represents a pressure, temperature condition under which all three states, solid, liquid, and gas, can exist in equilibrium.

The ice will sublime; i.e., it will warm up very slightly and then pass directly into the vapor state, without ever having melted, and become a liquid.

12.29 (a) Yes, if the temperature is at or below the critical temperature. In fact, it will be either liquefied or solidified, depending on the temperature. Below the critical temperature each vertical line (increasing P at constant T) crosses either a vapor/liquid boundary or a vapor/solid boundary.

(b) Yes. Sufficient cooling will liquefy any vapor. In fact, it will be condensed to liquid or solid, depending on the pressure, i.e., depending on which horizontal line is traveled from right to left.

12.31

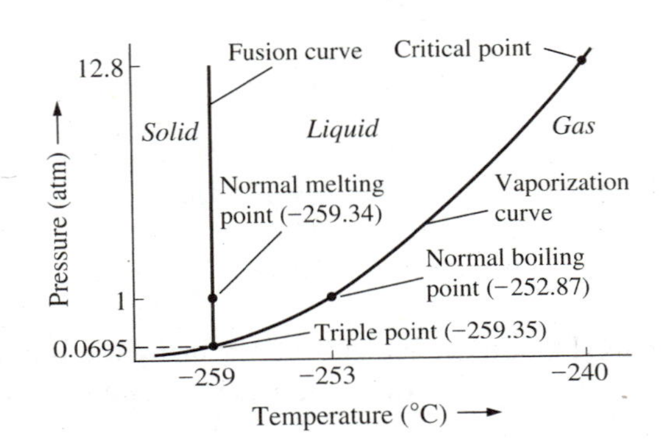

12.33 Greater. The upward fusion curve slants very slightly to the right. Therefore, a vertical line drawn in the direction of increasing pressure from any point on the fusion curve takes liquid hydrogen into solid hydrogen. This means that solid hydrogen is more compact than liquid hydrogen, so the density of solid hydrogen must be greater than that of liquid hydrogen.

12.35 (a) Three states. There are two triple points in the phase diagram. A solid-Liquid II–Liquid I triple point around 1.8 K and 30 atm, and a Liquid II–Liquid I-vapor triple point around 2.17 K and 0.05 atm.

Note: Each distinct liquid is a separate state.

(b) No. Helium cannot exist in the solid state below about 25 atm, no matter how much the temperature is lowered.

(c) Increases. The fusion curve slants upward to the right showing that at higher pressures the solid-liquid equilibrium or melting point occurs at higher temperatures.

Comment: According to Sears, Zemansky, and Young, *University Physics,* 6th ed., Addison-Wesley 1982 (page 339), Liquid I and Liquid II cannot coexist side by side; the liquid helium must be entirely in one state or entirely in the other. The Liquid I–Liquid II line in the phase diagram shows conditions of temperature and pressure at which the transition can occur; it is not a transition equilibrium line. Furthermore, the two points that would ordinarily be called triple points are not true triple points, since it is not possible at either point to have equilibrium with three states.

12.37 Liquid helium I. Increasing pressure tends to decrease the volume of a substance. Since increasing pressure changes liquid helium II into liquid helium I, liquid helium I must be more compact and, consequently, must have a greater density.

12.39 (a) 1. graphite 2. diamond
(b) 4000°C (the point where the graphite/vapor line intersects the horizontal 1-atm line).

12.41 (a) An intermolecular attractive force is any force that causes otherwise independent atoms and molecules to draw near each other. Chemical bonding forces cause atoms to join together to form molecules, covalent network solids, and metallic solids. They also cause ions to join in forming ionic solids.
(b) The chemical bonding forces are stronger.
(c) Refer to Tables 9.3 and 12.3.

12.43 See Figures 12.14 and 12.15 in the textbook for examples of hydrogen bonding. A hydrogen bond is an intermolecular attraction that results when a hydrogen atom bonded to a small, highly electronegative atom, principally F, O, or N, interacts with a lone pair of electrons situated on another small, highly electronegative atom. Hydrogen bonding can also occur intramolecularly.

The "hydrogen bond" is the strongest intermolecular attractive force, but it is still much weaker than chemical bonding forces. The forces or bonds holding hydrogen in H_2 and H_2S are chemical bonding forces, i.e., covalent bonds, and are not the much weaker intermolecular attractions called "hydrogen bonds". In conclusion, "hydrogen bonds" and covalent bonds involving hydrogen are not the same thing and, therefore, should not be confused or substituted for one another.

12.45 Usually linear. The partially positive, bonded hydrogen atom tends to interact to some extent with a lone pair of electrons on the other atom. So, the bonded hydrogen atom is involved in an interaction with the electron density contained in an orbital on the other atom, and this interaction will be maximized if the hydrogen atom and the orbital on the other atom interact or overlap with one another linearly. Hence, the three atomic centers involved in the hydrogen bond tend to lie in a straight line or be linear.

12.47 (a) A dipole-dipole interaction is the net electrostatic attraction in a collection of polar molecules. Dipole-dipole interactions bring and hold molecules together, and the strength of the attraction increases with the polarity of the participating molecules. A London dispersion force is the net attractive force that exists in a collection of mutually induced dipoles. London dispersion forces bring and hold molecules together, and the strength of the force depends on the magnitude of the induced dipole, which in turn depends on the polarizability of the substance.
(b) The London dispersion force.

12.49 (a) Both I_2 and CCl_4 are nonpolar molecules. Therefore, only London dispersion forces exist.
(b) Both CH_3Br and CH_3Cl are polar molecules. Therefore, both dipole-dipole forces and London dispersion forces exist.
(c) Both CH_3OH and CH_3NH_2 have very polar bonds involving the hydrogen atom. Therefore, both hydrogen bonding and London dispersion forces exist.

12.51 (a) Yes. (b) No. (c) No.
H–F– – –H–F

12.53 In general, when comparing similar substances, the substance with the strongest intermolecular forces has the highest boiling point. Both N_2 and O_2 are nonpolar molecules and, therefore, only London dispersion forces are operative in these substances. On the other hand, NO is a polar molecule and, therefore, both dipole-dipole and London dispersion forces exist between the NO molecules. Since NO has dipole-dipole forces in addition to London dispersion forces, NO should have a substantially higher boiling point than N_2 and O_2.

12.55 (a) CH_3OH (b) CH_3OH
CH_3OH can participate in hydrogen bonding while nonpolar CH_3CH_3 only has London dispersion forces holding its molecules together.

12.57 CH_3OCH_3. CH_3CH_2OH can participate in hydrogen bonding, the strongest intermolecular attractive force, while CH_3OCH_3 cannot. Therefore, the forces holding the CH_3OCH_3 molecules together should be considerably weaker than those holding the CH_3CH_2OH molecules together, and, consequently, CH_3OCH_3 should have a higher vapor pressure.

12.59 (a) Xe. Xe is larger than Ar and, therefore, should have stronger London forces.
(b) HF. HF participates in hydrogen bonding while HBr doesn't.
(c) $HOCH_2CH_2OH$. $HOCH_2CH_2OH$ can form twice as many hydrogen bonds as CH_3CH_2OH.

12.61 1. Liquids take the shape of their container. Gases also take the shape of their container, but the shape of solids is fixed.
2. Liquids have a definite volume. Solids also have a definite volume, but gases do not. Gases expand to fill the volume of their container.
3. Surface tension. The surface of a liquid resists stretching and maintains as small an area as possible.
4. Viscosity. Liquids exhibit an extremely wide range of resistances to flow.
5. Capillary action. Most liquids spontaneously rise in a narrow tube.

12.63 (a) The viscosity and surface tension of a liquid increases with increasing intermolecular attractions.
(b) The viscosity and surface tension of a liquid decreases with increasing temperature.

12.65 (a) A molecular crystal is held together by intermolecular forces. These always include the London dispersion forces, and possibly dipole-dipole forces and/or hydrogen bonds as well.
(b) A metallic crystal is held together by metallic bonding. This is a highly delocalized bond that encompasses the entire metallic crystal. The metal atoms or the positive cores of the metal atoms are held together at regular lattice sites by the delocalized, freely flowing, surrounding valence electrons of the metal atoms.
(c) An ionic crystal is held together by ionic bonds (ionic bonding). In an ionic crystal, the cations and anions are arranged such that the electrical attractions between oppositely charged ions exceed the electrical repulsions between like charged ions. The attractive forces between the oppositely charged ions (the ionic bonds) hold the ions at regular lattice sites.
(d) A network covalent crystal is held together by covalent bonds (covalent bonding). The atoms in a network covalent crystal are held together in large networks or chains by covalent bonds. Each atom in the network is covalently bonded to its neighboring atoms, and it's this continuous bonding scheme that maintains the crystal structure of the solid.
(e) An amorphous solid is held together by intermolecular forces. An amorphous solid can be looked upon as a supercooled liquid of very high viscosity. Amorphous solids are mainly held together by London dispersion forces. Certain amorphous solids, like glass, also contain or exhibit a degree of network covalent bonding.

12.67 For (a), (b), (c), and (d) refer to Table 12.7 in the textbook. (e) glass and tar. Glass and tar have no definite melting points, they can take on any shape, and their solids break into fragments that have curved surfaces. All this shows that they are not crystalline materials.

12.69 (a) Mothballs–molecular solid. Mothballs are volatile, low melting, and poor conductors of heat and electricity.
(b) Sand–network covalent solid. Sand is very hard and brittle. It has a very high melting point and very low volatility. It is also a poor conductor of electricity.
(c) Ice–molecular solid. Ice is low melting with a low heat of fusion. It is also a poor conductor of heat and electricity.
(d) Table salt–ionic solid. Table salt is hard and brittle. It has a high melting point and low volatility. Salt only conducts electricity in the molten state.
(e) Toaster filament–metallic solid. A toaster filament is a good conductor of heat and electricity.

12.71 (a) PrI_3–ionic solid (b) OsF_6–molecular solid (c) BN–network covalent solid

12.73 (a) The unit cell is a single stamp.
(b) The unit cell is a square containing two black and two white smaller squares arranged as follows:

(c) Fruits like grapefruit are usually stacked in hexagonal close-packed pyramids so that they will not slip. The unit cell is therefore the hexagonal close-packed unit cell depicted in Figure 12.37.

12.75 See Figures 12.31, 12.32, and 12.33 in textbook.

12.81 See "Lattice Imperfections" in Section 12.6 of the textbook.

12.83 Decreases. First-order scattering occurs when n = 1 in the Bragg equation. Therefore, for first-order scattering

$$\sin\theta = \lambda/2d$$

The $\sin\theta$ goes from 0 to +1 as θ goes from 0° to 90°. As d increases, the $\sin\theta$ decreases and the first-order scattering angle decreases also.

12.85 (a) Equation 12.5 applies to face-centered unit cells.

$$r = \frac{\text{edge}}{2\sqrt{2}} = \frac{407.86\text{ pm}}{2\sqrt{2}} = 144.20\text{ pm}$$

(b) The density of gold is the same as the density of the unit cell. A face-centered unit cell contains four atoms. The volume of the unit cell is

$$V = (\text{edge})^3 = (407.86 \times 10^{-10}\text{ cm})^3 = 6.7847 \times 10^{-23}\text{ cm}^3$$

The density of gold is

$$d = \frac{m}{V} = \frac{4\text{ Au atoms} \times 196.9665\text{ u/Au atom} \times 1.661 \times 10^{-24}\text{ g/u}}{6.7847 \times 10^{-23}\text{ cm}^3} = 19.29\text{ g/cm}^3$$

12.87 The density of iron is the same as the density of the unit cell. The volume of any cubic cell is equal to its edge cubed. The volume of the iron unit cell is therefore

$V = (\text{edge})^3 = (286 \times 10^{-10}\ \text{cm})^3 = 2.34 \times 10^{-23}\ \text{cm}^3$

The unit cell is a body-centered cube and therefore contains two iron atoms (Table 12.8). The density of iron is

$$d = \frac{m}{V} = \frac{2\ \text{Fe atoms} \times 55.847\ \text{u/Fe atom} \times 1.661 \times 10^{-24}\ \text{g/u}}{2.34 \times 10^{-23}\ \text{cm}^3} = 7.93\ \text{g/cm}^3$$

The calculated density is slightly larger than the observed density.

12.89 (a) The volume of the unit cell is $V = (\text{edge})^3 = (495.05 \times 10^{-10}\ \text{cm})^3 = 1.2132 \times 10^{-22}\ \text{cm}^3$.

The mass of the unit cell is $m = V \times d = 1.2132 \times 10^{-22}\ \text{cm}^3 \times 11.3\ \text{g/cm}^3 = 1.37 \times 10^{-21}\ \text{g}$.

One mole of lead, 207.2 g, contains 6.022×10^{23} atoms. The number of atoms in 1.37×10^{-21} g is

$$1.37 \times 10^{-21}\ \text{g} \times \frac{6.022 \times 10^{23}\ \text{atoms}}{207.2\ \text{g}} = 3.98 = 4\ \text{atoms}$$

Four atoms per unit cell corresponds to a face-centered cubic lattice (Table 12.8).

(b) Equation 12.5 applies to face-centered unit cells.

$$r = \frac{\text{edge}}{2\sqrt{2}} = \frac{495.05\ \text{pm}}{2\sqrt{2}} = 175.03\ \text{pm}$$

12.91 (a) Refer to Figure 12.37 in the textbook.

(b) The number of zinc atoms in the unit cell can either be calculated using the entire hexagonal arrangement and dividing by three or it can be calculated using a single unit cell. Using the entire hexagonal arrangement and dividing by three gives

2 face atoms (centers of hexagon) × 1/2 = 1 atom
12 corner atoms × 1/6 = 2 atoms
3 inside atoms × 1 = 3 atoms
Sum = 6 atoms

Therefore, the unit cell contains 6 atoms/3 = 2 zinc atoms.

(c) Each zinc atom is in contact with 12 other zinc atoms. Therefore, the coordination number is 12.

12.93 See Figure 12.39 in the textbook.

(a) Six (b) Six

Each Na^+ ion is surrounded by six Cl^- ions and each Cl^- ion is surrounded by six Na^+ ions.

12.95 $$d = \lambda/2 \sin\theta = \frac{0.154\ \text{nm}}{2 \sin 14.7°} = 0.303\ \text{nm} = 303\ \text{pm}$$

12.97 (a) The molecules leaving the liquid state are the ones that have a higher kinetic energy than the average. Therefore, as they escape, the average kinetic energy of the remaining molecules drops as well as the temperature of the remaining liquid, since the temperature depends on the average kinetic energy.

A more elaborate way of looking at this is the following:

Evaporation requires energy, because energy is required to overcome the intermolecular forces of attraction between the molecules. The energy used to overcome these attractive forces comes from the kinetic energy of the molecules. Therefore, evaporation causes the average kinetic energy of the molecules to decrease and, consequently, the temperature drops. (It is the higher kinetic energy molecules that expend their kinetic energy in overcoming the intermolecular attractive forces.)

(b) As the water on the skin evaporates, the remaining water drops in temperature. Body heat then flows to the cooler water and any excess body heat is removed in the process.

(c) The tree's respiration process produces O_2 and H_2O. The H_2O is removed by evaporation from the leaves and this cools down the tree.

12.99 On a dry day there is less water vapor in the air. This allows the perspiration to evaporate faster, since there is little condensation taking place. The faster the perspiration evaporates, the faster the excess heat is removed from the body. Consequently, perspiration is more effective in cooling down a body on a dry day.

12.101 Butane (C_4H_{10}). The London forces in butane, with its longer carbon chain, should be stronger than those in propane and methane. Hence, butane should have a higher boiling or condensation point.

12.103 (a) Plasma is the fluid portion of blood. Plasma contains many large complex molecules, such as carbohydrates, fatty acids, vitamins, etc. These complex molecules have strong London forces and that makes the blood plasma more viscous than water.

(b) Mercury is a poor wetting agent because it has a high surface tension. Therefore, when spilled, it forms spherical droplets, since a sphere provides the smallest surface area for a given volume. The surface tension that maintains the spherical shape is stronger than the gravitational force that would flatten the sphere. Also, the forces that hold the mercury surface together are stronger than the attractive forces between mercury and the surface on which it rests, therefore the mercury sphere is complete (water droplets on certain surfaces are more like hemispheres).

(c) See "Chemical Insight: How to Make Liquids Wetter" in Section 12.4 in textbook.

12.105 See the derivations in Figures 12.35, 12.36, and 12.37 in the textbook.

12.107 This problem can be broken down into two steps: (1) Find the equilibrium state that just the ice and liquid water would establish when mixed. (2) Find the final equilibrium temperature when the copper is added to the equilibrium state established in step 1.

Step 1: The heat lost by the liquid water is

$$q_{H_2O\ cooling} = m \times \text{specific heat} \times \Delta T = 250\ g \times 4.18\ J/g{\bullet}°C \times (T_f - 75°C)$$

The heat gained by the ice is

1. In melting: $q_{ice\ melting} = 50.0\ g \times \dfrac{1\ mol}{18.02\ g} \times \dfrac{6.01\ kJ}{1\ mol} = 16.7\ kJ$

2. In warming: $q_{H_2O\ warming} = 50.0\ g \times 4.18\ J/g{\bullet}°C \times (T_f - 0°C)$

Equating the heat gain to the heat loss we get

$$q_{H_2O\ cooling} = -q_{ice\ melting} - q_{H_2O\ warming}$$

$$1{,}045\ J/°C \times T_f - 78{,}375\ J = -16.7 \times 10^3\ J - 209\ J/°C \times T_f$$

$$1{,}254/°C \times T_f = 61{,}675$$

$$T_f = 49°C$$

Therefore, at equilibrium there is 300 g of liquid water at 49°C.

Step 2: The heat lost by the water is $q_{lost} = 300\ g \times 4.18\ J/g{\bullet}°C \times (T_f - 49°C)$.

The heat gained by the copper is $q_{gained} = 50.0\ g \times 0.385\ J/g{\bullet}°C \times (T_f - 25°C)$.

Equating the heat gain to the heat loss we get

$q_{lost} = -q_{gained}$

$1{,}254\ J/°C \times T_f - 61{,}446\ J = -19.25\ J/°C \times T_f + 481\ J$

$1{,}273/°C \times T_f = 61{,}927$

$T_f = 49°C$

Answer: The final equilibrium temperature is 49°C.

12.109 (a) The volume of the unit cell is $V = (\text{edge})^3 = (356.7 \times 10^{-10}\ cm)^3 = 4.538 \times 10^{-23}\ cm^3$.

The mass of the unit cell is $m = V \times d = 4.538 \times 10^{-23}\ cm^3 \times 3.513\ g/cm^3 = 1.594 \times 10^{-22}\ g$.

The number of carbon atoms in the unit cell is

$$1.594 \times 10^{-22}\ g\ C \times \frac{1\ u}{1.661 \times 10^{-24}\ g} \times \frac{1\ C\ atom}{12.011\ u\ C} = 7.99 \text{ or } 8\ C\ atoms$$

(b) In drawing the unit cell, keep in mind that:

1. Each corner atom contributes only one-eighth of an atom to the unit cell and there must be eight of them, since it is a cubic unit cell. The total contribution of the corner atoms is one atom.
2. Each face atom contributes one-half of an atom to the unit cell. There are six face atoms in the unit cell. The total contribution of the face atoms is three atoms.
3. An internal atom contributes one whole atom to the unit cell. There are four of them.
4. Each atom is bonded to four other atoms by covalent bonds.

We can draw this unit cell by bonding each internal atom to one corner atom and three face atoms. Alternating corners are thus bonded within one cell:

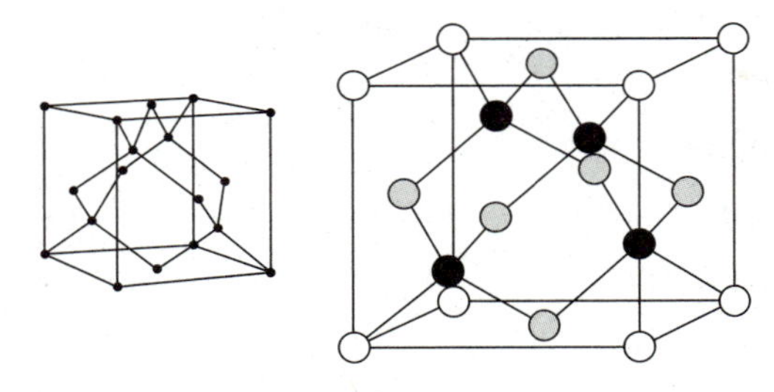

● Atoms inside unit cell
◐ Atoms on faces of unit cell
○ Atoms on corners of unit cell

(c) No.

12.111 The density of NaCl is the same as the unit cell density. Each unit cell contains the equivalent of four chlorine atoms and four sodium atoms. The unit cell is a cube; therefore, its volume is equal to the cube of its edge. From Figure 12.39 we see that the edge length is equal to twice the distance between the centers of the Na^+ and Cl^- ions, or edge = 2 × 282 pm = 564 pm. The density of NaCl is

$$d = \frac{m}{V} = \frac{(4\ \text{Na atoms} \times 22.99\ \text{u/Na atom} + 4\ \text{Cl atoms} \times 35.45\ \text{u/Cl atom}) \times 1.661 \times 10^{-24}\ g/u}{(564 \times 10^{-10}\ cm)^3} = 2.16\ g/cm^3$$

The calculated density is slightly lower than the observed density.

12.113 The density of the unit cell is the same as the density of CaO. The volume of the unit cell is

$V = (\text{edge})^3 = (480 \times 10^{-10}\ \text{cm})^3 = 1.11 \times 10^{-22}\ \text{cm}^3$

The mass of the unit cell is

$\text{mass unit cell} = V \times d = 1.11 \times 10^{-22}\ \text{cm}^3 \times 3.35\ \text{g/cm}^3 = 3.72 \times 10^{-22}\ \text{g}$

The mass of one CaO formula unit is

$\text{mass CaO formula unit} = (40.078\ \text{u} + 15.9994\ \text{u}) \times 1.661 \times 10^{-24}\ \text{g/u} = 9.314 \times 10^{-23}\ \text{g}$

The number of CaO formula units in the unit cell is

$$\frac{3.72 \times 10^{-22}\ \text{g/unit cell}}{9.314 \times 10^{-23}\ \text{g/CaO formula unit}} = 3.99 = 4\ \text{CaO formula units/unit cell}$$

This number of formula units in a unit cell indicates that the Ca^{2+} ions form a face-centered cubic lattice and the O^{2-} ions form a face-centered cubic lattice. The complete lattice is a superposition of the two separate lattices. The unit cell is a face-centered cube like that of NaCl, which is depicted in Figure 12.39.

CHAPTER 13

SOLUTIONS

Solutions To Practice Exercises

PE 13.1 $1.05 \times 10^{-4} \text{ mol/L} \times \dfrac{0.100 \text{ atm}}{0.993 \text{ atm}} = 1.06 \times 10^{-5} \text{ mol/L}$

PE 13.2 $P_{methanol} = X_{methanol} P^{\circ}_{methanol} = \dfrac{1.00}{3.00} \times 88.7 \text{ torr} = 29.6 \text{ torr}$

$P_{ethanol} = X_{ethanol} P^{\circ}_{ethanol} = \dfrac{2.00}{3.00} \times 44.5 \text{ torr} = 29.7 \text{ torr}$

$P_{total} = P_{methanol} + P_{ethanol} = 29.6 \text{ torr} + 29.7 \text{ torr} = 59.3 \text{ torr}$

PE 13.3 At 25°C the vapor pressure of pure water is 23.8 torr. The vapor pressure lowering is

$VPL = P^{\circ}_{solvent} - P_{solution} = 23.8 \text{ torr} - 22.0 \text{ torr} = 1.8 \text{ torr}$

The mole fraction of sucrose is

$X_{sucrose} = \dfrac{VPL}{P^{\circ}_{water}} = \dfrac{1.8 \text{ torr}}{23.8 \text{ torr}} = 0.076$

PE 13.4 The molar mass of ethanol is 46.07 g/mol. The number of moles of ethanol is

$n_{ethanol} = 6.00 \text{ g} \times \dfrac{1 \text{ mol}}{46.07 \text{ g}} = 0.130 \text{ mol}$

This number of moles is dissolved in 75.0 g = 0.0750 kg of water. The molality is

$\text{molality} = \dfrac{\text{number of moles of ethanol}}{\text{number of kilograms of water}} = \dfrac{0.130 \text{ mol}}{0.0750 \text{ kg}} = 1.73 \text{ mol/kg or } 1.73 \text{ m}$

PE 13.5 The molar mass of biphenyl is 154.21 g/mol. The number of moles of biphenyl is

$$n_{\text{biphenyl}} = 1.25 \text{ g} \times \frac{1 \text{ mol}}{154.21 \text{ g}} = 8.11 \times 10^{-3} \text{ mol}$$

The molality of the solution is

$$m_{\text{biphenyl}} = \frac{\text{number of moles of biphenyl}}{\text{number of kilograms of water}} = \frac{8.11 \times 10^{-3} \text{ mol}}{50.0 \times 10^{-3} \text{ kg}} = 0.162 \text{ mol/kg}$$

K_b for benzene is

$$K_b = \frac{\text{BPE}}{m_{\text{biphenyl}}} = \frac{0.411°\text{C}}{0.162 \text{ mol/kg}} = 2.54°\text{C•kg/mol}$$

PE 13.6 The normal freezing point and K_f of benzene are 5.5°C and 4.90°C•kg/mol, respectively (see Table 13.2). The freezing point depression is

$$\text{FPD} = T_f^\circ - T_f = 5.5°\text{C} - 3.41°\text{C} = 2.1°\text{C}$$

The molality is

$$m_{\text{benzene}} = \frac{\text{FPD}}{K_f} = \frac{2.1°\text{C}}{4.90°\text{C•kg/mol}} = 0.43 \text{ mol/kg}$$

PE 13.7

$$\pi = MRT = \frac{1.00 \text{ mol}}{1 \text{ L}} \times \frac{0.0821 \text{ L•atm}}{\text{mol•K}} \times 298 \text{ K} = 24.5 \text{ atm}$$

PE 13.8 Step 1. BPE = 0.376°C

Step 2.

$$m = \frac{\text{BPE}}{K_b} = \frac{0.376°\text{C}}{3.63°\text{C•kg/mol}} = 0.104 \text{ mol/kg}$$

Step 3. The number of moles of cholesterol in 25.0 g = 0.0250 kg of solvent is

$$n_{\text{cholesterol}} = 0.0250 \text{ kg solvent} \times \frac{0.104 \text{ mol}}{1 \text{ kg solvent}} = 2.60 \times 10^{-3} \text{ mol}$$

Step 4. 2.60×10^{-3} mol of cholesterol corresponds to 1.00 g. The molar mass of cholesterol is 1.00 g/2.60×10^{-3} mol = 385 g/mol.

PE 13.9 Step 1. The osmotic pressure of the solution is 12.0 torr × 1 atm/760 torr = 1.58×10^{-2} atm.
Steps 2 and 3. The number of moles of protein in 1 mL of solution is obtained by substituting $\pi = 1.58 \times 10^{-2}$ atm, $V_{\text{solution}} = 1.00 \times 10^{-3}$ L, R = 0.0821 L•atm/mol•K, and the temperature 25°C = 298 K into Equation 13.12a:

$$n_{\text{protein}} = \frac{\pi V_{\text{soln}}}{RT} = \frac{1.58 \times 10^{-2} \text{ atm} \times 1.00 \times 10^{-3} \text{ L}}{0.0821 \text{ L•atm/mol•K} \times 298 \text{ K}} = 6.46 \times 10^{-7} \text{ mol}$$

Step 4. The mass of 6.46×10^{-7} mol is 28.8 mg. The molar mass is 28.8×10^{-3} g/6.46×10^{-7} mol = 4.46×10^{4} g/mol.

PE 13.10 The freezing point depression of a 1.00 molal aqueous solution of any nonionized solute is 1.86°C.

1.00 m NaCl: $NaCl(aq) \rightarrow Na^+(aq) + Cl^-(aq)$

Assuming complete dissociation, a 1.00 m NaCl solution will be 2.00 m in solute particles. Therefore, the calculated freezing point depression is 2 × 1.86°C = 3.72°C.

1.00 m $MgSO_4$: $MgSO_4(aq) \rightarrow Mg^{2+}(aq) + SO_4^{2-}(aq)$

Assuming complete dissociation, a 1.00 m $MgSO_4$ solution will be 2.00 m in solute particles. Therefore, the calculated freezing point depression is 2 × 1.86°C = 3.72°C.

1.00 m H_2SO_4: $H_2SO_4(aq) \rightarrow 2H^+(aq) + SO_4^{2-}(aq)$

Assuming complete dissociation, a 1.00 m H_2SO_4 solution will be 3.00 m in solute particles. Therefore, the calculated freezing point depression is 3 × 1.86°C = 5.58°C.

1.00 m $ZnCl_2$: $ZnCl_2(aq) \rightarrow Zn^{2+}(aq) + 2Cl^-(aq)$

Assuming complete dissociation, a 1.00 m $ZnCl_2$ solution will be 3.00 m in solute particles. Therefore, the calculated freezing point depression is 3 × 1.86°C = 5.58°C.

0.100 m $MgSO_4$: The calculated freezing point depression of a 0.100 m $MgSO_4$ solution will be one-tenth that of a 1.00 m solution or 0.100 × 3.72°C = 0.372°C.

0.0100 m $MgSO_4$: The calculated freezing point depression of a 0.0100 m $MgSO_4$ solution will be one-hundredth that of a 1.00 m solution or 0.0100 × 3.72°C = 0.0372°C.

0.00100 m $MgSO_4$: The calculated freezing point depression of a 0.00100 m $MgSO_4$ solution will be one-thousandth that of a 1.00 m $MgSO_4$ solution or 0.00100 × 3.72°C = 0.00372°C.

PE 13.11

$$m = 0.100;\ i = \frac{0.225°C}{0.186°C} = 1.21$$

$$m = 0.0100;\ i = \frac{0.0285°C}{0.0186°C} = 1.53$$

$$m = 0.00100;\ i = \frac{0.00339°C}{0.00186°C} = 1.82$$

PE 13.12 The i factor for 1.00 m H_2SO_4 is

$$i = \frac{4.04°C}{1.86°C} = 2.17$$

If we assume no dissociation, the calculated BPE for a 1.00 m aqueous solution would be 0.512°C. The observed BPE is therefore

observed BPE = i × calculated BPE assuming no dissociation = 2.17 × 0.512°C = 1.11°C

The boiling point of the solution will be 100.00°C + 1.11°C = 101.11°C.

Solutions To Final Exercises

13.1 Intermolecular attractions between the solute particles tend to keep the solute particles attached to one another, and this holds for the solvent particles as well. On the other hand, the random thermal motions of all particles tend to disperse the solute and solvent particles in space. The amount that dissolves depends on the compromise reached between these two opposing tendencies and also on the attractive forces between solute and solvent molecules.

13.3 (a) Because water has the ability to stabilize ions in solution.
(b) By forming hydration layers around the ions. These hydration layers inhibit recombination of the ions.

13.5 (a) By hydrogen bonding between the ethanol and water molecules.
(b) By reacting with the hydrogen bromide gas to form hydrated hydrogen and bromide ions: $HBr(g) \rightarrow H^+(aq) + Br^-(aq)$.
(c) By forming hydration layers around the K^+ and NO_3^- ions.

13.7 Most solids become more soluble with increasing temperature because the dissolving of most solids near the saturation point is endothermic. (In fact, the heat of solution of most solids is endothermic, the solution becoming cooler as the solid dissolves.)

13.9 The solution process can be imagined as being made up of three steps; two endothermic and one exothermic (see "Heats of Solution" in Section 13.1 in textbook for details). The heat of solution is the sum of the heats of these three steps. For a gas the exothermic Step 3 usually predominates and, therefore, the heat of solution of a gas is usually negative.

13.11 Liquids and solids are not appreciably affected by moderate changes in pressure; therefore pressure does not influence the equilibrium between a liquid solution and the undissolved solid or liquid.

13.13 Blood plasma contains various dissolved salts that reduce the solubility of gases like N_2, which do not react with water. This effect is called "salting-out."

13.15 The partial pressure of CO_2 increases from 1.00 atm to 4.0 atm. Henry's law states that the solubility will increase by the same factor as the partial pressure. Hence, the new solubility will be

$$\frac{3.1 \times 10^{-2}\ \text{mol}}{1\text{L}} \times \frac{4.0\ \text{atm}}{1.00\ \text{atm}} = 0.12\ \text{mol/L}$$

13.17 An ideal solution is a postulated solution in which the forces between all particles of both solvent and solute are identical. The solvent–solvent attractive forces are equal to both the solute–solute and solvent–solute attractive forces.

1. Ideal solutions have a volume equal to the sum of its pure component values. Since the forces acting between the molecules are unchanged, there is no pulling closer together or spreading further apart.
2. The heat of solution is zero. Since the number and magnitude of the attractive forces between the molecules is unchanged, there is no change in the overall potential energy.
3. Each component of an ideal solution obeys Raoult's law. Since the forces acting on each molecule is the same as in the pure state, the differences in any properties is due entirely to the difference in the number of molecules. In particular, the forces holding the molecules in solution are identical to the pure state forces, so the vapor pressure of a given component should only depend on the number of molecules of that component present per unit area of solution surface. For example, if the number of molecules is half as many as in the pure solvent, then that component's vapor pressure should be half of that of the pure solvent.
4. The vapor above an ideal solution obeys Dalton's law. Since the molecules are behaving ideally in the solution, they also behave ideally in the vapor.

13.19 (a) Negative deviations from Raoult's law occur when the attractive forces between the component molecules A and B of the solution are greater than the A–A forces or the B–B forces.

(b) Positive deviations from Raoult's law occur when the attractive forces between the component molecules A and B of the solution are less than the A–A forces or the B–B forces.

13.21 Use the equations $P_{methanol} = X_{methanol} \times 89$ torr, $P_{ethanol} = X_{ethanol} \times 40$ torr, and $P_{total} = P_{methanol} + P_{ethanol}$ to calculate points at X = 0, 0.2, 0.4, 0.6, 0.8, and 1. Plot the results on graph paper using the same setup as that in Figure 13.9.

13.23 (a) To solve this problem we first have to calculate the mole fractions of CS_2 and $CH_3COOC_2H_5$ in the solution. The molar masses of CS_2 and $CH_3COOC_2H_5$ are 76.14 g/mol and 88.11 g/mol, respectively. The numbers of moles of CS_2 and $CH_3COOC_2H_5$ are

$$n_{CS_2} = 100\text{ g} \times \frac{1\text{ mol}}{76.14\text{ g}} = 1.31\text{ mol}$$

$$n_{CH_3COOC_2H_5} = 100\text{ g} \times \frac{1\text{ mol}}{88.11\text{ g}} = 1.13\text{ mol}$$

The mole fractions of CS_2 and $CH_3COOC_2H_5$ are

$$X_{CS_2} = \frac{n_{CS_2}}{n_{CS_2} + n_{CH_3COOC_2H_5}} = \frac{1.31\text{ mol}}{1.31\text{ mol} + 1.13\text{ mol}} = 0.537$$

$$X_{CH_3COOC_2H_5} = 1 - X_{CS_2} = 1 - 0.537 = 0.463$$

The partial pressures of CS_2 and $CH_3COOC_2H_5$ are

$$P_{CS_2} = X_{CS_2} P^{\circ}_{CS_2} = 0.537 \times 390\text{ torr} = 209\text{ torr}$$

$$P_{CH_3COOC_2H_5} = X_{CH_3COOC_2H_5} P^{\circ}_{CH_3COOC_2H_5} = 0.463 \times 100\text{ torr} = 46.3\text{ torr}$$

(b) The total pressure is obtained by adding the two partial pressures:

$$P_{total} = P_{CS_2} + P_{CH_3COOC_2H_5} = 209\text{ torr} + 46.3\text{ torr} = 255\text{ torr}$$

13.25

$$X_{water} = \frac{P_{water}}{P^{\circ}_{water}} = \frac{22.50\text{ torr}}{23.76\text{ torr}} = 0.9470$$

$$X_{urea} = 1 - X_{water} = 1 - 0.9470 = 0.0530$$

13.27 Yes. Positive. According to Raoult's law, the vapor pressure of ethanol above the solution should be

$$P_{ethanol} = X_{ethanol} P^{\circ}_{ethanol} = 0.282 \times 43.5\text{ torr} = 12.3\text{ torr}$$

This is considerably lower than the actual value of 23.4 torr. Therefore, the solution exhibits a positive deviation from Raoult's law.

13.29 The mole fraction of benzene in the solution is 0.25 (see Example 13.2). The mole fraction of benzene in the vapor above the solution is (see Example 13.2 for data)

$$X_{benzene\ vapor} = \frac{P_{benzene}}{P_{total}} = \frac{18.7\text{ torr}}{35.4\text{ torr}} = 0.528$$

The mole fraction of benzene in the vapor is more than twice the mole fraction of benzene in the solution.

13.31 Refer to definitions on page 604 of the textbook.

13.33 The molar mass of $C_6H_4Cl_2$ is 147.0 g/mol. The number of moles of $C_6H_4Cl_2$ is

$$n_{dichlorobenzene} = 0.615 \text{ g} \times \frac{1 \text{ mol}}{147.0 \text{ g}} = 4.18 \times 10^{-3} \text{ mol}$$

This number of moles is dissolved in 21.8 g = 0.0218 kg of benzene. The molality is

$$\text{molality} = \frac{\text{number of moles of dichlorobenzene}}{\text{number of kilograms of benzene}} = \frac{4.18 \times 10^{-3} \text{ mol}}{2.18 \times 10^{-2} \text{ kg}} = 0.192 \text{ mol/kg or } 0.192 \text{ m}$$

13.35

$$\text{mass } CaCl_2 = 1.50 \text{ mol} \times \frac{111.0 \text{ g}}{1 \text{ mol}} = 166 \text{ g}$$

total mass solution = mass water + mass $CaCl_2$ = 1000 g + 166 g = 1166 g

13.37 (a) The molar mass of H_2SO_4 is 98.08 g/mol. The molarity is

$$\text{molarity} = \frac{1.07 \text{ g soln}}{1 \text{ mL}} \times \frac{10.0 \text{ g } H_2SO_4}{100.0 \text{ g soln}} \times \frac{1000 \text{ mL}}{1 \text{ L}} \times \frac{1 \text{ mol } H_2SO_4}{98.08 \text{ g } H_2SO_4} = 1.09 \text{ mol } H_2SO_4/\text{L}$$

or 1.09 M H_2SO_4

(b)
$$X_{H_2SO_4} = \frac{n_{H_2SO_4}}{n_{H_2SO_4} + n_{H_2O}}$$

The number of moles of H_2SO_4 and H_2O in 100.0 g of solution are

$$n_{H_2SO_4} = 10.0 \text{ g } H_2SO_4 \times \frac{1 \text{ mol } H_2SO_4}{98.08 \text{ g } H_2SO_4} = 0.102 \text{ mol } H_2SO_4$$

$$n_{H_2O} = 90.0 \text{ g } H_2O \times \frac{1 \text{ mol } H_2O}{18.02 \text{ g } H_2O} = 4.99 \text{ mol } H_2O$$

The mole fraction of sulfuric acid in the solution is

$$X_{H_2SO_4} = \frac{0.102 \text{ mol}}{0.102 \text{ mol} + 4.99 \text{ mol}} = 0.0200$$

(c)
$$\text{molality} = \frac{0.102 \text{ mol } H_2SO_4}{0.0900 \text{ kg}} = 1.13 \text{ m } H_2SO_4$$

13.39 The molar mass of H_2SO_4 is 98.08 g/mol. The mass of H_2SO_4 in 1 L of solution is

$$\frac{2.45 \text{ mol}}{1 \text{ L}} \times \frac{98.08 \text{ g}}{1 \text{ mol}} = 240 \text{ g} = 0.240 \text{ kg}$$

The mass of 1 L of solution is 1000 mL × 1.15 g/mL = 1.15×10^3 g = 1.15 kg. The mass of H_2O in 1 L of solution is 1.15 kg – 0.240 kg = 0.91 kg. The molality is

$$\text{molality} = \frac{2.45 \text{ mol}}{0.91 \text{ kg}} = 2.7 \text{ m}$$

13.41 (a) In a solution, there are fractionally fewer solvent molecules per unit area at the surface that are capable of leaving than at the surface of the pure solvent, since a portion of the total volume of the solution is occupied by solute particles. Hence, the rate of escape or vaporization of the solvent molecules is less than it would be if no solute were present, and this causes the lowering or depression of the solvent vapor pressure, since an equal fraction of fewer molecules are needed in the vapor in order to obtain a rate of condensation equal to the rate of vaporization.

(b) Because the vapor pressure of the solvent is reduced by the presence of the solute, the solvent molecules have to be heated to a higher temperature in order to obtain a vapor pressure equal to the external or atmospheric pressure. Remember that the vapor pressure increases with temperature, because the average kinetic energy of the molecules increases with temperature and, consequently, more molecules have the required energy needed to escape the solution. Hence, the presence of the solute causes the boiling point of the solution to be elevated over that of the pure solvent.

13.43 (a) Increase. The colligative effects depend on the concentration of the solute. The greater the mass of solute, the greater the concentration.

(b) Decrease. The number of moles in a given mass of solute is less when the molar mass is greater, and the colligative effects depend on the number of moles of solute.

13.45 For a given solvent, the colligative properties of solutions of nonionic and nonvolatile solutes depend solely on the concentration of solute particles and not on their nature. The greater the concentration, the greater the effect.

(a) 0.10 m glucose (b) 0.010 m urea

13.47 (a) The number of moles of CCl_4 is

$$3.00\text{ g} \times \frac{1\text{ mol}}{153.82\text{ g}} = 0.0195\text{ mol}$$

The molality of the solution is

$$\text{molality} = \frac{0.0195\text{ mol}}{0.190\text{ kg}} = 0.103\text{ m}$$

The freezing point depression and freezing point of the solution are

$$\text{FPD} = K_f m = \frac{4.90°\text{C}\bullet\text{kg}}{1\text{ mol}} \times \frac{0.103\text{ mol}}{1\text{ kg}} = 0.505°\text{C}$$

$$T_f = T_f^\circ - \text{FPD} = 5.5°\text{C} - 0.505°\text{C} = 5.0°\text{C}$$

(b) The freezing point depression and freezing point of the solution are

$$\text{FPD} = K_f m = \frac{1.86°\text{C}\bullet\text{kg}}{1\text{ mol}} \times \frac{1.51\text{ mol}}{1\text{ kg}} = 2.81°\text{C}$$

$$T_f = T_f^\circ - \text{FPD} = 0.00°\text{C} - 2.81°\text{C} = -2.81°\text{C}$$

13.49 The number of moles of mannitol is

$$15.0\text{ g} \times \frac{1\text{ mol}}{182.17\text{ g}} = 8.23 \times 10^{-2}\text{ mol}$$

(a) The vapor pressure of the solution can be obtained from Raoult's law. The moles and mole fraction of water in the solution are

$$n_{water} = 500\text{ g} \times \frac{1\text{ mol}}{18.02\text{ g}} = 27.7\text{ mol}$$

$$X_{water} = \frac{27.7 \text{ mol}}{27.7 \text{ mol} + 0.0823 \text{ mol}} = 0.997$$

The vapor pressure is

$$P_{solution} = P_{water} = X_{water} P^{\circ}_{water} = 0.997 \times 55.3 \text{ torr} = 55.1 \text{ torr}$$

(b) The molality of the solution is

$$\text{molality} = \frac{8.23 \times 10^{-2} \text{ mol}}{0.500 \text{ kg}} = 0.165 \text{ mol/kg}$$

The boiling point elevation and boiling point of the solution are

$$BPE = K_b m = \frac{0.512°C \cdot kg}{1 \text{ mol}} \times \frac{0.165 \text{ mol}}{1 \text{ kg}} = 0.0845°C$$

$$T_b = T^{\circ}_b + BPE = 100.00°C + 0.0845°C = 100.08°C$$

(c) The freezing point depression and freezing point of the solution are

$$FPD = K_f m = \frac{1.86°C \cdot kg}{1 \text{ mol}} \times \frac{0.165 \text{ mol}}{1 \text{ kg}} = 0.307°C$$

$$T_f = T^{\circ}_f - FPD = 0.00°C - 0.307°C = -0.31°C$$

13.51 The approximate net molarity of the blood plasma is

$$M = \frac{\pi}{RT} = \frac{7.7 \text{ atm}}{0.0821 \text{ L} \cdot \text{atm/mol} \cdot \text{K} \times 310 \text{ K}} = 0.30 \text{ mol/L}$$

Assuming the molality is equal to the molarity, the FPD and freezing point of the blood plasma are

$$FPD = K_f m = \frac{1.86°C \cdot kg}{1 \text{ mol}} \times \frac{0.30 \text{ mol}}{1 \text{ kg}} = 0.56°C$$

$$T_f = T^{\circ}_f - FPD = 0.00°C - 0.56°C = -0.56°C$$

13.53 Chloroform. From Table 13.2 we see that the K_b value for chloroform is almost three times that for ethanol. Therefore, with chloroform the BPE obtained is three times as great as that obtained with ethanol and this enables us to make more accurate measurements.

13.55 (a) The number of moles of naphthalene is

$$5.00 \text{ g} \times \frac{1 \text{ mol}}{128.17 \text{ g}} = 0.0390 \text{ mol}$$

The molality is

$$\text{molality} = \frac{0.0390 \text{ mol}}{0.100 \text{ kg}} = 0.390 \text{ m}$$

The freezing point depression is

$$FPD = T^{\circ}_f - T_f = 6.5°C - (-1.4°C) = 7.9°C$$

The freezing point depression constant of cyclohexane is

$$K_f = \frac{FPD}{m} = \frac{7.9°C}{0.390\ mol/kg} = 20°C{\bullet}kg/mol$$

(b) Step 1. $FPD = T_f^{\circ} - T_f = 6.5°C - 3.4°C = 3.1°C$

Step 2. $m = \frac{FPD}{K_f} = \frac{3.1°C}{20°C{\bullet}kg/mol} = 0.155\ mol/kg$ (carry an extra significant figure)

Step 3. $n_{compound} = \frac{0.155\ mol}{1\ kg\ solvent} \times 0.100\ kg\ solvent = 0.0155\ mol$

Step 4. 0.0155 mol corresponds to 2.00 g. The molar mass is 2.00 g/0.0155 mol = 1.3×10^2 g/mol.

13.57 Steps 1 and 2.

$$m = \frac{BPE}{K_b} = \frac{0.893°C}{3.63°C{\bullet}kg/mol} = 0.246\ mol/kg$$

Step 3. $n_{compound} = \frac{0.246\ mol}{1\ kg} \times 0.0200\ kg = 4.92 \times 10^{-3}\ mol$

Step 4. 0.00492 mol corresponds to 3.00 g. The molar mass is 3.00 g/0.00492 mol = 610 g/mol.

13.59 Step 1. VPL = 6.35 torr

Step 2.

$$X_{compound} = \frac{VPL}{P^{\circ}_{benzene}} = \frac{6.35\ torr}{74.7\ torr} = 0.0850$$

Step 3. The number of moles of benzene is

$$250\ g \times \frac{1\ mol}{78.11\ g} = 3.20\ mol$$

The number of moles of compound is obtained as follows:

$$X_{compound} = \frac{n_{compound}}{n_{compound} + n_{benzene}}$$

Rearranging and solving for $n_{compound}$ gives

$$n_{compound} = \frac{X_{compound}\, n_{benzene}}{(1 - X_{compound})} = \frac{0.0850 \times 3.20\ mol}{(1 - 0.0850)} = 0.297\ mol$$

Step 4. The molar mass of the compound is

$$\mathcal{M}_{compound} = \frac{25.0\ g}{0.297\ mol} = 84.2\ g/mol$$

13.61 (a) boiling point: seawater > pure water
(b) freezing point: seawater < pure water
(c) vapor pressure: seawater < pure water
(d) osmotic pressure: seawater > pure water

13.63 (a) The magnitude of the colligative effects depends on the number of independent solute particles. In solutions of ionic solutes, however, the association of oppositely charged ions with one another reduces the number of independent solute particles and, consequently, the observed colligative effects are less than the effects calculated on the basis of complete dissociation.

(b) As the solution becomes more dilute, the amount of association between the oppositely charged ions decreases. Hence, the degree to which the solute particles behave independently increases with increasing dilution, and the observed effects approach the calculated values.

13.65 1.0 m $CaCl_2$. Assuming complete dissociation the 1.0 m $CaCl_2$ solution would have a solute particle concentration of 3.0 m, while the 1.0 m NaCl solution would only have a solute particle concentration of 2.0 m.

13.67 (a) $Al_2(SO_4)_3$ dissociates according to the equation

$$Al_2(SO_4)_3(aq) \rightarrow 2Al^{3+}(aq) + 3SO_4^{2-}(aq)$$

Hence, a 0.100 m $Al_2(SO_4)_3$ solution will be 0.500 m with respect to ions or solute particles. Assuming complete dissociation, the freezing point depression and freezing point of the solution will be

$$FPD = K_f m = \frac{1.86°C \cdot kg}{1\ mol} \times \frac{0.500\ mol}{1\ kg} = 0.93°C$$

$$T_f = T_f^\circ - FPD = 0.00°C - 0.93°C = -0.93°C$$

Assuming complete dissociation, the boiling point elevation and boiling point of the solution will be

$$BPE = K_b m = \frac{0.512°C \cdot kg}{1\ mol} \times \frac{0.500\ mol}{1\ kg} = 0.256°C$$

$$T_b = T_b^\circ + BPE = 100.000°C + 0.256°C = 100.256°C$$

(b) The actual freezing point will be greater than the estimated value and the actual boiling point will be less than the estimated value. Because of the ionic atmosphere surrounding the ions, the observed colligative effects in ionic solutions are always less than the effects calculated on the basis of complete dissociation.

13.69 (a) $m = \dfrac{FPD}{K_f} = \dfrac{2.00°C}{1.86°C \cdot kg/mol} = 1.08\ mol/kg$

(b) The boiling point elevation and boiling point of the sample is

$$BPE = K_b m = \frac{0.512°C \cdot kg}{1\ mol} \times \frac{1.08\ mol}{1\ kg} = 0.553°C$$

$$T_b = T_b^\circ + BPE = 100.000°C + 0.553°C = 100.553°C$$

13.71 If we assume no ionization, the calculated freezing point depression for a 1.00 m aqueous solution would be 1.86°C. Therefore, the i factor for the solution is

$$i = \frac{\text{observed colligative effect}}{\text{calculated effect assuming no ionization}} = \frac{4.04°C}{1.86°C} = 2.17$$

13.73 (a) If we assume no dissociation, the calculated freezing point depression for a 0.100 m aqueous solution would be 0.186°C. Therefore, the observed FPD is

observed FPD = i × calculated FPD assuming no dissociation = 1.87 × 0.186°C = 0.348°C

The freezing point of the solution will be 0.000°C – 0.348°C = –0.348°C.

(b) If we assume no dissociation, the calculated boiling point elevation for a 0.100 m aqueous solution would be 0.0512°C. Therefore, the observed BPE is

observed BPE = i × calculated BPE assuming no dissociation = 1.87 × 0.0512°C = 0.0957°C

The boiling point of the solution will be 100.000°C + 0.0957°C = 100.096°C.

(c) To estimate the osmotic pressure of the solution, we must first calculate its molarity. The molar mass of NaCl is 58.44 g/mol. Therefore, there is 0.100 mol of NaCl in 1000.00 g + 5.84 g = 1005.84 g of solution. The volume of 1005.84 g of solution is 1005.84 g × 1 mL/1.00 g = 1.01×10^3 mL or 1.01 L. The molarity is therefore 0.100 mol/1.01 L = 0.0990 mol/L to three significant figures. If we assume no dissociation, the calculated osmotic pressure of a 0.0990 M solution at 25°C is

$$\pi = MRT = \frac{0.0990 \text{ mol}}{1 \text{ L}} \times \frac{0.0821 \text{ L•atm}}{\text{mol•K}} \times 298 \text{ K} = 2.42 \text{ atm}$$

Therefore, the observed osmotic pressure is

observed osmotic pressure = i × calculated osmotic pressure assuming no dissociation = 1.87 × 2.42 atm = 4.53 atm

13.75 True solutions have particles with dimensions that are usually less than 1000 pm. Suspensions have particles with dimensions exceeding 100,000 pm. Colloidal dispersions have particles with dimensions greater than 1000 pm, but with at least one dimension smaller than 100,000 pm.

Solutions: salt water, sugar water, clean air, and gold-silver alloys.

Suspensions: AgCl precipitation, suspension of mud or fine sand in water, and the dirty exhaust from diesel trucks and buses.

Colloidal dispersions: mayonnaise, gems, fog, ink, and homogenized milk.

13.77 The Tyndall effect is the detection of a light beam being passed through a medium, when viewed from the side. The scattering of light at large angles by the particles in the medium causes the Tyndall effect.

Yes. Suspensions will give a Tyndall effect.

13.79 See Table 13.5 in textbook.

13.81 A sol is a colloidal dispersion of a solid in a liquid. A hydrophilic sol has colloidal particles made up of polar molecules that have an affinity for water. The polar molecules composing the colloidal particles are arranged so that the polar groups, ends, or parts of the molecules can form hydrogen bonds with water molecules. Starch dispersed in water and printing ink are two examples.

13.83 Hydrophilic colloids are stable because the hydrophilic colloidal particles form hydrogen bonds with water and these hydrogen bonds stabilize the colloid by preventing the particles from coagulating.

13.85 Adsorbed ions provide the colloidal particles with charges of the same sign. The colloidal particles do not coagulate and settle out because the mutual repulsion caused by the adsorbed ions keeps the particles away from each other.

13.87 It is the heat of solution at the saturation point and not the overall heat of solution that determines if a substance is more or less soluble with increasing temperature. Increased temperature makes for increased solubility when the solution process at the saturation point is endothermic, decreased solubility when the solution process is exothermic. For NaOH the process of dissolving at the saturation point is endothermic and, therefore, the solubility of NaOH increases with increasing temperature.

13.89 (a) Since the lettuce leaves and other vegetables in the salad are in contact with a concentrated salt solution, water will flow from the plant cells into the solution causing the salad to wilt.
(b) Salty foods cause loss of cell fluid into the plasma, since the salty foods increase the plasma's osmotic pressure. Drinking additional fluid restores the osmotic balance between the cells and the plasma.
(c) Pure water has zero osmotic pressure and, therefore, some water will flow into the wilted flowers and thereby freshen them up.
(d) Since oil and water do not mix, the oil coating will protect the vegetables from coming into direct contact with the vinegar solution and thereby reduce any loss of fluid from the vegetables (wilting) to the vinegar solution.

13.91 If the molality is significantly different from the molarity, then we cannot easily obtain the quantity (moles) of reactants consumed and products produced, or the molarity of an unknown solution, from the simple reactant volume measurements. We would also have to know the density of one or both of the solutions before we could calculate the desired quantities, depending on the experiment.

If the molarity is significantly different from the molality, then we cannot use the K_b and K_f constants to calculate the BPE and FPD, and the T_b and T_f of the solution. A BPE or T_b versus molarity curve and a FPD or T_f versus molarity curve would first have to be determined or worked out for that particular solute, and it would be good only for that particular solute, in order to use molarities instead of molalities.

13.93 (a) $50.0 \text{ g KNO}_3 \times \dfrac{100 \text{ g H}_2\text{O}}{247 \text{ g KNO}_3} \times \dfrac{1 \text{ mL H}_2\text{O}}{1 \text{ g H}_2\text{O}} = 20.2 \text{ mL H}_2\text{O}$

(b) The maximum amount of KNO_3 that $20.2 \text{ mL H}_2\text{O}$ can dissolve at 0°C is

$$\frac{13.3 \text{ g KNO}_3}{100 \text{ g H}_2\text{O}} \times \frac{1 \text{ g H}_2\text{O}}{1 \text{ mL H}_2\text{O}} \times 20.2 \text{ mL H}_2\text{O} = 2.69 \text{ g KNO}_3$$

The amount of KNO_3 that will recrystallize is 50.0 g – 2.69 g = 47.3 g.

13.95 (a) $P_{\text{benzene}} = X_{\text{benzene}} P^{\circ}_{\text{benzene}} = 0.451 \times 1108 \text{ torr} = 500 \text{ torr}$

$X_{\text{toluene}} = 1 - 0.451 = 0.549$

$P_{\text{toluene}} = X_{\text{toluene}} P^{\circ}_{\text{toluene}} = 0.549 \times 474 \text{ torr} = 260 \text{ torr}$

Alternately, P_{toluene} could be calculated from the fact that the solution boils at 94.8°C and, therefore, must have a total vapor pressure of 760 torr; i.e.,

$P_{\text{toluene}} = P_T - P_{\text{benzene}} = 760 \text{ torr} - 500 \text{ torr} = 260 \text{ torr}$

Hence, the composition of the vapor is

$$X_{\text{benzene}} = \frac{P_{\text{benzene}}}{P_T} = \frac{500 \text{ torr}}{760 \text{ torr}} = 0.658$$

$X_{\text{toluene}} = 1 - 0.658 = 0.342$

(b) The condensed liquid has the same mole fractions as the vapor. Therefore, the partial pressures of benzene and toluene above such a solution at 94.8°C are

$P_{\text{benzene}} = 0.658 \times 1108 \text{ torr} = 729 \text{ torr}$

$P_{\text{toluene}} = 0.342 \times 474 \text{ torr} = 162 \text{ torr}$

The total vapor pressure of the condensed liquid at 94.8°C is therefore

$$P_T = P_{benzene} + P_{toluene} = 729 \text{ torr} + 162 \text{ torr} = 891 \text{ torr}$$

13.97 –25°F is equivalent to –32°C. One liter of water has a mass of approximately 1 kg.

(a) To prevent freezing at –32°C, the freezing point depression would have to be

$$FPD = T_f^\circ - T_f = 0.00°C - (-32°C) = 32°C$$

The molality of the solution would have to be

$$m = \frac{FPD}{K_f} = \frac{32°C}{1.86°C \cdot kg/mol} = 17 \text{ mol/kg}$$

The grams of ethylene glycol needed for 1.0 L of water would be

$$1.0 \text{ L water} \times \frac{1 \text{ kg water}}{1 \text{ L water}} \times \frac{17 \text{ mol } C_2H_6O_2}{1 \text{ kg water}} \times \frac{62.07 \text{ g } C_2H_6O_2}{1 \text{ mol } C_2H_6O_2} = 1.1 \times 10^3 \text{ g } C_2H_6O_2$$

(b) $$BPE = K_b m = \frac{0.512°C \cdot kg}{1 \text{ mol}} \times \frac{17 \text{ mol}}{1 \text{ kg}} = 8.7°C$$

$$T_b = T_b^\circ + BPE = 100.00°C + 8.7°C = 108.7°C$$

(c) No. The FPD and BPE equations are not valid for such high concentrations of solute. For such solutions the actual freezing and boiling points must be obtained experimentally.

13.99 The osmotic pressure of the protein solution is

$$\pi = MRT = \frac{0.010 \text{ mol}}{1 \text{ L}} \times \frac{0.0821 \text{ L} \cdot \text{atm}}{\text{mol} \cdot \text{K}} \times 298 \text{ K} = 0.24 \text{ atm}$$

The maximum height is the height of a column of protein solution that exerts a pressure of 0.24 atm. The osmotic pressure of 0.24 atm can be converted into millimeters of mercury.

$$0.24 \text{ atm} \times \frac{760 \text{ torr}}{1 \text{ atm}} \times \frac{1 \text{ mm Hg}}{1 \text{ torr}} = 180 \text{ mm or } 18 \text{ cm Hg}$$

Therefore, a Hg column 18 cm in height provides a pressure of 0.24 atm. Since the density of Hg is 13.6 times greater than the density of the solution, it takes a column of solution $13.6 \times 18 \text{ cm} = 2.4 \times 10^2$ cm in height to provide a pressure of 0.24 atm.

13.101 (a) Step 1. $FPD = T_f^\circ - T_f = 5.5°C - 4.27°C = 1.23°C$ (carry an extra significant figure)

Step 2. $$m = \frac{FPD}{K_f} = \frac{1.23°C}{4.90°C \cdot kg/mol} = 0.251 \text{ mol/kg}$$

Step 3. $$n_{hormone} = \frac{0.251 \text{ mol}}{1 \text{ kg}} \times 5.00 \times 10^{-3} \text{ kg} = 1.26 \times 10^{-3} \text{ mol}$$

Step 4. $$\mathcal{M} = \frac{0.363 \text{ g}}{0.00126 \text{ mol}} = 288 \text{ g/mol or } 2.9 \times 10^2 \text{ g/mol to two significant figures}$$

(b) The numbers of moles of C, H, and O in the molar mass are

$$\text{C: } 2.9 \times 10^2 \text{ g hormone} \times \frac{79.12 \text{ g C}}{100 \text{ g hormone}} \times \frac{1 \text{ mol C atoms}}{12.01 \text{ g C}} = 19 \text{ mol C atoms}$$

$$\text{H: } 2.9 \times 10^2 \text{ g hormone} \times \frac{9.79 \text{ g H}}{100 \text{ g hormone}} \times \frac{1 \text{ mol H atoms}}{1.008 \text{ g H}} = 28 \text{ mol H atoms}$$

$$\text{O: } 2.9 \times 10^2 \text{ g hormone} \times \frac{11.09 \text{ g O}}{100 \text{ g hormone}} \times \frac{1 \text{ mol O atoms}}{16.00 \text{ g O}} = 2 \text{ mol O atoms}$$

The molecular formula for testosterone is therefore $C_{19}H_{28}O_2$. The precise molar mass of testosterone is $(19 \times 12.011) + (28 \times 1.00794) + (2 \times 15.9994) = 288.43$ g/mol.

CHAPTER 14

CHEMICAL KINETICS

Solutions To Practice Exercises

PE 14.1 The difference in the temperatures 35°C – 5°C = 30°C is equivalent to three temperature rises of ten degrees each. The growth rate of the colony doubles for each ten-degree rise in temperature. Therefore, the growth rate at 35°C should be 2 × 2 × 2 = 8 times faster than that at 5°C.

PE 14.2 The coefficients show that 1 mol of NO is produced for every 3 mol of NO_2 consumed. Hence,

$$\frac{\Delta[\text{NO}]}{\Delta t} = \frac{0.30 \text{ mol NO}_2}{\text{L•s}} \times \frac{1 \text{ mol NO}}{3 \text{ mol NO}_2} = 0.10 \text{ mol NO/L•s}$$

PE 14.3 From Figure 14.3 we see that the change in concentration over this interval is $\Delta[NO_2]$ = 0.0057 mol/L – 0.0071 mol/L = –0.0014 mol/L. The time interval is Δt = 70 s – 50 s = 20 s. Therefore, the average rate between 50 and 70 seconds is

$$\text{average rate} = \frac{-\Delta[\text{NO}_2]}{\Delta t} = \frac{-(-0.0014 \text{ mol/L})}{20 \text{ s}} = 7.0 \times 10^{-5} \text{ mol/L•s}$$

PE 14.4 Draw the tangent line to the $[NO_2]$ versus time curve at t = 40 s. The actual values for $\Delta[NO_2]$ and Δt depend on which right triangle is employed, i.e., on the length chosen for the tangent segment. One right triangle gives $\Delta[NO_2]$ = 0.006 mol/L – 0.00925 mol/L = –0.003 mol/L and Δt = 58.75 s – 30 s = 29 s. The slope of the tangent line is

$$\text{slope} = \frac{\Delta[\text{NO}_2]}{\Delta t} = \frac{-0.003 \text{ mol/L}}{29 \text{ s}} = -1 \times 10^{-4} \text{ mol/L•s}$$

Because concentration is decreasing with time, the instantaneous reaction rate is the negative of the slope, or 1×10^{-4} mol/L•s.

PE 14.5 Assume the rate law has the form

$$\text{rate} = k[\text{I}^-]^m[\text{S}_2\text{O}_8^{2-}]^n$$

We have to determine the values of m and n; m can be determined from runs 1 and 3, since the initial $[S_2O_8^{2-}]$ is the same in both runs.

$$\frac{\text{rate 3}}{\text{rate 1}} = \frac{8.6 \times 10^{-4}\ \text{mol/L•s}}{2.6 \times 10^{-4}\ \text{mol/L•s}} = \frac{k(0.50)^m(0.45)^n}{k(0.15)^m(0.45)^n} = \left[\frac{0.50}{0.15}\right]^m$$

or $3.3 = (3.3)^m$ and $m = 1$

In a like manner, n can be determined from runs l and 2, since the initial $[I^-]$ is the same in both runs.

$$\frac{\text{rate 1}}{\text{rate 2}} = \frac{2.6 \times 10^{-4}\ \text{mol/L•s}}{1.4 \times 10^{-4}\ \text{mol/L•s}} = \frac{k(0.15)^m(0.45)^n}{k(0.15)^m(0.25)^n} = \left[\frac{0.45}{0.25}\right]^n$$

or $1.9 = (1.8)^n$ and $n = 1$

Substituting m = 1 and n = 1 into the rate law gives

rate = $k\,[I^-][S_2O_8^{2-}]$

PE 14.6 Since the reaction is first order with respect to N_2O_5, the exponent of $[N_2O_5]$ in the rate law is 1. Hence, the rate law is

rate = $k[N_2O_5]$

PE 14.7 Since the reaction is first order with respect to both $S_2O_3^{2-}$ and H^+, the exponents of $[S_2O_3^{2-}]$ and $[H^+]$ in the rate law are both one. The reaction is zero order in O_2, so the rate law is independent of $[O_2]$. Hence, the rate law is

rate = $k[S_2O_3^{2-}][H^+]$

PE 14.8 The rate law for the reaction is rate = k[B].

(a) Using the given data gives 0.012 mol/L•s = k(0.60 mol/L). Hence, $k = 0.020\ s^{-1}$.

(b) The reaction rate when A and B are each 0.010 M is

rate = $k[B] = 0.020\ s^{-1} \times 0.010\ \text{mol/L} = 2.0 \times 10^{-4}$ mol/L•s

PE 14.9 $\ln([N_2O_5]/[N_2O_5]_0) = -kt$

$$\ln\frac{0.385\ \text{mol/L}}{0.400\ \text{mol/L}} = -k \times 1.0\ \text{min}$$

Rearranging and solving for k gives

$-k \times 1.0\ \text{min} = \ln(0.385/0.400) = -3.822 \times 10^{-2}$ and $k = 0.03822/1.0\ \text{min} = 0.038\ \text{min}^{-1}$

PE 14.10 Using the points (t = 0.0 min, $\ln[N_2O_5] = \ln 0.400$) and (t = 40.0 min, $\ln[N_2O_5] = \ln 0.090$) we get

$$m = \frac{\Delta \ln[N_2O_5]}{\Delta t} = \frac{\ln 0.090 - \ln 0.400}{(40.0 - 0.0)\text{min}} = -3.73 \times 10^{-2}\ \text{min}^{-1}$$

and $k = -m = 0.0373\ \text{min}^{-1}$

The value obtained is consistent with those obtained using Equation 14.4.

PE 14.11 The fraction of N_2O_5 remaining can be obtained by substituting $A = N_2O_5$, $k = 0.037\ min^{-1}$, and $t = 45$ min in Equation 14.3.

$$\frac{[N_2O_5]}{[N_2O_5]_0} = e^{-kt} = e^{-0.037\ min^{-1} \times 45\ min} = 0.19$$

PE 14.12 The rate constant, k, can be obtained by substituting $[sucrose]/[sucrose]_0 = 2/3$ and $t = 139$ min into Equation 14.4.

$\ln([sucrose]/[sucrose]_0) = -kt$

$\ln(2/3) = -k \times 139\ min$

$k = -\ln(2/3)/139\ min = 2.92 \times 10^{-3}\ min^{-1}$

PE 14.13 Substituting $t_{1/2} = 56.3$ min into Equation 14.6 gives

$k \times t_{1/2} = \ln 2$

$k \times 56.3\ min = 0.693$

and $k = 1.23 \times 10^{-2}\ min^{-1}$

PE 14.14 The time required can be obtained by substituting $[Fr]/[Fr]_0 = 0.10$ and $k = 0.0252\ s^{-1}$ into Equation 14.4.

$\ln 0.10 = -0.0252\ s^{-1} \times t$

and $t = 91.4$ s

PE 14.15 The NO_2 concentration can be obtained by substituting $A = NO_2$, $k = 10.1$ L/mol•s, $t = 30.0$ s, and $[NO_2]_0 = 0.0300$ mol/L into Equation 14.8.

$$\frac{1}{[NO_2]} = kt + \frac{1}{[NO_2]_0}$$

$$\frac{1}{[NO_2]} = 10.1\ L/mol{\bullet}s \times 30.0\ s + \frac{1}{0.300\ mol/L}$$

$$\frac{1}{[NO_2]} = 303\ L/mol + 33.3\ L/mol = 336\ L/mol$$

and $[NO_2] = 2.98 \times 10^{-3}$ mol/L

PE 14.16 (a) The length of the first half-life period is obtained by substituting $[NO_2]_0 = 0.0300$ mol/L and $k = 10.1$ L/mol•s into Equation 14.9.

$$t_{1/2} = \frac{1}{k[A]_0} = \frac{1}{10.1\ L/mol{\bullet}s \times 0.0300\ mol/L} = 3.30\ s$$

(b) The length of the second half-life period is obtained by substituting $[NO_2]_0 = 0.0150$ mol/L and $k = 10.1$ L/mol•s into Equation 14.9.

$$t_{1/2} = \frac{1}{10.1\ L/mol{\bullet}s \times 0.0150\ mol/L} = 6.60\ s$$

PE 14.17 $\text{rate} = \frac{-d[A]}{dt} = k[A]^2$

$$\frac{d[A]}{[A]^2} = -kdt$$

$$\int_{[A]_0}^{[A]} \frac{d[A]}{[A]^2} = -k\int_0^t dt$$

$$\frac{-1}{[A]} + \frac{1}{[A]_0} = -k(t-0)$$

$$\text{or } \frac{1}{[A]} = kt + \frac{1}{[A]_0} \qquad (14.8)$$

PE 14.18 (a) rate = $k[N_2O][O]$
(b) rate = $k[Br_2]$

PE 14.19 A slow first step starting with $NO_2(g) + F_2(g) \rightarrow$ products

would account for the observed rate law. Likely products might be NO_2F and F.

Step 1: $NO_2(g) + F_2(g) \rightarrow NO_2F(g) + F(g)$ (slow)

A second step to complete the mechanism brings in the second NO_2 molecule and eliminates the intermediate F.

Step 2: $NO_2(g) + F(g) \rightarrow NO_2F(g)$ (fast)

PE 14.20 The rate of the rate-determining step is

rate = $k_2[NO_3][NO]$

This rate law has the intermediate NO_3 in it. To eliminate the intermediate NO_3 and have a rate law only in terms of the initial reactants and final products, we employ the steady-state approximation.

rate of NO_3 formation = $k_1[NO][O_2]$

rate of NO_3 consumption = $k_{-1}[NO_3]$

Equating these rates gives $k_1[NO][O_2] = k_{-1}[NO_3]$

Rearranging gives the steady-state concentration of NO_3:

$$[NO_3] = \frac{k_1}{k_{-1}}[NO][O_2]$$

The rate law for the slow step can now be rewritten as follows:

$$\text{rate} = k_2[NO_3][NO] = k_2\frac{k_1}{k_{-1}}[NO][O_2][NO] = \frac{k_2k_1}{k_{-1}}[NO]^2[O_2] = k[NO]^2[O_2]$$

PE 14.21 $\Delta H°$ for the step can be obtained by substituting E_a (forward) = 42 kJ and E_a (reverse) = 32 kJ into Equation 14.12.

$\Delta H° = 42\text{ kJ} - 32\text{ kJ} = +10\text{ kJ}$

The step is endothermic because the enthalpy change is positive.

PE 14.22 Substituting $E_a = 18.7 \times 10^3$ J/mol, $k_1 = 0.0400\ s^{-1}$, $T_1 = 273$ K, $T_2 = 298$ K, and R = 8.314 J/mol•K into Equation 14.15 gives

$$\ln \frac{0.0400}{k_2} = \frac{18.7 \times 10^3 \text{ J/mol}}{8.314 \text{ J/mol•K}} \left[\frac{1}{298 \text{ K}} - \frac{1}{273 \text{ K}}\right]$$

$\ln 0.0400 - \ln k_2 = -0.69$

$\ln k_2 = -2.53$

and $k_2 = e^{-2.53} = 0.080\ s^{-1}$

PE 14.23 The frequency of bimolecular collisions is proportional to the concentration of each of the colliding molecules. Therefore, tripling the concentration of A will triple the collision frequency, while halving the concentration of B will cut the collision frequency in half. Consequently, the collision frequency will increase by a factor of 3/2 = 1.5.

PE 14.24 To solve the problem, first calculate the fraction of collisions with energies greater than or equal to 10.0 kJ/mol at 25°C and 35°C.

At 25°C: Substituting E_a, R, and T = 298 K in Equation 14.17 gives

$f_a = e^{-(10,000 \text{ J/mol})/(8.314 \text{ J/mol•K})(298 \text{ K})} = 1.77 \times 10^{-2}$

At 35°C: Substituting E_a, R and T = 308 K into Equation 14.17 gives $f_a = 2.01 \times 10^{-2}$

The reaction rate will increase by a factor of $2.01 \times 10^{-2}/1.77 \times 10^{-2} = 1.14$.

Solutions To Final Exercises

14.1 1. The nature of the reactants. 2. The reactant concentrations and state of subdivision. 3. The temperature. 4. Catalysts.

14.3
(a) Break the wood up into small pieces and fan the fire to increase oxygen supply.
(b) Increase the temperature by heating the iron and supply moisture, a catalyst.
(c) Increase the temperature and increase the concentrations by increasing the pressure.

14.5 Reactions between ions in solution tend to be rapid because of the high mobility of dissolved ions and the electrostatic attractions between them. Also, when an ionic solid dissolves, the ionic bond(s) is (are) broken; therefore there are no bonds to be broken in this type of reaction. Basically, just a rearrangement of ions takes place, and this can occur very rapidly. In general, the fewer the bonds there are to be broken during the reaction, and the weaker they are, the quicker the reaction will proceed.

14.7 In homogeneous catalysis, the catalyst is in solution with the reactants, while in heterogeneous catalysis, the catalyst is usually in a different state than the reactants and provides a surface on which the reaction takes place.

A contact catalyst is the name given to the catalyst in a heterogeneous catalysis reaction, i.e., a catalyst that provides a surface on which the reaction takes place.

14.9 By four. The difference in the temperatures 15.5°C – 35.5°C = –20.0°C is equivalent to two temperature drops of ten degrees each. The animal's oxygen consumption doubles for each ten-degree rise in temperature. Therefore, the animal's oxygen consumption at 15.5°C should be 2 × 2 = 4 times slower or less than that at 35.5°C.

14.11 Refer to "Measuring Reaction Rates," Section 14.2, in the textbook. The optical rotation of light can be used if one of the reactants or products is optically active, which means that it rotates the plane of polarized light. Nuclear magnetic resonance spectroscopy can be used to follow the creation or destruction of an NMR absorption band during a reaction. Radioactive emissions can be used to follow the disappearance or production of a radioactive substance. Techniques based on monitoring changes in density, viscosity, turbidity (cloudiness), and absorption of ultraviolet or infrared radiation can also be used.

14.13 (a) To determine the average rate, first calculate the concentration change during the time interval and then divide that by the time interval. Adjust signs to obtain a positive answer.

(b) Make a graph of the concentration versus time data and draw the tangent to the curve at the point representing the time of interest. The instantaneous rate will be the negative of the tangent's slope for a reactant, and the tangent's slope for a product.

For more explicit details, see the worked-out examples in the textbook and the accompanying text.

14.15 From the reaction equation we see that the rates of H_2O formation, NH_3 disappearance, and O_2 disappearance are related to the rate of NO formation as follows:

$$\frac{\Delta[H_2O]}{\Delta t} = \frac{6}{4}\frac{\Delta[NO]}{\Delta t};\quad \frac{-\Delta[NH_3]}{\Delta t} = \frac{\Delta[NO]}{\Delta t};\text{ and }\frac{-\Delta[O_2]}{\Delta t} = \frac{5}{4}\frac{\Delta[NO]}{\Delta t}$$

14.17 (a) Substituting P = 1.00 atm, V = 0.0100 L, T = 300 K, and R = 0.0821 L•atm/mol•K in the ideal gas equation and solving for n will give the number of moles of O_2 produced in a second's time.

$$n = \frac{PV}{RT} = \frac{1.00\text{ atm} \times 0.0100\text{ L}}{0.0821\text{ L•atm/mol•K} \times 300\text{ K}} = 4.06 \times 10^{-4}\text{ mol}$$

Hence, the rate in terms of moles of oxygen produced per second is 4.06×10^{-4} mol/s.

(b) From the reaction equation we see that

$$\frac{-\Delta n_{H_2O_2}}{\Delta t} = \frac{2\Delta n_{O_2}}{\Delta t} = 2 \times 4.06 \times 10^{-4}\text{ mol/s} = 8.12 \times 10^{-4}\text{ mol/s}$$

The volume of the solution in 0.0250 L. Therefore, the change in molarity of H_2O_2 per second is

$$\frac{-\Delta[H_2O_2]}{\Delta t} = \frac{8.12 \times 10^{-4}\text{ mol/s}}{0.0250\text{ L}} = 3.25 \times 10^{-2}\text{ mol/L•s}$$

14.19 (a) The graph shows the tangent line to the [sucrose] versus time curve at t = 0 min. A right triangle has been drawn with part of the tangent line as the hypotenuse. The legs of the triangle are Δ[sucrose] = 0.230 mol/L – 0.480 mol/L = –0.250 mol/L and Δt = 170 min – 50 min = 120 min. The slope of the tangent line is

$$\text{slope} = \frac{\Delta[\text{sucrose}]}{\Delta t} = \frac{-0.250\text{ mol/L}}{120\text{ min}} = -2.1 \times 10^{-3}\text{ mol/L•min}$$

The initial reaction rate is the negative of the slope, or 2.1×10^{-3} mol/L•min.

(b) The graph shows the tangent line to the [sucrose] versus time curve at t = 440 min. A right triangle has been drawn with part of the tangent line as the hypotenuse. The legs of the triangle are Δ[sucrose] = 0.070 mol/L – 0.210 mol/L = –0.140 mol/L and Δt = 600 min – 290 min = 310 min. The slope of the tangent line is

$$\text{slope} = \frac{\Delta[\text{sucrose}]}{\Delta t} = \frac{-0.140 \text{ mol/L}}{319 \text{ min}} = -4.5 \times 10^{-4} \text{ mol/L•min}$$

The rate at 440 min is the negative of the slope, or 4.5×10^{-4} mol/L•min.

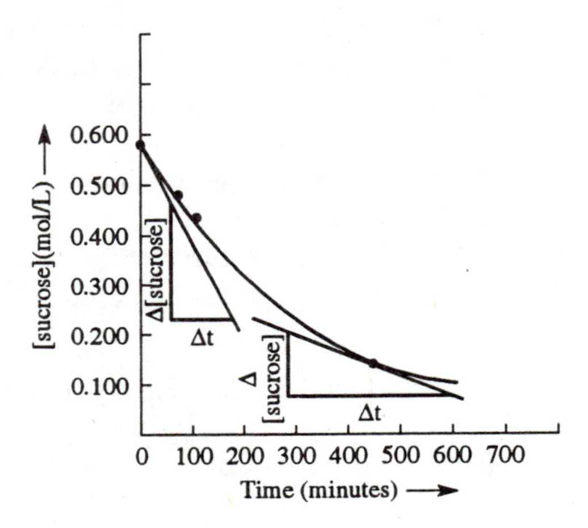

14.21 The average reaction rate over a time interval is equal to the change in concentration divided by the change in time. For the hydrolysis of sucrose, the average rate between 0 and 100 minutes is

$$\text{average rate} = \frac{-\Delta[\text{sucrose}]}{\Delta t} = \frac{-(0.427 \text{ mol/L} - 0.584 \text{ mol/L})}{100 \text{ min}} = \frac{-(-0.157 \text{ mol/L})}{100 \text{ min}} = 1.57 \times 10^{-3} \text{ mol/L•min}$$

14.23 A rate law or rate equation is an expression that relates the reaction rate to concentrations raised to various powers. The concentrations of reactants, catalyst, and products can appear in the rate law.

No. The exponents in a rate law can only be determined by experiment.

14.25 The order of a reaction is the sum of the exponents in its rate law.

14.27 A rate constant is the proportionality constant k in the rate law. Rate constants are determined by substituting experimental data into the rate law, or an equation involving k that is derived or obtained from the rate law, and solving for k.

14.29 (a) s^{-1} (b) L/mol•s

14.31 The rate is proportional to the square of the H_2 concentration and is proportional to the N_2 concentration. Combining both concentration effects gives

rate = $k[H_2]^2[N_2]$

14.33 (a) From runs (1) and (2), increasing [A] threefold while keeping [B] constant increases the rate by 0.108/0.012 or ninefold. Since $9 = 3^2$, the rate is proportional to $[A]^2$. From runs (1) and (3), cutting [B] in half while keeping [A] constant also cuts the rate in half. The rate is proportional to [B]. Therefore, rate = $k[A]^2[B]$. The reaction is third order.

(b) Using data from each run and averaging the results gives:

1st run: 0.012 mol/L•s = $k(0.10 \text{ mol/L})^2(0.60 \text{ mol/L})$ and k = 2.0 L^2/mol^2•s
2nd run: k = 2.0 L^2/mol^2•s; 3rd run: k = 2.0 L^2/mol^2•s

The value for k averaged over the three runs is k = 2.0 $L^2/mol^2 \cdot s$.

(c) Rate = 2.0 $L^2/mol^2 \cdot s \times (0.010\ mol/L)^2\ (0.010\ mol/L) = 2.0 \times 10^{-6}\ mol/L \cdot s$

14.35 (a) From runs (1) and (2), decreasing the $[C_2F_4]$ by 9.31/10.14 = 0.918 decreases the rate by $5.38 \times 10^{-12}/6.39 \times 10^{-12} = 0.842$. Since $0.842 = (0.918)^2$, the rate is proportional to $[C_2F_4]^2$. Therefore, rate = $k[C_2F_4]^2$. The reaction is second order.

(b) Using data from each run and averaging the results gives:

1st run: 6.39×10^{-12} mmol/L•s = k(10.14 mmol/L)^2 and k = 6.21×10^{-14} L/mmol•s
2nd run: k = 6.21×10^{-14} L/mmol•s; 3rd run: k = 6.21×10^{-14} L/mmol•s

The value for k averaged over the three runs is k = 6.21×10^{-14} L/mmol•s.

(c) rate = 6.21×10^{-14} L/mmol•s × (0.200 mmol/L)^2 = 2.48×10^{-15} mmol/L•s

14.37 (a) From runs (1) and (3), increasing $[N_2O_4]$ by $2.00 \times 10^{-4}/1.80 \times 10^{-4} = 1.11$ while keeping $[N_2]$ constant increases the rate by 7.69/6.93 = 1.11. The rate is proportional to $[N_2O_4]$. So, rate = $k[N_2O_4][N_2]^x$ and $k = rate/[N_2O_4][N_2]^x$. Using data from runs (1) and (2) and solving for x we get

$$\frac{6.93}{(1.80 \times 10^{-4})(0.0225)^x} = \frac{7.70}{(2.25 \times 10^{-4})(0.0200)^x} \quad \text{and } x = 1$$

The rate is proportional to $[N_2]$. Therefore, rate = $k[N_2O_4][N_2]$.

(b) Using data from each run and averaging the results gives:

1st run: 6.93 mol/L•s = k(1.80×10^{-4} mol/L)(0.0225 mol/L) and k = 1.71×10^6 L/mol•s
2nd run: k = 1.71×10^6 L/mol•s; 3rd run: k = 1.71×10^6 L/mol•s

The value for k averaged over the three runs is k = 1.71×10^6 L/mol•s.

(c) The reaction is first order with respect to N_2O_4 and first order with respect to N_2.

14.39 A usual first-order reaction is a reaction in which the rate is proportional to the first power of a single reactant concentration.

A plot of the logarithm of concentration versus time should yield a straight line for a first-order reaction. Therefore, if such a plot is a straight line, then that will identify the reaction as being a first-order reaction.

14.41 (a) The half-life ($t_{1/2}$) is the time required for a reactant's concentration to decrease to one-half of its value.
(b) The half-life of a first-order reaction is independent of concentration and, therefore, does not change as the reaction proceeds, while the half-life of a second-order reaction varies with concentration. For a second-order reaction the half-life periods become progressively larger as the reaction proceeds. A first-order reaction has a constant half-life.

14.43 (a) For a first-order reaction, a plot of the logarithm of the concentration versus time should yield a straight line.

t (min)	0	20.0	60.0	100	160	220	440	660
ln [sucrose]	−0.538	−0.600	−0.728	−0.851	−1.050	−1.234	−1.924	−2.62

The graph shows that this is essentially the case.

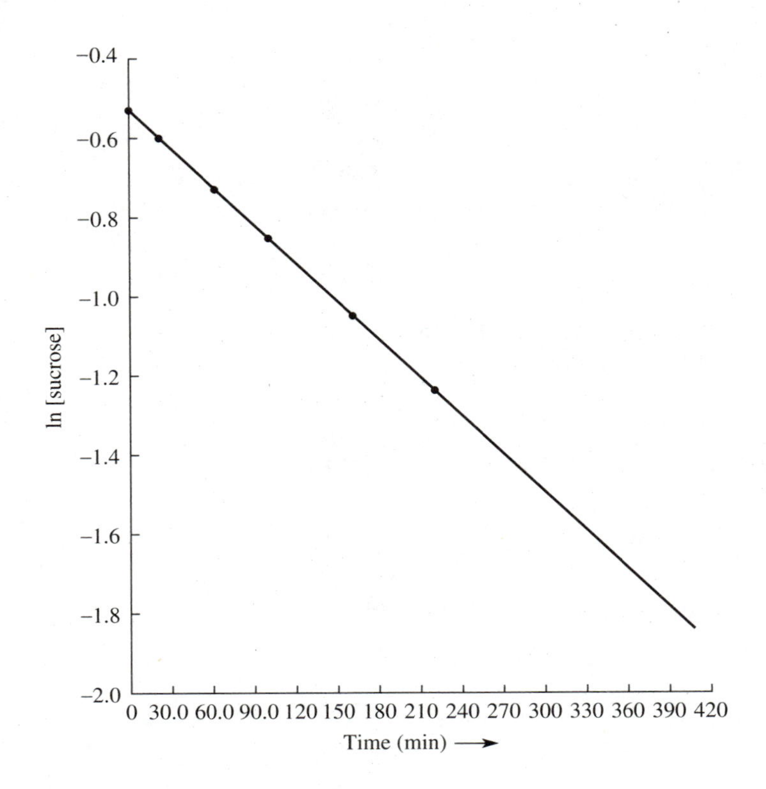

(b) Since $m = -k$, (see Equation 14.5) the rate constant for this first-order reaction can be obtained from the slope of the graph of ln [sucrose] versus time. It can also be obtained using Equation 14.4 and any two concentration values (data obtainable from Exercise 14.19).

From graph: Using the points (t = 100 min, ln [sucrose] = −0.851) and (t = 220 min, ln [sucrose] = −1.234) we get for the slope

$$\text{slope} = \frac{\Delta \ln[\text{sucrose}]}{\Delta t} = \frac{-0.383}{120 \text{ min}} = -3.19 \times 10^{-3} \text{ min}^{-1}$$

Hence, the rate constant is 3.19×10^{-3} min^{-1}.

(c) If 20% of the sucrose is hydrolyzed, then 80% remains: the fraction remaining is $[\text{sucrose}]/[\text{sucrose}]_0 =$ 0.80. This fraction and $k = 3.19 \times 10^{-3}$ min^{-1} are substituted in Equation 14.4.

$\ln([\text{sucrose}]/[\text{sucrose}]_0) = -kt$

$\ln 0.80 = -3.19 \times 10^{-3} \text{ min}^{-1} \times t$ and $t = 70$ min

14.45 (a) Substitute moles of N_2O_5 for [A] in Equation 14.4.

$\ln(n_{N_2O_5}/(n_{N_2O_5})_0) = -kt$

The time $t = 1.00 \text{ h} = 3.60 \times 10^3$ s, $k = 4.98 \times 10^{-4} \text{ s}^{-1}$, and the original number of moles of N_2O_5 is $(n_{N_2O_5})_0 = 1.00$ mol. Hence,

$\ln(n_{N_2O_5}/1.00 \text{ mol}) = -4.98 \times 10^{-4} \text{ s}^{-1} \times 3.60 \times 10^3 \text{ s} = -1.79$

and $n_{N_2O_5} = 0.17 \times 1.00 \text{ mol} = 0.17$ mol

(b) From the reaction equation we see that for every mole of N_2O_5 gas that decomposes, 2.5 mol of gas is produced. Therefore, after 1.00 h there is 0.17 mol + 0.83 mol × 2.5 = 2.2 mol. Hence, the pressure should increase by a factor of 2.2, or total gas pressure = 0.500 atm × 2.2 = 1.1 atm.

14.47 (a) $kt_{1/2} = \ln 2$

$0.000498\ s^{-1} \times t_{1/2} = 0.693$ and $t_{1/2} = 1.39 \times 10^3\ s$

(b) If 15.0% of the N_2O_5 has decomposed, then 85.0% remains: the fraction of N_2O_5 remaining is 0.850. This fraction and $k = 4.98 \times 10^{-4}\ s^{-1}$ are substituted in Equation 14.4.

$$\ln 0.850 = -4.98 \times 10^{-4}\ s^{-1} \times t \text{ and } t = 326\ s \times \frac{1\ min}{60\ s} = 5.43\ min$$

14.49 $k \times t_{1/2} = \ln 2$

$0.00319\ min^{-1} \times t_{1/2} = 0.693$ and $t_{1/2} = 217\ min$

14.51 First the rate constant k must be calculated.

$$k = \frac{\ln 2}{t_{1/2}} = \frac{0.693}{32.0\ min} = 0.0217\ min^{-1}$$

The time required can be obtained by substituting $[A]/[A]_0 = 0.0160\ M/0.256\ M$ and $k = 0.0217\ min^{-1}$ into Equation 14.4.

$$\ln \frac{0.0160\ mol/L}{0.256\ mol/L} = -0.0217\ min^{-1} \times t \text{ and } t = 2.773/0.0217\ min^{-1} = 128\ min$$

14.53 (a) First the rate constant k of the reaction must be calculated.

$$k = \frac{1}{t_{1/2}[A]_0} = \frac{1}{32.0\ min \times 0.256\ mol/L} = 0.122\ L/mol \bullet min$$

The time can be obtained by substituting $[A]_0 = 0.256\ mol/L$, $[A] = 0.0160\ mol/L$, and $k = 0.122\ L/mol \bullet min$ into Equation 14.8.

$$\frac{1}{0.0160\ mol/L} = 0.122\ L/mol \bullet min \times t + \frac{1}{0.256\ mol/L} \text{ and } t = \frac{58.6\ L/mol}{0.122\ L/mol \bullet min} = 480\ min$$

(b) Comparing the answers we see that the time required for the second-order reaction to reach the 0.0160 M concentration is considerably longer that required by the first-order reaction. This is because the half-life of a second-order reaction increases as the concentration decreases, while that of a first-order reaction remains the same.

14.55 (a) For a second-order reaction, a graph of 1/[A] versus t will be a straight line whose slope is k. Therefore, a plot of 1/[A] versus t can be used to show if a reaction is second order. The numbers needed are

t (minutes)	0	45.0	72.0	107.0	230.0
$[NH_4CNO]$	0.1000	0.0808	0.0716	0.0638	0.0463
$1/[NH_4CNO]$	10.00	12.38	14.0	15.7	21.6

A plot of these numbers (not shown) gives a straight line and shows that the reaction is second order. (Convince yourself of this by making the actual plot.)

(b) The slope = k can be calculated using any two points in the above data. Using t = 0 min and t = 45.0 min we get

$$k = \text{slope} = \frac{\Delta 1/[NH_4CNO]}{\Delta t} = \frac{12.38\ L/mol - 10.00\ L/mol}{45.0\ min} = 0.0529\ L/mol \bullet min$$

A somewhat more accurate value for the slope = k can be obtained from the graph.

(c) The time required can be obtained by substituting $[A]_0 = 0.1000$ mol/L, $[A] = 0.0750$ mol/L, and $k = 0.0529$ L/mol•min into Equation 14.8.

$$\frac{1}{0.0750 \text{ mol/L}} = 0.0529 \text{ L/mol•min} \times t + \frac{1}{0.1000 \text{ mol/L}}$$

$0.0529 \text{ L/mol•min} \times t = 3.33$ L/mol and $t = 62.9$ min

14.57 (a) A reaction mechanism is a postulated series of consecutive steps leading from reactants to products.
(b) 1. It must account for the observed products.
2. It must explain the experimental rate law.

14.59 The molecularity is the number of particles participating in an elementary step, while the reaction order is the sum of the concentration exponents in the rate law.

Reaction order is also defined in terms of individual species in the rate law, i.e., in terms of the exponent of the concentration of a given species.

14.61 An intermediate is a substance produced in one step of the reaction mechanism and used up or consumed in another, while a catalyst is a substance that increases the rate of a reaction without being consumed in the reaction. A catalyst may be used up in one step of the reaction, but is reproduced in another step.

14.63 The rate law for an elementary step can be deduced from the equation for the step.
(a) rate $= k[Cl_2]$ (b) rate $= k[Cl][Cl] = k[Cl]^2$

14.65 (a) rate $= k[ClO_2][F_2]$
(b) A two-step mechanism having a slow first step that is consistent with the rate law is:

Step 1. $ClO_2(g) + F_2(g) \rightarrow FClO_2(g) + F(g)$ (slow)
Step 2. $F(g) + ClO_2(g) \rightarrow FClO_2(g)$ (fast)

14.67 The rate law for the reaction is determined by the slow third step.

rate $= k_2[H_2][I]^2$

The intermediate I can be removed from the rate law by employing the steady-state approximation.

rate of I formation = rate of I consumption

$k_1[I_2] = k_{-1}[I]^2$

The steady-state concentration of atomic iodine is therefore $[I] = \sqrt{\frac{k_1}{k_{-1}}}\,[I_2]^{1/2}$

Substituting this expression for [I] into the rate law gives

$$\text{rate} = \frac{k_2 k_1}{k_{-1}}[H_2][I_2] = k[H_2][I_2]$$

where $k = k_2k_1/k_{-1}$

14.69 (a) $2NO(g) + O_2(g) \rightarrow 2NO_2(g)$

N_2O_2 is an intermediate.

(b) The rate law for the reaction is determined by the slow third step.

rate = $k_2[N_2O_2][O_2]$

The intermediate N_2O_2 can be removed from the rate law by employing the steady-state approximation.

rate of N_2O_2 formation = rate of N_2O_2 consumption

$k_1[NO]^2 = k_{-1}[N_2O_2]$

The steady state concentration of N_2O_2 is therefore

$$[N_2O_2] = \frac{k_1}{k_{-1}}[NO]^2$$

Substituting this expression for $[N_2O_2]$ into the rate law gives

$$\text{rate} = \frac{k_2k_1}{k_{-1}}[NO]^2[O_2] = k[NO]^2[O_2], \text{ where } k = k_2k_1/k_{-1}.$$

14.71 A chain initiation step produces free radicals, a chain propagation step uses up one free radical and forms another free radical, and a chain termination step removes free radicals from circulation.

14.73 (a) Chain initiation step:

$$Cl_2(g) \xrightarrow{h\nu} Cl\bullet(g) + Cl\bullet(g)$$

Chain propagation steps:

$$Cl\bullet(g) + CH_4(g) \rightarrow CH_3\bullet(g) + HCl(g)$$
$$CH_3\bullet(g) + Cl_2(g) \rightarrow CH_3Cl(g) + Cl\bullet(g)$$

Chain termination steps:

$$Cl\bullet(g) + Cl\bullet(g) \rightarrow Cl_2(g)$$
$$CH_3\bullet(g) + CH_3\bullet(g) \rightarrow C_2H_6(g)$$
$$CH_3\bullet(g) + Cl\bullet(g) \rightarrow CH_3Cl(g)$$

(b) The chain propagation steps.

14.75 As the magnitude of its activation energy increases, the rate of an elementary step decreases.

14.77 Potential energy profile for an endothermic bimolecular step.

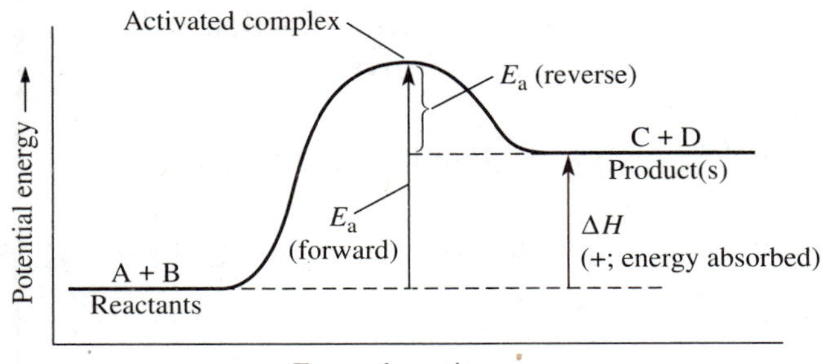

14.79

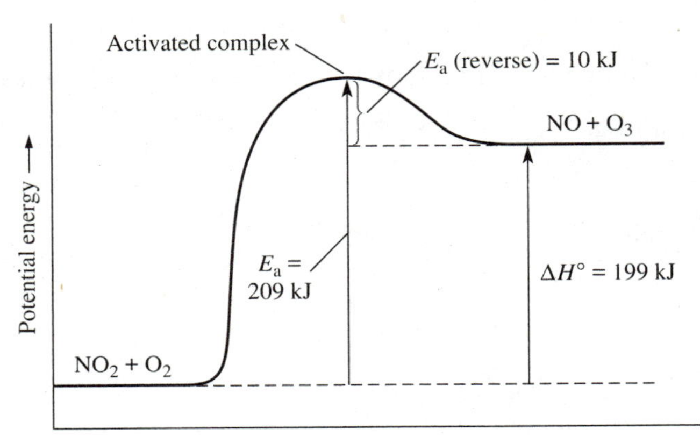

14.81 (a)

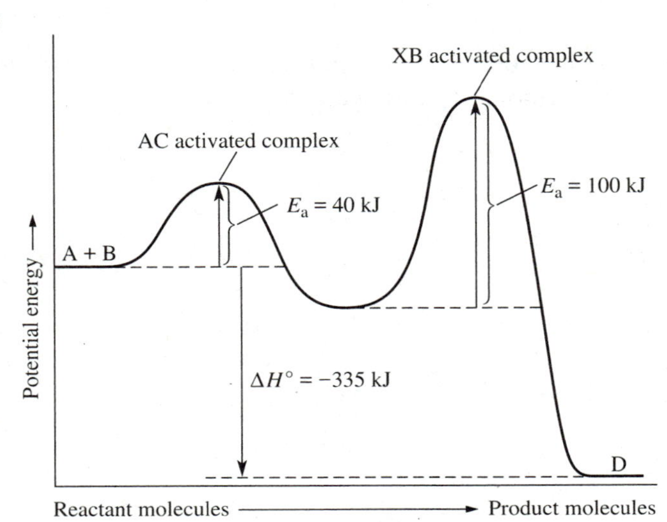

(b) The second step, since it has a greater activation energy.
(c) $\Delta H° = \Delta H°$ Step 1 + $\Delta H°$ Step 2 = –20 kJ – 315 kJ = –335 kJ.

14.83 $k = Ae^{-E_a/RT}$
(a) As E_a increases, the rate constant decreases.
(b) As T increases, the rate constant increases.

14.85 If the rate constant doubles, then the fraction $k_1/k_2 = 1/2$. Substituting this fraction and T_1 = 298 K, T_2 = 308 K, and R = 8.314 J/mol•K into Equation 14.15 gives

$$\ln(1/2) = \frac{E_a}{8.314\ \text{J/mol•K}}\left[\frac{1}{308\ \text{K}} - \frac{1}{298\ \text{K}}\right]$$ and $E_a = 5.3 \times 10^4$ J/mol = 53 kJ/mol

14.87 (a) Let k_1 and T_1 be $2.20 \times 10^{-5}\ \text{min}^{-1}$ and 457.6 K; let k_2 and T_2 be $3.07 \times 10^{-3}\ \text{min}^{-1}$ and 510.1 K. Substituting these values and R = 8.314 J/mol•K into Equation 14.15 gives

$$\ln\frac{2.20 \times 10^{-5}}{3.07 \times 10^{-3}} = \frac{E_a}{8.314\ \text{J/mol•K}}\left[\frac{1}{510.1\ \text{K}} - \frac{1}{457.6\ \text{K}}\right]$$

Solving gives $E_a = 1.83 \times 10^5$ J/mol = 183 kJ/mol.

(b) Substituting $E_a = 183{,}000$ J/mol, $k = 2.20 \times 10^{-5}\ \text{min}^{-1}$, $T = 457.6$ K, and $R = 8.314$ J/mol•K into Equation 14.13 and solving for A gives

$$A = ke^{E_a/RT} = 2.20 \times 10^{-5}\ \text{min}^{-1} \times e^{(183{,}000\ \text{J/mol})/(8.314\ \text{J/mol•K})(457.6\ \text{K})} = 1.71 \times 10^{16}\ \text{min}^{-1}$$

(c) Substituting $E_a = 183{,}000$ J/mol, $k_1 = 2.20 \times 10^{-5}\ \text{min}^{-1}$, $T_1 = 457.6$ K, $T_2 = 475.0$ K, and $R = 8.314$ J/mol•K into Equation 14.15 gives

$$\ln \frac{2.20 \times 10^{-5}}{k_2} = \ln 2.20 \times 10^{-5} - \ln k_2 = \frac{183{,}000\ \text{J/mol}}{8.314\ \text{J/mol•K}}\left[\frac{1}{475.0\ \text{K}} - \frac{1}{457.6\ \text{K}}\right] \text{ and } k_2 = 1.3 \times 10^{-4}\ \text{min}^{-1}$$

14.89 If a reaction follows the Arrhenius equation, then a graph of ln k versus 1/T should yield a straight line whose slope $m = -E_a/R$ and intercept $b = \ln A$. To prepare the graph we must first calculate ln k and 1/T (K^{-1}). The resulting numbers are given below.

k (L/mol•s)	ln k	T (K)	1/T (K^{-1})
1.6×10^{-20}	–45.58	298	3.36×10^{-3}
1.3×10^{-19}	–43.49	308	3.25×10^{-3}
8.7×10^{-19}	–41.59	318	3.14×10^{-3}
5.4×10^{-18}	–39.76	328	3.05×10^{-3}
3.0×10^{-17}	–38.05	338	2.96×10^{-3}

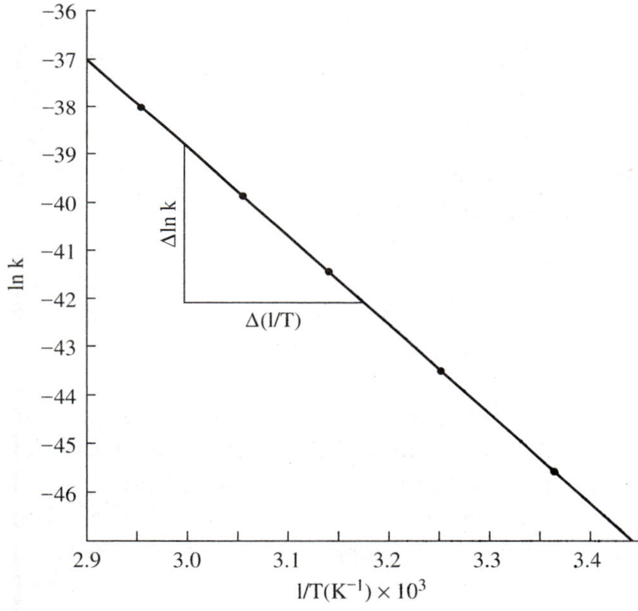

The graph shows the straight line obtained by plotting ln k versus 1/T. A right triangle has been drawn with part of the straight line as the hypotenuse. The legs of the triangle are $\Delta \ln k = -42.20 - (-38.80) = -3.40$ and $\Delta(1/T) = 3.18 \times 10^{-3}\ K^{-1} - 3.00 \times 10^{-3}\ K^{-1} = 0.18 \times 10^{-3}\ K^{-1}$. The slope of the line is

$$\text{slope} = \frac{\Delta \ln k}{\Delta(1/T)} = \frac{-3.40}{0.18 \times 10^{-3}\ K^{-1}} = -1.9 \times 10^4\ K$$

The activation energy for the reaction can be obtained from the slope.

$$E_a = -mR = 1.9 \times 10^4\ K \times 8.314\ \text{J/mol•K} = 1.6 \times 10^5\ \text{J/mol or 160 kJ/mol}$$

A can now be obtained by substituting the given data and E_a into the Arrhenius equation. Substituting $k = 1.6 \times 10^{-20}$ L/mol•s, T = 298 K, and E_a = 160,000 J/mol into Equation 14.13 gives

1.6×10^{-20} L/mol•s = $Ae^{-(160{,}000\ J/mol)/(8.314\ J/mol\cdot K)(298\ K)}$ and $A = 1.8 \times 10^{8}$ L/mol•s

14.91 The principle assumption underlying the collision theory of reaction rates is that in order for a reaction to occur, the reactant molecules must collide while at the same time possessing the necessary energy and suitable orientation. In other words, the principle assumption is that reacting particles must effectively collide in order to react, which means that (1) the collision must be sufficiently energetic, and (2) the colliding particles must have the proper orientation.

14.93 (a) The bimolecular collision frequency increases with increasing temperature.
(b) The bimolecular collision frequency increases with increasing concentration.

14.95 The orientation factor p, which is the fraction of collisions with suitable orientation for reaction, decreases with increasing molecular complexity.

14.97 (a) The collision frequencies are proportional to the square roots of these temperatures.

$$\frac{\text{collision frequency at 373 K}}{\text{collision frequency at 273 K}} = \sqrt{\frac{373\ \text{K}}{273\ \text{K}}} = 1.17$$

The collision frequency will increase by a factor of 1.17.

(b) $$\frac{e^{-(85{,}000\ J/mol)/(8.314\ J/mol\cdot K)(373\ K)}}{e^{-(85{,}000\ J/mol)/(8.314\ J/mol\cdot K)(273\ K)}} = 2.29 \times 10^{4}$$

$e^{-E_a/RT}$ increases by a factor of 2.29×10^{4}.

(c) The reaction rate will increase by a factor of $1.17 \times 2.29 \times 10^{4} = 2.68 \times 10^{4}$.

14.99 A catalyst provides a pathway or mechanism in which the rate-determining elementary step has a lower activation energy. The catalyzed reaction goes faster, since the slow step in the catalyzed reaction has a lower activation energy than the slow step in the uncatalyzed reaction.

14.101 The catalyst exists in a different state from the reacting species. Usually a solid catalyst is in contact with a liquid or gaseous solution. The reactant molecules diffuse from the liquid or gaseous solution to the catalyst surface where they are chemically adsorbed (the process is called chemisorption). The bonding in the reactant molecules is altered, bonds may actually be broken, by the forming of weak chemical bonds with the catalyst surface atoms. The altered molecules then undergo reaction by an alternative reaction mechanism with a lower activation energy than the uncatalyzed reaction. After the reaction is over, the products diffuse away from the catalyst surface.

14.103 (a) The activation energy for the reverse catalyzed reaction can be obtained by substituting ΔH° = +46 kJ and E_a (forward) = 165 kJ into Equation 14.12:

+ 46 kJ = 165 kJ – E_a (reverse) and E_a (reverse) = 165 kJ – 46 kJ = 119 kJ

(b) The activation energy for the reverse uncatalyzed reaction can be obtained by substituting ΔH° = +46 kJ and E_a (forward) = 335 kJ into Equation 14.12:

+46 kJ = 335 kJ – E_a (reverse) and E_a (reverse) = 335 kJ – 46 kJ = 289 kJ

14.105 Nitrogen (N_2) is unreactive because it has a strong triple bond holding the nitrogen atoms together. When subject to an electric discharge, the bonds of the N_2 molecules break. The resulting atoms are reactive, since they have incomplete octets.

14.107 rate = $k[H_2][I_2]$

$$\text{rate} = \frac{2.42\times10^{-2}\ \text{L}}{\text{mol}\cdot\text{s}}\times\frac{0.104\ \text{mol}}{\text{L}}\times\frac{0.0100\ \text{mol}}{\text{L}} = 2.52\times10^{-5}\ \text{mol/L}\cdot\text{s}$$

14.109 To solve the problem, we first have to obtain the rate constant for the reaction at 600°C.

$$k = \frac{0.693}{t_{1/2}} = \frac{0.693}{81\ \text{s}} = 8.6\times10^{-3}\ \text{s}^{-1}$$

The time can be obtained by substituting [acetone] = 0.45 atm, $[\text{acetone}]_0$ = 0.51 atm, and $k = 8.6\times10^{-3}\ \text{s}^{-1}$ into Equation 14.4.

$$\ln\frac{0.45}{0.51} = -8.6\times10^{-3}\ \text{s}^{-1}\times t \quad \text{and} \quad t = 15\ \text{s}$$

14.111 (a) $\text{rate} = \dfrac{-d[A]}{dt} = k$

Rearranging gives $d[A] = -k\,dt$

Integrating between time 0 and time t gives

$$\int_{[A]_0}^{[A]} d[A] = -k\int_0^t dt$$

$[A] - [A]_0 = -k(t-0)$ or $[A] = -kt + [A]_0$

(b) The derived equation has the form $y = mx + b$

$$\underset{y}{\underset{\uparrow}{[A]}} = \underset{mx}{\underset{\uparrow\uparrow}{-kt}} + \underset{b}{\underset{\uparrow}{[A]_0}}$$

which describes a straight line. Here [A] and t correspond to the variables y and x, respectively. Hence, a plot of [A] versus t will be linear.

(c) Substituting $t = t_{1/2}$ and $[A] = \frac{1}{2}[A]_0$ into the zero-order equation and solving for $t_{1/2}$ gives

$\frac{1}{2}[A]_0 = -kt_{1/2} + [A]_0$ and $t_{1/2} = [A]_0/2k$

14.113 $A + B \rightarrow$ products

rate = $-d[A]/dt = k[A][B]$

The rate equation can be integrated once the concentration of B is related to the concentration of A. Let $[A]_0$ and $[B]_0$ be the initial concentrations of A and B. The concentration of A can be written as follows

$[A] = [A]_0 - x$

where x is the drop in the concentration of A. From the reaction equation, we see that there is a one-to-one relationship between A and B and, therefore, the drop in concentration of B = the drop in concentration of A = x, and

$[B] = [B]_0 - x$. Hence, $-d[A]/dt = k[A][B] = k([A]_0 - x)([B]_0 - x)$

$d[A]/dt$ is converted to dx/dt as follows

$d[A]/dt = d([A]_0 - x)/dt = -dx/dt$

since $[A]_0$ is a constant. Therefore, the rate law becomes $dx/dt = k([A]_0 - x)([B]_0 - x)$

and can now be solved for x as a function of t. Rearranging gives

$$k\,dt = \frac{dx}{([A]_0 - x)([B]_0 - x)}$$

Integrating between time zero and time t gives (x goes from zero to x) (see comment below)

$$\int_0^t k\,dt = kt = \int_0^x \frac{dx}{([A]_0 - x)([B]_0 - x)}$$

$$= \int_0^x \left\{\frac{-1}{[A]_0 - [B]_0}\right\}\left\{\frac{1}{([A]_0 - x)} - \frac{1}{([B]_0 - x}\right\} dx$$

$$= \left\{\frac{-1}{[A]_0 - [B]_0}\right\}\left\{\ln\left(\frac{[A]_0}{[A]_0 - x}\right) - \ln\left(\frac{[B]_0}{[B]_0 - x}\right)\right\}$$

Finally, recalling that $[A] = [A]_0 - x$ and $[B] = [B]_0 - x$ and combining the two logarithms gives

$$kt = \left(\frac{1}{[A]_0 - [B]_0}\right) \ln \frac{[A][B]_0}{[A]_0[B]}$$

Comment: f(x) dx is simplified using the method of partial fractions. Using this method we can write f(x) as the sum of two simpler terms:

$$f(x) = \frac{1}{([A]_0 - x)([B]_0 - x)} = \frac{p}{([A]_0 - x)} + \frac{q}{([B]_0 - x)}$$

where p and q are constants. The terms p and q can be evaluated by using a common denominator and equating coefficients of like powers of x in the numerators:

$$\frac{1}{([A]_0 - x)([B]_0 - x)} = \frac{p([B]_0 - x) + q([A]_0 - x)}{([A]_0 - x)([B]_0 - x)}$$

Equating coefficients of x, $0 = -p - q$ (1)

Equating the constant terms, $1 = p[B]_0 + q[A]_0$ (2)

Solving (1) and (2) simultaneously for p and q gives

$$p = \frac{-1}{[A]_0 - [B]_0} \quad \text{and} \quad q = \frac{1}{[A]_0 - [B]_0}$$

Therefore, f(x) dx becomes

$$f(x)\,dx = \frac{dx}{([A]_0 - x)([B]_0 - x)} = \left(\frac{-1}{[A]_0 - [B]_0}\right)\left(\frac{1}{([A]_0 - x)} - \frac{1}{([B]_0 - x)}\right) dx$$

14.115 Yes. If the slow step involves an intermediate whose steady-state concentration depends on the product of two concentrations, then the rate law for the reaction will depend on the product of three concentrations and the reaction will be a third-order reaction.

14.117 (a) $CHCl_3(g) + Cl_2(g) \rightarrow CCl_4(g) + HCl(g)$

Cl and CCl_3 are intermediates.

(b) The rate law for the reaction is determined by the slow second step.

rate = $k_2[CHCl_3][Cl]$

The intermediate Cl can be removed from the rate law by employing the steady-state approximation. Since the second step is much slower than the first step, the steady-state approximation can be based just on the first step.

rate of Cl formation = rate of Cl consumption

$k_1[Cl_2] = k_{-1}[Cl]^2$ and $[Cl] = \sqrt{k_1/k_{-1}}\,[Cl_2]^{1/2}$

Substituting this expression for [Cl] in the rate law gives

rate = $k_2\sqrt{k_1/k_{-1}}\,[CHCl_3][Cl_2]^{1/2} = k[CHCl_3][Cl_2]^{1/2}$

14.119 (a) First we determine whether the reaction is first order or second order. A plot (not shown) of ln [protein] versus t gives a straight line; therefore the reaction is first order. Values for the slopes of the plots and the rate constants at the two temperatures can be calculated from the given data.

$$65.0°C:\ \text{slope} = \frac{\Delta \ln[\text{protein}]}{\Delta t} = \frac{\ln 5.22 - \ln 7.77}{40\ h} = -0.0099\ h^{-1}$$

and $k_{65.0°C} = -\text{slope} = 0.0099\ h^{-1}$

$$70.2°C:\ \text{slope} = \frac{\Delta \ln[\text{protein}]}{\Delta t} = \frac{\ln 2.76 - \ln 6.75}{300\ \text{min}} = -0.00298\ \text{min}^{-1}$$

and $k_{70.2°C} = -\text{slope} = 0.00298\ \text{min}^{-1} \times 60\ \text{min}\ h^{-1} = 0.179\ h^{-1}$

(b) The activation energy can be obtained by substituting $k_1 = 0.0099\ h^{-1}$, $T_1 = 338.2$ K, $k_2 = 0.179\ h^{-1}$, $T_2 = 343.4$ K, and R = 8.314 J/mol•K into Equation 14.15.

$$\ln\frac{0.0099}{0.179} = \frac{E_a}{8.314\ \text{J/mol•K}}\left[\frac{1}{343.4\ K} - \frac{1}{338.2\ K}\right]$$ and $E_a = 5.4 \times 10^5$ J/mol or 540 kJ/mol

14.121 The rate constants for the first-order decomposition at 25°C and 50°C are

25°C: $k = 0.693/20.0\ s = 3.46 \times 10^{-2}\ s^{-1}$

50°C: $k = 0.693/1.0\ s = 0.69\ s^{-1}$

The activation energy can be obtained by substituting $k_1 = 3.46 \times 10^{-2}\ s^{-1}$, $T_1 = 298$ K, $k_2 = 0.69\ s^{-1}$, $T_2 = 323$ K, and R = 8.314 J/mol•K into Equation 14.15.

$$\ln\frac{3.46\times10^{-2}}{0.69} = \frac{E_a}{8.314\ \text{J/mol•K}}\left[\frac{1}{323\ K} - \frac{1}{298\ K}\right]$$ and $E_a = 9.6 \times 10^4$ J/mol or 96 kJ/mol

The preexponential factor A can be obtained by substituting $k = 3.46 \times 10^{-2}\ s^{-1}$, T = 298 K, E_a = 96,000 J/mol, and R = 8.314 J/mol•K into Equation 14.13.

$$A = \frac{k}{e^{-E_a/RT}} = \frac{3.46\times10^{-2}\ s^{-1}}{e^{-(96{,}000\ \text{J/mol})/(8.314\ \text{J/mol•K})(298\ K)}} = 2.3 \times 10^{15}\ s^{-1}$$

14.123 From the reaction equation, we see that each reactive collision leads to the consumption of two NO_2 molecules. Therefore, there is only one-half a mole of reactive collisions per mole of NO_2 reacted or consumed. Consequently, converting the rate from mol/L•min to collisions/mL•s proceeds as follows:

$$\frac{4.8\times10^{-6}\text{ mol NO}_2}{\text{L}\bullet\text{min}}\times\frac{3.011\times10^{23}\text{ collisions}}{1\text{ mol NO}_2}\times\frac{1\text{ L}}{10^3\text{ mL}}\times\frac{1\text{ min}}{60\text{ s}}=2.4\times10^{13}\text{ collisions/mL}\bullet\text{s}$$

According to collision theory, the reaction rate is given by the following (Equation 14.16):

rate = $Z \times f_a \times p$ = frequency of reactive collisions × p

Substituting rate = 2.4×10^{13} collisions/mL•s and $Z \times f_a = 1.0 \times 10^{15}$ collisions/mL•s into Equation 14.16 gives

2.4×10^{13} collisions/mL•s = 1.0×10^{15} collisions/mL•s × p and $p = 2.4 \times 10^{-2}$

14.125 (a) Three possible mechanisms might be:

1. $H_2(g) \xrightarrow{Pt(s)} H_2(\text{on Pt})$ fast
   $H_2(\text{on Pt}) \rightarrow 2H(\text{on Pt})$ slow
   $2H(\text{on Pt}) + O_2(g) \rightarrow H_2O(g) + O(g)$ fast
   $O(g) + H_2(g) \rightarrow H_2O(g)$ fast

2. $O_2(g) \xrightarrow{Pt(s)} O_2(\text{on Pt})$ fast
   $O_2(\text{on Pt}) \rightarrow 2O(\text{on Pt})$ slow
   $O(\text{on Pt}) + H_2(g) \rightarrow H_2O(g)$ fast

3. $H_2(g) + O_2(g) \xrightarrow{Pt(s)} H_2(\text{on Pt}) + O_2(\text{on Pt})$ fast
   $H_2(\text{on Pt}) + O_2(\text{on Pt}) \rightarrow 2H(\text{on Pt}) + 2O(\text{on Pt})$ slow
   $2H(\text{on Pt}) + O(\text{on Pt}) \rightarrow H_2O(g)$ fast

(b) The second mechanism, since it just depends on the adsorption of oxygen.

Yes. The slow step is the break up of the O–O bond after the oxygen molecule is adsorbed on the platinum catalyst. The rate of this step is proportional to the fraction of the platinum surface that holds chemisorbed O_2 molecules. However, since this fraction is directly proportional to the pressure of O_2, the rate of the reaction is also directly proportional to the concentration of O_2. However, if the O_2 partial pressure or concentration is high enough so that the catalyst becomes saturated with O_2, then the mechanism or rate law would become independent of $[O_2]$; i.e., the reaction would then become a zero-order reaction.

CHAPTER 15

CHEMICAL EQUILIBRIUM

Solutions To Practice Exercises

PE 15.1 $Q_c = [Cl]^2/[Cl_2]$

PE 15.2 2nd: $K_c = \dfrac{(0.00823)^2}{(0.0146)} = 4.64 \times 10^{-3}$

3rd: $K_c = \dfrac{(0.00654)^2}{(0.00923)} = 4.63 \times 10^{-3}$

PE 15.3 Concentration Summary:

Equation:	$H_2O(g)$ +	$CO(g)$ $\rightleftharpoons$	$H_2(g)$ +	$CO_2(g)$
Initial concentrations, mol/L	1.00	1.00	0	0
Concentration changes, mol/L	–0.58	–0.58	+0.58	+0.58
Equilibrium concentrations, mol/L	0.422	0.422	0.58	0.58

$$K_c = (Q_c)_{eq} = \frac{[H_2]_{eq}\,[CO_2]_{eq}}{[H_2O]_{eq}\,[CO]_{eq}} = \frac{(0.58)(0.58)}{(0.422)(0.422)} = 1.9$$

PE 15.4 $Q_c = \dfrac{[ICl]^2}{[I_2]\,[Cl_2]} = \dfrac{(0.0025)^2}{(0.10)(0.10)} = 6.2 \times 10^{-4}$

Since $Q_c < K_c$, the reaction will move in the forward direction.

PE 15.5 There is no Br to begin with, so the reaction can move only in the forward direction.

Concentration Summary:

Equation:	$Br_2(g)$ $\rightleftharpoons$	$2Br(g)$
Initial concentrations, mol/L	1.00	0
Concentration changes, mol/L	$-x$	$+2x$
Equilibrium concentrations, mol/L	$1.00 - x$	$2x$

The K_c expression is $K_c = [Br]^2/[Br_2]$.

Substituting the equilibrium concentrations and $K_c = 2.64 \times 10^{-4}$ into the above expression gives

$$2.64 \times 10^{-4} = \frac{(2x)^2}{1.00 - x}$$

$4x^2 + 2.64 \times 10^{-4}x - 2.64 \times 10^{-4} = 0$

and $x = 8.09 \times 10^{-3}$ mol/L

The equilibrium concentration of Br is therefore $[Br] = 2x = 2 \times 0.00809$ mol/L = 0.0162 mol/L.

PE 15.6 Since only CO_2 is present to begin with, the reaction must proceed in the forward direction. The initial CO_2 concentration is 0.100 mol/2.00 L = 0.0500 mol/L.

Concentration Summary:

	$2CO_2(g)$ $\rightleftharpoons$	$2CO(g)$ +	$O_2(g)$
Equation:	$2CO_2(g)$	$2CO(g)$	$O_2(g)$
Initial concentrations, mol/L	0.0500	0	0
Concentration changes, mol/L	$-2x$	$+2x$	$+x$
Equilibrium concentrations, mol/L	$0.0500 - 2x$	$2x$	x

The K_c expression is $K_c = [CO]^2[O_2]/[CO_2]^2$.

Substituting the equilibrium concentrations and $K_c = 4.50 \times 10^{-23}$ into the above expression gives

$$4.50 \times 10^{-23} = \frac{(2x)^2(x)}{(0.0500 - 2x)^2}$$

Since K_c is so small we can simplify the calculation by assuming that $0.0500 - 2x$ is approximately equal to 0.0500. Replacing $0.0500 - 2x$ with 0.0500 in the above equation gives

$$4.50 \times 10^{-23} = \frac{(2x)^2(x)}{(0.0500)^2} = \frac{4x^3}{(0.0500)^2}$$

and $x = [O_2] = 3.04 \times 10^{-9}$ mol/L.

PE 15.7

$$8.00 \times 10^{-2} = \frac{x(1.00 + 4x)^4}{0.0500 - x}$$

First Approximation: $8.00 \times 10^{-2} = \dfrac{x(1.00)^4}{0.0500}$ and $x = 4.00 \times 10^{-3}$

Second Approximation: $8.00 \times 10^{-2} = \dfrac{x(1.00 + 0.0160)^4}{0.0500 - 0.00400}$ and $x = 3.45 \times 10^{-3}$

Third Approximation: $8.00 \times 10^{-2} = \dfrac{x(1.00 + 4(3.45 \times 10^{-3})^4}{0.0500 - 0.00345}$ and $x = 3.53 \times 10^{-3}$

Fourth Approximation: $x = 3.51 \times 10^{-3}$

Fifth Approximation: $x = 3.52 \times 10^{-3}$

Sixth Approximation: $x = 3.52 \times 10^{-3}$

Answer: $x = 3.52 \times 10^{-3}$

PE 15.8 The concentration of water is not included in the equilibrium constant expressions. Hence,
(a) $K_c = [HCN][OH^-]/[CN^-]$
(b) $K_c = [Be(OH)^+][H^+]/[Be^{2+}]$

PE 15.9 Q_c is initially zero, so the reaction can only move in the forward direction.

Concentration Summary:

Equation:	$NH_3(aq) + H_2O(l) \rightleftharpoons$	$NH_4^+(aq) +$	$OH^-(aq)$
Initial concentrations, mol/L	0.10	0	0
Concentration changes, mol/L	$-x$	$+x$	$+x$
Equilibrium concentrations, mol/L	$0.10 - x$	x	x

Substituting the equilibrium concentrations and $K_c = 1.77 \times 10^{-5}$ into the K_c expression gives

$K_c = [NH_4^+][OH^-]/[NH_3]$

$$1.77 \times 10^{-5} = \frac{(x)(x)}{(0.10 - x)}$$

$x^2 + 1.77 \times 10^{-5}x - 1.77 \times 10^{-6} = 0$ and $x = [OH^-] = 0.0013$ mol/L

PE 15.10 (a) The initial pressure of NOCl is 1.000 atm. The initial pressures of NO and Cl_2 are zero. Since Q_c is initially zero, the reaction moves in the forward direction.

Pressure Summary:

Equation:	$2NOCl(g) \rightleftharpoons$	$2NO(g) +$	$Cl_2(g)$
Initial pressures, atm	1.000	0	0
Pressure changes, atm	$-2x$	$+2x$	$+x$
Equilibrium pressures, atm	$1.000 - 2x$	$2x$	x

The total pressure at equilibrium, 1.135 atm, is the sum of the partial pressures. Therefore, 1.135 atm $= P_{NOCl} + P_{NO} + P_{Cl_2} = 1.000 - 2x + 2x + x = 1.000 + x$ and $x = 0.135$ atm.

The equilibrium pressures are

$P_{NOCl} = 1.000 - 2x = 1.000 \text{ atm} - 0.270 \text{ atm} = 0.730$ atm

$P_{NO} = 2x = 0.270$ atm

$P_{Cl_2} = x = 0.135$ atm

(b) Substituting the equilibrium pressures into the K_p expression gives

$$K_p = \frac{P_{NO}^2\, P_{Cl_2}}{P_{NOCl}^2} = \frac{(0.270)^2(0.135)}{(0.730)^2} = 1.85 \times 10^{-2}$$

PE 15.11 The reaction moves in the forward direction.

Pressure Summary:

Equation:	$C_2H_2(g) +$	$3H_2(g) \rightleftharpoons$	$2CH_4(g)$
Initial pressures, atm	1.00	1.00	0
Pressure changes, atm	$-x$	$-3x$	$+2x$
Equilibrium pressures, atm	$1.00 - x$	$1.00 - 3x$	$2x$

Substituting $K_p = 2.41 \times 10^{-4}$ and the equilibrium pressures into the K_p expression gives

$K_p = P^2_{CH_4}/P_{C_2H_2}P^3_{H_2}$

$$2.41 \times 10^{-4} = \frac{(2x)^2}{(1.00 - x)(1.00 - 3x)^3}$$

K_p is small, and we can assume that $2x$ and $3x$ will be small relative to 1.00. Substituting 1.00 for $(1.00 - 2x)$ and $(1.00 - 3x)$ gives

$$2.41 \times 10^{-4} = \frac{(2x)^2}{(1.00)(1.00)^3} = 4x^2$$

and $x = 0.00776$ atm. The approximation is justified because $3x = 3 \times 0.00776$ atm $= 0.0233$ atm is less than 5% of 1.00 atm. The partial pressure of methane is $P_{CH_4} = 2x = 2 \times 0.00776$ atm $= 1.55 \times 10^{-2}$ atm.

PE 15.12 (a) $2SO_3(g) \rightleftharpoons 2SO_2(g) + O_2(g)$
$\Delta n = 3 - 2 = +1$
(b) $K_c = K_p/RT$
$K_c = 1.79 \times 10^{-5}/0.0821 \times 623 = 3.50 \times 10^{-7}$

PE 15.13 (a) The equation in (a) is obtained by multiplying the original equation by 1/2, so its K_p is the square root of the original equilibrium constant:

$K_p = (7.6 \times 10^{-16})^{1/2} = 2.8 \times 10^{-8}$

(b) The equation in (b) is obtained by reversing the original equation and multiplying it by 1/2, so its K_p is the square root of the reciprocal of the original equilibrium constant:

$K_p = (1/7.6 \times 10^{-16})^{1/2} = 3.6 \times 10^7$

PE 15.14 The first equation is the third equation plus the reverse of the second equation. Hence,

$$K_p = \frac{1}{K_{p_2}} \times K_{p_3} = \frac{6.90 \times 10^{-19}}{4.27 \times 10^{-31}} = 1.62 \times 10^{12}$$

PE 15.15 (a) Br_2 is a liquid; $K_c = [NOBr]/[NO]$
(b) H_2O is the solvent and SbOCl is a solid; $K_c = [H^+]^2/[Sb^{3+}][Cl^-]$

PE 15.16 The reaction moves in the reverse direction.

Pressure Summary:

Equation:	C(graphite) + $S_2(g)$	$\rightleftharpoons CS_2(g)$
Initial pressures, atm	0	1.00
Pressure changes, atm	$+x$	$-x$
Equilibrium pressures, atm	x	$1.00 - x$

Substituting the equilibrium pressures and $K_p = 5.60$ into the equilibrium constant expression gives

$K_p = P_{CS_2}/P_{S_2}$

$$5.60 = \frac{1.00 - x}{x} \quad \text{and } x = 0.152 \text{ atm}$$

The equilibrium pressures are $P_{S_2} = x = 0.152$ atm

$P_{CS_2} = 1.00 - x = 1.00$ atm $- 0.152$ atm $= 0.85$ atm

PE 15.17 (a) More intense. Some of the added HI will be consumed producing more I_2 and H_2.
(b) More intense. The reaction will move in the reverse direction so as to restore some of the H_2. The new equilibrium mixture will contain more I_2 and less HI.

PE 15.18 Proportion of NO_2 will increase. Two moles of NO_2 occupy a greater volume than 1 mol of N_2O_4. Greater volume favors a larger proportion of NO_2 in the equilibrium mixture.

PE 15.19 (a) The forward reaction is exothermic. Increasing the temperature will cause the reaction to move in the reverse direction; thus the amounts of H_2 and F_2 increase and the amount of HF decreases.
(b) The partial pressures of H_2 and F_2 will increase; the partial pressure of HF will decrease. Hence, K_p will decrease.

Solutions To Final Exercises

15.1 A reversible reaction is a reaction that can proceed in both the forward and backward directions.

Dynamic equilibrium is a state of a system in which no net change takes place because two opposing processes are occurring at the same rate.

15.3 The magnitudes of the rates of the forward and reverse reactions approach each other as the system approaches equilibrium. In other words, the rate of one slows down, while the rate of the other continues to increase (which rate does what depends on how the reaction equation is written and on how the system is approaching equilibrium), until the both rates are identical.

15.5 (a) $Q_c = \dfrac{[CH_3OH]}{[CO][H_2]^2}$ $K_c = \dfrac{[CH_3OH]_{eq}}{[CO]_{eq}[H_2]^2_{eq}}$

(b) The Q_c expression is valid for all concentrations of the reactants and products; its value is variable. The K_c expression is only valid for the equilibrium concentrations of the reactants and products; its value is fixed at a given temperature.

15.7 (a) If $Q_c > K_c$, then the reaction will move in the reverse direction.
(b) If $Q_c < K_c$, then the reaction will move in the forward direction.

15.9 (a) $Q_c = \dfrac{[SO_2]^2[H_2O]^2}{[H_2S]^2[O_2]^3}$ (b) $Q_c = \dfrac{[Ag^+][CN^-]^2}{[Ag(CN)_2^-]}$ (c) $Q_c = \dfrac{[I_3^-]}{[I_2][I^-]}$

15.11 O_2. Since the equilibrium constant is so small, the equilibrium mixture will contain mostly reactant and very little product. (A very small value of K means the reaction lies predominantly to the left or reactant side).

15.13 (a) Since only PCl_5 is present to begin with, the reaction must proceed in the forward direction.

Concentration Summary:

Equation:	$PCl_5(g) \rightleftharpoons$	$PCl_3(g)$	$+ Cl_2(g)$
Initial concentrations, mol/L	0.100	0	0
Concentration changes, mol/L	$-x$	$+x$	$+x$
Equilibrium concentrations, mol/L	$0.100 - x$	x	x

The equilibrium concentrations are

$[Cl_2] = x = 0.470 \text{ mol}/10.0 \text{ L} = 0.0470 \text{ mol/L}$

$[PCl_3] = x = 0.0470 \text{ mol/L}$

$[PCl_5] = 0.100 - x = 0.100 \text{ mol/L} - 0.0470 \text{ mol/L} = 0.053 \text{ mol/L}$

(b) The K_c expression is $K_c = [PCl_3][Cl_2]/[PCl_5]$.

Substituting the above concentrations into the K_c expression gives

$$K_c = \frac{(0.0470)(0.0470)}{(0.053)} = 0.042$$

(c) The fraction of PCl_5 dissociated is

$$\frac{0.0470}{0.100} = 0.470$$

15.15 (a) To determine whether the reaction will move in the forward or reverse direction, we have to calculate Q_c and compare it with K_c. The Q_c expression is

$Q_c = [PCl_3][Cl_2]/[PCl_5]$

Substituting the initial concentrations into the Q_c expression gives

$$Q_c = \frac{(0.100)(0.100)}{(0.100)} = 0.100$$

Since Q_c (= 0.100) > K_c (= 0.042), the reaction will move in the reverse direction.

(b) Concentration Summary:

Equation:	$PCl_5(g)$ $\rightleftharpoons$	$PCl_3(g)$ +	$Cl_2(g)$
Initial concentrations, mol/L	0.100	0.100	0.100
Concentration changes, mol/L	$+x$	$-x$	$-x$
Equilibrium concentrations, mol/L	$0.100 + x$	$0.100 - x$	$0.100 - x$

The K_c expression is $K_c = [PCl_3][Cl_2]/[PCl_5]$.

Substituting the equilibrium concentrations and $K_c = 0.042$ into the above expression gives

$$0.042 = \frac{(0.100 - x)(0.100 - x)}{(0.100 + x)}$$

$x^2 - 0.242x + 0.0058 = 0$ and $x = 0.027$ mol/L

The solution was obtained by using the negative sign before the square root. The positive sign would have given a value for x greater than 0.100. The equilibrium concentrations are

$[PCl_5] = 0.100 + x = 0.100 \text{ mol/L} + 0.027 \text{ mol/L} = 0.127 \text{ mol/L}$

$[PCl_3] = 0.100 - x = 0.100 \text{ mol/L} - 0.027 \text{ mol/L} = 0.073 \text{ mol/L}$

$[Cl_2] = 0.100 - x = 0.073 \text{ mol/L}$

15.17 (a) The reaction moves in the forward direction.

Mole Summary:

Equation:	$H_2(g)$ +	$I_2(g) \rightleftharpoons$	$2HI(g)$
Initial moles	2.00	1.00	0
Change in moles	$-x$	$-x$	$+2x$
Equilibrium moles	$2.00 - x$	$1.00 - x$	$2x$

Substituting the equilibrium moles and $K_c = 50.3$ into the K_c expression gives

$$K_c = \frac{[HI]^2}{[H_2][I_2]} = \frac{(n_{HI}/V)^2}{(n_{H_2}/V)(n_{I_2}/V)} = \frac{n_{HI}^2}{n_{H_2}n_{I_2}}$$

$$50.3 = \frac{(2x)^2}{(2.00 - x)(1.00 - x)}$$

$46.3x^2 - 150.9x + 100.6 = 0$ and $x = 0.935$ mol

The number of moles of HI in the final mixture is

$n_{HI} = 2x = 2 \times 0.935 \text{ mol} = 1.87 \text{ mol}$

(b) The volume is not needed, since both the numerator and denominator in the K_c expression have the units mol^2/L^2 and therefore the volume cancels out.

15.19 The reaction moves in the forward direction.

Concentration Summary:

Equation:	$2NOCl(g) \rightleftharpoons$	$2NO(g)$ +	$Cl_2(g)$
Initial concentrations, mol/L	1.00	0	0
Concentration changes, mol/L	$-2x$	$+2x$	$+x$
Equilibrium concentrations, mol/L	$1.00 - 2x$	$2x$	x

Substituting the equilibrium concentrations and $K_c = 4.3 \times 10^{-4}$ into the K_c expression gives

$K_c = [NO]^2[Cl_2]/[NOCl]^2$

$$4.3 \times 10^{-4} = \frac{(2x)^2(x)}{(1.00 - 2x)^2}$$

Since K_c is small we will assume that $(1.00 - 2x)$ is approximately equal to 1.00. Replacing $(1.00 - 2x)$ by 1.00 in the above expression gives

$4.3 \times 10^{-4} = 4x^3$ and $x = 0.048$ mol/L

The equilibrium concentration of NO is $[NO] = 2x = 2 \times 0.048 \text{ mol/L} = 0.096 \text{ mol/L}$.

15.21 (a) $$4.3 \times 10^{-4} = \frac{4x^3}{(1.00 - 2x)^2}$$

First Approximation: $4.3 \times 10^{-4} = \dfrac{4x^3}{(1.00)^2}$ and $x = 0.048$

Second Approximation: $4.3 \times 10^{-4} = \dfrac{4x^3}{(1.00 - 2 \times 0.048)^2}$ and $x = 0.044$

Third Approximation: $x = 0.045$; Fourth Approximation: $x = 0.045$

The equilibrium concentration of NO is $[NO] = 2x = 2 \times 0.045$ mol/L = 0.090 mol/L.

(b) difference % $= \left(\dfrac{0.096 - 0.090}{0.090}\right) \times 100\% = 6.7\%$

The 5% approximation gave a value that is 6.7% higher than the value obtained by the method of successive approximations, which is a more accurate computation.

15.23 The reaction moves in the reverse direction. The initial NH_3 concentration is

$$25.0 \text{ g } NH_3 \times \frac{1 \text{ mol } NH_3}{17.03 \text{ g } NH_3} \times \frac{1}{10.0 \text{ L}} = 0.147 \text{ mol } NH_3\text{/L}$$

Concentration Summary:

Equation:	$N_2(g)$ +	$3H_2(g)$ ⇌	$2NH_3(g)$
Initial concentrations, mol/L	0	0	0.147
Concentration changes, mol/L	$+x$	$+3x$	$-2x$
Equilibrium concentrations, mol/L	x	$3x$	$0.147 - 2x$

Substituting the equilibrium concentrations and $K_c = 1.85$ into the K_c expression gives

$K_c = [NH_3]^2/[N_2][H_2]^3$

$$1.85 = \frac{(0.147 - 2x)^2}{(x)(3x)^3}$$

The equation is simplified by taking the square root of both sides

$$1.36 = \frac{(0.147 - 2x)}{\sqrt{27}\, x^2};\ 7.07x^2 + 2x - 0.147 = 0 \text{ and } x = 0.0605 \text{ mol/L}$$

The equilibrium concentrations are:

$[N_2] = x = 0.0605$ mol/L

$[H_2] = 3x = 3 \times 0.0605$ mol/L = 0.182 mol/L

$[NH_3] = 0.147 - 2x = 0.147$ mol/L $- 2 \times 0.0605$ mol/L = 0.026 mol/L

15.25 (a) $K_c = \dfrac{[CrO_4^{2-}]^2}{[Cr_2O_7^{2-}][OH^-]^2}$ (b) $K_c = \dfrac{[HCO_3^-][OH^-]}{[CO_3^{2-}]}$ (c) $K_c = [H^+][OH^-]$

15.27 The reaction moves in the forward direction.

Concentration Summary:

Equation:	$CH_3NH_2(aq)$ +	$H_2O(l)$ ⇌	$CH_3NH_3^+(aq)$ +	$OH^-(aq)$
Initial concentrations, mol/L	0.250		0	0
Concentration changes, mol/L	$-x$		$+x$	$+x$
Equilibrium concentrations, mol/L	$0.250 - x$		x	x

Substituting the equilibrium concentrations and $K_c = 3.70 \times 10^{-4}$ into the K_c expression gives

$K_c = [CH_3NH_3^+][OH^-]/[CH_3NH_2]$

$$3.70 \times 10^{-4} = \frac{(x)(x)}{(0.250 - x)}$$

Since K_c is small we will assume that $0.250 - x$ is approximately equal to 0.250. Replacing $0.250 - x$ by 0.250 in the above equation gives

$$3.70 \times 10^{-4} = \frac{x^2}{0.250} \quad \text{and } x = 9.62 \times 10^{-3} \text{ mol/L}$$

The equilibrium concentration of $CH_3NH_3^+$ is $[CH_3NH_3^+] = x = 9.62 \times 10^{-3}$ mol/L.

15.29 K_p is used for reactions involving gases (gas state reactions). The partial pressures of the gases involved are expressed in atmospheres (atm).

15.31 PV = nRT

$$P = \frac{n}{V}RT = MRT$$

15.33 (a) $Q_p = P^3_{O_2}/P^2_{O_3}$ (b) $Q_p = P_{C_2H_4}/P_{C_2H_2}P_{H_2}$

15.35

$$M = \frac{P}{RT} = \frac{(770 \text{ torr} \times 1 \text{ atm/760 torr})}{0.0821 \text{ L•atm/mol•K} \times 298 \text{ K}} = 4.14 \times 10^{-2} \text{ mol/L}$$

15.37 Increases. A larger K_p means that the reaction goes further to the right as the temperature increases and, therefore, more Cl_2 molecules are dissociated at the higher temperature.

15.39 (a) $P_{PCl_5} = [PCl_5]RT = \frac{0.053 \text{ mol}}{1 \text{ L}} \times \frac{0.0821 \text{ L•atm}}{\text{mol•K}} \times 523 \text{ K} = 2.3 \text{ atm}$

$$P_{PCl_3} = [PCl_3]RT = \frac{0.0470 \text{ mol}}{1 \text{ L}} \times \frac{0.0821 \text{ L•atm}}{\text{mol•K}} \times 523 \text{ K} = 2.02 \text{ atm}$$

$$P_{Cl_2} = P_{PCl_3} = 2.02 \text{ atm}$$

(b) Substituting the equilibrium pressures into the K_p expression gives

$$K_p = \frac{P_{PCl_3} P_{Cl_2}}{P_{PCl_5}} = \frac{(2.02)(2.02)}{(2.3)} = 1.8$$

15.41 The reaction moves in the forward direction.

Pressure Summary:

Equation:	$N_2O_4(g)$ ⇌	$2NO_2(g)$
Initial pressures, atm	0.500	0
Pressure changes, atm	$-x$	$+2x$
Equilibrium pressures, atm	$0.500 - x$	$2x$

Substituting $K_p = 0.113$ and the equilibrium pressures into the K_p expression gives

$$K_p = P^2_{NO_2}/P_{N_2O_4}$$

$$0.113 = \frac{(2x)^2}{0.500 - x}$$

$4x^2 + 0.113x - 0.0565 = 0$ and $x = 0.106$ atm

The partial pressures at equilibrium are

$P_{NO_2} = 2x = 2 \times 0.106 \text{ atm} = 0.212 \text{ atm}$

$P_{N_2O_4} = 0.500 - x = 0.500 \text{ atm} - 0.106 \text{ atm} = 0.394 \text{ atm}$

15.43 (a) The reaction moves in the forward direction.

Pressure Summary:

Equation:	$Br_2(g)$ $\rightleftharpoons$	$2Br(g)$
Initial pressures, atm	2.00	0
Pressure changes, atm	$-x$	$+2x$
Equilibrium pressures, atm	$2.00 - x$	$2x$

Substituting $K_p = 0.255$ and the equilibrium pressures into the K_p expression gives

$K_p = P_{Br}^2 / P_{Br_2}$

$$0.255 = \frac{(2x)^2}{2.00 - x}$$

$4x^2 + 0.255x - 0.510 = 0$ and $x = 0.327$ atm

The partial pressures at equilibrium are

$P_{Br_2} = 2.00 - x = 2.00 \text{ atm} - 0.327 \text{ atm} = 1.67 \text{ atm}$

$P_{Br} = 2x = 2 \times 0.327 \text{ atm} = 0.654 \text{ atm}$

(b) $\frac{0.327}{2.00} = 0.164$

15.45 $$1.79 \times 10^{-5} = \frac{(2x)^2(x)}{(1.00 - 2x)^2}$$

First Approximation: $1.79 \times 10^{-5} = \frac{4x^3}{(1.00)^2}$ and $x = 0.0165$

Second Approximation: $1.79 \times 10^{-5} = \frac{4x^3}{(1.00 - 2 \times 0.0165)^2}$ and $x = 0.0161$

Third Approximation: $x = 0.0161$

The equilibrium pressures are:

$P_{SO_3} = 1.00 - 2x = 1.00 \text{ atm} - 0.0322 \text{ atm} = 0.97 \text{ atm}$

$P_{SO_2} = 2x = 0.0322 \text{ atm}$

$P_{O_2} = x = 0.0161 \text{ atm}$

The total pressure is the sum of the partial pressures

$P_{total} = 0.97 \text{ atm} + 0.0322 \text{ atm} + 0.0161 \text{ atm} = 1.02 \text{ atm}$

15.47 (a) $\Delta n = 3 - 2 = 1$

$K_c = K_p(RT)^{-\Delta n} = K_p/RT$

(b) $\Delta n = 1 - 2 = -1$

$K_c = K_p(RT)^{-\Delta n} = K_pRT$

15.49 $\Delta n = 1 - 3/2 = -1/2$

$K_p = K_c(RT)^{\Delta n} = K_c/(RT)^{1/2}$

$$K_p = \frac{2.5 \times 10^{15}}{(0.0821 \times 773)^{1/2}} = 3.1 \times 10^{14}$$

15.51 (a) $\Delta n = 3 - 2 = 1$; $K_p = K_c(RT)^{\Delta n} = K_cRT$

$K_p = 4.3 \times 10^{-4} \times 0.0821 \times 773 = 2.7 \times 10^{-2}$

(b) $\Delta n = 1 - 3 = -2$; $K_p = K_c(RT)^{\Delta n} = K_c/(RT)^2$

$$K_p = \frac{6.6}{(0.0821 \times 523)^2} = 3.6 \times 10^{-3}$$

15.53 (a) The equation in (a) is obtained by reversing the original equation, so its K_p is the reciprocal of the original equilibrium constant:

$$K_p = \frac{1}{5.2 \times 10^{-7}} = 1.9 \times 10^6$$

(b) The equation in (b) is obtained by reversing the original equation and multiplying it by 1/2, so its K_p is the square root of the reciprocal of the original equilibrium constant:

$$K_p = \left[\frac{1}{5.2 \times 10^{-7}}\right]^{1/2} = 1.4 \times 10^3$$

15.55 The desired equation is obtained by reversing the second equation and adding the result to the first equation. Therefore, its K_c is equal to the product of the first equilibrium constant and the reciprocal of the second equilibrium constant:

$$K_c = \frac{1.0 \times 10^{-14}}{3.1 \times 10^{-7}} = 3.2 \times 10^{-8}$$

15.57 In homogeneous equilibria the reactants and products are in the same state (the same solution), while in heterogeneous equilibria the reactants and products are in more than one state. In heterogeneous equilibria the opposing reactions occur at the surface of a solid or liquid.

Refer to examples in Section 15.7 of the textbook.

15.59 (a) $K_p = P_{H_2}P_{CO}/P_{H_2O}$ (b) $K_p = P^4_{H_2O}/P^4_{H_2}$ (c) $K_p = 1/P_{Cl_2}$

15.61 (a) $K_c = [H_2][CO]/[H_2O]$ (b) $K_c = [H_2O]^4/[H_2]^4$ (c) $K_c = 1/[Cl_2]$

15.63 (a) $K_c = [Bi^{3+}]^2[S^{2-}]^3$ (b) $K_c = [H^+][Br^-][HOBr]$

15.65 The reaction moves in the forward direction.

Pressure Summary:

	$NH_2COONH_4(s) \rightleftharpoons$	$2NH_3(g)$	+ $CO_2(g)$
Equation:			
Initial pressures, atm		0	0
Pressure changes, atm		$+2x$	$+x$
Equilibrium pressures, atm		$2x$	x

Substituting $K_p = 2.87 \times 10^5$ and the equilibrium pressures into the K_p expression gives

$K_p = P^2_{NH_3} P_{CO_2}$

$2.87 \times 10^5 = (2x)^2 x$ and $x = 41.6$ atm

The equilibrium pressures are

$P_{CO_2} = x = 41.6$ atm

$P_{NH_3} = 2x = 2 \times 41.6$ atm $= 83.2$ atm

The total pressure is $P_{total} = P_{NH_3} + P_{CO_2} = 41.6$ atm $+ 83.2$ atm $= 124.8$ atm.

15.67 The total pressure coming from the reaction is: $P_{total} = P_{NH_3} + P_{H_2S}$

0.66 atm $= x + x = 2x$ and $x = 0.33$ atm

Substituting the equilibrium partial pressures into the K_p expression gives

$K_p = P_{NH_3} P_{H_2S} = (0.33)(0.33) = 0.11$

15.69 The final equation is the sum of two times the reverse of the first equation and two times the second equation. Hence,

$$K_p = \frac{1}{K^2_{p_1}} \times K^2_{p_2} = \frac{(1.25 \times 10^{69})^2}{(1.13 \times 10^{24})^2} = 1.22 \times 10^{90}$$

15.71 (a) An equilibrium mixture disturbed by a concentration change will move towards a new equilibrium state which diminishes the concentration change to some extent.

(b) An equilibrium mixture disturbed by a volume change will move towards a new equilibrium state that diminishes the change in the system's pressure. So, if the volume decreases, the system's pressure increases, and the reaction mixture moves in the direction of fewer moles of gas, and vice versa.

(c) An equilibrium mixture disturbed by a temperature change will move towards a new equilibrium state such that the change in the system's temperature is diminished. So, if the temperature increases, the reaction mixture moves in the direction that absorbs heat (endothermic direction); and if the temperature decreases, the reaction mixture moves in the direction that releases heat (exothermic direction).

15.73 (a) Adding NaOH drives the reaction to the right and, therefore, should make the color less orange and more yellow.

(b) Adding HCl drives the reaction to the left by reducing the OH^- concentration and, therefore, should make the color more orange and less yellow.

15.75 (a) The color will increase; more intense. From the K_c values, we see that the reaction is driven to the right by increasing temperature. Therefore, there will be a greater proportion of NO_2 in the equilibrium mixture at higher temperatures.

(b) The color will increase; more intense. Higher N_2O_4 concentrations move the reaction to the right.
(c) The color will increase; more intense. Higher NO_2 partial pressures move the reaction to the left. However, the partial pressure of NO_2 after the increase will always be greater than what it was initially, since only some of the additional NO_2 introduced will be consumed in the reverse reaction.
(d) The color will remain the same; no change. Helium is a nonparticipating substance. Helium does not appear in the reaction quotient and, therefore, does not affect the equilibrium.

15.77 Endothermic. When heated, the reaction will move in the direction that absorbs heat. Since the equilibrium constant increases with increasing temperature, the reaction moves in the forward direction when heated and, therefore, the forward reaction must be endothermic.

15.79 (a) Increase. Compression moves the reaction in the forward direction, since there are 3 mol of gas on the left side of the equation and only 2 mol of gas on the right side of the equation. Therefore, compression should increase the equilibrium yield of SO_3.
(b) Decrease. Increasing the temperature moves the reaction in the reverse direction, since the forward reaction is exothermic. Therefore, increasing temperature should decrease the equilibrium yield of SO_3.

15.81 (a) Decrease. Since the reaction is exothermic, increasing the temperature will drive the reaction in the reverse direction.
(b) Increase. Higher Cl_2 partial pressures move the reaction to the right.
(c) No change. There are 2 mol of gas on each side of the equation.
(d) No change. Catalysts do not change the position of an equilibrium, only the rate of attainment of equilibrium.

15.83 Substituting $K_1 = 0.90$, $T_1 = 393$ K, $K_2 = 3.2$, $T_2 = 423$ K, and $R = 8.314 \times 10^{-3}$ kJ/mol•K into Equation 15.4 gives

$$\ln \frac{3.2}{0.90} = \frac{-\Delta H°}{8.314 \times 10^{-3}\ \text{kJ/mol•K}}\left[\frac{1}{423\ \text{K}} - \frac{1}{393\ \text{K}}\right]$$

Solving gives $\Delta H° = +58$ kJ/mol.

15.85 Substituting $K_1 = 6.9 \times 10^{-5}$, $T_1 = 298$ K, $T_2 = 373$ K, $\Delta H° = -92.2$ kJ/mol, and $R = 8.314 \times 10^{-3}$ kJ/mol•K into Equation 15.4 gives

$$\ln \frac{K_2}{6.9 \times 10^{-5}} = \frac{+92.2\ \text{kJ/mol}}{8.314 \times 10^{-3}\ \text{kJ/mol•K}}\left[\frac{1}{373\ \text{K}} - \frac{1}{298\ \text{K}}\right]$$

Solving gives $K_2 = 3.9 \times 10^{-8}$.

15.87 (a) $Ag(S_2O_3)_2^{3-}$. The $Ag(S_2O_3)_2^{3-}$ equilibrium constant is much smaller than the $Ag(NH_3)_2^+$ equilibrium constant. Therefore, much less $Ag(S_2O_3)_2^{3-}$ dissociates.
(b) A 1.0 M solution of $Ag(NH_3)_2^+$ will have a greater concentration of free Ag^+ ion, since it has a greater K_c. Note that the reaction equations have similar forms.

15.89 (a) The Q_c expression is $Q_c = [H^+][CH_3COO^-]/[CH_3COOH]$.

Substituting the initial concentrations into the Q_c expression gives

$$Q_c = \frac{(1.0 \times 10^{-7})(0.50)}{(0.50)} = 1.0 \times 10^{-7}$$

Since $Q_c < K_c$, the reaction will move in the forward direction.

(b) Concentration Summary:

	$CH_3COOH(aq) \rightleftharpoons$	$H^+(aq)$ +	$CH_3COO^-(aq)$
Equation:			
Initial concentrations, mol/L	0.50	1.0×10^{-7}	0.50
Concentration changes, mol/L	$-x$	$+x$	$+x$
Equilibrium concentrations, mol/L	$0.50 - x$	$1.0 \times 10^{-7} + x$	$0.50 + x$

Substituting the equilibrium concentrations and $K_c = 1.76 \times 10^{-5}$ into the K_c expression gives

$K_c = [H^+][CH_3COO^-]/[CH_3COOH]$

$$1.76 \times 10^{-5} = \frac{(1.0 \times 10^{-7} + x)(0.50 + x)}{(0.50 - x)}$$

Since K_c is small we will assume that both $0.50 - x$ and $0.50 + x$ are approximately equal to 0.50. Making these replacements in the above equation gives

$$1.76 \times 10^{-5} = \frac{(1.0 \times 10^{-7} + x)(0.50)}{(0.50)} = 1.0 \times 10^{-7} + x$$

and $x = 1.75 \times 10^{-5}$ mol/L

The equilibrium concentrations are

$[CH_3COOH] = 0.50 - x = 0.50 \text{ mol/L} - 1.75 \times 10^{-5} \text{ mol/L} = 0.50 \text{ mol/L}$

$[H^+] = 1.0 \times 10^{-7} + x = 1.0 \times 10^{-7} \text{ mol/L} + 1.75 \times 10^{-5} \text{ mol/L} = 1.76 \times 10^{-5} \text{ mol/L}$

$[CH_3COO^-] = 0.50 + x = 0.50 \text{ mol/L} + 1.75 \times 10^{-5} \text{ mol/L} = 0.50 \text{ mol/L}$

15.91 The reaction moves in the forward direction.

Pressure Summary:

	$CO(g)$ +	$H_2O(g) \rightleftharpoons$	$CO_2(g)$ +	$H_2(g)$
Equation:				
Initial pressures, atm	12.5	12.5	0	0
Pressure changes, atm	$-x$	$-x$	$+x$	$+x$
Equilibrium pressures, atm	$12.5 - x$	$12.5 - x$	x	x

Substituting $K_p = 0.227$ and the equilibrium pressures into the K_p expression gives

$K_p = P_{CO_2} P_{H_2}/P_{CO} P_{H_2O}$

$$0.227 = \frac{(x)(x)}{(12.5 - x)(12.5 - x)}$$

The equation is simplified by taking the square root of both sides

$$0.476 = \frac{x}{12.5 - x} \quad \text{and } x = P_{H_2} = 4.03 \text{ atm}$$

Since there is an equal number of moles of gas on both sides of the equation, the total pressure doesn't change during the reaction; $P_{total} = 25.0$ atm. By Dalton's law, the mole fraction of a gas in a mixture is equal to its pressure fraction. Therefore, the mol % hydrogen in the equilibrium mixture is

$$\text{mol\% } H_2 = X_{H_2} \times 100\% = \frac{P_{H_2}}{P_{total}} \times 100\% = \frac{4.30 \text{ atm}}{25.0 \text{ atm}} \times 100\% = 16.1\%$$

15.93 $Q_c = \frac{[I_2][Cl_2]}{[ICl]^2} = \frac{(0.250)(0.250)}{(0.250)^2} = 1.00$

Since $Q_c > K_c$, the reaction moves in the reverse direction.

Concentration Summary:

Equation:	$2ICl(g)$	$\rightleftharpoons$ $I_2(g)$	+ $Cl_2(g)$
Initial concentrations, mol/L	0.250	0.250	0.250
Concentration changes, mol/L	$+2x$	$-x$	$-x$
Equilibrium concentrations, mol/L	$0.250 + 2x$	$0.250 - x$	$0.250 - x$

Substituting the equilibrium concentrations and $K_c = 0.110$ into the K_c expression gives

$K_c = [I_2][Cl_2]/[ICl]^2$

$$0.110 = \frac{(0.250 - x)(0.250 - x)}{(0.250 + 2x)^2}$$

The equation is simplified by taking the square root of both sides

$$0.332 = \frac{(0.250 - x)}{(0.250 + 2x)}; \; 1.664x = 0.167 \text{ and } x = 0.100 \text{ mol/L}$$

The equilibrium concentration of ICl is $[ICl] = 0.250 + 2x = 0.250$ mol/L + 2 × 0.100 mol/L = 0.450 mol/L.

15.95 First we have to convert the pressure to atm:

$$17.5 \text{ torr} \times \frac{1 \text{ atm}}{760 \text{ torr}} = 0.0230 \text{ atm}$$

Substituting the vapor pressure into the K_p expression gives

$K_p = P_{H_2O} = 0.0230$

15.97 (a) Increase. Increasing the temperature moves the reaction in the forward direction, since the forward reaction is endothermic. Therefore, increasing temperature should lead to an increase in the fraction of $COCl_2$ dissociated.

(b) Decrease. Compression moves the reaction in the reverse direction, since there are 2 mol of gas on the right side of the equation and only 1 mol of gas on the left side of the equation. Therefore, compression should lead to a decrease in the fraction of $COCl_2$ dissociated.

(c) Decrease. Higher Cl_2 partial pressures move the reaction to the left. Therefore, increasing partial pressure of Cl_2 should lead to a decrease in the fraction of $COCl_2$ dissociated.

15.99 The reaction moves in the reverse direction.

Concentration Summary:

Equation:	$2ICl(g)$	$\rightleftharpoons$ $I_2(g)$	+ $Cl_2(g)$
Initial concentrations, mol/L	0	0.500	0.250
Concentration changes, mol/L	$+2x$	$-x$	$-x$
Equilibrium concentrations, mol/L	$2x$	$0.500 - x$	$0.250 - x$

$K_c = [I_2][Cl_2]/[ICl]^2$

$$0.110 = \frac{(0.500 - x)(0.250 - x)}{(2x)^2}; \; 0.560x^2 - 0.750x + 0.125 = 0 \text{ and } x = 0.195 \text{ mol/L}$$

The equilibrium concentration of ICl is $[ICl] = 2x = 2 \times 0.195$ mol/L = 0.390 mol/L.

CHAPTER 16

ACIDS AND BASES

Solutions To Practice Exercises

PE 16.1 (a) $CH_3COOH(aq) + H_2O(l) \rightleftharpoons H_3O^+(aq) + CH_3COO^-(aq)$
(b) $H_3O^+(aq) + OH^-(aq) \rightleftharpoons 2H_2O(l)$

PE 16.2

$$\underset{\text{base}}{CH_3-\overset{H}{\overset{|}{\ddot{N}}}-H(l)} \quad + \quad \underset{\text{acid}}{:\overset{H}{\overset{|}{N}}-CH_3(l)} \rightleftharpoons CH_3NH_3^+ + CH_3NH^-$$

or $2CH_3NH_2(l) \rightleftharpoons CH_3NH_3^+ + CH_3NH^-$

PE 16.3 (a) $H_2PO_4^-$ (b) O^{2-}

PE 16.4 $HC_2O_4^-(aq) + HS^-(aq) \rightarrow C_2O_4^{2-}(aq) + H_2S(g)$

$HC_2O_4^-$ gives up a proton and is an acid; its conjugate base is $C_2O_4^{2-}$. HS^- accepts a proton and is a base; its conjugate acid is H_2S.

PE 16.5 (a) $HClO_4(l) + H_2O(l) \rightarrow H_3O^+(aq) + ClO_4^-(aq)$
(b) $NH_2^- + H_2O(l) \rightarrow NH_3(g) + OH^-(aq)$

PE 16.6 (a) $2H_2S(l) \rightleftharpoons H_3S^+ + HS^-$

The strongest base that can exist in liquid hydrogen sulfide is HS^-.

(b) $2CH_3NH_2(l) \rightleftharpoons CH_3NH_3^+ + CH_3NH^-$

The strongest acid that can exist in liquid methylamine is $CH_3NH_3^+$.

PE 16.7 (a) $HF(aq) + OH^-(aq) \rightarrow F^-(aq) + H_2O(l)$
(b) $CN^-(aq) + H_3O^+(aq) \rightarrow HCN(aq) + H_2O(l)$

PE 16.8 (a) $HF(aq) + CN^-(aq) \rightleftharpoons F^-(aq) + HCN(aq)$
(b) $NH_3(aq) + CH_3COOH(aq) \rightleftharpoons NH_4^+(aq) + CH_3COO^-(aq)$

PE 16.9 $Cu(H_2O)_4^{2+}(aq) + H_2O(l) \rightleftharpoons Cu(H_2O)_3(OH)^+(aq) + H_3O^+(aq)$

PE 16.10 $H_3PO_4(aq) + H_2O(l) \rightleftharpoons H_3O^+(aq) + H_2PO_4^-(aq)$
$H_2PO_4^-(aq) + H_2O(l) \rightleftharpoons H_3O^+(aq) + HPO_4^{2-}(aq)$
$HPO_4^{2-}(aq) + H_2O(l) \rightleftharpoons H_3O^+(aq) + PO_4^{3-}(aq)$

PE 16.11 A strong base can remove up to two protons:

$Zn(H_2O)_4^{2+}(aq) + 2OH^-(aq) \rightarrow Zn(H_2O)_2(OH)_2(s) + 2H_2O(l)$

or $Zn^{2+}(aq) + 2OH^-(aq) \rightarrow Zn(OH)_2(s)$

PE 16.12 $Ca(OH)_2$ is an ionic compound; 1.0×10^{-3} mol of $Ca(OH)_2$ furnishes $2 \times 1.0 \times 10^{-3}$ mol $= 2.0 \times 10^{-3}$ mol of $OH^-(aq)$. The hydroxide ion furnished by the water is negligible compared to that furnished by the $Ca(OH)_2$, so $[OH^-] = 2.0 \times 10^{-3}$ M. The $H^+(aq)$ ions come only from the water. Their concentration is

$$[H^+] = \frac{K_W}{[OH^-]} = \frac{1.00 \times 10^{-14}}{2.0 \times 10^{-3}} = 5.0 \times 10^{-12} \text{ M}$$

PE 16.13 $pH = -\log [H^+] = -\log (1.5 \times 10^{-2}) = -(-1.82) = 1.82$

PE 16.14 Seawater is basic, since its pH is greater than 7.0.

PE 16.15 The H^+ concentration in Example 16.3(b) is 2.0×10^{-14} M; the OH^- concentration is 0.50 M. Hence,

$pH = -\log [H^+] = -\log (2.0 \times 10^{-14}) = 13.70$

$pOH = -\log [OH^-] = -\log 0.50 = 0.30$

PE 16.16 In a neutral solution $[OH^-] = 1.0 \times 10^{-7}$ M. Hence,

$pOH = -\log [OH^-] = -\log (1.0 \times 10^{-7}) = -(-7.00) = 7.00$

PE 16.17 $pH = 4.3 = -\log [H^+]$

$\log [H^+] = -4.3$

and $[H^+] = \text{antilog} (-4.3) = 10^{-4.3} = 5 \times 10^{-5}$ M

The answer has one significant figure because its logarithm, –4.3, has one digit after the decimal place.

PE 16.18 For the pH 8.0 solution we have

$pOH = 14.00 - pH = 14.00 - 8.0 = 6.0$

$\log [OH^-] = -pOH = -6.0$

and $[OH^-] = \text{antilog} (-6.0) = 10^{-6.0} = 1 \times 10^{-6}$ M

The pH 10.0 solution has an $[OH^-]$ of 1×10^{-4} M (see Example 16.8). Therefore, the pH 8.0 solution has an $OH^-(aq)$ concentration 100 times smaller than that of the pH 10.0 solution.

PE 16.19 (a) H_2S. Acid strength increases with increasing positive charge; hence, $H_2S > HS^-$.

(b) O^{2-}. Base strength increases with increasing negative charge; hence, $O^{2-} > OH^-$.

PE 16.20 (a) $CHCl_2COOH$. $CHCl_2COOH$ has two electronegative chlorine atoms drawing electron density away from the COOH group, while $CH_2ClCOOH$ only has one.

(b) OI^-. Since chlorine is more electronegative than iodine, HOCl is a stronger acid than HOI and, consequently, OCl^- is a weaker base than OI^-.

PE 16.21 (a) HBr. The strength of binary acids increases from top to bottom within a periodic group; hence, HBr > HCl.

(b) OH^-. Since H_2S is a stronger acid than H_2O, OH^- is a stronger base than HS^-. Also, since S (third period) is larger than O (second period), HS^- is larger than OH^- and its charge is more delocalized, which results in HS^- being a weaker base than OH^-.

PE 16.22 $Cr(H_2O)_6^{3+}$. The chromium ion is a trivalent cation (3+), while the nickel ion is only divalent (2+).

PE 16.23

$$AlCl_3 + :\ddot{\underset{..}{Cl}}:^- \rightarrow [AlCl_4]^-$$

(Lewis structures: Cl–Al(–Cl)–Cl + :Cl:⁻ → [Cl–Al(–Cl)(–Cl)–Cl]⁻)

From the reaction written in terms of Lewis structures we see that $AlCl_3$ is the Lewis acid and Cl^- is the Lewis base.

Solutions To Final Exercises

16.1 An Arrhenius acid is any substance that produces $H^+(aq)$ ions in aqueous solution.

An Arrhenius base is any substance that produces $OH^-(aq)$ ions in aqueous solution.

Neutralization is the reaction of $H^+(aq)$ ions with $OH^-(aq)$ ions to form water. $H^+(aq) + OH^-(aq) \rightarrow H_2O(l)$.

16.3 (a) All Brønsted acids can donate a proton (H^+) to a base; i.e., they all have an acidic hydrogen.

(b) Ammonia (NH_3), methylamine (CH_3NH_2), and water (H_2O) are three examples. They can donate protons to stronger bases, but cannot produce $H^+(aq)$ ions in aqueous solution.

16.5 (a) Using $H_3O^+(aq)$ instead of $H^+(aq)$ emphasizes the fact that the ionization of acids in water is actually a Brønsted-Lowry proton transfer reaction with water acting as the Brønsted base.

(b) 1. The hydronium ion (H_3O^+) in solution is further hydrated and, therefore, $H_3O^+(aq)$ doesn't represent the actual situation.

2. Since the $[H_2O]$ is constant, the H_2O concentration is omitted from equilibrium constant expressions and, therefore, the use of H_2O is not really necessary when writing the acid ionization reaction equations. In other words, $H^+(aq)$ has the advantage of simplicity over $H_3O^+(aq)$ in situations where the role of H_2O doesn't have to be emphasized or can be ignored.

16.7 An amphiprotic substance is a substance that can act either as an acid or a base; i.e., it can donate or accept a proton in different reactions.

Amphiprotic substances contain both hydrogen atoms and lone electron pairs.

16.9 Both OH^- and NH_3 are generally thought of as bases because their tendency or ability to accept a proton is considerably greater than their tendency to part with or donate a proton. From Table 16.2 we see that both OH^- and NH_3 are classified as very weak acids, while OH^- is classified as a strong base and NH_3 a weak base. In other words, since they most often act as bases, they are generally thought of as bases. Also, both OH^- and NH_3 are *basic toward water,* and since life exists in an aqueous medium we naturally see them as bases.

16.11 (a) $HCO_3^-(aq) + HCl(aq) \rightarrow H_2O(l) + CO_2(g) + Cl^-(aq)$
$HCO_3^-(aq) + OH^-(aq) \rightarrow CO_3^{2-}(aq) + H_2O(l)$
(b) $H_2PO_4^-(aq) + HCl(aq) \rightarrow H_3PO_4(aq) + Cl^-(aq)$
$H_2PO_4^-(aq) + OH^-(aq) \rightarrow HPO_4^{2-}(aq) + H_2O(l)$

16.13 (a) $2CH_3OH(l) \rightleftharpoons CH_3OH_2^+ + CH_3O^-$
(b) $2HCl(l) \rightleftharpoons H_2Cl^+ + Cl^-$

16.15 (a) In the Arrhenius theory a strong acid is an acid that ionizes completely, is 100% ionized, in aqueous solution. Strong bases are soluble ionic hydroxides.
(b) In the Brønsted-Lowry theory a strong acid is any acid stronger than the conjugate acid of the solvent. For aqueous solutions, the acid must be stronger than the hydronium ion, $H_3O^+(aq)$, in order to be categorized as strong. Strong acids react completely with the solvent to form the conjugate acid of the solvent; hence, the conjugate acid of the solvent is the strongest acid that can exist in its solutions. The acids above H_3O^+ in Table 16.2 are labeled strong acids.

In the Brønsted-Lowry theory a strong base is any base stronger than the conjugate base of the solvent. For aqueous solutions, the base must be stronger than the hydroxyl ion, $OH^-(aq)$, in order to be categorized as strong. Like the strong acids, strong bases react completely with the solvent to form the conjugate base of the solvent; hence, the conjugate base of the solvent is the strongest base that can exist in its solutions. The bases below OH^- in Table 16.2 are labeled strong bases.

Comment: Soluble ionic hydroxides are also classified as strong bases in the Brønsted-Lowry theory, since they ionize completely in aqueous solution, directly supplying hydroxyl ions.

16.17 Conjugate acid-base pairs are species whose formulas differ by a single proton. One species can be formed from the other by the gain or loss of a proton.

Examples are: H_3O^+ and H_2O, H_2O and OH^-, NH_4^+ and NH_3, HBr and Br^-, O^{2-} and OH^-, and $HClO_4$ and ClO_4^-.

16.19 The stronger the base, the weaker its conjugate acid. $NH_4^+ > C_2H_5NH_3^+$.

16.21 The leveling effect of water refers to the fact that all strong acids appear to be equally strong in water. The leveling effect of water has to do with the fact that the strongest acid that can exist in aqueous solution is the hydronium ion. All acids stronger than the hydronium ion appear to be equally strong in water, since they are all 100% ionized. Therefore, the differences in the strengths of these acids have been rendered inconsequential or leveled out by the action of the water. In other words, water demonstrates or has a leveling effect on the strengths of strong acids, since it makes them all appear to be equally strong in water.

Strongest acid: $H_3O^+(aq)$; strongest base: $OH^-(aq)$

A stronger acid will be completely deprotonated with the concurrent formation of an equal amount of the conjugate acid of water, i.e., $H_3O^+(aq)$. A stronger base will be completely protonated with the concurrent formation of an equal amount of the conjugate base of water, i.e., $OH^-(aq)$.

16.23 A differentiating solvent is a solvent in which differences in acid and base strengths can be distinguished.

Water is a differentiating solvent for weak acids and weak bases. In particular, water is a differentiating solvent for acids weaker than H_3O^+ and bases weaker than OH^-. Therefore, the acids below H_3O^+ in Table 16.2 and the bases above OH^- in Table 16.2 can be differentiated in water.

Water can differentiate between these weak acids and bases, because these acids and bases only partially deprotonate and protonate, respectively, in water.

16.25 Water is a differentiating solvent for weak acids and weak bases. Therefore, a determination of the $H^+(aq)$ concentrations in equimolar aqueous solutions of HCN and HF should show which of these acids is stronger.

16.27

1. In a reaction between HBr and HI, the weaker acid will act as a base and accept protons from the stronger acid. Therefore, an examination of the products of this reaction will determine which is the stronger acid.
2. Strong acids can also be differentiated by the use of a solvent that is not easily protonated. For example, determining how extensively HBr and HI are ionized in liquid acetic acid should show which one is the stronger acid.

16.29 (a) HSO_4^- (b) SO_4^{2-} (c) PH_3 (d) H_2O (e) OH^- (f) $Cr(H_2O)_5(OH)^{2+}$ (g) $C_2H_5O^-$ (h) HS^-

16.31
(a) NH_3 and NH_4^+; HCO_3^- and CO_3^{2-}
(b) CH_3NH_2 and $CH_3NH_3^+$; HCl and Cl^-
(c) H_3O^+ and H_2O; OH^- and H_2O

16.33 OH^- and H_2O; H_3O^+ and H_2O

16.35 (a) $HI(aq) + H_2O(l) \rightarrow H_3O^+(aq) + I^-(aq)$

$$\underset{\text{base}}{H-\ddot{O}(H):(l)} + \underset{\text{acid}}{(H)-\ddot{\underset{..}{I}}:(aq)} \rightarrow H_3O^+(aq) + I^-(aq)$$

(b) $NaNH_2(s) + H_2O(l) \rightarrow NaOH(aq) + NH_3(g)$

$$\underset{\text{base}}{[H-\ddot{N}(H):]^-} + \underset{\text{acid}}{(H)-\ddot{O}(H):(l)} \rightarrow OH^-(aq) + NH_3(g)$$

16.37
(a) $HNO_2(aq) + H_2O(l) \rightleftharpoons H_3O^+(aq) + NO_2^-(aq)$
(b) $HCN(aq) + H_2O(l) \rightleftharpoons H_3O^+(aq) + CN^-(aq)$
(c) $HF(aq) + H_2O(l) \rightleftharpoons H_3O^+(aq) + F^-(aq)$
(d) $CH_3COOH(aq) + H_2O(l) \rightleftharpoons H_3O^+(aq) + CH_3COO^-(aq)$

16.39
(a) $H_2SO_4(aq) + H_2O(l) \rightarrow H_3O^+(aq) + HSO_4^-(aq)$
$HSO_4^-(aq) + H_2O(l) \rightleftharpoons H_3O^+(aq) + SO_4^{2-}(aq)$
(b) $H_2SeO_3(aq) + H_2O(l) \rightleftharpoons H_3O^+(aq) + HSeO_3^-(aq)$
$HSeO_3^-(aq) + H_2O(l) \rightleftharpoons H_3O^+(aq) + SeO_3^{2-}(aq)$

16.41 (a) $HCOOH(aq) + H_2O(l) \rightleftharpoons H_3O^+(aq) + HCOO^-(aq)$
(b) $H_2C_2O_4(aq) + H_2O(l) \rightleftharpoons H_3O^+(aq) + HC_2O_4^-(aq)$
$HC_2O_4^-(aq) + H_2O(l) \rightleftharpoons H_3O^+(aq) + C_2O_4^{2-}(aq)$

16.43 (a) $NH_4^+(aq) + H_2O(l) \rightleftharpoons H_3O^+(aq) + NH_3(aq)$
(b) $O^{2-} + H_2O(l) \rightarrow 2OH^-(aq)$
(c) $Cd(H_2O)_4^{2+}(aq) + H_2O(l) \rightleftharpoons H_3O^+(aq) + Cd(H_2O)_3(OH)^+(aq)$
(d) $H_2PO_4^-(aq) + H_2O(l) \rightleftharpoons H_3O^+(aq) + HPO_4^{2-}(aq)$
and $H_2PO_4^-(aq) + H_2O(l) \rightleftharpoons H_3PO_4(aq) + OH^-(aq)$

16.45 (a) $Zn(s) + 2HCl(aq) \rightarrow H_2(g) + Zn^{2+}(aq) + 2Cl^-(aq)$
(b) $Zn(H_2O)_2(OH)_2(s) + 2HNO_3(aq) \rightarrow Zn(H_2O)_4^{2+}(aq) + 2NO_3^-(aq)$
or $Zn(OH)_2(s) + 2HNO_3(aq) \rightarrow Zn(NO_3)_2(aq) + 2H_2O(l)$
(c) $2HClO_4(aq) + Ba(OH)_2(aq) \rightarrow Ba(ClO_4)_2(aq) + 2H_2O(l)$
(d) $NaOH(aq) + H_2C_2O_4(aq) \rightarrow NaHC_2O_4(aq) + H_2O(l)$

16.47 (a) $C_6H_5COOH(aq) + OH^-(aq) \rightarrow C_6H_5COO^-(aq) + H_2O(l)$
(b) $NH_3(aq) + H_3O^+(aq) \rightarrow NH_4^+(aq) + H_2O(l)$
(c) $HCOOH(aq) + F^-(aq) \rightleftharpoons HCOO^-(aq) + HF(aq)$

16.49 The $\Delta H°$ for the reaction

$HNO_2(aq) + H_2O(l) \rightarrow H_3O^+(aq) + NO_2^-(aq) \quad \Delta H° = ?$

is the sum of the enthalpy changes for the following steps:
$HNO_2(aq) + OH^-(aq) \rightarrow NO_2^-(aq) + H_2O(l) \quad \Delta H° = -41.2$ kJ
$2H_2O(l) \rightarrow H_3O^+(aq) + OH^-(aq) \quad \Delta H° = +55.8$ kJ
Therefore, $\Delta H° = -41.2$ kJ $+ 55.8$ kJ $= +14.6$ kJ

16.51 No. There will always be some OH^- present from the autoionization of water.

16.53 Refer to Equations 16.3 and 16.4 in the textbook.

16.55 All these acids are strong acids and, therefore, ionize completely.

(a) $[H^+] = 1.25$ M; $[OH^-] = \frac{K_w}{[H^+]} = \frac{1.00 \times 10^{-14}}{1.25} = 8.00 \times 10^{-15}$ M

(b) $[H^+] = 3.4$ M; $[OH^-] = \frac{K_w}{[H^+]} = \frac{1.00 \times 10^{-14}}{3.4} = 2.9 \times 10^{-15}$ M

(c) $[H^+] = 0.0500$ M; $[OH^-] = \frac{K_w}{[H^+]} = \frac{1.00 \times 10^{-14}}{0.0500} = 2.00 \times 10^{-13}$ M

16.57 HCl is a strong acid and, therefore, ionizes completely.
(a) $pH = -\log [H^+] = -\log 0.015 = 1.82$
(b) $pH = -\log [H^+] = -\log 0.150 = 0.824$
(c) $pH = -\log [H^+] = -\log 0.15000 = 0.82391$
(d) $pH = -\log [H^+] = -\log 1.50 = -0.176$

16.59 $\log [H^+] = -2.3$ and $[H^+] = \text{antilog}(-2.3) = 10^{-2.3} = 5 \times 10^{-3}$ M

The lemon juice is acidic, since the pH < 7.0.

16.61 $pOH = 14.00 - pH = 14.00 - 13.0 = 1.0$
$\log [OH^-] = -1.0$ and $[OH^-] = \text{antilog}(-1.0) = 10^{-1.0} = 1 \times 10^{-1}$ M

The solution is basic, since the pH > 7.0.

16.63 (a) $pH = -\log [H^+] = -\log 1.25 = -0.097$
$pOH = 14.00 - pH = 14.00 - (-0.097) = 14.10$ or $pOH = -\log [OH^-] = -\log (8.00 \times 10^{-15}) = 14.097$
(b) $pH = -\log [H^+] = -\log 3.4 = -0.53$
$pOH = 14.00 - pH = 14.00 - (-0.53) = 14.53$ or $pOH = -\log [OH^-] = -\log (2.9 \times 10^{-15}) = 14.54$
(c) $pH = -\log [H^+] = -\log 0.0500 = 1.301$
$pOH = 14.00 - pH = 14.00 - 1.301 = 12.70$ or $pOH = -\log [OH^-] = -\log (2.00 \times 10^{-13}) = 12.699$

16.65 (a) $[OH^-] = \frac{K_w}{[H^+]} = \frac{1.00 \times 10^{-14}}{0.015} = 6.7 \times 10^{-13}$ M

$pOH = -\log [OH^-] = -\log (6.7 \times 10^{-13}) = 12.17$

(b) $[OH^-] = 1.00 \times 10^{-14}/0.150 = 6.67 \times 10^{-14}$ M
$pOH = -\log (6.67 \times 10^{-14}) = 13.176$

(c) $[OH^-] = 1.00 \times 10^{-14}/0.15000 = 6.67 \times 10^{-14}$ M
$pOH = 13.176$

(d) $[OH^-] = 1.00 \times 10^{-14}/1.50 = 6.67 \times 10^{-15}$ M
$pOH = -\log (6.67 \times 10^{-15}) = 14.176$

16.67 Trichloracetic acid is the stronger acid. The strength of an acid depends on how readily it is willing to part with its protons. Since the trichloracetic acid solution has a lower pH, it is more willing to give up its protons and, therefore, must be the stronger acid of the two. Remember, the lower the pH, the greater the hydrogen ion concentration.

16.69 0.10 M HF: $\log [H^+] = -2.23$ and $[H^+] = \text{antilog}(-2.23) = 10^{-2.23} = 5.9 \times 10^{-3}$ M

$$[OH^-] = \frac{K_w}{[H^+]} = \frac{1.00 \times 10^{-14}}{5.9 \times 10^{-3}} = 1.7 \times 10^{-12} \text{ M}$$

0.10 M trichloracetic acid: $\log [H^+] = -1.13$ and $[H^+] = \text{antilog}(-1.13) = 10^{-1.13} = 7.4 \times 10^{-2}$ M

$$[OH^-] = \frac{K_w}{[H^+]} = \frac{1.00 \times 10^{-14}}{7.4 \times 10^{-2}} = 1.4 \times 10^{-13} \text{ M}$$

16.71 (a) Acidic. Solutions with pH values below 7.0 are acidic.
(b) $\log [H^+] = -5.6$

and $[H^+] = \text{antilog}(-5.6) = 10^{-5.6} = 3 \times 10^{-6}$ M

16.73 Oxo acids are acids that contain hydrogen, oxygen, and at least one other element. In these acids the oxygen atoms are bonded to a central atom, in three atom oxyacids (X–O–H) oxygen is the central atom, and the acidic hydrogens are at the positive ends of polar O–H bonds.

The O–H bond polarity is the principal factor affecting the acidity of an oxo acid; the more polar the O–H bond, the stronger the oxo acid.

16.75 Charge delocalization is the spreading of ionic charge over a greater volume or larger surface area. In other words, a given charge is more delocalized or spread out on a large ion than on a small ion.

The strength of a base depends on its ability to attract and hold on to protons. The concentrated charge on a small anion is more effective in attracting protons than the spread out or delocalized charge on a large anion and, consequently, larger anions are weaker bases than smaller anions with the same charge and same central atom.

16.77 Acid strength increases with increasing positive charge for species differing only in the number of protons.
(a) $PH_4^+ > PH_3$ (b) $H_2S > HS^-$ (c) $H_2PO_4^- > HPO_4^{2-}$

16.79
(a) The charge on Br^- is more delocalized than that on Cl^-, since bromine is larger than chlorine. Therefore, Br^- is a weaker base than Cl^- and, consequently, HBr is a stronger acid than HCl.
(b) Cl is more electronegative than Br. Because Cl is more electronegative, the H–O bond in $HClO_3$ is more polar than that in $HBrO_3$ and, consequently, $HClO_3$ is a stronger acid than $HBrO_3$.
(c) HNO_3 has an additional electronegative atom attached to the central nitrogen atom. This additional electronegative atom increases the O–H bond polarity and makes for a larger oxo anion, which is less effective in attracting protons because its charge is delocalized over a greater volume. Therefore, the additional electronegative atom makes HNO_3 a stronger acid than HNO_2.
(d) Aluminum with its 3+ positive charge is better able to attract the electron clouds of the bonded water molecules than zinc with only its 2+ positive charge. Consequently, the O–H bond polarity is greater in $Al(H_2O)_6^{3+}$ than in $Zn(H_2O)_4^{2+}$, and $Al(H_2O)_6^{3+}$ is a stronger acid than $Zn(H_2O)_4^{2+}$.

16.81
(a) HI. I^- is a weaker base than Br^-, because iodine is larger than bromine.
(b) H_3PO_4. P is more electronegative than As.
(c) H_2S. HS^- is a larger anion than HO^-.
(d) HCl. Cl is more electronegative than S and HS^- is an anion, while HCl is neutral.
(e) $Cr(H_2O)_6^{3+}$. Chromium in $Cr(H_2O)_6^{3+}$ has a 3+ charge, while in $Cr(H_2O)_6^{2+}$ it only has a 2+ charge.
(f) $HClO_4$. $HClO_4$ has an additional electronegative oxygen atom.
(g) $CH_3CH_2ClCOOH$. Cl is more electronegative than Br.

16.83 A Lewis acid is any substance capable of forming a covalent bond by accepting a share in an electron pair from another substance.

A Lewis base is any substance capable of forming a covalent bond by providing or donating a share in an electron pair with another substance.

Neutralization in the Lewis theory is the formation of a covalent bond, the sharing of an electron pair, between an acid and a base.

A Brønsted-Lowry acid has to have a proton to donate to a base, while Lewis acids are not limited to species with acidic protons. The Lewis definition encompasses all the Brønsted-Lowry acids, as well as nonprotonic acids. A Brønsted-Lowry base must accept a proton, while Lewis bases are not limited to just forming covalent bonds with the hydrogen ion (H^+). However, all Brønsted-Lowry bases are Lewis bases, and vice versa, since the origin of the base character in both theories is the existence of a lone pair of electrons on the base species.

16.85 A Lewis acid is called an electrophile, because it seeks out and accepts a share in an electron pair.

The corresponding term for a Lewis base is a nucleophile, because it is seeking out a positive nucleus (or site) or an electron-deficient species (or site). The term literally means "nucleus-lover."

16.87
(a) All Lewis acids can form a covalent bond by accepting a share in an electron pair.
(b) $AlCl_3$, BF_3, most cations (Al^{3+}, Zn^{2+}, Ag^+, etc.), $SnCl_4$, and CO_2 are examples of Lewis acids that are not Brønsted acids.

16.89 (a) $BF_3 + F^- \rightarrow [BF_4]^-$

Lewis acid   Lewis base   (new covalent bond)

(b) $CH_3NH_2 + HCl \rightarrow [CH_3NH_3]^+ + Cl^-$

Lewis base   Lewis acid   (this bond dissociates; new covalent bond)

16.91 (a) $CH_3NH_2(aq) + H_2O(l) \rightleftharpoons CH_3NH_3^+(aq) + OH^-(aq)$
$HCl(aq) + H_2O(l) \rightarrow H_3O^+(aq) + Cl^-(aq)$

Arrenhius neutralization:

$H_3O^+(aq) + Cl^-(aq) + CH_3NH_3^+(aq) + OH^-(aq) \rightarrow 2H_2O(l) + CH_3NH_3Cl(aq)$
or $H_3O^+(aq) + OH^-(aq) \rightarrow 2H_2O(l)$

In the Arrhenius system, neutralization is interpreted as the combination of hydrogen and hydroxide ions to form water.

(b) $HCl(aq) + H_2O(l) \rightarrow H_3O^+(aq) + Cl^-(aq)$
$acid_1$ $base_2$ $acid_2$ $base_1$

Brønsted-Lowry neutralization:

$CH_3NH_2(aq) + H_3O^+(aq) \rightarrow CH_3NH_3^+(aq) + H_2O(l)$
$base_1$ $acid_2$ $acid_1$ $base_2$

In the Brønsted-Lowry system, a neutralization or acid-base reaction is interpreted as a proton transfer; the acid is the proton donor and the base is the proton acceptor. For the reaction of weak bases with strong acids in aqueous solution, the weak base accepts protons directly from $H_3O^+(aq)$

(c) $H_2\ddot{O}:(l) + H{-}Cl(aq) \rightarrow H_3O^+(aq) + :Cl:^-(aq)$

Lewis neutralization:

$CH_3{-}\ddot{N}H_2 + [H{-}OH_2]^+ \rightarrow [CH_3{-}NH_3]^+ + H_2O$

Lewis base   Lewis acid

In the Lewis system, a neutralization or acid-base reaction is interpreted as the formation of a covalent bond, with the base donating the electron pair that is shared between the acid and base.

(d) The Brønsted-Lowry and Lewis interpretations seem simpler, because they don't require the reaction between the weak base and water.

16.93 No. The exact number of water molecules associated with the proton or H^+ ion is not known. In fact, it is possible that several types of hydrated ions exist simultaneously in aqueous solution.

16.95
(a) $3CaO(s) + 2H_3PO_4(aq) \rightarrow Ca_3(PO_4)_2(s) + 3H_2O(l)$
(b) $12NaOH(aq) + P_4O_{10}(s) \rightarrow 4Na_3PO_4(aq) + 6H_2O(l)$
(c) $CaO(s) + SO_2(g) \rightarrow CaSO_3(s)$
(d) $MgO(s) + 2HNO_3(aq) \rightarrow Mg(NO_3)_2(aq) + H_2O(l)$
(e) $Ba(OH)_2(aq) + SO_2(aq) \rightarrow BaSO_3(aq) + H_2O(l)$
(f) $Cr_2O_3(s) + 3H_2SO_4(aq) \rightarrow Cr_2(SO_4)_3(aq) + 3H_2O(l)$
(g) $2KOH(aq) + CO_2(aq) \rightarrow K_2CO_3(aq) + H_2O(l)$

16.97
These bases are ionic compounds. Doubling the volume halves the initial concentrations. So,

$[OH^-] = 0.250\text{ M} + 0.500\text{ M} = 0.750\text{ M}$
$[H^+] = K_w/[OH^-] = 1.00 \times 10^{-14}/0.750 = 1.33 \times 10^{-14}\text{ M}$
$pH = -\log[H^+] = -\log(1.33 \times 10^{-14}) = 13.876$

16.99
Brand A: The numbers of moles of $Al(OH)_3$ and $Mg(OH)_2$ are

$$0.300\text{ g }Al(OH)_3 \times \frac{1\text{ mol }Al(OH)_3}{78.00\text{ g }Al(OH)_3} = 0.00385\text{ mol }Al(OH)_3$$

$$0.200\text{ g }Mg(OH)_2 \times \frac{1\text{ mol }Mg(OH)_2}{58.32\text{ g }Mg(OH)_2} = 0.00343\text{ mol }Mg(OH)_2$$

The number of moles of OH^- available for neutralization is

$3 \times 0.00385 + 2 \times 0.00343 = 0.0184\text{ mol }OH^-$

Therefore, Brand A can neutralize 0.0184 mol of $H^+(aq)$ per tablet.

Brand B: The number of moles of $Al(OH)_3$ is

$$0.600\text{ g }Al(OH)_3 \times \frac{1\text{ mol }Al(OH)_3}{78.00\text{ g }Al(OH)_3} = 0.00769\text{ mol }Al(OH)_3$$

The number of moles of OH^- available for neutralization is $3 \times 0.00769 = 0.0231\text{ mol }OH^-$.

Therefore, Brand B can neutralize 0.0231 mol of $H^+(aq)$ per tablet.

Brand C: $CaCO_3(s) + 2H^+(aq) \rightarrow Ca^{2+}(aq) + CO_2(g) + H_2O(l)$
500 mg ?

The number of moles of $H^+(aq)$ reacting with 500 mg of $CaCO_3$ can be calculated as follows:
g $CaCO_3 \rightarrow$ mol $CaCO_3 \rightarrow$ mol $H^+(aq)$.

$$0.500\text{ g }CaCO_3 \times \frac{1\text{ mol }CaCO_3}{100.09\text{ g }CaCO_3} \times \frac{2\text{ mol }H^+(aq)}{1\text{ mol }CaCO_3} = 0.00999\text{ mol }H^+(aq)$$

Therefore, Brand C can neutralize 0.00999 mol of $H^+(aq)$ per tablet.

The order of ability to neutralize $H^+(aq)$ per tablet is: Brand B > Brand A > Brand C.

16.101
$Mg(OH)_2(s) + 2HNO_3(aq) \rightarrow Mg(NO_3)_2(aq) + 2H_2O(l)$
1.00 g mL = ?

The calculation proceeds as follows: g $Mg(OH)_2 \rightarrow$ mol $Mg(OH)_2 \rightarrow$ mol $HNO_3 \rightarrow$ mL HNO_3.

$$1.00\text{ g }Mg(OH)_2 \times \frac{1\text{ mol }Mg(OH)_2}{58.32\text{ g }Mg(OH)_2} \times \frac{2\text{ mol }HNO_3}{1\text{ mol }Mg(OH)_2} = 0.0343\text{ mol }HNO_3$$

The number of milliliters of 0.100 M HNO_3 required is

$$0.0343 \text{ mol } HNO_3 \times \frac{1 \text{ L}}{0.100 \text{ mol } HNO_3} \times \frac{10^3 \text{ mL}}{1 \text{ L}} = 343 \text{ mL}$$

16.103 $Al(OH)_3(s) + 3HCl(aq) \rightarrow AlCl_3(aq) + 3H_2O(l)$

(a) The number of grams of $Al(OH)_3$ neutralized is calculated as follows:

$V \times M$ (of HCl) $\rightarrow$ mol HCl $\rightarrow$ mol $Al(OH)_3$ $\rightarrow$ g $Al(OH)_3$.

$$0.100 \text{ L HCl} \times \frac{0.150 \text{ mol HCl}}{1 \text{ L HCl}} \times \frac{1 \text{ mol } Al(OH)_3}{3 \text{ mol HCl}} \times \frac{78.00 \text{ g } Al(OH)_3}{1 \text{ mol } Al(OH)_3} = 0.390 \text{ g } Al(OH)_3$$

(b) The percentage of $Al(OH)_3$ in the tablet is

$$\% \ Al(OH)_3 = \frac{\text{mass } Al(OH)_3}{\text{mass tablet}} \times 100\% = \frac{0.390 \text{ g}}{0.800 \text{ g}} \times 100\% = 48.8\%$$

CHAPTER 17

ACID-BASE EQUILIBRIA IN AQUEOUS SOLUTIONS

Solutions To Practice Exercises

PE 17.1 Hydrofluoric acid. The smaller the pK_a value, the stronger the acid.

PE 17.2 $[H^+] = \text{antilog}(-2.23) = 10^{-2.23} = 5.9 \times 10^{-3}$ M

Concentration Summary:

Equation:	$HCOOH(aq)$	$\rightleftharpoons$	$H^+(aq)$	+	$HCOO^-(aq)$
Initial concentrations, mol/L	0.200		0		0
Concentration changes, mol/L	-5.9×10^{-3}		$+5.9 \times 10^{-3}$		$+5.9 \times 10^{-3}$
Equilibrium concentrations, mol/L	0.194		5.9×10^{-3}		5.9×10^{-3}

Substituting the equilibrium concentrations in the K_a expression gives

$$K_a = \frac{[H^+][HCOO^-]}{[HCOOH]} = \frac{(5.9 \times 10^{-3})^2}{0.194} = 1.8 \times 10^{-4}$$

PE 17.3 Concentration Summary:

Equation:	$HCN(aq)$	$\rightleftharpoons$	$H^+(aq)$	+	$CN^-(aq)$
Initial concentrations, mol/L	0.050		0		0
Concentration changes, mol/L	$-x$		$+x$		$+x$
Equilibrium concentrations, mol/L	$0.050 - x$		x		x
Approximate concentrations, mol/L	0.050		x		x

Substituting the approximate concentrations and $K_a = 4.93 \times 10^{-10}$ (Table 17.1) into the K_a expression gives

$K_a = [H^+][CN^-]/[HCN]$

$$4.93 \times 10^{-10} = \frac{x^2}{0.050} \quad \text{and} \quad x = [H^+] = [CN^-] = 5.0 \times 10^{-6} \text{ M}$$

PE 17.4 (a) 0.100 M acetic acid:

Concentration Summary:

Equation:	$CH_3COOH(aq) \rightleftharpoons$	$H^+(aq) +$	$CH_3COO^-(aq)$
Initial concentrations, mol/L	0.100	0	0
Concentration changes, mol/L	$-x$	$+x$	$+x$
Equilibrium concentrations, mol/L	$0.100 - x$	x	x
Approximate concentrations, mol/L	0.100	x	x

1. With approximation: Substituting the approximate concentrations and $K_a = 1.76 \times 10^{-5}$ (Table 17.1) into the K_a expression gives

$K_a = [H^+][CH_3COO^-]/[CH_3COOH]$

$$1.76 \times 10^{-5} = \frac{x^2}{0.100} \text{ and } x = [H^+] = 1.33 \times 10^{-3}\text{ M}$$

2. Without approximation: Substituting the equilibrium concentrations and $K_a = 1.76 \times 10^{-5}$ into the K_a expression gives

$$1.76 \times 10^{-5} = \frac{x^2}{0.100 - x}$$

$x^2 + 1.76 \times 10^{-5}x - 1.76 \times 10^{-6} = 0$ and $x = [H^+] = 1.32 \times 10^{-3}$ M

(b) 1.00×10^{-4} M acetic acid (replace 0.100 with 1.00×10^{-4} in the above concentration summary):

1. With approximation: Substituting the approximate concentrations and $K_a = 1.76 \times 10^{-5}$ into the K_a expression gives

$$1.76 \times 10^{-5} = \frac{x^2}{1.00 \times 10^{-4}} \text{ and } x = [H^+] = 4.20 \times 10^{-5}\text{ M}$$

2. Without approximation: Substituting the equilibrium concentrations and K_a value into the K_a expression gives

$$1.76 \times 10^{-5} = \frac{x^2}{1.00 \times 10^{-4} - x}$$

$x^2 + 1.76 \times 10^{-5}x - 1.76 \times 10^{-9} = 0$ and $x = [H^+] = 3.41 \times 10^{-5}$ M

PE 17.5 Trimethylamine. The lower the pK_b value, the higher the K_b and the stronger the base.

PE 17.6 Concentration Summary:

Equation:	$C_6H_5NH_2(aq) + H_2O(l) \rightleftharpoons$	$C_6H_5NH_3^+(aq) +$	$OH^-(aq)$
Initial concentrations, mol/L	0.15	0	0
Concentration changes, mol/L	$-x$	$+x$	$+x$
Equilibrium concentrations, mol/L	$0.15 - x$	x	x
Approximate concentrations, mol/L	0.15	x	x

Substituting the approximate concentrations and $K_b = 4.3 \times 10^{-10}$ (Table 17.2) into the K_b expression gives

$K_b = [C_6H_5NH_3^+][OH^-]/[C_6H_5NH_2]$

$$4.3 \times 10^{-10} = \frac{x^2}{0.15} \text{ and } x = [OH^-] = 8.0 \times 10^{-6}\text{ M}$$

The pOH of the solution is pOH = $-\log [OH^-] = -\log (8.0 \times 10^{-6}) = 5.10$.

Answer: pH = 14.00 – pOH = 14.00 – 5.10 = 8.90

PE 17.7

$$K_a = \frac{K_w}{K_b} = \frac{1.00 \times 10^{-14}}{1.77 \times 10^{-5}} = 5.65 \times 10^{-10}$$

PE 17.8 Concentration Summary:

	$NH_3(aq) + H_2O(l) \rightleftharpoons$	$NH_4^+(aq)$ +	$OH^-(aq)$
Equation:			
Initial concentrations, mol/L	0.200	0	0
Concentration changes, mol/L	$-x$	$+x$	$+x$
Equilibrium concentrations, mol/L	$0.200 - x$	x	x
Approximate concentrations, mol/L	0.200	x	x

Substituting the approximate concentrations and $K_b = 1.77 \times 10^{-5}$ (Table 17.2) into the K_b expression gives

$K_b = [NH_4^+][OH^-]/[NH_3]$

$$1.77 \times 10^{-5} = \frac{x^2}{0.200} \text{ and } x = [NH_4^+] = [OH^-] = 1.88 \times 10^{-3} \text{ M}$$

The number of moles of NH_3 ionized per liter is equal to the hydroxide ion concentration. The percent ionization is

$$\% \text{ ionization} = 100\% \times \frac{\text{moles of } NH_3 \text{ ionized per liter}}{\text{initial moles of } NH_3 \text{ per liter}} = 100\% \times \frac{1.88 \times 10^{-3} \text{ M}}{0.200 \text{ M}} = 0.940\%$$

PE 17.9 $NH_3(aq) + H_2O(l) \rightleftharpoons NH_4^+(aq) + OH^-(aq)$

(a) Decreases. In addition to decreasing all concentrations, dilution also drives the reaction to the right.
(b) No change. The K_b value depends only on temperature as long as the solution is dilute.
(c) Increases. Dilution drives the reaction to the right; therefore the total number of moles of OH^- increases.
(d) Increases. Dilution drives the reaction to the right; therefore the fraction of molecules ionized increases.

PE 17.10 $H_2AsO_4^-$. For acids that differ only in the number of protons, acid strength increases with increasing positive charge; hence, $H_2AsO_4^- > HAsO_4^{2-}$.

PE 17.11 (a) The $[H^+]$ can be calculated from the first step alone.

Concentration Summary:

	$H_2C_3H_2O_4(aq) \rightleftharpoons$	$H^+(aq)$ +	$HC_3H_2O_4^-(aq)$
Equation:			
Initial concentrations, mol/L	0.80	0	0
Concentration changes, mol/L	$-x$	$+x$	$+x$
Equilibrium concentrations, mol/L	$0.80 - x$	x	x
Approximate concentrations, mol/L	0.80	x	x

Substituting the approximate concentrations and $K_{a_1} = 1.49 \times 10^{-3}$ (Table 17.1) into the K_{a_1} expression gives

$K_{a_1} = [H^+][HC_3H_2O_4^-]/[H_2C_3H_2O_4]$

$$1.49 \times 10^{-3} = \frac{x^2}{0.80} \text{ and } x = [H^+] = 0.035 \text{ M}$$

The approximation is valid, since x is less than 5% of 0.80. The pH is

$pH = -\log [H^+] = -\log (0.035) = 1.46$

(b) The concentration of the doubly charged anion is approximately equal to K_{a_2}. Therefore,

$[C_3H_2O_4^{2-}] = K_{a_2} = 2.0 \times 10^{-6}$ M

PE 17.12 $S^{2-}(aq) + H_2O(l) \rightleftharpoons HS^-(aq) + OH^-(aq)$

$K_b = [HS^-][OH^-]/[S^{2-}]$

PE 17.13 $CH_3NH_3^+(aq) \rightleftharpoons CH_3NH_2(aq) + H^+(aq)$

$K_a = [CH_3NH_2][H^+]/[CH_3NH_3^+]$

PE 17.14 (a) Basic. The solution contains $Na^+(aq)$ and $PO_4^{3-}(aq)$. The hydrated sodium ion does not react with water and has no effect on the pH of the solution. The phosphate ion is the anion of the weak acid HPO_4^{2-} and is therefore basic with respect to water. The solution will be basic.

(b) Acidic. The solution contains $Cu^{2+}(aq)$ and $NO_3^-(aq)$. The hydrated copper(II) ion is acidic with respect to water. The nitrate ion is the anion of a strong acid and will not react with water. The solution will be acidic.

(c) Neutral. The solution contains $Li^+(aq)$ and $NO_3^-(aq)$. The hydrated lithium ion is monovalent and will not react with water. The nitrate ion is the anion of a strong acid and will not react with water. The solution will be neutral.

(d) Slightly acidic. The solution contains $(CH_3)_3NH^+(aq)$ and $F^-(aq)$. K_a for the $(CH_3)_3NH^+$ ion is

$$K_a = \frac{K_w}{K_b} = \frac{1.00 \times 10^{-14}}{6.3 \times 10^{-5}} = 1.6 \times 10^{-10}$$

K_b for the fluoride ion is

$$K_b = \frac{K_w}{K_a} = \frac{1.00 \times 10^{-14}}{3.53 \times 10^{-4}} = 2.83 \times 10^{-11}$$

K_a for $(CH_3)_3NH^+$ is slightly greater than K_b for F^-; hence, the acid strength of $(CH_3)_3NH^+$ is slightly greater than the base strength of F^-. The solution should be slightly acidic.

PE 17.15 Barium acetate provides the ions $Ba^{2+}(aq)$ and $CH_3COO^-(aq)$. The barium ion does not react with water and does not affect the pH of the solution. The acetate ion is basic and will form hydroxide ions. The concentration summary will be based on this reaction.

Concentration Summary:

Equation:	$CH_3COO^-(aq) + H_2O(l) \rightleftharpoons$	$CH_3COOH(aq)$	$+ OH^-(aq)$
Initial concentrations, mol/L	0.400	0	0
Concentration changes, mol/L	$-x$	$+x$	$+x$
Equilibrium concentrations, mol/L	$0.400 - x$	x	x
Approximate concentrations, mol/L	0.400	x	x

Substituting the approximate concentrations and $K_b = 5.68 \times 10^{-10}$ (Example 17.5) into the K_b expression gives

$K_b = [CH_3COOH][OH^-]/[CH_3COO^-]$

$$5.68 \times 10^{-10} = \frac{x^2}{0.400}$$

$x = [OH^-] = 1.51 \times 10^{-5}$ M

$pOH = -\log [OH^-] = -\log (1.51 \times 10^{-5}) = 4.821$ and $pH = 14.00 - pOH = 14.00 - 4.821 = 9.18$

PE 17.16 Basic. The reactions of HCO_3^- with water are

1. $HCO_3^-(aq) \rightleftharpoons H^+(aq) + CO_3^{2-}(aq)$ $K_{a_2} = 5.61 \times 10^{-11}$
2. $HCO_3^-(aq) + H_2O(l) \rightleftharpoons H_2CO_3(aq) + OH^-(aq)$

K_b for the second reaction is

$$K_b = \frac{K_w}{K_{a_1}} = \frac{1.00 \times 10^{-14}}{4.30 \times 10^{-7}} = 2.33 \times 10^{-8}$$

K_b is greater than K_{a_2}, showing that HCO_3^- is a stronger base than acid. A solution containing HCO_3^- ions will therefore be basic.

PE 17.17 (a) Sodium benzoate (C_6H_5COONa); the pH is raised since the $[H^+]$ is lowered by the addition of the benzoate ion.

(b) $C_6H_5NH_3Cl$; lowered since the $[OH^-]$ decreases and the $[H^+]$ increases.

PE 17.18 Concentration Summary:

	$CH_3NH_2(aq) + H_2O(l) \rightleftharpoons$	$CH_3NH_3^+(aq)$ +	$OH^-(aq)$
Equation:			
Initial concentrations, mol/L	0.200	0.200	0
Concentration changes, mol/L	$-x$	$+x$	$+x$
Equilibrium concentrations, mol/L	$0.200 - x$	$0.200 + x$	x
Approximate concentrations, mol/L	0.200	0.200	x

Substituting the approximate concentrations and $K_b = 3.70 \times 10^{-4}$ (Table 17.2) into the K_b expression gives

$K_b = [CH_3NH_3^+][OH^-]/[CH_3NH_2]$

$$3.70 \times 10^{-4} = \frac{(0.200)(x)}{0.200} \quad \text{and } x = [OH^-] = 3.70 \times 10^{-4} \text{ M}$$

PE 17.19 Concentration Summary:

	$HC_3H_2O_4^-(aq) \rightleftharpoons$	$H^+(aq)$ +	$C_3H_2O_4^{2-}(aq)$
Equation:			
Initial concentrations, mol/L	0.10	0	0.10
Concentration changes, mol/L	$-x$	$+x$	$+x$
Equilibrium concentrations, mol/L	$0.10 - x$	x	$0.10 + x$
Approximate concentrations, mol/L	0.10	x	0.10

Substituting the approximate concentrations and $K_a = 2.03 \times 10^{-6}$ into the K_a expression gives

$K_a = [H^+][C_3H_2O_4^{2-}]/[HC_3H_2O_4^-]$

$$2.03 \times 10^{-6} = \frac{(x)(0.10)}{0.10}$$

$x = 2.03 \times 10^{-6}$ M $= [H^+]$ and $pH = -\log [H^+] = -\log (2.03 \times 10^{-6}) = 5.69$

PE 17.20 pH = 1.20. HCl is a strong acid and, therefore, the pH of the gastric juice after ingestion of the aspirin tablet should be almost identical to what it was originally.

PE 17.21 NaOH, a strong base, will repress the ionization of NH_3. Therefore, virtually all of the hydroxide ions in the mixture will come from the NaOH. The $[OH^-]$ concentration is

$[OH^-] = 0.010 \text{ mol}/0.250 \text{ L} = 0.040 \text{ M}$

$pOH = -\log [OH^-] = -\log 0.040 = 1.40$ and $pH = 14.00 - pOH = 14.00 - 1.40 = 12.60$

PE 17.22 HCOOH is the acid and $HCOO^-$ is the base. Equation 17.4 becomes

$$pH = pK_a + \log \frac{[HCOO^-]}{[HCOOH]}$$

$$\log \frac{[HCOO^-]}{[HCOOH]} = pH - pK_a = 3.95 + \log (1.9 \times 10^{-4}) = 0.23$$

$$\text{and } \frac{[HCOO^-]}{[HCOOH]} = \text{antilog } 0.23 = 10^{+0.23} = 1.7$$

PE 17.23 Since NaOH is a strong base, it will react completely with the acid component of the buffer; 1.00×10^{-3} mol NaOH will convert 1.00×10^{-3} mol C_6H_5COOH to 1.00×10^{-3} mol $C_6H_5COO^-$.

$C_6H_5COOH(aq) + OH^-(aq) \rightarrow C_6H_5COO^-(aq) + H_2O(l)$

The new quantities are:

$C_6H_5COO^-$: 0.0100 mol + 0.00100 mol = 0.0110 mol

C_6H_5COOH: 0.0100 mol – 0.00100 mol = 0.0090 mol

The final pH is calculated from Equation 17.4:

$$pH = pK_a + \log \frac{[C_6H_5COO^-]}{[C_6H_5COOH]} = 4.190 + \log \frac{0.0110}{0.0090} = 4.28$$

Note: The mole ratio, which is the same as the concentration ratio, was used in this calculation.

PE 17.24 Look in Table 17.1 for acids with pK_a values within one unit of pH 3.52. Formic acid with $pK_a = 3.72$ and its salt are the practical choice among several candidates.

PE 17.25 The pK_a for the formic acid-sodium formate buffer is 3.72 (Table 17.1).

$$pH = pK_a + \log \frac{[HCOO^-]}{[HCOOH]}$$

$$3.52 = 3.72 + \log \frac{[HCOO^-]}{[HCOOH]}$$

$$\log \frac{[HCOO^-]}{[HCOOH]} = 3.52 - 3.72 = -0.20; \text{ and } \frac{[HCOO^-]}{[HCOOH]} = \text{antilog } (-0.20) = 10^{-0.20} = 0.63$$

The concentration of HCOOH is 0.10 M. Hence, the concentration of $HCOO^-$ must be

$[HCOO^-] = 0.63 \times [HCOOH] = 0.63 \times 0.10 \text{ M} = 0.063 \text{ M}$

The number of moles of sodium formate that must be added to 500 mL of the solution is

$$0.500 \text{ L} \times \frac{0.063 \text{ mol}}{1 \text{ L}} = 0.032 \text{ mol}$$

PE 17.26 The pH of this solution at equivalence will be less than 7 because $C_6H_5NH_3^+$ is a weak acid.

$C_6H_5NH_3^+(aq) \rightleftharpoons H^+(aq) + C_6H_5NH_2(aq)$

The number of moles of aniline at the beginning of the titration is

$$0.0350\ L \times \frac{0.0500\ mol}{1\ L} = 0.00175\ mol$$

From the neutralization equation

$C_6H_5NH_2(aq) + HCl(aq) \rightleftharpoons C_6H_5NH_3Cl(aq)$

we see that 0.00175 mol of aniline will react with 0.00175 mol of hydrochloric acid to form 0.00175 mol of $C_6H_5NH_3^+$. The volume of 0.0500 M HCl required to reach equivalence is

$$0.00175\ mol\ HCl \times \frac{1\ L}{0.0500\ mol\ HCl} = 0.0350\ L = 35.0\ mL$$

The total volume of the solution at the equivalence point is

35.0 mL + 35.0 mL = 70.0 mL = 0.0700 L

The molarity of $C_6H_5NH_3^+$ at equivalence is initially

$\frac{0.00175\ mol}{0.0700\ L} = 0.0250$ M, but some of this ionizes to form hydrogen ions.

Concentration Summary:

Equation:	$C_6H_5NH_3^+(aq) \rightleftharpoons$	$H^+(aq)$ +	$C_6H_5NH_2(aq)$
Initial concentrations, mol/L	0.0250	0	0
Concentration changes, mol/L	$-x$	$+x$	$+x$
Equilibrium concentrations, mol/L	$0.0250 - x$	x	x
Approximate concentrations, mol/L	0.0250	x	x

Substituting $K_a = K_w/K_b = 1.00 \times 10^{-14}/4.3 \times 10^{-10} = 2.3 \times 10^{-5}$ and the approximate concentrations into the K_a expression gives

$K_a = [H^+][C_6H_5NH_2]/[C_6H_5NH_3^+]$

$$2.3 \times 10^{-5} = \frac{x^2}{0.0250}$$

$x = [H^+] = 7.6 \times 10^{-4}$ M and $pH = -\log [H^+] = -\log (7.6 \times 10^{-4}) = 3.12$

PE 17.27 initial moles of HCl = 0.02500 L × 0.1000 mol/L = 2.500×10^{-3} mol

moles of HCl neutralized = moles of NaOH added = 0.02400 L × 0.1000 mol/L = 2.400×10^{-3} mol

moles of HCl remaining = $2.500 \times 10^{-3} - 2.400 \times 10^{-3} = 1.00 \times 10^{-4}$ mol

concentration of HCl = moles of HCl remaining/final solution volume

= 1.00×10^{-4} mol/(0.02500 L + 0.02400 L)

= 2.04×10^{-3} M and $pH = -\log [H^+] = -\log (2.04 \times 10^{-3}) = 2.69$

PE 17.28 At points between the initial point and the equivalence point, the pH can be calculated using Equation 17.4.

initial moles of acetic acid = 0.02500 L × 0.1000 mol/L = 2.500×10^{-3} mol

moles of acetic acid neutralized = moles of NaOH added = 0.02000 L × 0.1000 mol/L = 2.000×10^{-3} mol moles of acetic acid remaining = $2.500 \times 10^{-3} - 2.000 \times 10^{-3} = 5.00 \times 10^{-4}$ mol

The number of moles of acetic ion formed is 2.00×10^{-3} mol, and the buffer ratio is

$$\frac{[CH_3COO^-]}{[CH_3COOH]} = \frac{2.00 \times 10^{-3}\,mol}{5.00 \times 10^{-4}\,mol} = 4.00$$

Substituting $pK_a = 4.754$ and 4.00 for the buffer ratio into Equation 17.4 gives

$$pH = pK_a + \log\frac{[CH_3COO^-]}{[CH_3COOH]} = 4.754 + \log 4.00 = 5.356$$

PE 17.29 (a) Using a concentration summary similar to that in Practice Exercise 17.8 and substituting the approximate concentrations and $K_b = 1.77 \times 10^{-5}$ into the K_b expression gives

$$1.77 \times 10^{-5} = \frac{x^2}{0.1000}$$

$x = [OH^-] = 1.33 \times 10^{-3}$ M

$pOH = -\log[OH^-] = -\log(1.33 \times 10^{-3}) = 2.876$ and $pH = 14.00 - pOH = 14.00 - 2.876 = 11.1$

(b) At the half-neutralization point the buffer ratio is unity. Substituting $pK_a = 9.25$ and one for the buffer ratio into Equation 17.4 gives

$$pH = pK_a + \log\frac{[NH_3]}{[NH_4^+]} = 9.25 + \log 1 = 9.25$$

(c) At the equivalence point 25.00 mL of HCl has been added and 0.02500 L × 0.1000 mol/L = 2.500×10^{-3} mol NH_4^+ is initially present in 50.00 mL of solution. Using a concentration summary similar to that in Example 17.10 and substituting the approximate concentrations and $K_a = 5.65 \times 10^{-10}$ into the K_a expression gives

$$5.65 \times 10^{-10} = \frac{x^2}{0.05000}$$

$x = [H^+] = 5.32 \times 10^{-6}$ M and $pH = -\log[H^+] = -\log(5.32 \times 10^{-6}) = 5.3$

PE 17.30 The pH at the equivalence point is 3.1. A suitable indicator would be any indicator whose transition interval includes pH 3.1, for example, congo red.

Solutions To Final Exercises

17.1 (a) HOI: $K_a = [H^+][OI^-]/[HOI]$; CCl_3COOH: $K_a = [H^+][CCl_3COO^-]/[CCl_3COOH]$

(b) K_a $CCl_3COOH(2 \times 10^{-1})$ is greater than K_a $HOI(2.3 \times 10^{-11})$; hence, $CCl_3COOH > HOI$.

17.3 Stronger. The lower pK_b value at 50°C means that the ammonia ionization reaction is driven further to the right as the temperature goes up.

17.5 (a) $[H^+] = \text{antilog}(-4.0) = 10^{-4.0} = 1 \times 10^{-4}$ M

$$[OH^-] = \frac{K_w}{[H^+]} = \frac{1.00 \times 10^{-14}}{1 \times 10^{-4}} = 1 \times 10^{-10}\text{ M}$$

(b) The $[H^+]$ from the autoionization of water is repressed by the presence of the weak acid and, therefore, is usually several orders of magnitude smaller than that obtained from the ionization of the weak acid.

17.7 $NH_4^+(aq) \rightleftharpoons H^+(aq) + NH_3(aq)$

$$K_a = \frac{[H^+][NH_3]}{[NH_4^+]}$$

$NH_3(aq) + H_2O(l) \rightleftharpoons NH_4^+(aq) + OH^-(aq)$

$$K_b = \frac{[NH_4^+][OH^-]}{[NH_3]}$$

$$K_a \times K_b = \frac{[H^+][NH_3]}{[NH_4^+]} \times \frac{[NH_4^+][OH^-]}{[NH_3]} = [H^+][OH^-] = K_w$$

17.9 $pK_a = -\log K_a = 4.92$

$K_a = \text{antilog}(-4.92) = 10^{-4.92} = 1.2 \times 10^{-5}$

17.11 $$K_a = \frac{K_w}{K_b} = \frac{1.00 \times 10^{-14}}{6.41 \times 10^{-4}} = 1.56 \times 10^{-11}$$

17.13 (a) $[H^+] = \text{antilog}(-2.96) = 10^{-2.96} = 1.1 \times 10^{-3}$ M

(b) Concentration Summary:

Equation:	$CH_3CH_2COOH(aq) \rightleftharpoons$	$H^+(aq)$ +	$CH_3CH_2COO^-(aq)$
Initial concentrations, mol/L	0.100	0	0
Concentration changes, mol/L	-1.1×10^{-3}	$+1.1 \times 10^{-3}$	$+1.1 \times 10^{-3}$
Equilibrium concentrations, mol/L	0.099	1.1×10^{-3}	1.1×10^{-3}

Substituting the equilibrium concentrations in the K_a expression gives

$$K_a = \frac{[H^+][CH_3CH_2COO^-]}{[CH_3CH_2COOH]} = \frac{(1.1 \times 10^{-3})^2}{0.099} = 1.2 \times 10^{-5}$$

17.15 (a) Concentration Summary:

Equation:	$HC_3H_5O_3(aq) \rightleftharpoons$	$H^+(aq)$ +	$C_3H_5O_3^-(aq)$
Initial Concentrations, mol/L	0.180	0	0
Concentration changes, mol/L	$-x$	$+x$	$+x$
Equilibrium concentrations, mol/L	$0.180 - x$	x	x
Approximate concentrations, mol/L	0.180	x	x

Substituting the approximate concentrations and $K_a = 1.37 \times 10^{-4}$ (Table 17.1) into the K_a expression gives

$K_a = [H^+][C_3H_5O_3^-]/[HC_3H_5O_3]$

$$1.37 \times 10^{-4} = \frac{x^2}{0.180} \quad \text{and } x = [H^+] = 4.97 \times 10^{-3} \text{ M}$$

The approximation is acceptable, since x is less than 5% of 0.180. The pH is

$pH = -\log [H^+] = -\log(4.97 \times 10^{-3}) = 2.304$

(b) Exchanging 0.0180 for 0.180 in the above concentration summary and substituting the approximate concentrations and K_a value into the K_a expression gives

$$1.37 \times 10^{-4} = \frac{x^2}{0.0180} \quad \text{and } x = [H^+] = 1.57 \times 10^{-3}$$

The approximation is unacceptable or not valid, since x is more than 5% of 0.0180. The correct value of x is obtained by substituting the equilibrium concentrations and K_a value into the K_a expression and using the quadratic formula.

$$1.37 \times 10^{-4} = \frac{x^2}{0.0180 - x}$$

$$x^2 + 1.37 \times 10^{-4}x - 2.47 \times 10^{-6} = 0$$

$$x = [H^+] = 1.50 \times 10^{-3}\text{ M and pH} = -\log [H^+] = -\log (1.50 \times 10^{-3}) = 2.824$$

17.17 $CH_3COOH(aq) \rightleftharpoons H^+(aq) + CH_3COO^-(aq)$

(a) Decreases. Dilution decreases all the concentrations and also drives the reaction to the right.
(b) Decreases. Even though the overall number of moles of CH_3COO^- increases, the $[CH_3COO^-]$ must decrease in order to satisfy the K_a expression.
(c) Increases. Even though the overall number of moles of H^+ increases, the $[H^+]$ must decrease to satisfy the K_a expression.
(d) Increases. Since the $[H^+]$ decreases, the $[OH^-]$ must increase to satisfy the K_w expression.
(e) Remains the same. In dilute solution the K_a value depends only on the temperature.
(f) Increases. Dilution drives the reaction to the right. Therefore, the fraction of molecules ionized increases.

17.19 Use of a concentration summary similar to the one in Exercise 17.16 for the acid, and the one in Practice Exercise 17.8 for the base, and substituting the approximate concentrations and K_a and K_b values from Tables 17.1 and 17.2 into the K_a and K_b expressions gives

(a) $K_a = [H^+][C_3H_5O_3^-]/[HC_3H_5O_3]$

$$1.37 \times 10^{-4} = \frac{x^2}{0.10} \quad \text{and } x = [C_3H_5O_3^-] = [H^+] = 3.7 \times 10^{-3}\text{ M}$$

The number of moles of $HC_3H_5O_3$ ionized per liter is equal to x, so the percent ionization is

$$\%\text{ ionization} = 100\% \times \frac{3.7 \times 10^{-3}\text{ M}}{0.10\text{ M}} = 3.7\%$$

(b) $K_b = [CH_3NH_3^+][OH^-]/[CH_3NH_2]$

$$3.70 \times 10^{-4} = \frac{x^2}{0.25} \quad \text{and } x = [OH^-] = [CH_3NH_3^+] = 9.6 \times 10^{-3}\text{ M}$$

The number of moles of CH_3NH_2 ionized per liter is equal to x, so the percent ionization is

$$\%\text{ ionization} = 100\% \times \frac{9.6 \times 10^{-3}\text{ M}}{0.25\text{ M}} = 3.8\%$$

17.21 (a) moles of formic acid ionized per liter

$$= \frac{\%\text{ ionization}}{100\%} \times \text{initial moles of formic acid per liter} = \frac{34\%}{100\%} \times 0.0010\text{ M} = 0.00034\text{ M}$$

The hydrogen ion concentration is equal to the number of moles of formic acid ionized per liter, or

$[H^+] = 0.00034$ M. The pH is

$pH = -\log [H^+] = -\log 0.00034 = 3.47$

(b) The equilibrium concentrations are

$[H^+] = [HCOO^-] = 0.00034$ M

$[HCOOH] = 0.0010 - 0.00034 = 0.0007$ M

Substituting these equilibrium concentrations into the K_a expression gives

$$K_a = [H^+][HCOO^-]/[HCOOH] = \frac{(0.00034)^2}{(0.0007)} = 2 \times 10^{-4}$$

17.23 (a) Concentration Summary:

Equation:	$HCN(aq) \rightleftharpoons$	$H^+(aq) +$	$CN^-(aq)$
Initial concentrations, mol/L	0.250	0	0
Concentration changes, mol/L	$-x$	$+x$	$+x$
Equilibrium concentrations, mol/L	$0.250 - x$	x	x
Approximate concentrations, mol/L	0.250	x	x

Substituting the approximate concentrations and $K_a = 4.93 \times 10^{-10}$ (Table 17.1) into the K_a expression gives

$K_a = [H^+][CN^-]/[HCN]$

$$4.93 \times 10^{-10} = \frac{x^2}{0.250} \quad \text{and} \quad x = [H^+] = 1.11 \times 10^{-5} \text{ M}$$

$pH = -\log [H^+] = -\log (1.11 \times 10^{-5}) = 4.955$

(b) The number of moles of HCN ionized per liter is equal to x, so the fraction of HCN ionized is

$$\frac{1.11 \times 10^{-5} \text{ M}}{0.250 \text{ M}} = 4.44 \times 10^{-5}$$

17.25 For acids that differ only in the number of protons, acid strength decreases with decreasing positive charge; hence, in order of decreasing K_a: $H_3A > H_2A^- > HA^{2-}$. In other words, the more negative the species, the more difficult it is to remove a positive proton.

17.27 The $[H^+]$ and $[HSO_3^-]$ can be calculated from the first step alone. Since K_{a_1} is large, the exact solution will be used.

Concentration Summary:

Equation:	$H_2SO_3(aq) \rightleftharpoons$	$H^+(aq) +$	$HSO_3^-(aq)$
Initial concentrations, mol/L	0.150	0	0
Concentration changes, mol/L	$-x$	$+x$	$+x$
Equilibrium concentrations, mol/L	$0.150 - x$	x	x

Substituting the equilibrium concentrations and $K_{a_1} = 1.71 \times 10^{-2}$ (Table 17.1) into the K_{a_1} expression and using the quadratic formula gives

$K_{a_1} = [H^+][HSO_3^-]/[H_2SO_3]$

$$1.71 \times 10^{-2} = \frac{x^2}{0.150 - x}$$

$x^2 + 1.71 \times 10^{-2}x - 2.565 \times 10^{-3} = 0$ and $x = [H^+] = [HSO_3^-] = 0.0428$ M

The concentration of SO_3^{2-} is approximately equal to K_{a_2}. Therefore,

$[SO_3^{2-}] = K_{a_2} = 6.0 \times 10^{-8}$ M

17.29 (a) The $[H^+]$ and $[HCO_3^-]$ can be calculated from the first step alone.

Concentration Summary:

Equation:	$CO_2(aq) + H_2O(l) \rightleftharpoons$	$H^+(aq)$ +	$HCO_3^-(aq)$
Initial concentrations, mol/L	0.033	0	0
Concentration changes, mol/L	$-x$	$+x$	$+x$
Equilibrium concentrations, mol/L	$0.033 - x$	x	x
Approximate concentrations, mol/L	0.033	x	x

Substituting the approximate concentrations and $K_{a_1} = 4.30 \times 10^{-7}$ (Table 17.1) into the K_{a_1} expression gives

$K_{a_1} = [H^+][HCO_3^-]/[CO_2]$

$$4.30 \times 10^{-7} = \frac{x^2}{0.033}$$

$x = [H^+] = [HCO_3^-] = 1.2 \times 10^{-4}$ M and $pH = -\log [H^+] = -\log (1.2 \times 10^{-4}) = 3.92$

The approximation is valid, since x is less than 5% of 0.033.

(b) From (a) we have $[HCO_3^-] = 1.2 \times 10^{-4}$ M. The concentration of CO_3^{2-} is approximately equal to K_{a_2}. Therefore,

$[CO_3^{2-}] = K_{a_2} = 5.6 \times 10^{-11}$ M

17.31
1. The anions of strong acids, excluding the amphiprotic anions like HSO_4^-. These anions are weaker bases than water. Examples are: ClO_4^-, Cl^-, NO_3^-, I^-, Br^-, and ClO_3^-.
2. Monovalent metal cations and divalent alkaline earth metal cations other than beryllium. These hydrated cations are weaker acids than water. Examples are: Li^+, Na^+, K^+, Mg^{2+}, Ca^{2+}, and Ba^{2+}.

17.33
1. Small, highly charged metal cations. These hydrated cations are stronger acids than water. Examples are: Al^{3+}, Be^{2+}, Cu^{2+}, Fe^{3+}, Cr^{3+}, and Zn^{2+}.

$Zn(H_2O)_4^{2+}(aq) + H_2O(l) \rightleftharpoons Zn(H_2O)_3(OH)^+(aq) + H_3O^+(aq)$
$Be(H_2O)_4^{2+}(aq) + H_2O(l) \rightleftharpoons Be(H_2O)_3(OH)^+(aq) + H_3O^+(aq)$
$Cr(H_2O)_6^{3+}(aq) + H_2O(l) \rightleftharpoons Cr(H_2O)_5(OH)^{2+}(aq) + H_3O^+(aq)$

2. Ammonium ion (NH_4^+) and protonated amines. These cations are stronger acids than water. Examples are: NH_4^+, $CH_3NH_3^+$, and $C_6H_5NH_3^+$.

$NH_4^+(aq) + H_2O(l) \rightleftharpoons NH_3(aq) + H_3O^+(aq)$
$CH_3NH_3^+(aq) + H_2O(l) \rightleftharpoons CH_3NH_2(aq) + H_3O^+(aq)$
$C_6H_5NH_3^+(aq) + H_2O(l) \rightleftharpoons C_6H_5NH_2(aq) + H_3O^+(aq)$

Note: Some amphiprotic anions may be acidic; examples are $H_2PO_4^-$, HSO_4^-, and $HC_2O_4^-$.

17.35 (a) $Al(NO_3)_3$: acidic (b) KOBr: basic (c) $Ca(HCO_3)_2$ basic (see Practice Exercise 17.16 for details) (d) NH_4ClO_4: acidic

17.37 $CN^-(aq) + H_2O(l) \rightleftharpoons HCN(aq) + OH^-(aq)$

$K_b = [HCN][OH^-]/[CN^-]$

$$K_b = \frac{K_w}{K_a} = \frac{1.00 \times 10^{-14}}{4.93 \times 10^{-10}} = 2.03 \times 10^{-5}$$

$F^-(aq) + H_2O(l) \rightleftharpoons HF(aq) + OH^-(aq)$

$K_b = [HF][OH^-]/[F^-]$

$$K_b = \frac{K_w}{K_a} = \frac{1.00 \times 10^{-14}}{3.53 \times 10^{-4}} = 2.83 \times 10^{-11}$$

Since the K_b for CN^- is considerably greater than the K_b for F^-, the KCN solution should be more basic than the KF solution.

17.39 $Al(H_2O)_6^{3+}(aq) + H_2O(l) \rightleftharpoons Al(H_2O)_5(OH)^{2+}(aq) + H_3O^+(aq)$

$$K_a = \frac{[Al(H_2O)_5(OH)^{2+}][H_3O^+]}{[Al(H_2O)_6^{3+}]} = 1.4 \times 10^{-5}$$

$Fe(H_2O)_6^{3+}(aq) + H_2O(l) \rightleftharpoons Fe(H_2O)_5(OH)^{2+}(aq) + H_3O^+(aq)$

$$K_a = \frac{[Fe(H_2O)_5(OH)^{2+}][H_3O^+]}{[Fe(H_2O)_6^{3+}]} = 7.9 \times 10^{-3}$$

Since the K_a for $Fe(H_2O)_6^{3+}$ is greater than the K_a for $Al(H_2O)_6^{3+}$, the $FeCl_3$ solution should be more acidic than the $AlCl_3$ solution.

17.41 (a) $K_a = 5.65 \times 10^{-10}$ (Practice Exercise 17.7)

(b) Using a concentration summary similar to that in Example 17.10 and substituting the equilibrium concentrations and K_a value into the K_a expression, making the usual assumption that $0.15 - x = 0.15$, gives

$K_a = [NH_3][H^+]/[NH_4^+]$

$$5.65 \times 10^{-10} = \frac{x^2}{0.15}$$

$x = [H^+] = 9.2 \times 10^{-6}$ M and $pH = -\log [H^+] = -\log (9.2 \times 10^{-6}) = 5.04$

17.43 Potassium cyanide provides the ions $K^+(aq)$ and $CN^-(aq)$. The potassium ion does not react with water and does not affect the pH of the solution. The cyanide ion is basic and will form hydroxide ions. The concentration summary will be based on this reaction.

Concentration Summary:

Equation:	$CN^-(aq) + H_2O(l) \rightleftharpoons$	$HCN(aq)$	$+ OH^-(aq)$
Initial concentrations, mol/L	0.075	0	0
Concentration changes, mol/L	$-x$	$+x$	$+x$
Equilibrium concentrations, mol/L	$0.075 - x$	x	x
Approximate concentrations, mol/L	0.075	x	x

Substituting the approximate concentrations and $K_b = 2.03 \times 10^{-5}$ (Exercise 17.37) into the K_b expression gives

$K_b = [HCN][OH^-]/[CN^-]$

$$2.03 \times 10^{-5} = \frac{x^2}{0.075} \text{ and } x = [OH^-] = 1.2 \times 10^{-3} \text{ M}$$

$pOH = -\log [OH^-] = -\log (1.2 \times 10^{-3}) = 2.92$ and $pH = 14.00 - pOH = 14.00 - 2.92 = 11.08$

17.45 NaOBr provides the ions $Na^+(aq)$ and $OBr^-(aq)$. The sodium ion does not react with water and does not affect the pH of the solution. OBr^- is basic and will form hydroxide ions.

The K_a for HOBr is 2.06×10^{-9}. Therefore,

$K_b = K_w/K_a = 1.00 \times 10^{-14}/2.06 \times 10^{-9} = 4.85 \times 10^{-6}$

Using the usual concentration summary and approximations we get

$K_b = [HOBr][OH^-]/[OBr^-]$

$$4.85 \times 10^{-6} = \frac{x^2}{0.500} \text{ and } x = [OH^-] = 1.56 \times 10^{-3} \text{ M}$$

$pOH = -\log [OH^-] = -\log (1.56 \times 10^{-3}) = 2.807$ and $pH = 14.00 - pOH - 14.00 - 2.807 = 11.19$

17.47 Only the first hydrolysis step is important in determining the pH of this solution.

Concentration Summary:

	$PO_4^{3-}(aq) + H_2O(l) \rightleftharpoons$	$HPO_4^{2-}(aq) +$	$OH^-(aq)$
Equation:			
Initial concentrations, mol/L	0.50	0	0
Concentration changes, mol/L	$-x$	$+x$	$+x$
Equilibrium concentrations, mol/L	$0.50 - x$	x	x

K_b for the reaction is

$$K_b = \frac{K_w}{K_{a_3}} = \frac{1.00 \times 10^{-14}}{4.5 \times 10^{-13}} = 2.2 \times 10^{-2}$$

Since the K_b value is not small, we must perform an exact solution. Substituting the equilibrium concentrations and K_b value into the K_b expression and using the quadratic formula gives

$$K_b = \frac{[HPO_4^{2-}][OH^-]}{[PO_4^{3-}]}$$

$$2.2 \times 10^{-2} = \frac{x^2}{0.50 - x}; \quad x^2 + 2.2 \times 10^{-2}x - 1.1 \times 10^{-2} = 0$$

$x = [OH^-] = 9.4 \times 10^{-2}$ M

$pOH = -\log [OH^-] = -\log (9.4 \times 10^{-2}) = 1.03$ and $pH = 14.00 - pOH = 14.00 - 1.03 = 12.97$

17.49 $CH_3NH_2(aq) + H_2O(l) \rightleftharpoons CH_3NH_3^+(aq) + OH^-(aq)$

(a) Decreases. The addition of $CH_3NH_3^+$ drives the reaction to the left.
(b) Increases. Since the $[OH^-]$ decreases, the $[H^+]$ must increase to satisfy the K_w expression.
(c) Increases. The addition of $CH_3NH_3^+$ drives the reaction to the left.
(d) Decreases. The $[H^+]$ increases, so the pH decreases.
(e) Increases. The $[OH^-]$ decreases, so the pOH increases.
(f) Remains unchanged.

17.51 The sodium formate concentration is

$$\frac{10.0 \text{ g HCOONa}}{1.00 \text{ L}} \times \frac{1 \text{ mol HCOONa}}{68.01 \text{ g HCOONa}} = \frac{0.147 \text{ mol HCOONa}}{1.00 \text{ L}} = 0.147 \text{ M HCOONa}$$

Concentration Summary:

Equation:	$HCOOH(aq) \rightleftharpoons$	$HCOO^-(aq) +$	$H^+(aq)$
Initial concentrations, mol/L	0.0800	0.147	0
Concentration changes, mol/L	$-x$	$+x$	$+x$
Equilibrium concentrations, mol/L	$0.0800 - x$	$0.147 + x$	x
Approximate concentrations, mol/L	0.0800	0.147	x

Substituting the approximate concentrations and $K_a = 1.9 \times 10^{-4}$ (Table 17.1) into the K_a expression gives

$K_a = [HCOO^-][H^+]/[HCOOH]$

$$1.9 \times 10^{-4} = \frac{(0.147)(x)}{0.0800}$$

$x = [H^+] = 1.0 \times 10^{-4}$ M and $pH = -\log [H^+] = -\log (1.0 \times 10^{-4}) = 4.00$

The pH without the 10.0 g of HCOONa is 2.41. The added HCOONa represses the ionization of HCOOH, lowering the $[H^+]$ and, consequently, increasing the pH of the solution.

17.53 The solution contains CO_3^{2-}, which is basic:

$$CO_3^{2-}(aq) + H_2O(l) \rightleftharpoons HCO_3^-(aq) + OH^-(aq)$$

$$\frac{[HCO_3^-][OH^-]}{[CO_3^{2-}]} = K_b = \frac{K_W}{K_{a_2}} = \frac{1.00 \times 10^{-14}}{5.61 \times 10^{-11}} = 1.78 \times 10^{-4}$$

The solution also contains amphoteric HCO_3^-:

$$HCO_3^-(aq) + H_2O(l) \rightleftharpoons H_2CO_3(aq) + OH^-(aq)$$

$$\frac{[H_2CO_3][OH^-]}{[HCO_3^-]} = K_b = \frac{K_W}{K_{a_1}} = \frac{1.00 \times 10^{-14}}{4.30 \times 10^{-7}} = 2.33 \times 10^{-8}$$

$$HCO_3^-(aq) \rightleftharpoons H^+(aq) + CO_3^{2-}(aq)$$

$$\frac{[H^+][CO_3^{2-}]}{[HCO_3^-]} = K_{a_2} = 5.61 \times 10^{-11}$$

Since the K_b constant for CO_3^{2-} is considerably greater than the HCO_3^- K_b and HCO_3^- K_{a_2} values, we can confine our attention to the first hydrolysis step of CO_3^{2-}.

Concentration Summary:

Equation:	$CO_3^{2-}(aq) + H_2O(l) \rightleftharpoons$	$HCO_3^-(aq) +$	$OH^-(aq)$
Initial concentrations, mol/L	0.050	0.050	0
Concentration changes, mol/L	$-x$	$+x$	$+x$
Equilibrium concentrations, mol/L	$0.050 - x$	$0.050 + x$	x
Approximate concentrations, mol/L	0.050	0.050	x

Substituting the approximate concentrations and $K_b = 1.78 \times 10^{-4}$ into the K_b expression gives

$$1.78 \times 10^{-4} = \frac{(0.050)(x)}{(0.050)}$$

$x = 1.78 \times 10^{-4}$ M = [OH⁻]

pOH = –log [OH⁻] = –log (1.78×10^{-4}) = 3.75 and pH = 14.00 – pOH = 14.00 – 3.75 = 10.25

Comment: This problem can also be considered as a HCO_3^- – CO_3^{2-} buffer, and the pH calculated using Equation 17.4. The pK_a for this buffer system is 10.25 (Table 17.1).

$$pH = pK_a + \log \frac{[CO_3^{2-}]}{[HCO_3^-]} = 10.25 + \log \frac{0.050}{0.050} = 10.25 + \log 1 = 10.25$$

Notice that the result is the same. This shows that buffers are really a special case of the common ion effect.

17.55 (a) The K_a's of ascorbic acid are: $K_{a_1} = 7.9 \times 10^{-5}$, $K_{a_2} = 1.6 \times 10^{-12}$. Since K_{a_1} is considerably greater than K_{a_2}, the concentration of H^+ ion can be calculated from the first ionization step alone.

Concentration Summary:

Equation:	$H_2C_6H_6O_6(aq)$ ⇌	$H^+(aq)$ +	$HC_6H_6O_6^-(aq)$
Initial concentrations, mol/L	0.10	0	0
Concentration changes, mol/L	$-x$	$+x$	$+x$
Equilibrium concentrations, mol/L	$0.10 - x$	x	x
Approximate concentrations, mol/L	0.10	x	x

Substituting the approximate concentrations and K_{a_1} value in the K_{a_1} expression gives

$K_{a_1} = [H^+][HC_6H_6O_6^-]/[H_2C_6H_6O_6]$

$$7.9 \times 10^{-5} = \frac{(x)(x)}{0.10} \quad \text{and } x = [H^+] = 2.8 \times 10^{-3}\text{ M}$$

Since ascorbic acid is a weak diprotic acid, the concentration of ascorbate ion is equal to K_{a_2}, or $[C_6H_6O_6^{2-}] = K_{a_2} = 1.6 \times 10^{-12}$ M.

(b) Since HCl is a strong acid, virtually all of the H^+ ion in solution comes from the HCl. Therefore, the $[H^+] = 0.010$ M. Remember, the H^+ from the HCl ionization represses the ascorbic acid ionization.

The concentration of ascorbate ion can be best or most conveniently calculated using the expression that results when the K_{a_1} expression is multiplied by the K_{a_2} expression, since then all the concentrations except that of the ascorbate ion are known. The calculation proceeds as follows:

The K_a expressions for the two ionization steps are:

$$\frac{[H^+][HC_6H_6O_6^-]}{[H_2C_6H_6O_6]} = K_{a_1} = 7.9 \times 10^{-5}$$

$$\frac{[H^+][C_6H_6O_6^{2-}]}{[HC_6H_6O_6^-]} = K_{a_2} = 1.6 \times 10^{-12}$$

The $[HC_6H_6O_6^-]$ is unknown and not wanted, and can be eliminated by multiplying the two K_a expressions together:

$$\frac{[H^+]^2[C_6H_6O_6^{2-}]}{[H_2C_6H_6O_6]} = K_{a_1} \times K_{a_2} = (7.9 \times 10^{-5})(1.6 \times 10^{-12}) = 1.3 \times 10^{-16}$$

Solving for $[C_6H_6O_6^{2-}]$ and substituting $[H_2C_6H_6O_6] = 0.10$ M and $[H^+] = 0.010$ M into the resulting expression gives

$$[C_6H_6O_6^{2-}] = \frac{(1.3\times10^{-16})[H_2C_6H_6O_6]}{[H^+]^2} = \frac{(1.3\times10^{-16})(0.10)}{(0.010)^2} = 1.3\times10^{-13}\text{ mol/L}$$

Comment: Notice that $[C_6H_6O_6^{2-}]$ is not equal to K_{a_2} in this case. That result applies only when the weak diprotic acid is the only acid in solution. Here, the solution also contains the strong acid HCl.

17.57 A buffer system is a common ion system that maintains a steady $H^+(aq)$ concentration by shifting the equilibrium between a conjugate acid-base pair.

(a) Added hydrogen ions will be consumed by formate ions.

$H^+(aq) + HCOO^-(aq) \rightarrow HCOOH(aq)$

Added hydroxide ions will be consumed by molecular formic acid.

$HCOOH(aq) + OH^-(aq) \rightarrow HCOO^-(aq) + H_2O(l)$

(b) Added hydrogen ions will be consumed by CO_3^{2-}.

$CO_3^{2-}(aq) + H^+(aq) \rightarrow HCO_3^-(aq)$

Added hydroxide ions will be consumed by HCO_3^-.

$HCO_3^-(aq) + OH^-(aq) \rightarrow CO_3^{2-}(aq) + H_2O(l)$

17.59 $HA(aq) \rightleftharpoons H^+(aq) + A^-(aq)$

$K_a = [H^+][A^-]/[HA]$

$$[H^+] = K_a \times \frac{[HA]}{[A^-]}$$

Taking the negative logarithm of both sides gives

$$-\log[H^+] = -\log K_a - \log\frac{[HA]}{[A^-]} \quad\text{and}\quad pH = pK_a + \log\frac{[A^-]}{[HA]}$$

17.61 The buffer ratio is the concentration ratio, [base]/[acid], in the Henderson-Hasselbalch equation.

(a) No change occurs. (b) No change occurs.
Dilution changes the concentrations of the acid and its conjugate base by the same factor; therefore, there is no change in the buffer ratio or pH of the solution.

17.63 (a) $[H^+]$ should be equal to K_a.
(b) The pH should be equal to pK_a.
(c) $[OH^-]$ should be equal to K_b.
(d) The pOH should be equal to pK_b.
(e) The concentrations of conjugate acid and conjugate base should be equal.

17.65 Look in Table 17.1 for acids with pK_a values within one unit of the desired pH and in Table 17.2 for bases with pK_b values within one unit of 14.00 – pH. The possible candidates are: CH_3COOH with $pK_a = 4.754$, $H_2C_6H_6O_6$ with $pK_{a_1} = 4.10$, $HC_4H_3N_2O_3$ with $pK_a = 4.01$, C_6H_5COOH with $pK_a = 4.190$, $HC_6H_4NO_2$ with $pK_a = 4.85$, $H_2C_2O_4$ with $pK_{a_2} = 4.194$, and $C_6H_5NH_2$ with $pK_b = 9.37$. Utilizing acetic acid (CH_3COOH), the buffer can be prepared as follows: Dissolve equal numbers of moles of acetic acid (CH_3COOH) and sodium acetate (CH_3COONa) in a beaker. This will give us a buffer with pH = 4.754. To adjust the buffer to pH 4.50, add additional acetic acid or HCl until the pH meter reads 4.50. Transfer the solution to a 500-mL volumetric flask and add water up to mark.

17.67 (a) Since the buffer ratio is unity, the pH equals the pK_a of barbituric acid; pH = pK_a = 4.01 (Table 17.1).

(b) The buffer ratio is 1/2. Substituting 1/2 for the buffer ratio and pK_a = 4.01 into Equation 17.4 gives

$$pH = pK_a + \log \frac{[C_4H_3N_2O_3^-]}{[HC_4H_3N_2O_3]} = 4.01 + \log \tfrac{1}{2} = 3.71$$

17.69 The pK_a for this ammonia-ammonium chloride buffer is 9.248 (Example 17.13).

$$pH = pK_a + \log \frac{[NH_3]}{[NH_4^+]}; \quad 8.50 = 9.248 + \log \frac{[NH_3]}{[NH_4^+]}$$

$$\log \frac{[NH_3]}{[NH_4^+]} = 8.50 - 9.248 = -0.75$$

and $$\frac{[NH_3]}{[NH_4^+]} = \text{antilog}(-0.75) = 10^{-0.75} = 0.18$$

The concentration of NH_3 is 0.200 M. Hence, the concentration of NH_4^+ must be

$$[NH_4^+] = \frac{[NH_3]}{0.18} = \frac{0.200\text{ M}}{0.18} = 1.1\text{ M}$$

The number of grams of NH_4Cl that must be added to 1.00 L of the solution is

$$1.00\text{ L} \times \frac{1.1\text{ mol}}{1\text{ L}} \times \frac{53.49\text{ g } NH_4Cl}{1\text{ mol}} = 58.8\text{ g } NH_4Cl$$

17.71 (a) The pK_a for this formic acid–sodium formate buffer is 3.72 (Table 17.1).

$$pH = pK_a + \log \frac{[HCOO^-]}{[HCOOH]}$$

$$3.52 = 3.72 + \log \frac{[HCOO^-]}{[HCOOH]}$$

$$\log \frac{[HCOO^-]}{[HCOOH]} = 3.52 - 3.72 = -0.20$$

and $$\frac{[HCOO^-]}{[HCOOH]} = \text{antilog}(-0.20) = 10^{-0.20} = 0.63$$

The concentration of HCOOH is 0.100 M. Hence, the concentration of $HCOO^-$ must be

$[HCOO^-] = 0.63 \times [HCOOH] = 0.63 \times 0.100\text{ M} = 0.063\text{ M}$

The number of grams of HCOONa that must be added to 1.00 L of the solution is

$$1.00\text{ L} \times \frac{0.63\text{ mol}}{1\text{ L}} \times \frac{68.01\text{ g HCOONa}}{1\text{ mol}} = 4.3\text{ g HCOONa}$$

(b) pH = 3.52. Dilution doesn't change the buffer ratio or the pH of the solution.

17.73 The number of moles of HCl added to each solution is

$$1.00\text{ mL} \times \frac{1\text{ L}}{1000\text{ mL}} \times \frac{10.0\text{ mol HCl}}{1\text{ L}} = 0.0100\text{ mol HCl}$$

(a) The pH of the original HCl solution is

$pH = -\log [H^+] = -\log (1.76 \times 10^{-5}) = 4.754$

The HCl ionizes completely. The new H^+ concentration is

$(0.0100 \text{ mol} + 0.100 \text{ L} \times 1.76 \times 10^{-5} \text{ mol/L})/0.101 \text{ L} = 0.0990 \text{ M}$

The new pH is $pH = -\log [H^+] = -\log 0.0990 = 1.004$

The change in pH is from 4.754 to 1.004, a decrease of 3.750 units.

(b) Since the buffer ratio is initially unity, the pH of the original acetic acid–sodium acetate buffer is equal to pK_a, or $pH = pK_a = 4.754$ (Table 17.1). The added HCl reacts completely with the basic component of the buffer; 0.0100 mol HCl will convert 0.0100 mol CH_3COO^- to 0.0100 mol CH_3COOH:

$CH_3COO^-(aq) + H^+(aq) \rightarrow CH_3COOH(aq)$

The new mole quantities are:

CH_3COO^-: 0.100 mol – 0.0100 mol = 0.090 mol
CH_3COOH: 0.100 mol + 0.0100 mol = 0.110 mol

The new pH calculated from Equation 17.4 is

$$pH = pK_a + \log \frac{[CH_3COO^-]}{[CH_3COOH]} = 4.754 + \log \frac{0.090}{0.110} = 4.67$$

The change in pH is from 4.754 to 4.67, a decrease of 0.08 units.

17.75 (a) The pK_a for this ammonia–ammonium chloride buffer is 9.248 (Example 17.13).

$$pH = pK_a + \log \frac{[NH_3]}{[NH_4^+]} = 9.248 + \log \frac{0.10}{0.25} = 8.85$$

(b) The number of moles of HCl added to the solution is (P = 1.00 atm, V = 0.500 L, and T = 298 K)

$$n = \frac{PV}{RT} = \frac{1.00 \text{ atm} \times 0.500 \text{ L}}{0.0821 \text{ L•atm/mol•K} \times 298 \text{ K}} = 0.0204 \text{ mol}$$

The added HCl reacts completely with the basic component of the buffer; 0.0204 mol HCl will convert 0.0204 mol NH_3 to 0.0204 mol NH_4^+:

$NH_3(aq) + H^+(aq) \rightarrow NH_4^+(aq)$

The new mole quantities are:

NH_3: 0.10 mol – 0.0204 mol = 0.08 mol
NH_4^+: 0.25 + 0.0204 mol = 0.27 mol

The new pH is calculated from Equation 17.4:

$$pH = pK_a + \log \frac{[NH_3]}{[NH_4^+]} = 9.248 + \log \frac{0.08}{0.27} = 8.72$$

17.77 (a) The equivalence point is the point in a titration at which neither reactant is in excess. The endpoint is an experimentally determined point, which may be close to the equivalence point.
(b) No. For instance, the poor choice of an indicator or experimental error will cause the two points to differ.

17.79 Because only a few drops of indicator are required to produce color changes visible to the eye. Also, an indicator is a weak acid or base in itself. If too much is present, it will then be in direct competition with the main reactants, and a titration error may occur; i.e., the endpoint and the equivalence point may differ significantly, or a false endpoint may be recorded.

17.81 The pH of this solution at equivalence will be greater than 7, because CN^- is a weak base:

$$CN^-(aq) + H_2O(l) \rightleftharpoons HCN(aq) + OH^-(aq)$$

The number of moles of HCN at the beginning of the titration is

$$0.0250 \text{ L} \times \frac{0.150 \text{ mol}}{1 \text{ L}} = 3.75 \times 10^{-3} \text{ mol}$$

From the neutralization equation

$$HCN(aq) + NaOH(aq) \rightarrow NaCN(aq) + H_2O(l)$$

we see that 3.75×10^{-3} mol of HCN will react with 3.75×10^{-3} mol of NaOH to form 3.75×10^{-3} mol of CN^-. The volume of 0.100 M NaOH required to reach equivalence is

$$3.75 \times 10^{-3} \text{ mol NaOH} \times \frac{1 \text{ L}}{0.100 \text{ mol NaOH}} = 0.0375 \text{ L} = 37.5 \text{ mL}$$

The total volume of the solution at the equivalence point is

$$25.0 \text{ mL} + 37.5 \text{ mL} = 62.5 \text{ mL} = 0.0625 \text{ L}$$

The molarity of CN^- at equivalence is initially

$$\frac{3.75 \times 10^{-3} \text{ mol}}{0.0625 \text{ L}} = 0.0600 \text{ M}$$

but some of this reacts with water to form hydroxide ions. Using a concentration summary similar to the one in Exercise 17.43 and substituting the approximate concentrations and $K_b = 2.03 \times 10^{-5}$ (Exercise 17.43) into the K_b expression gives

$$K_b = [HCN][OH^-]/[CN^-]$$

$$2.03 \times 10^{-5} = \frac{x^2}{0.0600}$$

$$x = [OH^-] = 1.10 \times 10^{-3} \text{ M}$$

$$pOH = -\log [OH^-] = -\log (1.10 \times 10^{-3}) = 2.959 \text{ and } pH = 14.00 - pOH = 14.00 - 2.959 = 11.04$$

17.83 The pH's at the various points are:

(a) Before acid is added:

$$pOH = -\log [OH^-] = -\log 0.200 = 0.699 \text{ and } pH = 14.00 - pOH = 14.00 - 0.699 = 13.30$$

(b) At the equivalence point: All of the NaOH is neutralized. The solution contains NaCl, a neutral salt, so its pH is 7.

(c) At the half-neutralization point: The pH depends on the concentration of unreacted NaOH remaining in the solution.

initial moles of NaOH = 0.0500 L × 0.200 mol/L = 0.0100 mol

moles of NaOH neutralized = moles of HCl added = 0.0100 mol/2 = 0.0050 mol

$$\text{volume of HCl added} = 0.0050 \text{ mol} \times \frac{1 \text{ L}}{0.100 \text{ mol}} = 0.050 \text{ L}$$

concentration of NaOH = moles of NaOH remaining/final solution volume

$= 0.0050 \text{ mol}/(0.0500 \text{ L} + 0.050 \text{ L}) = 0.050 \text{ M}$

$\text{pOH} = -\log [\text{OH}^-] = -\log 0.050 = 1.30$ and $\text{pH} = 14.00 - \text{pOH} = 14.00 - 1.30 = 12.70$

(d) At some point before equivalence: The volume of HCl needed for complete neutralization is

$$0.0100 \text{ mol} \times \frac{1 \text{ L}}{0.100 \text{ mol}} = 0.100 \text{ L} = 100 \text{ mL}$$

Therefore, we have to calculate the concentration of unreacted NaOH remaining in the solution after 99 mL of HCl has been added.

moles of NaOH neutralized = moles of HCl added = $0.099 \text{ L} \times 0.100 \text{ mol/L} = 0.0099 \text{ mol}$

moles of NaOH remaining = initial moles – moles neutralized = $0.0100 - 0.0099 = 0.0001 \text{ mol}$

concentration of NaOH = moles of NaOH remaining/final solution volume

$= 0.0001 \text{ mol}/(0.0500 \text{ L} + 0.099 \text{ L}) = 7 \times 10^{-4} \text{ M}$

$\text{pOH} = -\log [\text{OH}^-] = -\log (7 \times 10^{-4}) = 3.2$ and $\text{pH} = 14.00 - \text{pOH} = 14.00 - 3.2 = 10.8$

(e) After the equivalence point: The NaOH is gone and only the excess HCl determines the pH. 101 mL of HCl is added, and the pH is calculated as follows:

moles of excess HCl = volume of excess HCl × molarity = $0.001 \text{ L} \times 0.100 \text{ mol/L} = 1 \times 10^{-4} \text{ mol}$

concentration of excess HCl = moles of excess HCl/final solution volume

$= 1 \times 10^{-4} \text{ mol}/(0.0500 \text{ L} + 0.101 \text{ L}) = 7 \times 10^{-4} \text{ M}$ and $\text{pH} = -\log [\text{H}^+] = -\log (7 \times 10^{-4}) = 3.2$

Make a sketch of the pH versus milliliters added HCl using these pH values.

17.85 (a) The pH at the equivalence point is 7. Since this is a strong acid–strong base titration, any of the following indicators whose transition intervals fall between pH 4.0 and pH 10 are acceptable: methyl red, litmus, bromocresol purple, bromthymol blue, phenol red, thymol blue (basic range), and phenolphthalein.

(b) The pH at the equivalence point is 8.51. Since this is a weak acid–strong base titration, the pH rise in the steeply rising portion of the titration curve is smaller than that for a strong acid–strong base titration. The following indicators that change about pH 8.51 are acceptable: phenol red, thymol blue (basic range), and phenolphthalein.

17.87 (a) This is a strong acid–strong base titration. Therefore, any indicator from methyl red down to phenolphthalein in Table 17.3 is suitable.

(b) At the equivalence point we have a 0.050 M sodium benzoate solution. The pH of the weak base $C_6H_5COO^-$ is calculated in the usual manner (see Exercise 17.42). The pH of the solution is 8.44. Table 17.3 shows that phenolphthalein (8.2-10.0) is acceptable.

(c) At the equivalence point we have a 0.333 M NH_4Cl solution. The pH of the weak acid NH_4^+ is calculated in the usual manner (see Exercise 17.41). The pH of the solution is 4.86. Table 17.3 shows that methyl red (4.4-6.2) is acceptable.

17.89 (a) Use of a concentration summary similar to the one in Exercise 17.16 and substitution of the approximate concentrations and $K_a = 6.46 \times 10^{-5}$ (Table 17.1) into the K_a expression gives

1. 0.50 M: $K_a = [H^+][C_6H_5COO^-]/[C_6H_5COOH]$

$$6.46 \times 10^{-5} = \frac{x^2}{0.50} \quad \text{and} \quad x = [H^+] = 5.7 \times 10^{-3}\ M$$

The number of moles of C_6H_5COOH ionized per liter is equal to x, so the percent ionization is

$$\%\ \text{ionization} = 100\% \times \frac{5.7 \times 10^{-3}\ M}{0.50\ M} = 1.1\%$$

2. 0.10 M: $6.46 \times 10^{-5} = \dfrac{x^2}{0.10}$

$$x = 2.5 \times 10^{-3}\ M = [H^+] \quad \text{and} \quad \%\ \text{ionization} = 100\% \times \frac{2.5 \times 10^{-3}\ M}{0.10\ M} = 2.5\%$$

3. 0.050 M: $6.46 \times 10^{-5} = \dfrac{x^2}{0.050}$

$$x = 1.8 \times 10^{-3}\ M = [H^+] \quad \text{and} \quad \%\ \text{ionization} = 100\% \times \frac{1.8 \times 10^{-3}\ M}{0.050\ M} = 3.6\%$$

4. 0.010 M: $6.46 \times 10^{-5} = \dfrac{x^2}{0.010}$ and $x = 8.0 \times 10^{-4}\ M = [H^+]$

Since x is greater than 5% of 0.010, the approximation is not valid. The correct value of x is obtained by substituting the equilibrium concentrations and K_a value into the K_a expression and using the quadratic formula.

$$6.46 \times 10^{-5} = \frac{x^2}{0.010 - x}$$

$$x^2 + 6.46 \times 10^{-5}x - 6.46 \times 10^{-7} = 0$$

$$x = 7.7 \times 10^{-4}\ M = [H^+]$$

$$\text{and}\ \%\ \text{ionization} = 100\% \times \frac{7.7 \times 10^{-4}\ M}{0.010\ M} = 7.7\%$$

(b) See Figure 17.1 for a similar plot of the percent ionization of $NH_3(aq)$. The general trend is for the % ionization → 100% as concentration → zero.

17.91

$HA + B \rightleftharpoons HB^+ + A^-$; $K_{eq} = [HB^+][A^-]/[HA][B]$

$HA + H_2O \rightleftharpoons H_3O^+ + A^-$; $K_a = [H_3O^+][A^-]/[HA]$

$B + H_2O \rightleftharpoons HB^+ + OH^-$; $K_b = [HB^+][OH^-]/[B]$

$$\frac{K_a \times K_b}{K_w} = \frac{\dfrac{[H_3O^+][A^-][HB^+][OH^-]}{[HA][B]}}{[H_3O^+][OH^-]} = \frac{[HB^+][A^-]}{[HA][B]} = K_{eq}$$

17.93

$HNO_2(aq) \rightleftharpoons H^+(aq) + NO_2^-(aq)$

(a) Increases. The $[H^+]$ decreases, so the pH increases.

(b) Increases. Since the $[H^+]$ decreases, the $[OH^-]$ must increase to satisfy the K_w expression.

(c) Decreases. The addition of NO_2^- drives the reaction to the left.

(d) Remains unchanged.

17.95 This problem can be treated as a $Fe(H_2O)_6^{3+} - Fe(H_2O)_5(OH)^{2+}$ buffer with $pK_a = 2.10$, and solved using Equation 17.4.

$$pH = pK_a + \log \frac{[Fe(H_2O)_5(OH)^{2+}]}{[Fe(H_2O)_6^{3+}]}$$

$$\log \frac{[Fe(H_2O)_5(OH)^{2+}]}{[Fe(H_2O)_6^{3+}]} = pH - pK_a = 3.00 - 2.10 = 0.90$$

and $$\frac{[Fe(H_2O)_5(OH)^{2+}]}{[Fe(H_2O)_6^{3+}]} = \text{antilog } 0.90 = 10^{+0.90} = 7.9$$

Since this ratio holds for any initial amount of $Fe(H_2O)_6^{3+}$, let's choose the initial amount as 1 mol, and let x equal the number of moles of $Fe(H_2O)_5(OH)^{2+}$ formed. Then

$$\frac{x}{1-x} = 7.9 \text{ and } x = 7.9/8.9 = 0.89 \text{ mol } Fe(H_2O)_5(OH)^{2+}$$

Answer: The fraction of ions in the form of $Fe(H_2O)_5(OH)^{2+}$ is 0.89.

17.97 $F^-(aq) + H_2O(l) \rightleftharpoons HF(aq) + OH^-(aq)$

Using the usual concentration summary and approximations ($K_b = 1.00 \times 10^{-14}/3.53 \times 10^{-4} = 2.83 \times 10^{-11}$), we get

$K_b = [HF][OH^-]/[F^-]$

$$2.83 \times 10^{-11} = \frac{x^2}{0.250} \text{ and } x = [OH^-] = 2.66 \times 10^{-6} \text{ M}$$

$pOH = -\log [OH] = -\log (2.66 \times 10^{-6}) = 5.575$ and $pH = 14.00 - pOH = 14.00 - 5.575 = 8.42$

17.99 The neutralization reaction is

$NaHCO_3(aq) + HCl(aq) \rightarrow NaCl(aq) + H_2O(l) + CO_2(g)$

The initial number of moles of HCl is

$[H^+] = \text{antilog } (-0.80) = 10^{-0.80} = 0.16 \text{ mol/L}$

$$0.200 \text{ L} \times \frac{0.16 \text{ mol HCl}}{1 \text{ L}} = 0.032 \text{ mol HCl}$$

The final number of moles of HCl is

$[H^+] = \text{antilog } (-1.20) = 10^{-1.20} = 6.3 \times 10^{-2} \text{ M}$

$$0.400 \text{ L} \times \frac{6.3 \times 10^{-2} \text{ mol HCl}}{1 \text{ L}} = 0.025 \text{ mol HCl}$$

The number of moles of HCl that has to be neutralized is 0.032 – 0.025 = 0.007 mol. The grams of $NaHCO_3$ required is

$$0.007 \text{ mol HCl} \times \frac{1 \text{ mol } NaHCO_3}{1 \text{ mol HCl}} \times \frac{84.01 \text{ g } NaHCO_3}{1 \text{ mol } NaHCO_3} = 0.6 \text{ g } NaHCO_3$$

17.101 $H_2C_8H_4O_4(aq) \rightleftharpoons H^+(aq) + HC_8H_4O_4^-(aq) \quad pK_a = 2.89$

(a) $pH = pK_a + \log ([HC_8H_4O_4^-]/[H_2C_8H_4O_4])$

$2.75 = 2.89 + \log ([HC_8H_4O_4^-]/0.200)$

$\log ([HC_8H_4O_4^-]/0.200) = 2.75 - 2.89 = -0.14$

$[HC_8H_4O_4^-]/0.200 = \text{antilog} (-0.14) = 10^{-0.14} = 0.72$

and $[HC_8H_4O_4^-] = 0.200 \times 0.72 = 0.14$ M

(b) $H_2C_8H_4O_4(aq) + NaOH(s) \rightarrow NaHC_8H_4O_4(aq) + H_2O(l)$

Originally, there is 0.200 L × 0.200 mol/L = 0.0400 mol of $H_2C_8H_4O_4$ and 0.200 L × 0.14 mol/L = 0.028 mol of $HC_8H_4O_4^-$. After the addition of 0.0100 mol of solid NaOH, there is 0.0400 mol – 0.0100 mol = 0.0300 mol of $H_2C_8H_4O_4$ and 0.028 mol + 0.0100 mol = 0.038 mol of $HC_8H_4O_4^-$.

$$pH = 2.89 + \log \frac{\text{mol } HC_8H_4O_4^-}{\text{mol } H_2C_8H_4O_4} = 2.89 + \log \frac{0.038}{0.0300} = 2.99$$

17.103 To solve the problem we must first find the buffer ratio at pH = 5.6.

$$pH = pK_a + \log \frac{[HPO_4^{2-}]}{[H_2PO_4^-]}$$

$$\log \frac{[HPO_4^{2-}]}{[H_2PO_4^-]} = pH - pK_a = 5.6 - 7.21 = -1.6$$

and $\dfrac{[HPO_4^{2-}]}{[H_2PO_4^-]} = \text{antilog} (-1.6) = 10^{-1.6} = 0.03$

$[HPO_4^{2-}] = 0.03 \times [H_2PO_4^-]$

$[H_2PO_4^-] + [HPO_4^{2-}] = 0.0466$ M

$[H_2PO_4^-] + 0.03[H_2PO_4^-] = 0.0466$ M and $[H_2PO_4^-] = \dfrac{0.0466 \text{ M}}{1.03} = 0.0452$ M

$[HPO_4^{2-}] = 0.0466 \text{ M} - 0.0452 \text{ M} = 0.0014$ M

17.105 The $C_8H_{11}N_2O_3^-$ is produced by adding NaOH to the $HC_8H_{11}N_2O_3$.

$HC_8H_{11}N_2O_3(aq) + NaOH(aq) \rightarrow NaC_8H_{11}N_2O_3(aq) + H_2O(l)$

Let x be the moles of NaOH added. Then x also equals the moles of $C_8H_{11}N_2O_3^-$ produced, and $1.00 - x$ = moles of $HC_8H_{11}N_2O_3$ remaining. The pK_a for this buffer system is 7.43 (Table 17.1). Then

$$pH = pK_a + \log \frac{[C_8H_{11}N_2O_3^-]}{[HC_8H_{11}N_2O_3]} = pK_a + \log \frac{\text{mol } C_8H_{11}N_2O_3^-}{\text{mol } HC_8H_{11}N_2O_3}$$

$$8.00 = 7.43 + \log \frac{x}{1.00 - x}$$

$$\log \frac{x}{1.00 - x} = 8.00 - 7.43 = 0.57$$

$$\frac{x}{1.00 - x} = \text{antilog } 0.57 = 10^{+0.57} = 3.7$$

and $x = 3.7/4.7 = 0.79$ mol $C_8H_{11}N_2O_3^- = 0.79$ mol NaOH

0.79 mol NaOH must be added to 1.00 L of solution.

17.107 The $(CH_2OH)_3CNH_2$ is produced by adding KOH to the $(CH_2OH)_3CNH_3^+$.

$(CH_2OH)_3CNH_3^+(aq) + OH^-(aq) \rightarrow (CH_2OH)_3CNH_2(aq) + H_2O(l)$

Let x be the moles of KOH added. Then x also equals the moles of $(CH_2OH)_3CNH_2$ produced, and $0.100 - x$ = moles of $(CH_2OH)_3CNH_3^+$ remaining. Then

$$pH = pK_a + \log \frac{[(CH_2OH)_3CNH_2]}{[(CH_2OH)_3CNH_3^+]} = pK_a + \log \frac{\text{mol }(CH_2OH)_3CNH_2}{\text{mol }(CH_2OH)_3CNH_3^+}$$

$$7.80 = 8.08 + \log \frac{x}{0.100 - x}$$

$$\log \frac{x}{0.100 - x} = 7.80 - 8.08 = -0.28$$

$$\frac{x}{0.100 - x} = \text{antilog}(-0.28) = 10^{-0.28} = 0.52$$

and $x = 0.052/1.52 = 0.034$ mol $(CH_2OH)_3CNH_2 = 0.034$ mol KOH

0.034 mol KOH must be added to 1.00 L of solution.

17.109 (a) When a weak acid is titrated with a strong base producing a basic salt.
(b) When a strong acid is titrated with a strong base producing a neutral salt.
(c) When a weak base is titrated with a strong acid producing an acidic salt.

17.111 The calculation proceeds as follows: V × M (of NaOH) → mol NaOH → mol veronal → concentration veronal.

$$0.02345 \text{ L} \times \frac{0.0700 \text{ mol NaOH}}{1 \text{ L}} \times \frac{1 \text{ mol veronal}}{1 \text{ mol NaOH}} = 1.64 \times 10^{-3} \text{ mol veronal}$$

The original concentration of veronal was

$$[\text{veronal}] = \frac{1.64 \times 10^{-3} \text{mol}}{0.02500 \text{ L}} = 6.56 \times 10^{-2} \text{ M}$$

17.113 (a) $P_{CO_2} = X_{CO_2} P_T = 0.00030 \times 1 \text{ atm} = 3.0 \times 10^{-4} \text{ atm}$

The partial pressure of CO_2 decreases from 1 atm to 3.0×10^{-4} atm. Henry's law states that the solubility will decrease by the same factor as the partial pressure. Hence, the $[CO_2]$ will be

$$0.033 \text{ M} \times \frac{3.0 \times 10^{-4} \text{atm}}{1 \text{ atm}} = 9.9 \times 10^{-6} \text{ M}$$

(b) The $[HCO_3^-]$ can be calculated from the first step alone. Using a concentration summary like the one used in Exercise 17.29 and substituting the equilibrium concentrations and $K_{a_1} = 4.30 \times 10^{-7}$ into the K_{a_1} expression gives

$$4.30 \times 10^{-7} = \frac{x^2}{9.9 \times 10^{-6} - x}$$

$x^2 + 4.30 \times 10^{-7}x - 4.257 \times 10^{-12} = 0$ and $x = [HCO_3^-] = 1.9 \times 10^{-6}$ M

The $[CO_3^{2-}]$ is approximately equal to K_{a_2}. Therefore,

$[CO_3^{2-}] = K_{a_2} = 5.6 \times 10^{-11}$ M

A slightly more accurate calculation would take into account the $[H^+]$ coming from the autoionization of water.

17.115 The half-neutralization of a diprotic acid (H_2A) produces an amphoteric anion (HA^-). HA^-, being an amphoteric anion, acts as both an acid and base in aqueous solution. The two reactions are:

$HA^-(aq) \rightleftharpoons H^+(aq) + A^{2-}(aq) \quad K_{a_2} = [H^+][A^{2-}]/[HA^-]$

$HA^-(aq) + H_2O(l) \rightleftharpoons H_2A(aq) + OH^-(aq) \quad K_b = K_w/K_{a_1} = [H_2A][OH^-]/[HA^-]$

Since the constants K_{a_2} and K_b generally are not considerably different in magnitude, both reactions have to be considered when calculating the pH of the solution. An expression for $[H^+]$ can be obtained as follows:

Dividing K_{a_2} by K_b gives

$$\frac{K_{a_2}}{K_b} = \frac{[H^+][A^{2-}]}{[H_2A][OH^-]} \qquad (1)$$

From the autoionization of water we have

$H_2O(l) \rightleftharpoons H^+(aq) + OH^-(aq)$

$[H^+][OH^-] = 1.00 \times 10^{-14} = K_w$ and $[OH^-] = K_w/[H^+]$

Substituting this expression for $[OH^-]$ in (1) gives

$$\frac{K_{a_2}}{K_b} = \frac{[H^+]^2[A^{2-}]}{[H_2A]K_w} \qquad (2)$$

Since HA^- is both a very weak acid and a very weak base, it is likely that $[HA^-]$ will not change significantly due to these reactions. A more important reaction in this regard is the reaction of HA^- with itself

$HA^-(aq) + HA^-(aq) \rightleftharpoons A^{2-}(aq) + H_2A(aq)$

which has a much greater equilibrium constant, though $[HA^-]$ still does not change significantly. Focusing our attention on this reaction, we see that equal molar amounts of A^{2-} and H_2A are formed in the reaction, and that to a good approximation we can assume that $[A^{2-}] = [H_2A]$ overall. Using this approximation greatly simplifies Equation (2), since these concentrations now cancel out of the expression. Expressions for $[H^+]$ and the pH are now obtained as follows:

$$\frac{K_{a_2}}{K_b} = \frac{[H^+]^2}{K_w}$$

$$[H^+]^2 = K_{a_2}\frac{K_w}{K_b} = K_{a_2}K_{a_1}$$

$$[H^+] = \sqrt{K_{a_1}K_{a_2}} \qquad (3)$$

$$\text{and pH} = \tfrac{1}{2}(pK_{a_1} + pK_{a_2}) \qquad (4)$$

In general, Equations (3) and (4) are reliable for concentrations of at least 0.010 M.

Comment: Equations (3) and (4) are general equations that can be used for any amphoteric ion. A similar $[H^+]$ expression holds for salts in which both the cation and anion undergo hydrolysis, e.g., NH_4CH_3COO, NH_4CN, and NH_4NO_2. The $[H^+]$ expression for these salts of weak acids and weak bases is

$$[H^+] = \sqrt{K_a \times (K_w/K_b)}$$

Notice that K_w/K_b is equal to the K_a of the conjugate acid of the base. Therefore, this expression and (3) are identical. This result is interesting, since it shows that the pH of an amphoteric ion or salt of a weak acid and a weak base is independent of the concentration, so long as the concentration is not too small.

CHAPTER 18

SOLUBILITY AND COMPLEX ION EQUILIBRIA

Solutions To Practice Exercises

PE 18.1 (a) Calcium oxalate (CaC_2O_4) contains one Ca^{2+} ion for every $C_2O_4^{2-}$ ion. The solubility product constant is

$K_{sp} = [Ca^{2+}][C_2O_4^{2-}]$

(b) Bismuth sulphide (Bi_2S_3) contains two Bi^{3+} ions for every three S^{2-} ions. The solubility product constant is

$K_{sp} = [Bi^{3+}]^2[S^{2-}]^3$

PE 18.2 $PbCl_2$. $PbCl_2$ with a K_{sp} of 1.7×10^{-5} is more soluble than $PbBr_2$ with a K_{sp} of 2.1×10^{-6}.

PE 18.3 The K_{sp} for AgCl is $K_{sp} = [Ag^+][Cl^-]$.

The molar mass of AgCl is 143.3 g/mol; therefore 8.9×10^{-4} g/L is

$$\frac{8.9 \times 10^{-4}\text{ g}}{1\text{ L}} \times \frac{1\text{ mol}}{143.3\text{ g}} = 6.2 \times 10^{-6}\text{ mol/L}$$

One mole of AgCl provides 1 mol of Ag^+ ions and 1 mol of Cl^- ions. The ionic concentrations in 6.2×10^{-6} M AgCl are therefore

$[Ag^+] = 6.2 \times 10^{-6}$ mol/L and $[Cl^-] = 6.2 \times 10^{-6}$ mol/L

Substituting the concentrations into the K_{sp} expression gives

$K_{sp} = [Ag^+][Cl^-] = (6.2 \times 10^{-6})(6.2 \times 10^{-6}) = 3.8 \times 10^{-11}$

PE 18.4 $K_{sp} = [Ag^+]^2[SO_4^{2-}]$

$1.4 \times 10^{-5} = (2x)^2(x) = 4x^3$ and $x = 1.5 \times 10^{-2}$ mol/L $= [SO_4^{2-}]$

The equilibrium concentration of SO_4^{2-}, 1.5×10^{-2} mol/L, is also the molar solubility of Ag_2SO_4. The molar mass of Ag_2SO_4 is 311.8 g/mol; hence, its solubility in grams per liter is

$$\frac{1.5 \times 10^{-2}\text{ mol}}{1\text{ L}} \times \frac{311.8\text{ g}}{1\text{ mol}} = 4.7\text{ g/L}$$

The amount of Ag_2SO_4 dissolved in 500 mL = 0.500 L of water is 0.500 L × 4.7 g/L = 2.4 g. The grams of Ag_2SO_4 that remain undissolved: 3.00 g – 2.4 g = 0.6 g.

PE 18.5 (a) Less soluble. The SO_4^{2-} ions present will repress the solubility of Ag_2SO_4.

(b) Let x stand for the molarity of the sodium sulfate solution. One mole of Ag_2SO_4 supplies 2 mol of Ag^+, so the equilibrium $[Ag^+]$ is twice the Ag_2SO_4 solubility, or 0.020 M.

Concentration Summary:

	$Ag_2SO_4(s) \rightleftharpoons$	$2Ag^+(aq)$	$+ SO_4^{2-}(aq)$
Equation:			
Initial concentrations, mol/L		0	x
Concentration changes, mol/L		+0.020	+0.010
Equilibrium concentrations, mol/L		0.020	$0.010 + x$

Substituting the equilibrium concentrations into the K_{sp} expression gives

$K_{sp} = [Ag^+]^2[SO_4^{2-}]$ $\quad K_{sp} = 1.4 \times 10^{-5}$

$1.4 \times 10^{-5} = (0.020)^2(0.010 + x)$ and $x = [Na_2SO_4] = 0.025$ M

Answer: A 0.025 M Na_2SO_4 solution will dissolve 0.010 mol Ag_2SO_4 per liter.

PE 18.6 The quantities of each ion are:

Ca^{2+}: 0.050 L × 0.050 mol/L= 0.0025 mol

CrO_4^{2-}: 0.050 L × 0.050 mol/L = 0.0025 mol

The final volume is 50 mL + 50 mL = 100 mL = 0.100 L. The concentrations after mixing are:

Ca^{2+}: 0.0025 mol/0.100 L = 0.025 mol/L

CrO_4^{2-}: 0.0025 mol/0.100 L = 0.025 mol/L

Substituting the concentrations into the ion product expression gives

$Q = [Ca^{2+}]\,[CrO_4^{2-}] = (0.025)(0.025) = 6.2 \times 10^{-4}$

The ion product Q is less than the K_{sp} value of 7.1×10^{-4}, so $CaCrO_4$ will not precipitate.

PE 18.7 $K_{sp} = [Pb^{2+}][Cl^-]^2 = 1.7 \times 10^{-5}$

$(1.0 \times 10^{-3})[Cl^-]^2 = 1.7 \times 10^{-5}$ and $[Cl^-] = 0.13$ M

If the HCl, which is completely ionized, is maintained at 0.13 M or greater, the concentration of lead ion will not exceed 1.0×10^{-3} M. However, a high $[Cl^-]$ must be avoided, since the soluble complex ion $PbCl_4^{2-}$ forms in the presence of a large excess of chloride ions.

PE 18.8 The K_{sp}'s are: $Mg(OH)_2$, 7.1×10^{-12}; $Ca(OH)_2$, 6.5×10^{-6}. The optimum hydroxide concentration occurs when the solution is just barely saturated with respect to $Ca(OH)_2$. This concentration is obtained by substituting $[Ca^{2+}] = 0.10$ M into the K_{sp} expression for $Ca(OH)_2$:

$K_{sp} = [Ca^{2+}][OH^-]^2$

$6.5 \times 10^{-6} = (0.10)[OH^-]^2$ and $[OH^-] = 8.1 \times 10^{-3}$ M

The maximum separation occurs when the NaOH concentration is 8.1×10^{-3} M.

PE 18.9 (a) $K_{sp} = [Mg^{2+}][OH^-]^2$

$7.1 \times 10^{-12} = [Mg^{2+}](8.1 \times 10^{-3})^2$ and $[Mg^{2+}] = 1.1 \times 10^{-7}$ M

(b) Yes. Most of the magnesium ions are in the precipitate; all of the calcium ions are in the solution.

PE 18.10 $K_{sp} = [Al^{3+}][OH^-]^3$

$3 \times 10^{-34} = (0.10)[OH^-]^3$ and $[OH^-] = 1 \times 10^{-11}$ M

$pOH = -\log [OH^-] = -\log (1 \times 10^{-11}) = 11.0$ and $pH = 14.00 - 11.0 = 3.0$

Answer: The pH would have to be lowered to 3.0.

PE 18.11 The concentration of hydroxide ions in a saturated solution of $Mg(OH)_2$ containing 0.20 M Mg^{2+} is

$K_{sp} = [Mg^{2+}][OH^-]^2 = 7.1 \times 10^{-12}$

$(0.20)[OH^-]^2 = 7.1 \times 10^{-12}$ and $[OH^-] = 6.0 \times 10^{-6}$ M

Substituting $[OH^-] = 6.0 \times 10^{-6}$ M and $[NH_3] = 0.050$ M into the K_b expression for ammonia gives

$K_b = [NH_4^+][OH^-]/[NH_3]$

$$1.8 \times 10^{-5} = \frac{[NH_4^+](6.0 \times 10^{-6})}{(0.050)} \quad \text{and} \quad [NH_4^+] = 0.15 \text{ M}$$

Magnesium hydroxide will not precipitate if the concentration of ammonium ions is greater than or equal to 0.15 M .

PE 18.12 Only mixture (b). Mixture (a) cannot be separated, since neither $Ca(OH)_2$ nor $Fe(OH)_2$ is amphoteric. Mixture (c) cannot be separated, since both $Pb(OH)_2$ and $Zn(OH)_2$ are amphoteric.

PE 18.13 $Al_2O_3(s) + 6H^+(aq) \rightarrow 2Al^{3+}(aq) + 3H_2O(l)$

$Al_2O_3(s) + 2OH^-(aq) + 3H_2O(l) \rightarrow 2Al(OH)_4^-(aq)$

PE 18.14 (a) $PbCO_3$ dissolves, since $PbCO_3$ is the salt of carbonic acid (H_2CO_3), a weak, unstable acid.

$PbCO_3(s) + 2HNO_3(aq) \rightarrow Pb(NO_3)_2(aq) + H_2O(l) + CO_2(g)$

(b) PbI_2 does not dissolve, since PbI_2 is the salt of the strong acid HI.

(c) $PbCl_2$ does not dissolve, since $PbCl_2$ is the salt of the strong acid HCl.

(d) PbF_2 dissolves, since PbF_2 is the salt of the weak acid HF.

$PbF_2(s) + 2HNO_3(aq) \rightarrow Pb(NO_3)_2(aq) + 2HF(aq)$

PE 18.15 (a) $CdS(s) + 2H^+(aq) \rightleftharpoons Cd^{2+}(aq) + H_2S(aq)$

$$\frac{[Cd^{2+}][H_2S]}{[H^+]^2} = K_{spa} = 8 \times 10^{-7}$$

$$[H^+]^2 = \frac{(1.0 \times 10^{-8})(0.10)}{8 \times 10^{-7}}; \; [H^+] = 0.04 \text{ M and pH} = 1.4$$

A pH of 1.4 or less.

(b) Increase it. Reducing the hydrogen ion concentration will shift the equilibrium in the reverse direction causing CdS to precipitate.

PE 18.16 No. The K_{spa}'s are: CuS, 6×10^{-16}; HgS, 2×10^{-32}. HgS, with a much smaller K_{spa}, is less soluble than CuS in acid solution. At maximum separation, the solution will be saturated with respect to 0.10 M Cu^{2+}. The $[H^+]$ of the maximum separation solution is obtained by substituting $[Cu^{2+}] = 0.10$ M and $[H_2S] = 0.10$ M into the K_{spa} expression of CuS. So,

$$\frac{[Cu^{2+}][H_2S]}{[H^+]^2} = K_{spa} = 6 \times 10^{-16}$$

$$\frac{(0.10)(0.10)}{[H^+]^2} = 6 \times 10^{-16} \text{ and } [H^+] = 4 \times 10^6 \text{ M}$$

Such a high concentration of HCl cannot be realistically obtained, and, therefore, it is not possible to precipitate the HgS by itself. Therefore, the answer is no.

PE 18.17 The K_f's are: $Cd(NH_3)_4^{2+}$, 1×10^7; $Fe(SCN)^{2+}$, 1.2×10^2; HgI_4^{2-}, 1.9×10^{30}. The larger the formation constant, the more stable the ion; therefore, HgI_4^{2-} with $K_f = 1.9 \times 10^{30}$ is the most stable and $Fe(SCN)^{2+}$ with $K_f = 1.2 \times 10^2$ is the least stable.

PE 18.18 Because the K_f of $Zn(OH)_4^{2-}$ is large, 2.2×10^{16}, and because hydroxide ion is present in excess, most of the zinc will be in the form of $Zn(OH)_4^{2-}$. We will therefore assume that all of the zinc is initially in the form of the complex, which then breaks down to form x moles of zinc ion at equilibrium. So, initially $[Zn(OH)_4^{2-}] = 0.10$ M. The formation of this complex reduces the concentration of OH^- by 0.40 M, since each $Zn(OH)_4^{2-}$ ion contains four OH^- ions. Therefore, the initial $[OH^-]$ is

$$[OH^-] = 2.0 \text{ M} - 0.40 \text{ M} = 1.6 \text{ M}$$

Concentration Summary:

Equation:	$Zn(OH)_4^{2-}(aq)$ $\rightleftharpoons$	$Zn^{2+}(aq)$ +	$4OH^-(aq)$
Initial concentrations, mol/L	0.10	0	1.6
Concentration changes, mol/L	$-x$	$+x$	$+4x$
Equilibrium concentrations, mol/L	$0.10 - x$	x	$1.6 + 4x$
Approximate concentrations, mol/L	0.10	x	1.6

Substituting the approximate concentrations and $K_d = 4.5 \times 10^{-17}$ into the K_d expression gives

$$\frac{[Zn^{2+}][OH^-]^4}{[Zn(OH)_4^{2-}]} = K_d$$

$$\frac{(x)(1.6)^4}{(0.10)} = 4.5 \times 10^{-17} \text{ and } x = 6.9 \times 10^{-19} \text{ M} = [Zn^{2+}]$$

PE 18.19 Substituting $[Ag^+] = 1.6 \times 10^{-10}$ M and $[Br^-] = 1.0$ M into the ion product expression gives

$$Q = [Ag^+][Br^-] = (1.6 \times 10^{-10})(1.0) = 1.6 \times 10^{-10}$$

The ion product Q is greater than the K_{sp} value of 5.0×10^{-13}, so AgBr will precipitate.

PE 18.20 Let x be the molar solubility of AgCl.

Concentration Summary:

Equation:	$AgCl(s) + 2NH_3(aq)$ $\rightleftharpoons$	$Ag(NH_3)_2^+(aq)$ +	$Cl^-(aq)$
Initial concentrations, mol/L	2.0	0	0
Concentration changes, mol/L	$-2x$	$+x$	$+x$
Equilibrium concentrations, mol/L	$2.0 - 2x$	x	x

Substituting $K_{sp} = 1.8 \times 10^{-10}$, $K_f = 1.6 \times 10^7$, and the equilibrium concentrations into the combined equilibrium constant expression gives

$$\frac{[Ag(NH_3)_2^+][Cl^-]}{[NH_3]^2} = K_{sp} \times K_f$$

$$\frac{(x)(x)}{(2.0 - 2x)^2} = (1.8 \times 10^{-10})(1.6 \times 10^7) = 2.9 \times 10^{-3}$$

The equation is simplified by taking the square root of both sides:

$$\frac{x}{2.0 - 2x} = 5.4 \times 10^{-2} \text{ and } x = 9.7 \times 10^{-2}\text{ M}$$

The solubility of AgCl in 2.0 M NH_3 is 9.7×10^{-2} M.

Solutions To Final Exercises

18.1 Solubility product constants are used only for sparingly soluble solids, because the K_{sp}'s are reliable only when the ionic concentrations are low enough for interionic attractions to be ignored.

18.3 An ionic solid (precipitate) exists in equilibrium with its aqueous ions. The addition of one of the ions comprising the precipitate to the equilibrium solution moves the reaction to the left (Le Châtelier's principle), forming more precipitate. Consequently, the solubility of the substance decreases; i.e., the addition of the common ion represses the solubility of the substance.

18.5 (a) $PbSO_4$: $K_{sp} = [Pb^{2+}][SO_4^{2-}]$ (b) $Sb_2(SO_4)_3$: $K_{sp} = [Sb^{3+}]^2[SO_4^{2-}]^3$ (c) $Fe(OH)_3$: $K_{sp} = [Fe^{3+}][OH^-]^3$

18.7 (a) $[Co^{2+}] = 0.011$ M and $[IO_3^-] = 2 \times 0.011$ M $= 0.022$ M

$K_{sp} = [Co^{2+}][IO_3^-]^2 = (0.011)(0.022)^2 = 5.3 \times 10^{-6}$

(b) $[Ba^{2+}] = 0.008$ M and $[S_2O_3^{2-}] = 0.008$ M

$K_{sp} = [Ba^{2+}][S_2O_3^{2-}] = (0.008)(0.008) = 6 \times 10^{-5}$

(c) $[Tl^+] = 2 \times 7.8 \times 10^{-5}$ M $= 1.6 \times 10^{-4}$ M and $[PtCl_6^{2-}] = 7.8 \times 10^{-5}$ M

$K_{sp} = [Tl^+]^2[PtCl_6^{2-}] = (1.6 \times 10^{-4})^2(7.8 \times 10^{-5}) = 2.0 \times 10^{-12}$

18.9 (a) $K_{sp} = [Sr^{2+}][CO_3^{2-}]$

$9.3 \times 10^{-10} = (x)(x) = x^2$ and $x = 3.0 \times 10^{-5}$ M $= [Sr^{2+}] = [CO_3^{2-}]$

The equilibrium concentration of Sr^{2+}, 3.0×10^{-5} mol/L, is also the molar solubility of $SrCO_3$. The molar mass of $SrCO_3$ is 147.6 g/mol; hence, its solubility in grams per 100 mL is

$$\frac{3.0 \times 10^{-5}\text{ mol}}{1\text{ L}} \times \frac{0.10\text{ L}}{100\text{ mL}} \times \frac{147.6\text{ g}}{1\text{ mol}} = 4.4 \times 10^{-4}\text{ g/100 mL}$$

(b) $K_{sp} = [Pb^{2+}][I^-]^2$

$7.9 \times 10^{-9} = (x)(2x)^2 = 4x^3$ and $x = 1.3 \times 10^{-3}$ M $= [Pb^{2+}]$

The equilibrium concentration of Pb^{2+}, 1.3×10^{-3} mol/L, is also the molar solubility of PbI_2. The molar mass of PbI_2 is 461.0 g/mol; hence, its solubility in grams per 100 mL is

$$\frac{1.3 \times 10^{-3}\,\text{mol}}{1\text{ L}} \times \frac{0.10\text{ L}}{100\text{ mL}} \times \frac{461.0\text{ g}}{1\text{ mol}} = 6.0 \times 10^{-2}\text{ g/100 mL}$$

Note, that in part (a), the hydrolysis of CO_3^{2-} has not been taken into account.

18.11 AgCl: $K_{sp} = [Ag^+][Cl^-]$

$1.8 \times 10^{-10} = (x)(x) = x^2$ and $x = 1.3 \times 10^{-5}$ M = $[Ag^+]$

Ag_2CO_3: $K_{sp} = [Ag^+]^2[CO_3^{2-}]$

$8.1 \times 10^{-12} = (2x)^2(x) = 4x^3$ and $x = 1.3 \times 10^{-4}$ M = $[CO_3^{2-}]$

$[Ag^+] = 2 \times [CO_3^{2-}] = 2 \times 1.3 \times 10^{-4}$ M $= 2.6 \times 10^{-4}$ M

Answer: The saturated Ag_2CO_3 solution has the highest Ag^+ concentration.

18.13 (a) The molar solubility of Ag_2SO_4 in water is 1.5×10^{-2} mol/L (Practice Exercise 18.4).

(b) Let x equal the molar solubility of Ag_2SO_4.

Concentration Summary:

Equation:	$Ag_2SO_4(s) \rightleftharpoons$	$2Ag^+(aq)$ +	$SO_4^{2-}(aq)$
Initial concentrations, mol/L		0	0.20
Concentration changes, mol/L		$+2x$	$+x$
Equilibrium concentrations, mol/L		$2x$	$0.20 + x$
Approximate concentrations, mol/L		$2x$	0.20

Substituting the approximate concentrations and $K_{sp} = 1.4 \times 10^{-5}$ into the K_{sp} expression gives

$K_{sp} = [Ag^+]^2[SO_4^{2-}]$

$1.4 \times 10^{-5} = (2x)^2(0.20)$ and $x = 4.2 \times 10^{-3}$ mol/L

The molar solubility of Ag_2SO_4 in 0.20 M Na_2SO_4 is 4.2×10^{-3} mol/L.

(c) Let x equal the molar solubility of Ag_2SO_4.

Concentration Summary:

Equation:	$Ag_2SO_4(s) \rightleftharpoons$	$2Ag^+(aq)$ +	$SO_4^{2-}(aq)$
Initial concentrations, mol/L		0.10	0
Concentration changes, mol/L		$+2x$	$+x$
Equilibrium concentrations, mol/L		$0.10 + 2x$	x
Approximate concentrations, mol/L		0.10	x

Substituting the approximate concentrations and K_{sp} value into the K_{sp} expression gives

$1.4 \times 10^{-5} = (0.10)^2(x)$ and $x = 1.4 \times 10^{-3}$ mol/L

The molar solubility of Ag_2SO_4 in 0.10 M $AgNO_3$ is 1.4×10^{-3} mol/L.

18.15 (a) $K_{sp} = [Ca^{2+}][CO_3^{2-}]$

$4.5 \times 10^{-9} = (x)(x) = x^2$ and $x = 6.7 \times 10^{-5}$ mol/L = $[Ca^{2+}] = [CO_3^{2-}]$

The equilibrium concentration of Ca^{2+}, 6.7×10^{-5} mol/L, is also the molar solubility of $CaCO_3$. The molar mass of $CaCO_3$ is 100.1 g/mol; hence, the number of grams of $CaCO_3$ that will dissolve in 10.0 L of pure water is

$$10.0\ \text{L} \times \frac{6.7 \times 10^{-5}\ \text{mol}}{1\ \text{L}} \times \frac{100.1\ \text{g}}{1\ \text{mol}} = 6.7 \times 10^{-2}\ \text{g}$$

(b) Let x equal the molar solubility of $CaCO_3$.

Concentration Summary:

Equation:	$CaCO_3(s) \rightleftharpoons Ca^{2+}(aq) + CO_3^{2-}(aq)$
Initial concentrations, mol/L	0.50 0
Concentration changes, mol/L	$+x$ $+x$
Equilibrium concentrations, mol/L	$0.50 + x$ x
Approximate concentrations, mol/L	0.50 x

Substituting the approximate concentrations and K_{sp} value into the K_{sp} expression gives

$4.5 \times 10^{-9} = (0.50)(x)$ and $x = 9.0 \times 10^{-9}$ mol/L

The molar solubility of $CaCO_3$ in 0.50 M $CaCl_2$ is 9.0×10^{-9} mol/L. The number of grams of $CaCO_3$ that will dissolve in 10.0 L of this solution is

$$10.0\ \text{L} \times \frac{9.0 \times 10^{-9}\ \text{mol}}{1\ \text{L}} \times \frac{100.1\ \text{g}}{1\ \text{mol}} = 9.0 \times 10^{-6}\ \text{g}$$

18.17 (a) The molar mass of Ag_2CrO_4 is 331.7 g/mol; therefore 0.014 g/L is

$$\frac{0.014\ \text{g}}{1\ \text{L}} \times \frac{1\ \text{mol}}{331.7\ \text{g}} = 4.2 \times 10^{-5}\ \text{M}$$

$[Ag^+] = 2 \times 4.2 \times 10^{-5}\ \text{M} = 8.4 \times 10^{-5}\ \text{M}$ and $[CrO_4^{2-}] = 4.2 \times 10^{-5}\ \text{M}$

$K_{sp} = [Ag^+]^2[CrO_4^{2-}] = (8.4 \times 10^{-5})^2(4.2 \times 10^{-5}) = 3.0 \times 10^{-13}$

(b) Let x equal the molar solubility of Ag_2CrO_4. Employing the usual concentration summary and approximations gives

$3.0 \times 10^{-13} = (2x)^2(0.10)$ and $x = 8.7 \times 10^{-7}\ \text{M} = [Ag_2CrO_4]$

18.19 $K_{sp} = [Ca^{2+}][OH^-]^2$

$6.5 \times 10^{-6} = (x)(2x)^2 = 4x^3$ and $x = 1.2 \times 10^{-2}\ \text{M} = [Ca^{2+}]$

$[OH^-] = 2x = 2 \times 1.2 \times 10^{-2}\ \text{M} = 2.4 \times 10^{-2}\ \text{M}$

$\text{pOH} = -\log [OH^-] = -\log (2.4 \times 10^{-2}) = 1.62$ and $\text{pH} = 14.00 - \text{pOH} = 14.00 - 1.62 = 12.38$

18.21 The molar mass of TlCl is 239.8 g/mol. The concentration of Tl^+ in 0.10 M NaCl solution is

$$\frac{0.98\ \text{g TlCl}}{0.500\ \text{L}} \times \frac{1\ \text{mol Tl}^+}{239.8\ \text{g TlCl}} = 8.2 \times 10^{-3}\ \text{mol Tl}^+/\text{L} = [Tl^+]$$

The concentration of Cl^- in solution is 0.10 M + 0.0082 M = 0.11 M. Substituting these concentrations into the K_{sp} expression gives

$K_{sp} = [Tl^+][Cl^-] = (8.2 \times 10^{-3})(0.11) = 9.0 \times 10^{-4}$

18.23 The ion product is the reaction quotient Q for an ionic solid in solution. It has the same form as the K_{sp} expression.

(a) equal to (b) less than (c) greater than

18.25 (a) The concentrations just after mixing are:

mol Pb^{2+} = 0.050 L × 0.20 mol/L = 0.010 mol

mol Cl^- = 0.080 L × 0.020 mol/L = 0.0016 mol

$[Pb^{2+}]$ = 0.010 mol/(0.050 L + 0.080 L) = 0.077 M

$[Cl^-]$ = 0.0016 mol/(0.050 L + 0.080 L) = 0.012 M

and $Q = [Pb^{2+}][Cl^-]^2 = (0.077)(0.012)^2 = 1.1 \times 10^{-5}$

The ion product Q is smaller than the K_{sp} value of 1.7×10^{-5}, so $PbCl_2$ will not precipitate.

(b) The concentrations just after mixing are:

mol Pb^{2+} = 0.075 L × 0.0010 mol/L = 7.5×10^{-5} mol

mol I^- = 0.050 L × 0.0020 mol/L = 1.0×10^{-4} mol

$[Pb^{2+}]$ = 7.5×10^{-5} mol/(0.075 L + 0.050 L) = 6.0×10^{-4} M

$[I^-]$ = 1.0×10^{-4} mol/(0.075 L + 0.050 L) = 8.0×10^{-4} M

and $Q = [Pb^{2+}][I^-]^2 = (6.0 \times 10^{-4})(8.0 \times 10^{-4})^2 = 3.8 \times 10^{-10}$

The ion product Q is smaller than the K_{sp} value of 7.9×10^{-9}, so PbI_2 will not precipitate.

(c) The concentrations just after mixing are:

mol Pb^{2+} = 1.0 L × 0.10 mol/L = 0.10 mol

mol OH^- = 0.0010 L × 0.0010 mol/L = 1.0×10^{-6} mol

$[Pb^{2+}]$ = 0.10 mol/(1.0 L + 0.001 L) = 0.10 M

$[OH^-]$ = 1.0×10^{-6} mol/(1.0 L + 0.001 L) = 1.0×10^{-6} M

and $Q = [Pb^{2+}][OH^-]^2 = (0.10)(1.0 \times 10^{-6})^2 = 1.0 \times 10^{-13}$

The ion product Q is greater than the K_{sp} value of 1.2×10^{-15}, so $Pb(OH)_2$ will precipitate.

18.27 (a) The concentrations just after mixing are:

mol Ca^{2+} = 0.100 L × 0.020 mol/L = 2.0×10^{-3} mol

mol CrO_4^{2-} = 0.100 L × 0.020 mol/L = 2.0×10^{-3} mol

$[Ca^{2+}]$ = 2.0×10^{-3} mol/(0.100 L + 0.100 L) = 0.010 M

$[CrO_4^{2-}]$ = 2.0×10^{-3} mol/(0.100 L + 0.100 L) = 0.010 M

and $Q = [Ca^{2+}][CrO_4^{2-}] = (0.010)(0.010) = 1.0 \times 10^{-4}$

The ion product Q is smaller than the K_{sp} value of 7.1×10^{-4}, so $CaCrO_4$ will not precipitate. The concentrations remaining in the solution are therefore

$[Ca^{2+}]$ = 0.010 M; $[Cl^-]$ = 0.020 M; $[K^+]$ = 0.020 M; and $[CrO_4^{2-}]$ = 0.010 M

(b) The concentrations just after mixing are:

mol Ba^{2+} = 0.100 L × 0.015 mol/L = 1.5×10^{-3} mol

mol F^- = 0.200 L × 0.0045 mol/L = 9.0×10^{-4} mol

$[Ba^{2+}]$ = 1.5×10^{-3} mol/(0.100 L + 0.200 L) = 5.0×10^{-3} M

$[F^-]$ = 9.0×10^{-4} mol/(0.100 L + 0.200 L) = 3.0×10^{-3} M and $Q = [Ba^{2+}][F^-]^2 = (5.0 \times 10^{-3})(3.0 \times 10^{-3})^2 = 4.5 \times 10^{-8}$

The ion product Q is smaller than the K_{sp} value of 1.3×10^{-6}, so BaF_2 will not precipitate. The concentrations remaining in solution are therefore

$[Ba^{2+}] = 5.0 \times 10^{-3}$ M; $[Cl^-] = 1.0 \times 10^{-2}$ M; $[Na^+] = 3.0 \times 10^{-3}$ M; and $[F^-] = 3.0 \times 10^{-3}$ M

18.29 Since the volume doubles on mixing, the initial concentrations just after mixing are half what they were originally. If we assume that all the sulfate ion initially precipitates, then the concentration of Ba^{2+} remaining in the mixed solution is

$[Ba^{2+}] = 0.100\text{ M} - 0.045\text{ M} = 0.055\text{ M}$

We now have to determine the molar solubility of $BaSO_4$ in 0.055 M Ba^{2+} solution.

Concentration Summary:

Equation:	$BaSO_4(s) \rightleftharpoons$	$Ba^{2+}(aq)$ +	$SO_4^{2-}(aq)$
Initial concentrations, mol/L		0.055	0
Concentration changes, mol/L		$+x$	$+x$
Equilibrium concentrations, mol/L		$0.055 + x$	x
Approximate concentrations, mol/L		0.055	x

Substituting the approximate concentrations and $K_{sp} = 1.1 \times 10^{-10}$ into the K_{sp} expression gives

$K_{sp} = [Ba^{2+}][SO_4^{2-}]$

$1.1 \times 10^{-10} = (0.055)(x)$ and $x = 2.0 \times 10^{-9}\text{ M} = [SO_4^{2-}]$

The percent of sulfate ion that will not precipitate is

$$\frac{2.0 \times 10^{-9}}{0.045} \times 100\% = 4.4 \times 10^{-6}\,\%$$

18.31 (a) The K_{sp}'s are: BaF_2, 1.3×10^{-6}; CaF_2, 3.9×10^{-11}. The $[F^-]$ that achieves maximum separation occurs when the solution is just barely saturated with respect to BaF_2. This concentration is obtained by substituting $[Ba^{2+}] = 0.225$ M into the K_{sp} expression for BaF_2:

$K_{sp} = [Ba^{2+}][F^-]^2$

$1.3 \times 10^{-6} = (0.225)[F^-]^2$ and $[F^-] = 2.4 \times 10^{-3}$ M

(b) $[Ba^{2+}] = 0.225$ M

$K_{sp} = [Ca^{2+}][F^-]^2$

$3.9 \times 10^{-11} = [Ca^{2+}](0.0024)^2$ and $[Ca^{2+}] = 6.8 \times 10^{-6}$ M

18.33 (a) The K_{sp}'s are: $CaCrO_4$, 7.1×10^{-4}; $BaCrO_4$, 1.2×10^{-10}. The $[CrO_4^{2-}]$ needed to attain saturation for each ion is

$CaCrO_4$: $K_{sp} = [Ca^{2+}][CrO_4^{2-}]$

$7.1 \times 10^{-4} = (0.050)[CrO_4^{2-}]$ and $[CrO_4^{2-}] = 0.014$ M

$BaCrO_4$: $K_{sp} = [Ba^{2+}][CrO_4^{2-}]$

$1.2 \times 10^{-10} = (0.0050)[CrO_4^{2-}]$ and $[CrO_4^{2-}] = 2.4 \times 10^{-8}$ M

The maximum separation occurs with the $[CrO_4^{2-}] = 0.014$ M.

(b) All of the Ca^{2+} ions are still in solution, therefore $[Ca^{2+}] = 0.050$ M. The concentration of Ba^{2+} can be obtained from the K_{sp} expression.

$1.2 \times 10^{-10} = [Ba^{2+}](0.014)$ and $[Ba^{2+}] = 8.6 \times 10^{-9}$ M

18.35 Increasing the hydroxyl ion (OH^-) concentration drives the reaction to the left, producing precipitate. On the other hand, acid lowers the hydroxyl ion concentration, which drives the reaction to the right, causing the precipitate to dissolve.

18.37 (a) The H^+ and OH^- concentrations are each 1.0×10^{-7} M at pH 7.00.

Fe^{3+}: $K_{sp} = [Fe^{3+}][OH^-]^3$

$1.6 \times 10^{-39} = [Fe^{3+}](1.0 \times 10^{-7})^3$ and $[Fe^{3+}] = 1.6 \times 10^{-18}$ M

Pb^{2+}: $K_{sp} = [Pb^{2+}][OH^-]^2$

$1.2 \times 10^{-15} = [Pb^{2+}](1.0 \times 10^{-7})^2$ and $[Pb^{2+}] = 0.12$ M

(b) $K_{sp} = [Al^{3+}][OH^-]^3$

$3 \times 10^{-34} = (0.12)[OH^-]^3$ and $[OH^-] = 1 \times 10^{-11}$

$pH = -\log [H^+] = -\log (1 \times 10^{-3}) = 3.0$

The pH must be lowered to 3.0 from 7.00.

18.39 If 95% of the barium precipitated, then we are left with $1.500\text{ g} - 0.95 \times 1.500\text{ g} = 0.075\text{ g } Ba^{2+}$ in solution. The $[Ba^{2+}]$ after precipitation is

$$\frac{0.075\text{ g }Ba^{2+}}{100\text{ mL}} \times \frac{1000\text{ mL}}{1\text{ L}} \times \frac{1\text{ mol }Ba^{2+}}{137.33\text{ g }Ba^{2+}} = 5.5 \times 10^{-3}\text{ M }Ba^{2+}$$

The pH can be obtained from the K_{sp} expression.

$K_{sp} = [Ba^{2+}][OH^-]^2$

$5 \times 10^{-3} = (5.5 \times 10^{-3})[OH^-]^2$

$[OH^-] = 1$ M and $pH = -\log [H^+] = -\log (1.00 \times 10^{-14}/1) = 14.0$

18.41 The hydroxide ions come from the ionization of ammonia

$NH_3(aq) + H_2O(l) \rightleftharpoons NH_4^+(aq) + OH^-(aq)$

The $[OH^-]$ before mixing is obtained by substituting $[NH_3] = 0.50$ M and $[NH_4^+] = 0.30$ M into the K_b expression for ammonia.

$$[OH^-] = \frac{K_b[NH_3]}{[NH_4^+]} = \frac{(1.8 \times 10^{-5})(0.50)}{(0.30)} = 3.0 \times 10^{-5}\text{ M}$$

Since this is an ammonia–ammonium chloride buffer system, the $[OH^-]$ does not change upon dilution. Therefore, the concentrations initially after mixing are

$[Mn^{2+}] = [Fe^{2+}] = [Mg^{2+}] = 0.0010$ M and $[OH^-] = 3.0 \times 10^{-5}$ M

To determine if any of these ions precipitate as hydroxides, we have to calculate their ion products and compare them to their K_{sp} values.

Mn^{2+}: $Q = [Mn^{2+}][OH^-]^2 = (0.0010)(3.0 \times 10^{-5})^2 = 9.0 \times 10^{-13}$

The ion product Q is greater than the K_{sp} value of 6×10^{-14}, so $Mn(OH)_2$ will precipitate.

Fe^{2+}: $Q = [Fe^{2+}][OH^-]^2 = 9.0 \times 10^{-13}$

The ion product Q is greater than the K_{sp} value of 7.9×10^{-16}, so $Fe(OH)_2$ will precipitate.

Mg^{2+}: $Q = [Mg^{2+}][OH^-]^2 = 9.0 \times 10^{-13}$

The ion product Q is smaller than the K_{sp} value of 7.1×10^{-12}, so $Mg(OH)_2$ will not precipitate.

18.43 First calculate the $[OH^-]$ that would make the solution saturated with respect to $Pb(OH)_2$. Precipitation will not occur if the actual hydroxide ion concentration is less than or equal to this value.

$K_{sp} = [Pb^{2+}][OH^-]^2$

$(0.0010)[OH^-]^2 = 1.2 \times 10^{-15}$ and $[OH^-] = 1.1 \times 10^{-6}$ M

The ratio of NH_4^+ to NH_3 corresponding to this $[OH^-]$ is obtained by substituting $[OH^-] = 1.1 \times 10^{-6}$ M into the K_b expression for ammonia

$$\frac{[NH_4^+]}{[NH_3]} = \frac{K_b}{[OH^-]} = \frac{1.8 \times 10^{-5}}{1.1 \times 10^{-6}} = 16$$

An NH_4^+/NH_3 ratio of 16 or greater would prevent precipitation.

18.45 An amphoteric hydroxide is a metal hydroxide that has the ability to neutralize both acids and bases.

Refer to Table 18.2 in the textbook.

18.47 The ionic and molecular equations are:

(a) $Sn(OH)_2(s) + 2H^+(aq) \rightarrow Sn^{2+}(aq) + 2H_2O(l)$
$Sn(OH)_2(s) + 2HCl(aq) \rightarrow SnCl_2(aq) + 2H_2O(l)$

(b) $Cr(OH)_3(s) + OH^-(aq) \rightarrow Cr(OH)_4^-(aq)$
$Cr(OH)_3(s) + NaOH(aq) \rightarrow NaCr(OH)_4(aq)$

18.49 Iron(III) hydroxide is not amphoteric; it dissolves in acids, but not in bases. Aluminum hydroxide, on the other hand, is amphoteric. Therefore, the addition of a strong base solution will dissolve the aluminum hydroxide, but not the iron(III) hydroxide.

18.51 Weak acids have strong conjugate bases. Therefore, the anions of sparingly soluble salts of weak acids partake in hydrolysis reactions that produce hydroxyl ions; i.e., they are basic. Addition of acid reduces the $[OH^-]$ and drives the hydrolysis reaction to the right. This, in turn, reduces the anion concentration and drives the solubility reaction to the right. Consequently, more precipitate dissolves.

18.53 CuCN and Cu_2S. CuCN and Cu_2S are salts of the weak acids HCN and H_2S, respectively. CuCl and CuI are not more soluble, since CuCl and CuI are salts of the strong acids HCl and HI, respectively.

18.55 Dilute HCl. The reaction of $BaSO_3$ with HCl would produce a choking SO_2 odor:

$BaSO_3(s) + 2HCl(aq) \rightarrow BaCl_2(aq) + SO_2(g) + H_2O(l)$

18.57 (a) $CaC_2O_4(s) + 2HCl(aq) \rightarrow CaCl_2(aq) + H_2C_2O_4(aq)$
(b) $CaCO_3(s) + 2HCl(aq) \rightarrow CaCl_2(aq) + H_2O(l) + CO_2(g)$
(c) $MnS(s) + 2HCl(aq) \rightarrow MnCl_2(aq) + H_2S(aq)$

18.59 The reactions that have to be considered are

$CaCO_3(s) \rightleftharpoons Ca^{2+}(aq) + CO_3^{2-}(aq)$

$CO_3^{2-}(aq) + H_2O(l) \rightleftharpoons HCO_3^-(aq) + OH^-(aq)$

The Ca^{2+} ion remains unchanged in solution. The molar solubility of $CaCO_3$, x, is equal to $[Ca^{2+}]$, which is also equal to the sum of the concentrations of the species containing the carbonate ion, or

$x = [Ca^{2+}] = [CO_3^{2-}] + [HCO_3^-]$

The ratio of $[HCO_3^-]$ to $[CO_3^{2-}]$ at this pH is ($[OH^-] = 1.0 \times 10^{-5}$ M and $K_b = 1.78 \times 10^{-4}$)

$$\frac{[HCO_3^-][OH^-]}{[CO_3^{2-}]} = K_b \text{ and } \frac{[HCO_3^-]}{[CO_3^{2-}]} = \frac{K_b}{[OH^-]} = \frac{1.78 \times 10^{-4}}{1.0 \times 10^{-5}} = 18$$

Therefore, $[HCO_3^-] = 18[CO_3^{2-}]$

$[CO_3^{2-}] + 18[CO_3^{2-}] = x$ and $[CO_3^{2-}] = x/19$

Substituting in the K_{sp} expression gives

$[Ca^{2+}][CO_3^{2-}] = K_{sp}$

$$(x)\left[\frac{x}{19}\right] = 4.5 \times 10^{-9} \text{ and } x = 2.9 \times 10^{-4}\text{ M}$$

18.61 Less soluble. Sulfides dissolve in acid because H^+ converts the S^{2-} to H_2S. Raising the pH reduces the concentrations of H^+ and reduces the extent of this conversion.

18.63 K_{spa} ("solubility product in acid solution") is the equilibrium constant governing the equilibrium reaction of a sparingly soluble salt in acid media.

$MnS(s) + 2H^+(aq) \rightleftharpoons Mn^{2+}(aq) + H_2S(aq)$

$K_{spa} = [Mn^{2+}][H_2S]/[H^+]^2$

18.65 $CdS(s) + 2H^+(aq) \rightleftharpoons Cd^{2+}(aq) + H_2S(aq)$

$[Cd^{2+}][H_2S]/[H^+]^2 = K_{spa} = 8 \times 10^{-7}$

$$[Cd^{2+}] = \frac{(8 \times 10^{-7})(0.75)^2}{(0.10)} = 4 \times 10^{-6}\text{ M}$$

18.67 $CuS(s) + 2H^+(aq) \rightleftharpoons Cu^{2+}(aq) + H_2S(aq)$
$MnS(s) + 2H^+(aq) \rightleftharpoons Mn^{2+}(aq) + H_2S(aq)$

A precipitate will form if $Q > K_{spa}$.

$$Cu^{2+}:\ Q = \frac{[Cu^{2+}][H_2S]}{[H^+]^2} = \frac{(0.10)(0.10)}{(0.10)^2} = 1.0 > K_{spa} = 6 \times 10^{-16}$$

$$Mn^{2+}:\ Q = \frac{[Mn^{2+}][H_2S]}{[H^+]^2} = \frac{(0.10)(0.10)}{(0.10)^2} = 1.0 < K_{spa} = 3 \times 10^{10}$$

Answer: A precipitate of CuS will form.

18.69 (a) The K_{spa}'s are: PbS, 3×10^{-7}; SnS, 1×10^{-5}. Maximum separation occurs when the $[H^+]$ is such that the solution is just barely saturated with respect to SnS. This concentration is obtained by substituting $[Sn^{2+}] = 0.0010$ M and $[H_2S] = 0.10$ M into the K_{spa} expression for SnS:

$K_{spa} = [Sn^{2+}][H_2S]/[H^+]^2$

$1 \times 10^{-5} = (0.0010)(0.10)/[H^+]^2$ and $[H^+] = 3$ M

(b) All of the Sn^{2+} remains in the solution; therefore $[Sn^{2+}] = 0.0010$ M. The $[Pb^{2+}]$ remaining in solution can be obtained by substituting $[H^+] = 3$ M and $[H_2S] = 0.10$ M into the K_{spa} expression.

$$[Pb^{2+}] = \frac{K_{spa}[H^+]^2}{[H_2S]} = \frac{(3 \times 10^{-7})(3)^2}{(0.10)} = 3 \times 10^{-5}\text{ M}$$

18.71 (a) If the pH is less than 0.2, then the $[H^+]$ would be greater than the maximum separation value of 0.6 M, and a maximum separation of Zn^{2+} and Cd^{2+} will not have been achieved. However, the precipitate would only contain CdS.

(b) If the pH is greater than 0.2, then the $[H^+]$ would be less than the maximum separation value of 0.6 M, and some ZnS will have precipitated.

The pH change in (b) is more likely to spoil the separation, since it leads to a mixed precipitate of CdS and ZnS.

18.73 For complex ions that contain several ligands, a separate equilibrium equation may be written for the addition of each successive ligand. A stepwise formation constant is an equilibrium constant for one of these separate equilibrium equations or steps.

An overall formation constant is the equilibrium constant for the single overall equation that shows the formation of a complex ion containing several ligands from the free cation and ligands. The overall formation constant is the product of the stepwise formation constants.

The overall formation constant is the one usually used in calculations involving complex ions.

18.75 The K_f's are: $Ag(NH_3)_2^+$, 1.6×10^7; $Ag(CN)_2^-$, 1×10^{21}; and $Ag(S_2O_3)_2^{3-}$, 1.7×10^{13}. $Ag(CN)_2^-$ will form in the greatest concentration, since its K_f value is considerably greater than those of $Ag(NH_3)_2^+$ and $Ag(S_2O_3)_2^{3-}$.

18.77 $K_d = 1/K_f = 1/(6.7 \times 10^{19}) = 1.5 \times 10^{-20}$

18.79 (a) $Sn(OH)_2(s) + 2OH^-(aq) \rightarrow Sn(OH)_4^{2-}(aq)$

(b) $AgC_2H_3O_2(s) + 2NH_3(aq) \rightarrow Ag(NH_3)_2^+(aq) + C_2H_3O_2^-(aq)$

(c) $Cu_2O(s) + 6CN^-(aq) + H_2O(l) \rightarrow 2Cu(CN)_3^{2-}(aq) + 2OH^-(aq)$

18.81 Substituting $[SCN^-] = 2.0$ M, $[Fe(SCN)^{2+}] = 0.25$ M, and $K_f = 1.2 \times 10^2$ into the K_f expression for $Fe(SCN)^{2+}$ gives

$$\frac{[Fe(SCN)^{2+}]}{[Fe^{3+}][SCN^-]} = K_f$$

$$\frac{(0.25)}{[Fe^{3+}](2.0)} = 1.2 \times 10^2 \text{ and } [Fe^{3+}] = 1.0 \times 10^{-3}\text{ M}$$

18.83 Because the K_f of $Cu(NH_3)_4^{2+}$ is large, 1.1×10^{13}, and because ammonia is present in excess, most of the copper will be in the form of $Cu(NH_3)_4^{2+}$. To solve the problem we will therefore assume that all of the copper is initially in the form of the complex, which then breaks down to form *x* moles of copper ion at equilibrium. So, initial $[Cu(NH_3)_4^{2+}] = 0.150$ M. The formation of the complex requires 0.600 mol NH_3 per liter, since each $Cu(NH_3)_4^{2+}$ ion contains four NH_3 molecules. Therefore, the initial $[NH_3]$ is

$$\frac{(3.00 \text{ mol } NH_3 - 0.600 \text{ mol } NH_3)}{1.00 \text{ L}} = \frac{2.40 \text{ mol } NH_3}{1.00 \text{ L}} = 2.40 \text{ M } NH_3$$

Concentration Summary:

Equation:	$Cu(NH_3)_4^{2+}(aq) \rightleftharpoons$	$Cu^{2+}(aq)$ +	$4NH_3(aq)$
Initial concentrations, mol/L	0.150	0	2.40
Concentration changes, mol/L	$-x$	$+x$	$+4x$
Equilibrium concentrations, mol/L	$0.150 - x$	x	$2.40 + 4x$
Approximate concentrations, mol/L	0.150	x	2.40

Substituting $K_d = 1/K_f = 1/(1.1 \times 10^{13}) = 9.1 \times 10^{-14}$ and the approximate concentrations into the K_d expression gives

$$\frac{[Cu^{2+}][NH_3]^4}{[Cu(NH_3)_4^{2+}]} = K_d$$

$$\frac{(x)(2.40)^4}{0.150} = 9.1 \times 10^{-14} \text{ and } x = 4.1 \times 10^{-16} \text{ M} = [Cu^{2+}]$$

18.85 Assume that all of the Hg^{2+} forms $HgCl_4^{2-}$, which then dissociates slightly. The Cl^- concentration will have been reduced by 4×0.010 M, or 0.040 M, to 0.16 M.

Concentration Summary:

Equation:	$HgCl_4^{2-}(aq) \rightleftharpoons$	$Hg^{2+}(aq)$ +	$4Cl^-(aq)$
Initial concentrations, mol/L	0.010	0	0.16
Concentration changes, mol/L	$-x$	$+x$	$+4x$
Equilibrium concentrations, mol/L	$0.010 - x$	x	$0.16 + 4x$
Approximate concentrations, mol/L	0.010	x	0.16

Substituting $K_d = 1/K_f = 1/(1.2 \times 10^{15}) = 8.3 \times 10^{-16}$ and the approximate concentrations into the K_d expression gives

$$\frac{[Hg^{2+}][Cl^-]^4}{[HgCl_4^{2-}]} = K_d$$

$$\frac{(x)(0.16)^4}{(0.010)} = 8.3 \times 10^{-16} \text{ and } x = 1.3 \times 10^{-14} \text{ M} = [Hg^{2+}]$$

$$[Cl^-] = 0.16 + 4x = 0.16 \text{ M} + 5.2 \times 10^{-14} \text{ M} = 0.16 \text{ M}$$

18.87 Substituting $[SCN^-] = 0.010$ M, $[Fe(SCN)^{2+}] = 1 \times 10^{-4}$ M, and $K_f = 1.2 \times 10^2$ into the K_f expression for $Fe(SCN)^{2+}$ gives

$$\frac{[Fe(SCN)^{2+}]}{[Fe^{3+}][SCN^-]} = K_f$$

$$\frac{(1 \times 10^{-4})}{[Fe^{3+}](0.010)} = 1.2 \times 10^2 \text{ and } [Fe^{3+}] = 8 \times 10^{-5} \text{ M}$$

This concentration is added to the Fe^{3+} that entered the complex.

$[Fe^{3+}] + [Fe(SCN)^{2+}] = 0.8 \times 10^{-4}\ M + 1 \times 10^{-4}\ M = 2 \times 10^{-4}\ M$

Fe^{3+} can be detected down to a concentration of 2×10^{-4} M.

18.89

Because the K_f of $Zn(NH_3)_4^{2+}$ is large, 2.9×10^9, and because ammonia is present in excess, most of the zinc will be in the form of $Zn(NH_3)_4^{2+}$. To solve the problem we will therefore assume that all the zinc is initially in the form of the complex, which then breaks down to form x moles of zinc ion at equilibrium. The initial concentration of the complex is

$$\frac{1.0\text{ g }ZnCl_2}{0.500\text{ L}} \times \frac{1\text{ mol }ZnCl_2}{136.3\text{ g }ZnCl_2} \times \frac{1\text{ mol }Zn(NH_3)_4^{2+}}{1\text{ mol }ZnCl_2} = 0.015\text{ M }Zn(NH_3)_4^{2+}$$

The formation of this complex reduces the concentration of NH_3 by 0.060 M, since each $Zn(NH_3)_4^{2+}$ ion contains four NH_3 molecules. Therefore, the initial $[NH_3]$ is

$[NH_3] = 0.50\ M - 0.060\ M = 0.44\ M$

Concentration Summary:

Equation:	$Zn(NH_3)_4^{2+}(aq) \rightleftharpoons$	$Zn^{2+}(aq) +$	$4NH_3(aq)$
Initial concentrations, mol/L	0.015	0	0.44
Concentration changes, mol/L	$-x$	$+x$	$+4x$
Equilibrium concentrations, mol/L	$0.015 - x$	x	$0.44 + 4x$
Approximate concentrations, mol/L	0.015	x	0.44

Substituting $K_d = 1/K_f = 1/(2.9 \times 10^9) = 3.4 \times 10^{-10}$ and the approximate concentrations into the K_d expression gives

$$\frac{[Zn^{2+}][NH_3]^4}{[Zn(NH_3)_4^{2+}]} = K_d$$

$$\frac{(x)(0.44)^4}{(0.015)} = 3.4 \times 10^{-10} \text{ and } x = 1.4 \times 10^{-10}\ M = [Zn^{2+}]$$

The equilibrium concentrations are:

$[Zn(NH_3)_4^{2+}] = 0.015 - x = 0.015\ M - 1.4 \times 10^{-10}\ M = 0.015\ M$

$[Zn^{2+}] = x = 1.4 \times 10^{-10}\ M$

$[NH_3] = 0.44 + 4x = 0.44\ M + 5.6 \times 10^{-10}\ M = 0.44\ M$

18.91

(1) $AgCl(s) \rightleftharpoons Ag^+(aq) + Cl^-(aq) \quad K_{sp} = 1.8 \times 10^{-10}$
(2) $Ag^+(aq) + 2Cl^-(aq) \rightleftharpoons AgCl_2^-(aq) \quad K_f = 2.5 \times 10^5$

The net reaction is the sum of steps (1) and (2)

$AgCl(s) + Cl^-(aq) \rightleftharpoons AgCl_2^-(aq)$

and the combined equilibrium constant is

$$\frac{[AgCl_2^-]}{[Cl^-]} = K_{sp} \times K_f = (1.8 \times 10^{-10})(2.5 \times 10^5) = 4.5 \times 10^{-5}$$

Let x be the molar solubility of AgCl.

Concentration Summary:

Equation:	$AgCl(s) + Cl^-(aq) \rightleftharpoons$	$AgCl_2^-(aq)$
Initial concentrations, mol/L	1.0	0
Concentration changes, mol/L	$-x$	$+x$
Equilibrium concentrations, mol/L	$1.0 - x$	x

Substituting the equilibrium concentrations into the combined equilibrium constant expression gives

$$\frac{x}{1.0 - x} = 4.5 \times 10^{-5} \text{ and } x = 4.5 \times 10^{-5} \text{ M}$$

The molar solubility of AgCl in 1.0 M HCl is 4.5×10^{-5} M.

18.93 The overall reaction is $AgBr(s) + 2S_2O_3^{2-}(aq) \rightleftharpoons Ag(S_2O_3)_2^{3-}(aq) + Br^-(aq)$.

The combined equilibrium constant is

$$\frac{[Ag(S_2O_3)_2^{3-}][Br^-]}{[S_2O_3^{2-}]^2} = K_{sp} \times K_f = (5.0 \times 10^{-13})(1.7 \times 10^{13}) = 8.5$$

The moles of dissolved AgBr per liter are

$$20 \text{ g AgBr} \times \frac{1 \text{ mol AgBr}}{187.8 \text{ g AgBr}} = 0.11 \text{ mol AgBr}$$

0.11 mol of Ag goes into solution where it is present as either free Ag^+ or $Ag(S_2O_3)_2^{3-}$. Because K_f is very large, virtually all of the silver will be present as the complex; therefore $[Ag(S_2O_3)_2^{3-}] = 0.11$ M. Substituting $[Ag(S_2O_3)_2^{3-}] = 0.11$ M and $[Br^-] = 0.11$ M into the combined equilibrium constant expression gives

$$\frac{(0.11)(0.11)}{[S_2O_3^{2-}]^2} = 8.5 \text{ and } [S_2O_3^{2-}] = 3.8 \times 10^{-2} \text{ M}$$

Consequently, at equilibrium the solution would have to be 3.8×10^{-2} M in $S_2O_3^{2-}$ for 20 g of AgBr per liter to dissolve. The original $S_2O_3^{2-}$ concentration would be 3.8×10^{-2} M + 2×0.11 M = 0.26 M. The second term (0.22 M) is the amount of $S_2O_3^{2-}$ consumed in making the complex.

18.95 (a) The molar mass of $Ca(OH)_2$ is 74.09 g/mol. The molar solubility of $Ca(OH)_2$ at 0°C is therefore

$$\frac{0.185 \text{ g}}{100 \text{ mL}} \times \frac{1000 \text{ mL}}{1 \text{ L}} \times \frac{1 \text{ mol}}{74.09 \text{ g}} = 2.50 \times 10^{-2} \text{ mol/L}$$

$[Ca^{2+}] = 2.50 \times 10^{-2}$ M and $[OH^-] = 2 \times 2.50 \times 10^{-2}$ M $= 5.00 \times 10^{-2}$ M. Substituting these concentrations into the K_{sp} expression gives

$$K_{sp} = [Ca^{2+}][OH^-]^2 = (2.50 \times 10^{-2})(5.00 \times 10^{-2})^2 = 6.25 \times 10^{-5}$$

(b) Decreases. The K_{sp} at 25°C, 6.5×10^{-6}, is smaller than that at 0°C, 6.25×10^{-5}.

18.97 $K_{sp} = [Fe^{3+}][OH^-]^3$

$1.6 \times 10^{-39} = (x)(3x)^3 = 27x^4$ and $x = 8.8 \times 10^{-11} \text{ M} = [Fe^{3+}]$

The equilibrium concentration of Fe^{3+}, 8.8×10^{-11} mol/L, is also the molar solubility of $Fe(OH)_3$.

18.99 Some of the possible reagents are: Na_2CO_3, NaOH, KI, and $Na_2C_2O_4$. The concentrations required can be calculated by substituting $[Pb^{2+}] = 1.0 \times 10^{-7}$ M and the K_{sp} value into the appropriate K_{sp} expression and solving for the unknown concentration of the anion.

18.101 The molar mass of Ag_2S is 247.80 g/mol. The number of moles of Ag in 3.00 g of Ag_2S are

$$3.00 \text{ g } Ag_2S \times \frac{1 \text{ mol } Ag_2S}{247.80 \text{ g } Ag_2S} \times \frac{2 \text{ mol Ag}}{1 \text{ mol } Ag_2S} = 0.0242 \text{ mol Ag}$$

Therefore, the concentration of Ag^+ in the original 500 mL = 0.500 L portion of saturated CH_3COOAg was

$$[Ag^+] = \frac{0.0242 \text{ mol}}{0.500 \text{ L}} = 0.0484 \text{ M}$$

$$[CH_3COO^-] = [Ag^+] = 0.0484 \text{ M and } K_{sp} = [CH_3COO^-][Ag^+] = (0.0484)(0.0484) = 2.3 \times 10^{-3}$$

18.103 The molarity of the calcium ion is

$$\frac{11.8 \text{ mg } Ca^{2+}}{0.100 \text{ L}} \times \frac{1 \text{ g}}{10^3 \text{ mg}} \times \frac{1 \text{ mol } Ca^{2+}}{40.08 \text{ g } Ca^{2+}} = 2.94 \times 10^{-3} \text{ M } Ca^{2+}$$

The molarity of the oxalate ion can be obtained from the K_{sp} expression

$$[Ca^{2+}][C_2O_4^{2-}] = K_{sp}$$

$$(2.94 \times 10^{-3})[C_2O_4^{2-}] = 1.3 \times 10^{-9} \text{ and } [C_2O_4^{2-}] = 4.4 \times 10^{-7} \text{ M}$$

18.105 (a) Since the chloride ion is present in excess, we can assume that essentially all of the Ag^+ and Pb^{2+} will precipitate out of solution. Therefore, the numbers of millimoles of AgCl and $PbCl_2$ that precipitate are

$$\text{millimoles AgCl} = \text{millimoles } PbCl_2 = 5.0 \text{ mL} \times \frac{0.10 \text{ mmol}}{1 \text{ mL}} = 0.50 \text{ mmol}$$

(b) The molar solubilities of AgCl and $PbCl_2$ at 100°C are:

AgCl: $K_{sp} = [Ag^+][Cl^-]$

$2.2 \times 10^{-10} = (x)(x) = x^2$ and $x = 1.5 \times 10^{-5}$ M

The molar solubility of AgCl at 100°C is 1.5×10^{-5} mol/L or 1.5×10^{-5} mmol/mL.

$PbCl_2$: $K_{sp} = [Pb^{2+}][Cl^-]^2$

$2.1 \times 10^{-4} = (x)(2x)^2 = 4x^3$ and $x = 3.7 \times 10^{-2}$ M

The molar solubility of $PbCl_2$ at 100°C is 3.7×10^{-2} mol/L or 3.7×10^{-2} mmol/mL.

The number of millimoles of each dissolved in 10 mL of water at 100°C is

AgCl: 10 mL $\times$ 1.5×10^{-5} mmol/mL = 1.5×10^{-4} mmol

$PbCl_2$: 10 mL $\times$ 0.037 mmol/mL = 0.37 mmol

18.107 (a) A high pH favors the formation of kidney stones, since then there will be more oxalate ion in solution, namely,

$$H_2C_2O_4(aq) \rightleftharpoons 2H^+(aq) + C_2O_4^{2-}(aq)$$

A low $[H^+]$ drives the reaction to the right; a high $[H^+]$ drives the reaction to the left, reducing the oxalate ion concentration.

(b) The low and high $[Ca^{2+}]$ are:

Low $[Ca^{2+}]$: 1.25 mmol/1500 mL = 8.3×10^{-4} M

High $[Ca^{2+}]$: 3.75 mmol/600 mL = 6.25×10^{-3} M

The low and high $[C_2O_4^{2-}]$ that could possibly cause calcium oxalate precipitation are:

Low $[C_2O_4^{2-}]$: $[Ca^{2+}][C_2O_4^{2-}] = K_{sp}$

$(6.25 \times 10^{-3})[C_2O_4^{2-}] = 1.3 \times 10^{-9}$ and $[C_2O_4^{2-}] = 2.1 \times 10^{-7}$ M

High $[C_2O_4^{2-}]$: $(8.3 \times 10^{-4})[C_2O_4^{2-}] = 1.3 \times 10^{-9}$ and $[C_2O_4^{2-}] = 1.6 \times 10^{-6}$ M

Using the oxalic acid ionization equations, it can be easily shown that at pH 7.0 a large proportion of the oxalate is in the form of the oxalate ion, approximately 99.8%. Therefore, we will take these low and high oxalate ion concentrations to essentially represent the low and high total oxalate concentrations in the wine. Consequently, the range of oxalate intake that might cause calcium oxalate to precipitate is

Low oxalate intake:

$$\frac{600\text{ mL}}{1\text{ day}} \times \frac{2.1 \times 10^{-7}\text{ mmol oxalate}}{1\text{ mL}} = 1.3 \times 10^{-4}\text{ mmol oxalate/day}$$

High oxalate intake:

$$\frac{1500\text{ mL}}{1\text{ day}} \times \frac{1.6 \times 10^{-6}\text{ mmol oxalate}}{1\text{ mL}} = 2.4 \times 10^{-3}\text{ mmol oxalate/day}$$

18.109 The reactions that have to be considered are

$BaF_2(s) \rightleftharpoons Ba^{2+}(aq) + 2F^-(aq)$

$F^-(aq) + H^+(aq) \rightleftharpoons HF(aq)$

The Ba^{2+} ion remains unchanged in solution. The molar solubility of BaF_2, x, is equal to $[Ba^{2+}]$, which is also equal to half the sum of the concentrations of the species containing the fluoride ion, or

$$x = \frac{[F^-] + [HF]}{2} \quad \text{and} \quad [F^-] + [HF] = 2x$$

The ratio of [HF] to $[F^-]$ at this pH is ($[H^+] = 3.5 \times 10^{-3}$ M and $K_a = 3.53 \times 10^{-4}$)

$$\frac{[HF]}{[F^-][H^+]} = \frac{1}{K_a} \quad \text{and} \quad \frac{[HF]}{[F^-]} = \frac{[H^+]}{K_a} = \frac{3.5 \times 10^{-3}}{3.53 \times 10^{-4}} = 9.9$$

Therefore, $[HF] = 9.9[F^-]$; $[F^-] + 9.9[F^-] = 2x$ and $[F^-] = 2x/10.9$

Substituting in the K_{sp} expression gives

$[Ba^{2+}][F^-]^2 = K_{sp}$

$$(x)\left[\frac{2x}{10.9}\right]^2 = 1.3 \times 10^{-6} \quad \text{and} \quad x = 3.4 \times 10^{-2}\text{ M}$$

The molar solubility of BaF_2 in a solution buffered at pH 2.46 is 3.4×10^{-2} mol/L.

18.111 The reactions that have to be considered are

$MgC_2O_4(s) \rightleftharpoons Mg^{2+}(aq) + C_2O_4^{2-}(aq)$ (1)

$C_2O_4^{2-}(aq) + H^+(aq) \rightleftharpoons HC_2O_4^-(aq)$ (2)

$HC_2O_4^-(aq) + H^+(aq) \rightleftharpoons H_2C_2O_4(aq)$ (3)

The Mg^{2+} ion remains unchanged in solution. The $[Mg^{2+}]$ is equal to the molar solubility of MgC_2O_4, 0.010 M, which is also equal to the sum of the concentrations of the species containing the oxalate ion, or

$[Mg^{2+}] = [C_2O_4^{2-}] + [HC_2O_4^-] + [H_2C_2O_4] = 0.010$ M (4)

Since the solution is saturated with MgC_2O_4, the $[C_2O_4^{2-}]$ can be obtained from the K_{sp} expression

$[Mg^{2+}][C_2O_4^{2-}] = 8.6 \times 10^{-5}$

$(0.010)[C_2O_4^{2-}] = 8.6 \times 10^{-5}$ and $[C_2O_4^{2-}] = 8.6 \times 10^{-3}$ M

To solve the problem we will first establish a relationship between H^+ and $HC_2O_4^-$ using the equilibrium constant expression for Equation (2), and then substitute this relationship, and the relationship between H^+ and $H_2C_2O_4$ obtained using Equation (4), into the equilibrium constant expression for Equation (3):

$K_{a_2} = [H^+][C_2O_4^{2-}]/[HC_2O_4^-]$

$$\frac{[H^+]}{[HC_2O_4^-]} = \frac{K_{a_2}}{[C_2O_4^{2-}]} = \frac{6.40 \times 10^{-5}}{8.6 \times 10^{-3}} = 7.44 \times 10^{-3}$$

and $[H^+] = 0.00744[HC_2O_4^-]$ or $[HC_2O_4^-] = [H^+]/0.00744$

From Equation (4) we have

$$[H_2C_2O_4] = -[HC_2O_4^-] + 0.010 - 0.0086 = \frac{-[H^+]}{0.00744} + 0.0014 = -134.4[H^+] + 0.0014$$

Substituting these expressions for $[H_2C_2O_4]$ and $[HC_2O_4^-]$ into the equilibrium constant expression for Equation (3) gives

$K_{a_1} = [H^+][HC_2O_4^-]/[H_2C_2O_4]$

$$5.90 \times 10^{-2} = \frac{[H^+] \times 134.4[H^+]}{-134.4[H^+] + 0.0014}$$

$134.4[H^+]^2 + 7.93[H^+] - 8.26 \times 10^{-5} = 0$ and $[H^+] = 1.0 \times 10^{-5}$ M

Comment: This problem can also be worked out by assuming that only the first protonation of $C_2O_4^{2-}$ (Equation (2)) has to be considered, solving the problem using this assumption, and checking to see if the assumption is valid. If the assumption is valid, nothing further has to be done. If the assumption is invalid, then the complete treatment used above has to be employed. Using this assumption, the problem would have the following solution:

$[Mg^{2+}] = [C_2O_4^{2-}] + [HC_2O_4^-] = 0.010$ M

$[C_2O_4^{2-}] = 8.6 \times 10^{-3}$ M (as before) and $[HC_2O_4^-] = 0.010 - [C_2O_4^{2-}]$

$= 0.010 - 0.0086 = 0.0014$ M (carry an extra significant figure)

Substituting $[C_2O_4^{2-}] = 0.0086$ M and $[HC_2O_4^-] = 0.0014$ M into the equilibrium constant expression for Equation (2) gives

$$[H^+] = K_{a_2} \times \frac{[HC_2O_4^-]}{[C_2O_4^{2-}]} = \frac{(6.40 \times 10^{-5})(0.0014)}{(0.0086)} = 1.0 \times 10^{-5} \text{ M}$$

The assumption can be tested by calculating the amount of $H_2C_2O_4$ present at equilibrium. The amount of $H_2C_2O_4$ can be obtained from the combined equilibrium constant expression for oxalic acid.

$$[H_2C_2O_4] = \frac{[H^+]^2[C_2O_4^{2-}]}{K_{a_1} \times K_{a_2}} = \frac{(1.0 \times 10^{-5})^2(8.6 \times 10^{-3})}{(5.90 \times 10^{-2})(6.40 \times 10^{-5})} = 2.3 \times 10^{-7}\ M$$

The amount of $H_2C_2O_4$ is indeed quite small and the assumption is valid. This shows that assumptions, hopefully based on sound chemical principles, can possibly simplify the solutions to problems considerably. However, the validity of the assumptions must always be confirmed; exceptions being those cases where the standard approximations are always made. As it turns out, normally the reaction of a polyprotic base, the conjugate base of a polyprotic acid, with water takes place for the most part in the first step and, therefore, only the first hydrolysis or protonation step need be considered in most calculations.

18.113 $Ag^+(aq) + 2CN^-(aq) \rightleftharpoons Ag(CN)_2^-$ $K_f = 1 \times 10^{21}$

Because K_f is very large, virtually all of the silver will be present as the complex; therefore it is reasonable to take the $[Ag(CN)_2^-] = 0.0750$ M. Substituting $[Ag(CN)_2^-] = 0.0750$ M, $[Ag^+] = 1.0 \times 10^{-10}$ M, and the K_f value into the K_f expression gives

$$\frac{[Ag(CN)_2^-]}{[Ag^+][CN^-]^2} = K_f$$

$$\frac{(0.0750)}{(1.0 \times 10^{-10})[CN^-]^2} = 1 \times 10^{21} \text{ and } [CN^-] = 9 \times 10^{-7}\ M$$

The number of moles of free CN^- is $0.250\ L \times 9 \times 10^{-7}$ mol/L $= 2 \times 10^{-7}$ mol. However, we also have

$$0.250\ L \times \frac{0.0750 \text{ mol } Ag(CN)_2^-}{1\ L} \times \frac{2 \text{ mol } CN^-}{1 \text{ mol } Ag(CN)_2^-} = 0.0375 \text{ mol } CN^-$$

combined in the complex. Consequently, the total number of moles of KCN that must be added to the solution is $0.0375 + 2 \times 10^{-7} = 0.0375$ mol, to three significant figures.

SURVEY OF THE ELEMENTS 3
THE REMAINING NONMETALS

Solutions To Practice Exercises

PE S3.1 Draw a horizontal line from left to right at 1 atm pressure to intersect the rhombic-monoclinic transformation curve. Then draw a vertical line down to the temperature scale. The phases are in equilibrium under 1 atm pressure at approximately 95°C.

PE S3.2 $Fe_2O_3(s) + 3H_2SO_4(aq) \rightarrow Fe_2(SO_4)_3(aq) + 3H_2O(l)$

PE S3.3 (a) $SO_4^{2-}(aq) + 10H^+(aq) + 8e^- \rightarrow H_2S(g) + 4H_2O(l)$ reduction

(b) $4(2I^-(aq) \rightarrow I_2(s) + 2e^-)$ oxidation

The net ionic equation is the sum of the adjusted oxidation and reduction half-reactions:

$SO_4^{2-}(aq) + 8I^-(aq) + 10H^+(aq) \rightarrow H_2S(g) + 4I_2(s) + 4H_2O(l)$

PE S3.4 Refer to Figure S3.13.

(a) Each oxygen atom has four VSEPR pairs (two bonding and two lone pairs) and, therefore, is sp^3 hybridized. According to this model one would expect the P–O–P bond angles to be slightly less than 109.5°, because of some lone pair–lone pair repulsion. (Note: The observed angle is 127°.)

(b) Predicted angle same as (a). (Note: The observed angle is 123°.)

Comment: VSEPR predictions are not universally correct.

PE S3.5 $H_3PO_4(aq) + 3NaOH(aq) \rightarrow Na_3PO_4(aq) + 3H_2O(l)$

PE S3.6 $H_2SO_4(aq) + NaHCO_3(aq) \rightarrow NaHSO_4(aq) + CO_2(g) + H_2O(l)$

PE S3.7 (a) $[:C \equiv C:]^{2-}$ (b) $\ddot{\underset{\cdot\cdot}{S}}=C=\ddot{\underset{\cdot\cdot}{S}}$ (c) $H-C \equiv N:$

PE S3.8 $B(OH)_4^-$: formal charge = 3 – 0 – 4 = –1

Solutions To Final Exercises

S3.1
(a) oxygen, sulfur, selenium, tellurium, and polonium
(b) tellurium
(c) selenium and tellurium

S3.3 Refer to Figure S3.2 in the textbook.

S3.5 Sulfur in the form of the free element is found in vast deposits lying on domes of salt, usually located several hundred feet underground. Much of the sulfur occurs in the form of metal sulfide minerals, pyrite (FeS_2) is an example, and to a lesser extent in the sulfate minerals, anhydrite ($CaSO_4$) is an example.

S3.7
(a) Refer to "Occurrence, Preparation, Properties and Uses of the Elements" in Section S3.1 for details about each type of sulfur.

$$S_8(\text{rhombic}) \overset{95.5^\circ C}{\rightleftharpoons} S_8(\text{monoclinic}) \overset{119.0^\circ C}{\rightleftharpoons} S_8(l) \underset{\text{to } 190^\circ C}{\overset{\text{heat}}{\rightleftharpoons}} \text{viscous sulfur} \underset{>190^\circ C}{\overset{\text{heat}}{\rightleftharpoons}} \text{mobile sulfur} \overset{447.7^\circ C}{\rightleftharpoons}$$

$$\text{gaseous sulfur} \overset{1000^\circ C}{\rightleftharpoons} S_2(g)$$

Sudden cooling of liquid sulfur heated to 160°C or more produces plastic sulfur. Sudden cooling of gaseous sulfur at 1000°C produces a paramagnetic solid made up of S_2 molecules.

(b) Refer to Figure S3.5 in the textbook.

In the change from rhombic to monoclinic sulfur, the S_8 ring molecules in the rhombic crystal lattice rearrange themselves into a more stable arrangement, which is the monoclinic crystal lattice. When monoclinic sulfur melts, a liquid composed of S_8 ring molecules forms. Upon further heating, above 159°C, the S_8 ring molecules become unstable, open up, and combine to form less mobile long chains, which is viscous sulfur. Further heating still, above 200°C, brakes down these long chains into smaller chains, which is mobile sulfur. At the boiling point, gaseous sulfur composed mainly of S_8 ring molecules forms. As the temperature rises, these S_8 ring molecules break down and S_2 diatomic molecules are formed. At 1000°C the vapor consists entirely of S_2 molecules. Sudden cooling of the vapor at 1000°C produces a paramagnetic solid, unstable above –80°C, that is made up of S_2 molecules.

S3.9 Refer to The "+6 Oxidation State: Sulfuric Acid and Sulfates" in Section S3.1 in the textbook.

S3.11
(a) See Figure S3.4 and answer to Final Exercise 10.107.
(b) Each S atom has four VSEPR pairs (two bonding and two lone pairs) and, therefore, is sp^3 hybridized. Consequently, the bond angles should be approximately 109.5°.
(c) The S–S–S tetrahedral bond angles force the ring to take on a puckered shape, since that shape has the least potential energy associated with it.

S3.13 Some of the resonance structures are:

(a) O=S–O ⟷ O–S=O ⟷ O=S=O

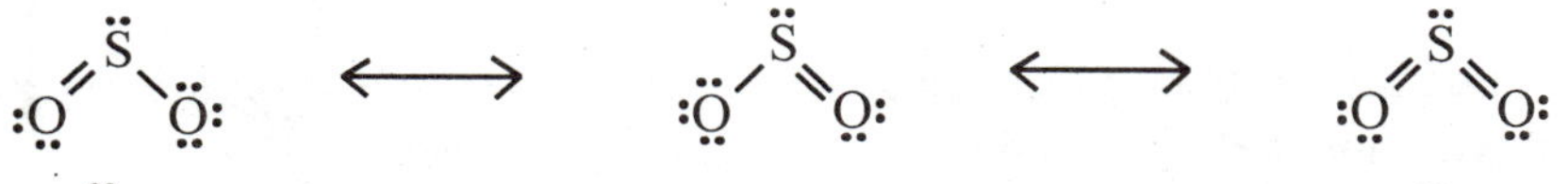

(b) ⟷ ⟷ ⟷ (resonance structures of SO_3 with S bonded to three O atoms)

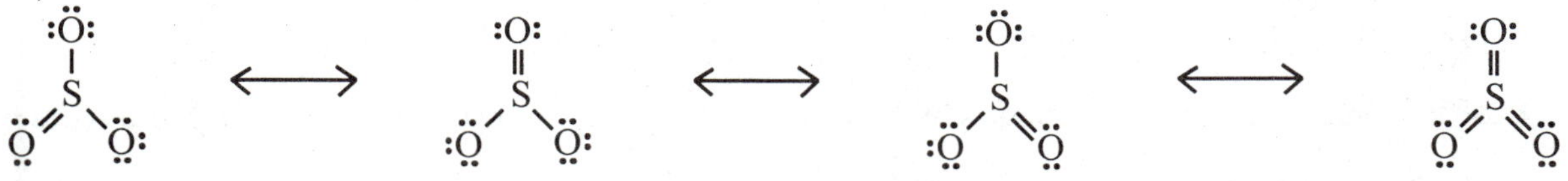

(c) H–O–S–O–H with –O: single bond ⟷ H–O–S–O–H with =O: double bond

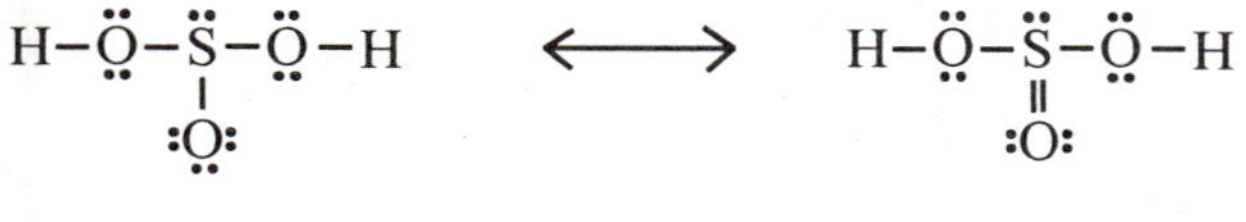

(d) $\left[\begin{array}{c} :\ddot{O}: \\ | \\ :\ddot{O}-S-\ddot{O}: \\ | \\ :\ddot{O}: \end{array} \right]^{2-}$ or $\left[\begin{array}{c} :O: \\ \| \\ :\ddot{O}-S-\ddot{O}: \\ \| \\ :O: \end{array} \right]^{2-}$ resonance hybrid

Comment: In Lewis structures any of the $S-\ddot{O}:$ bonds can be written as $S=\ddot{O}$ and vice versa. Some chemists prefer one form over the other, depending on their viewpoint.

S3.15 (a) SF_6: S bonded to six :F̤: atoms (b) $:\ddot{Cl}-\ddot{S}-\ddot{S}-\ddot{Cl}:$ (c) $:\ddot{Cl}-\ddot{S}-\ddot{Cl}:$

S3.17 Sulfur, being in the third period, has empty d orbitals in the valence shell that can be involved in bonding and it is therefore known to form as many as six bonds. Oxygen, on the other hand, being in the second period has only two vacancies in its valence shell and can form only two bonds.

S3.19 Refer to Table S3.3 in the textbook.

S3.21 (a) $S_2O_3^{2-}$: $x_O = -2$ (Rule 4); therefore $x_S = +2$ (Rule 2).

(b) $S_2O_6^{2-}$: $x_O = -2$ (Rule 4); therefore $x_S = +5$ (Rule 2).

(c) $S_4O_6^{2-}$: $x_O = -2$ (Rule 4); therefore $x_S = +2.5$ (Rule 2).

S3.23 (a) $Na_2SO_3(aq) + 2HCl(aq) \rightarrow 2NaCl(aq) + SO_2(g) + H_2O(l)$

(b) $2S_2O_3^{2-}(aq) + I_2(s) \rightarrow S_4O_6^{2-}(aq) + 2I^-(aq)$

(c) $3(S^{2-}(aq) \rightarrow S(s) + 2e^-)$ oxidation

$2(CrO_4^{2-}(aq) + 4H_2O(l) + 3e^- \rightarrow Cr(OH)_4^-(aq) + 4OH^-(aq))$ reduction

$3S^{2-}(aq) + 2CrO_4^{2-}(aq) + 8H_2O(l) \rightarrow 3S(s) + 2Cr(OH)_4^-(aq) + 8OH^-(aq)$

(d) $Na_2SO_3(aq) + S(s) \xrightarrow{\text{heat}} Na_2S_2O_3(aq)$

S3.25 (a) $SO_2(g) + H_2O(l) \rightleftharpoons H_2SO_3(aq)$

(b) $SO_2(g) + OH^-(aq) \rightarrow HSO_3^-(aq)$
limited base — hydrogen sulfite ion

$SO_2(g) + 2OH^-(aq) \rightarrow SO_3^{2-}(aq) + H_2O(l)$
excess base — sulfite ion

(c) $CaCO_3(s) + SO_2(g) \rightarrow CaSO_3(s) + CO_2(g)$

S3.27 (a) Wood is made up largely of cellulose $(C_6H_{10}O_5)_n$, a polymer in which the unit $C_6H_{10}O_5$ is repeated n times. H_2SO_4 chars wood by removing the hydrogen and oxygen from the cellulose.

$(C_6H_{10}O_5)_n(s) + 5nH_2SO_4(98\%) \rightarrow 6nC(s) + 5nH_2SO_4 \cdot H_2O(l)$

(b) The concentrated sulfuric acid dehydrates the $CuSO_4 \cdot 5H_2O$ forming $CuSO_4$, which is a white powder.

$CuSO_4 \cdot 5H_2O(s) + 5H_2SO_4(98\%) \rightarrow CuSO_4(s) + 5H_2SO_4 \cdot H_2O(l)$

(c) The concentrated sulfuric acid oxidizes the bromide ion to bromine, which is a red brown liquid (see "Hydrogen Halides and Their Salts" in Section S2.3 in the textbook).

$$2NaBr(s) + 3H_2SO_4(98\%) \rightarrow Br_2(l) + 2NaHSO_4(s) + SO_2(g) + 2H_2O(l)$$

(d) The concentrated sulfuric acid oxidizes the copper.

$$Cu(s) + 2H_2SO_4(aq)(concentrated) \rightarrow CuSO_4(aq) + SO_2(g) + 2H_2O(l)$$

S3.29 (a) Sulfur has a similar outer or valence electronic configuration as oxygen. They are both Group 6A elements. Therefore, sulfur can participate in the same kind of bonding as oxygen.
(b) Sulfur replacement should diminish the strength of an oxo acid, since sulfur is less electronegative than oxygen and, therefore, the O–H bond should not be as polar.

S3.31 (a) nitrogen, phosphorus, arsenic, antimony, and bismuth
(b) arsenic and antimony

S3.33 Refer to Figure S3.7 and the accompanying text in the textbook.

S3.35 Refer to Figure S3.11 in the textbook.

S3.37 (a) Refer to Figures S3.8 and S3.13 in the textbook.
(b) P_4O_6 is related to P_4 by having an oxygen atom inserted in each of the six edges of the P_4 tetrahedron. P_4O_{10} is related to P_4O_6 by having another oxygen atom added or bonded to each of the four phosphorus atoms.

S3.39 Refer to the Lewis structures in Figure S3.14. In H_3PO_4 each H atom is attached to an oxygen atom and is therefore acidic. In H_3PO_3, on the other hand, only two of the H atoms are attached to oxygen atoms; the third H atom is attached to phosphorus, and therefore, only two of the H atoms are acidic.

S3.41 (a) $P_4(s) + 3O_2(g) \xrightarrow{\text{limited oxygen}} P_4O_6(s)$

(b) $P_4(s) + 5O_2(g) \xrightarrow{\text{excess oxygen}} P_4O_{10}(s)$

(c) $P_4(l) + 3S(l) \xrightarrow{\text{heat}} P_4S_3(s)$ (+ $P_4S_5(s)$, $P_4S_7(s)$, etc.)

(d) $P_4(s) + 12Na(s) \xrightarrow{\text{heat}} 4Na_3P(s)$

S3.43 (a) $P_4O_6(s) + 6H_2O(l) \rightarrow 4H_3PO_3(aq)$ (b) $P_4O_{10}(s) + 6H_2O(l) \rightarrow 4H_3PO_4(aq)$

S3.45 (a) 1. $P_4O_{10}(s) + 6H_2O(l \text{ or } g) \rightarrow 4H_3PO_4(l)$
2. $Ca_3(PO_4)_2(s) + 3H_2SO_4(aq) \rightarrow 2H_3PO_4(85\%) + 3CaSO_4(s)$
phosphate rock

(b) Method (1) gives the purer product, since phosphate rock is not just an impure $Ca_3(PO_4)_2$, but is a mineral composed of several different phosphorus compounds, the main ones being fluorapatite ($3Ca_3(PO_4)_2{\cdot}CaF_2$), chlorapatite ($3Ca_3(PO_4)_2{\cdot}CaCl_2$), and hydroxyapatite ($3Ca_3(PO_4)_2{\cdot}Ca(OH)_2$).

S3.47 (a) $PCl_5(s) + 4H_2O(l) \rightarrow H_3PO_4(aq) + 5HCl(aq)$
(b) $AsCl_3(s \text{ or } l) + 3H_2O(l) \rightarrow As(OH)_3(aq) + 3HCl(aq)$

S3.49 (a) $Ca_3(PO_4)_2(s) + H_2SO_4(aq) \rightarrow 2CaHPO_4(aq) + CaSO_4(aq)$
(b) $Ca_3(PO_4)_2(s) + 2H_2SO_4(aq) \rightarrow Ca(H_2PO_4)_2(s) + 2CaSO_4(s)$

These products make better fertilizers because they are more soluble than the original phosphate rock.

S3.51 This illustrates the trend in main periodic groups of favoring the lower of two positive oxidation states toward the bottom of the group. Both As and P are in +5 oxidation states. But As lies below P in Group 5A and, therefore, its +5 oxidation state will be somewhat less stable and, consequently, more reactive than the +5 oxidation state of phosphorous, everything else being equal.

S3.53 (a) carbon, silicon, germanium, tin and lead
(b) silicon and germanium
(c) silicon and germanium

S3.55 Refer to Figure 2.18(b) in the textbook.

Graphite is a shiny, black, soft solid. It conducts electricity, has a high melting point, and lubricating properties when it absorbs air and moisture.

Graphite consists of layers of fused, six-membered rings. A delocalized pi orbital exists over each entire layer. These delocalized pi electrons account for graphite's ability to conduct electricity, and also its shiny, black appearance. The layers themselves are spaced relatively far apart and are only held together by van der Waals forces. This accounts for graphite's softness and ability to absorb air and moisture and thereby take on lubricating properties. Graphite has a high melting point, because the sigma covalent bonds of the fused rings are strong and the delocalized pi bond stabilizes the entire layer.

S3.57 Refer to "Carbon, the Element" and "Chemical Insight: Buckyballs and Other Fullerenes" in Section S3.3 in the textbook.

S3.59 Step 1: Crushed limestone is decomposed to CaO and CO_2.

$$CaCO_3(s) \xrightarrow{heat} CaO(s) + CO_2(g)$$

Step 2: The CO_2 is passed into concentrated salt water that has been made basic with ammonia. Sparingly soluble $NaHCO_3$ precipitates out and is separated from the solution by filtration.

$$CO_2(g) + NaCl(aq) + H_2O(l) + NH_3(aq) \rightarrow NaHCO_3(s) + NH_4Cl(aq)$$

Step 3: The costly ammonia is recovered, and then recycled, by adding CaO from Step 1 to the NH_4Cl solution from Step 2.

$$CaO(s) + 2NH_4Cl(aq) \rightarrow CaCl_2(aq) + H_2O(l) + 2NH_3(g)$$

S3.61 1. Manufacture of carbonated beverages. 2. Used to cause dough to rise. 3. Used to smother or put out fires. 4. Solid carbon dioxide (Dry Ice) is sometimes used as a refrigerant.

S3.63 A carbide is a binary compound of carbon in which carbon is combined with a less electronegative element. By convention, the compounds of carbon with hydrogen are not considered in this category. Also, the familiar binary compounds of carbon with oxygen, sulfur, the halogens, etc., are not carbides. Carbides are usually divided into three types: (1) ionic or saltlike carbides, compounds formed for the most part with elements of Groups 1A, 2A, and 3A; (2) interstitial carbides, compounds formed for the most part with transition metals; and (3) covalent carbides, compounds formed with nonmetals and semimetals.

Ionic carbides like CaC_2 are used as a portable fuel source.

The interstitial carbides WC and TiC are used to make tougher and longer lasting cutting tools.

The covalent carbides B_4C and SiC are used as abrasives. SiC is also used to make resistance elements for electric furnaces.

S3.65 (a) $\ddot{\underset{\cdot\cdot}{O}}=C=\ddot{\underset{\cdot\cdot}{O}}$ (b) $H-\ddot{\underset{\cdot\cdot}{O}}-\underset{\underset{:O:}{\|}}{C}-\ddot{\underset{\cdot\cdot}{O}}-H$ (c) $H-C\equiv N:$ (d) $\ddot{\underset{\cdot\cdot}{S}}=C=\ddot{\underset{\cdot\cdot}{S}}$

S3.67 (a) $2C(s) + O_2(g) \rightarrow 2CO(g)$

(b) $C(s) + O_2(g) \rightarrow CO_2(g)$

Limited oxygen supply and/or temperatures above 500°C favor the formation of CO.

S3.69 Cyanide solutions must be kept basic, because otherwise highly toxic hydrogen cyanide gas (HCN(g)) would be released into the air.

$KCN(aq) + H^+(aq) \rightarrow K^+(aq) + HCN(g)$

S3.71 An acetylide is a carbide containing the C_2^{2-} ion.

(a) $CaO(s) + 3C(s) \xrightarrow{\text{heat}} CaC_2(l) + CO(g)$

(b) $CaC_2(s) + 2H_2O(l) \rightarrow Ca(OH)_2(s) + C_2H_2(g)$

S3.73 Silicon is a major constituent of the earth's crust, comprising 27.72% by mass and, in combination with oxygen, provides the structure of most minerals and rocks. Germanium, on the other hand, is a rare element found as a sulfide mixed with the sulfide ores of arsenic, copper, and zinc.

S3.75 (a) The crystal lattices of silicon and germanium are the same as the diamond lattice of carbon. However, since the silicon and germanium atoms are larger than carbon atoms, their tetrahedrally directed sigma bonds are weaker and their crystals are not as hard as diamond. Also, since the bonds in silicon and germanium are larger and not so strong as those in diamond, their sigma bonding electrons are not so firmly localized in the crystal lattice, and, consequently, silicon and germanium are semiconductors, while diamond is an insulator.

(b) Silicon and germanium do not form graphite-type lattices because their atoms are too large to form strong pi bonds.

S3.77 A silicate is a compound containing the tetrahedral SiO_4^{4-} anion or extended anions resulting from the joining together of two or more SiO_4 tetrahedra by shared oxygen atoms. Up to three oxygen atoms of each SiO_4 tetrahedron can be used as bridge atoms. The nonsilicate crystalline network covalent compound SiO_2 is produced if all four oxygen atoms of each SiO_4 tetrahedron are used as bridge atoms. Aluminosilicates are compounds in which aluminum atoms have replaced some of the silicon atoms in a silicate structure. They can be thought of as compounds containing extended anions resulting from the joining together of SiO_4 tetrahedra with some (hypothetical) AlO_4 tetrahedra by shared oxygen atoms. The aluminosilicate extended anions have a greater negative charge than a corresponding extended silicate anion, since AlO_4 would have a 5– charge if it existed.

Feldspars, certain micas (for example, muscovite, $KAl_3Si_3O_{10}(OH)_2$), and zeolites are aluminosilicates.

S3.79 (a) $\left[\begin{array}{c} :\ddot{O}: \\ | \\ :\ddot{\underset{\cdot\cdot}{O}}-Si-\ddot{\underset{\cdot\cdot}{O}}: \\ | \\ :\underset{\cdot\cdot}{O}: \end{array}\right]^{4-}$

orthosilicate ion (SiO_4^{4-})

(b) $\left[\begin{array}{c} :\ddot{O}: \quad :\ddot{O}: \\ | \quad\quad | \\ :\ddot{\underset{\cdot\cdot}{O}}-Si-\ddot{\underset{\cdot\cdot}{O}}-Si-\ddot{\underset{\cdot\cdot}{O}}: \\ | \quad\quad | \\ :\underset{\cdot\cdot}{O}: \quad :\underset{\cdot\cdot}{O}: \end{array}\right]^{6-}$

pyrosilicate ion ($Si_2O_7^{6-}$)

(c) Water glass is a saturated solution of sodium silicate (Na_4SiO_4).

Comment: Water glass is really more complicated than the formula Na_4SiO_4 suggests. Water glass contains a variety of relatively small linear extended anions of SO_4 tetrahedra linked together by oxygen bridges, i.e., a variety of linear anion chains each containing multiple SiO_3^{2-} repeat units. These extended chains are called pyroxene chains. So, pyroxene chains of varying length occur in water glass. This is why the formula for water glass is often given as Na_2SiO_3. In conclusion, water glass is a complicated mixture of several different chain sodium silicates called pyroxenes. The formulas Na_4SiO_4 and Na_2SiO_3 are just rough approximations of the actual elemental composition. It is also one of the few soluble silicates.

S3.81 (a) See Figure S3.31a in the textbook.
(b) The thermal stability of the silicon-oxygen bonds.
(c) Silicones are used to make water repellent fabrics, lubricants, heat and electrical insulators, releasing agents, and polishes.

S3.83 Boron occurs in the mineral borax, which is found in dried lake beds in certain desert areas. Boron is prepared by reducing B_2O_3, obtained from the chemical treatment of borax, with magnesium at high temperature:

$$B_2O_3(s) + 3Mg(s) \xrightarrow{\text{heat}} 2B(s) + 3MgO(s)$$

Elemental boron is a hard, brittle, network covalent crystalline solid. It is a semimetal and a semiconductor. It is the only nonmetal in Group 3A.

S3.85 (a) Boron crystals are composed of B_{12} groups interconnected by covalent bonds. Each B_{12} group is in the form of a regular icosahedron, a twelve-cornered figure with 20 triangular sides.
(b) See Figure S3.34 in the textbook.
(c) Boric acid is flat or planar with an equilateral triangle shape that has the boron atom in the center of the equilateral triangle and OH groups occupying the three corners of the equilateral triangle.

S3.87 Refer to Figure S3.37 and accompanying discussion in Section S3.5 in the textbook.

S3.89 $B(OH)_3(aq) + H_2O(l) \rightleftharpoons B(OH)_4^-(aq) + H^+(aq)$

or $(H_2O)B(OH)_3(aq) \rightleftharpoons B(OH)_4^-(aq) + H^+(aq)$

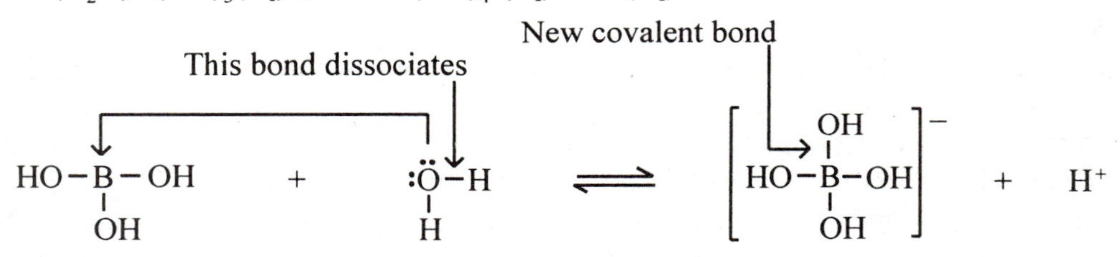

No. Boric acid is a Lewis acid, not a Brønsted acid. Also, the reason why boric acid is not a triprotic acid, as you might expect it to be from looking at its structure, is because it acts as a Lewis acid and not a Brønsted acid.

S3.91 (a) The reaction in terms of the tetraborate ion, $B_4O_7^{2-}$, portion of borax, $Na_2B_4O_7 \cdot 10H_2O$, is

$$B_4O_7^{2-} + 2H^+(aq) + 5H_2O(l) \rightleftharpoons 4H_3BO_3(aq)$$

(strong acid)

(b) $B(OH)_3(aq) + H_2O(l) \rightleftharpoons B(OH)_4^-(aq) + H^+(aq)$

(c) $2H_3BO_3(s) \xrightarrow{\text{heat}} B_2O_3(s) + 3H_2O(g)$

S3.93 (a) brimstone: S, sulfur
(b) epsom salts: $MgSO_4 \cdot 7H_2O$, magnesium sulfate heptahydrate
(c) gypsum: $CaSO_4 \cdot 2H_2O$, calcium sulfate dihydrate
(d) oil of vitriol: H_2SO_4, sulfuric acid
(e) oleum: $H_2S_2O_7$, pyrosulfuric acid
(f) phosphoric anhydride: P_4O_{10}, phosphorus pentoxide
(g) silica gel: SiO_2, silicon dioxide (with some hydration)
(h) talc: $Mg_2(Si_2O_5)_2 \cdot Mg(OH)_2$, sheet silicate
(i) borax: $Na_2B_4O_7 \cdot 10H_2O$, hydrated sodium tetraborate

S3.95 (a) Refer to "Chemical Insight: Phosphorus, the Limiting Element" in Section 3.8 in the textbook. Phosphorus detergents are a major cause of lake eutrophication.
(b) Phosphate rock is insoluble and, therefore, does not enter the water system.

S3.97 We have to show that the ion product Q exceeds the solubility product constant K_{sp}. If $Q > K_{sp}$, then a precipitate forms. If $Q \leq K_{sp}$, then all the ions remain in solution.

(a) $[Pb^{2+}] = \dfrac{10 \text{ mL} \times 0.10 \text{ mmol/mL}}{11 \text{ mL}} = 0.091 \text{ M}$

$[SO_4^{2-}] = \dfrac{1.0 \text{ mL} \times 0.10 \text{ mmol/mL}}{11 \text{ mL}} = 0.0091 \text{ M}$

Substituting these concentrations into the ion product expression gives

$Q = [Pb^{2+}][SO_4^{2-}] = (0.091)(0.0091) = 8.3 \times 10^{-4}$

The ion product Q is greater than the K_{sp} value of 6.3×10^{-7}, so $PbSO_4$ will precipitate.

(b) Since the K_{sp} value, 6.3×10^{-7}, is small and Pb^{2+} is present in excess, we can assume that all the SO_4^{2-} precipitates out. So, in the final equilibrium mixture we have

$[Pb^{2+}] = 0.091 \text{ M} - 0.0091 \text{ M} = 0.082 \text{ M}$

Check: The minimum $[Pb^{2+}]$ after precipitation is 0.082 M. Therefore, the maximum $[SO_4^{2-}]$ after precipitation is

$[SO_4^{2-}] = K_{sp}/[Pb^{2+}] = 6.3 \times 10^{-7}/0.082 = 7.7 \times 10^{-6} \text{ M}$

Consequently, the maximum $[Pb^{2+}]$ after precipitation is

$[Pb^{2+}] = 0.082 \text{ M} + 7.7 \times 10^{-6} \text{ M} = 0.082 \text{ M}$

Therefore, the assumption is valid.

S3.99 (a) $Ca(OH)_2(s \text{ or } aq) + SO_2(g) \rightarrow CaSO_3(s) + H_2O(l)$
(b) $MgO(s) + SO_2(g) \rightarrow MgSO_3(s)$

Note: Both sulfites soon oxidize to sulfates in the air.

S3.101 (a) $Si_3H_8(g) + 5O_2(g) \rightarrow 3SiO_2(s) + 4H_2O(l)$

The enthalpy change is the sum of the enthalpy changes for the following steps:

		$\Delta H°$
1. Breaking 2 mol of Si–Si bonds:	$2 \times +222$ kJ =	+444 kJ
2. Breaking 8 mol of Si–H bonds:	$8 \times +318$ kJ =	+2540 kJ
3. Breaking 5 mol of O=O bonds:	$5 \times +498.3$ kJ =	+2492 kJ
4. Forming 12 mol of Si–O bonds:	12×-452 kJ =	–5420 kJ
5. Forming 8 mol of O–H bonds:	8×-463 kJ =	–3700 kJ
	ΔH_c° = Sum =	–3644 kJ

The enthalpy of combustion of Si_3H_8 is –3644 kJ/mol. The $\Delta H°$ for the combustion of 1.00 mol of Si_3H_8 is 1.00 mol × –3644 kJ/mol = –3640 kJ.

(b) The enthalpy of combustion of C_3H_8 is –2219.9 kJ/mol (see Table 6.2). The $\Delta H°$ for the combustion of 1.00 mol of C_3H_8 is 1.00 mol × –2219.9 kJ/mol = –2220 kJ. The enthalpy of combustion of Si_3H_8 is considerably greater than that of C_3H_8.

(c) Propane is much more stable than Si_3H_8. Therefore, propane can be safely handled, transported, and stored, while Si_3H_8 cannot.

S3.103 (a) S_2: $(\sigma_{3s})^2(\sigma^*_{3s})^2(\pi_{3px})^2(\pi_{3py})^2(\sigma_{3pz})^2(\pi^*_{3px})^1(\pi^*_{3py})^1$

O_2: $(\sigma_{2s})^2(\sigma^*_{2s})^2(\pi_{2px})^2(\pi_{2py})^2(\sigma_{2pz})^2(\pi^*_{2px})^1(\pi^*_{2py})^1$

They differ mainly in that sulfur uses orbitals from the third shell, while oxygen uses orbitals from the second shell.

(b) The size of sulfur does not allow the 3p orbitals to overlap sufficiently to form strong pi bonds.

S_2 has essentially a single σ bond, the σ_{3pz} bond, holding it together, since the net pi bond is weak. Therefore, any structure, like the S_8 ring structure, that has two σ bonds per sulfur atom would be more stable at room temperature.

S3.105 (a) Substituting d = 1.09 g/L, T = 298 K, and P = 0.500 atm into Equation 5.10 gives

$$\mathcal{M} = \frac{dRT}{P} = \frac{1.09 \text{ g}}{1 \text{ L}} \times \frac{0.0821 \text{ L•atm/mol•K} \times 298 \text{ K}}{0.500 \text{ atm}} = 53.3 \text{ g/mol}$$

(b) The molar mass of boron is 10.811 g/mol. Therefore, there can be at most four boron atoms in the molecule. Three boron atoms is not acceptable, however, since then we would have to exceed the maximum number of hydrogen atoms allowed, which is nine. Therefore, the molecule must contain four boron atoms. The number of hydrogen atoms in the molecule is

$$(53.3 - 43.2) \text{ g H} \times \frac{1 \text{ mol H atoms}}{1.008 \text{ g H}} = 10 \text{ mol H atoms}$$

or 10 H atoms per molecule. A reasonable molecular formula is therefore B_4H_{10}.

S3.107 All of the $KMnO_4$ originally present is consumed in converting the iodide ion to iodine. Therefore, the amount of iodine produced is a measure of the amount of $KMnO_4$ originally present. The reaction between $Na_2S_2O_3$ and I_2 is:

$$I_2(aq) + 2S_2O_3^{2-}(aq) \rightarrow 2I^-(aq) + S_4O_6^{2-}(aq)$$

The steps in the calculation are:

$V \times M\ (Na_2S_2O_3) \rightarrow$ mol $Na_2S_2O_3 \rightarrow$ mol $I_2 \rightarrow$ mol $KMnO_4 \rightarrow$ molarity $KMnO_4$.

$$0.02970 \text{ L} \times \frac{0.1010 \text{ mol Na}_2\text{S}_2\text{O}_3}{1 \text{ L}} \times \frac{1 \text{ mol I}_2}{2 \text{ mol Na}_2\text{S}_2\text{O}_3} \times \frac{2 \text{ mol KMnO}_4}{5 \text{ mol I}_2} = 5.999 \times 10^{-4} \text{ mol KMnO}_4$$

The molarity of the $KMnO_4$ solution is

$$\frac{5.999 \times 10^{-4} \text{ mol}}{0.02500 \text{ L}} = 0.02400 \text{ M}$$

Comment: In the presence of iodide ion, molecular iodine is mostly in the form of the triiodide ion and, therefore, the equation for the iodine-thiosulfate reaction is oftentimes written as follows:

$$I_3^-(aq) + 2S_2O_3^{2-}(aq) \rightarrow 3I^-(aq) + S_4O_6^{2-}(aq)$$

Both equations give the same results.

CHAPTER 19

FREE ENERGY, ENTROPY AND THE SECOND LAW OF THERMODYNAMICS

Solutions To Practice Exercises

PE 19.1 (a) The system is at equilibrium; therefore $\Delta G = 0$.
(b) No. maximum useful work $= -\Delta G = 0$

PE 19.2 (a) maximum useful work $= 2.00 \times (-\Delta G) = 2.00 \times [-(-817.9 \text{ kJ})] = 1.64 \times 10^3$ kJ
(b) heat evolved = (–enthalpy) – work done = 2.00×890.4 kJ – 1640 kJ = 141 kJ

PE 19.3 Substituting $K_p = 50.3$, T = 731 K, and R = 8.314 J/mol•K into Equation 19.4 gives

$$\Delta G^\circ = -RT \ln K_p = -8.314 \text{ J/mol•K} \times 731 \text{ K} \times \ln 50.3 = -23{,}800 \text{ J/mol} = -23.8 \text{ kJ/mol}$$

PE 19.4 (a)

$$H_2O(l) \rightarrow H_2O(g)$$

ΔG_f°(kJ): –237.129   –228.572

$$\Delta G^\circ = -228.572 \text{ kJ} - (-237.129 \text{ kJ}) = +8.6 \text{ kJ}$$

(b) The water vapor condenses to liquid water.

PE 19.5

$$\Delta G^\circ = 2 \times \Delta G_f^\circ(NO) + \Delta G_f^\circ(Cl_2) - 2 \times \Delta G_f^\circ(NOCl) = 2 \times 86.55 \text{ kJ} + 0 \text{ kJ} - 2 \times 66.08 \text{ kJ} = +40.94 \text{ kJ}$$

Substituting $\Delta G_f^\circ = 40.94$ kJ/mol = 40,940 J/mol, R = 8.314 J/mol•K, and T = 298 K into Equation 19.4 gives

$$\ln K_p = \frac{\Delta G^\circ}{-RT} = \frac{40{,}940 \text{ J/mol}}{-8.314 \text{ J/mol•K} \times 298 \text{ K}} = -16.5$$

and $K_p = \text{antilog}(-16.5) = e^{-16.5} = 7 \times 10^{-8}$

PE 19.6 (a) $H_2O(l) \rightleftharpoons H_2O(g)$

$$\Delta G = \Delta G^\circ + RT \ln Q = \Delta G^\circ + RT \ln P_{H_2O}$$

$\Delta G = 0 + 8.314 \times 10^{-3}$ kJ/mol•K × 373 K × ln 2.00 = +2.15 kJ/mol

(b) ΔG is positive, so condensation is thermodynamically favored.

PE 19.7 (a) Warming increases the entropy; the entropy increases.
(b) Two moles of gas are more disordered than 1 mol of gas; the entropy decreases.
(c) Two moles of gas are more disordered than 1 mol of gas and 1 mol of liquid; the entropy increases.

PE 19.8 The vaporization of liquid water at 100°C and 1 atm is an equilibrium process; therefore Equation 19.8 applies. The change in enthalpy is (the vaporization takes place under standard state conditions):

$\Delta H° = 1$ mol × 40.7 kJ/mol = 40.7 kJ

Substituting ΔH° = 40.7 kJ and T = 373 K into Equation 19.8 gives

$$\Delta S° = \frac{\Delta H°}{T} = \frac{40.7 \text{ kJ}}{373 \text{ K}} = 0.109 \text{ kJ/K} = 109 \text{ J/K}$$

PE 19.9 $1/2N_2(g) + 3/2H_2(g) \rightarrow NH_3(g)$

Substituting $\Delta G_f^\circ = -16.45$ kJ/mol, $\Delta H_f^\circ = -46.11$ kJ/mol, and T = 298 K into Equation 19.7 gives

$$\Delta G_f^\circ = \Delta H_f^\circ - T\Delta S_f^\circ$$

-16.45 kJ/mol = -46.11 kJ/mol – 298 K × ΔS_f° and $\Delta S_f^\circ = -0.0995$ kJ/mol•K = -99.5 J/mol•K

PE 19.10 (a)

	C(diamond) →	C(graphite)
S°(J/K):	2.377	5.740

ΔS° = 5.740 J/K – 2.377 J/K = +3.363 J/K

(b) Diamond has a lower entropy value; therefore the diamond lattice is more ordered.

PE 19.11
$$\Delta S° = \frac{\Delta H° - \Delta G°}{T} = \frac{-2816 \text{ kJ} + 2870 \text{ kJ}}{298 \text{ K}} = 0.181 \text{ kJ/K}$$

Assume that the values of ΔH° and ΔS° do not change significantly between 25°C and 37°C. Substituting ΔH° = –2816 kJ, ΔS° = 0.181 kJ/K, and T = 310 K into Equation 19.7 gives

ΔG° = ΔH° – TΔS° = –2816 kJ – 310 K × 0.181 kJ/K = –2872 kJ

PE 19.12 $\Delta G = \Delta G° + RT \ln Q = \Delta G° + RT \ln P_{CO_2}$ (1)

We want to find the temperature at which ΔG is zero. The ΔG° in Equation (1) is the ΔG° at that temperature. Using Equation 19.7 and assuming that the values of ΔH° and ΔS° do not change significantly with temperature gives

ΔG° = ΔH° – TΔS° = 178.3 kJ/mol – T × 0.1606 kJ/mol•K

Substituting ΔG = 0, ΔG° = ΔH° – TΔS°, ΔH° = +178.3 kJ/mol, ΔS° = +0.1606 kJ/mol•K, $R = 8.314 \times 10^{-3}$ kJ/mol•K, and $P_{CO_2} = 0.10$ atm into Equation (1), and solving for T gives

$0 = \Delta H° - T\Delta S° + RT \ln P_{CO_2}$; $-\Delta H° = T \times (-\Delta S° + R \ln P_{CO_2})$

$$\text{and } T = \frac{-\Delta H°}{-\Delta S° + R \ln P_{CO_2}} = \frac{-178.3 \text{ kJ/mol}}{-0.1606 \text{ kJ/mol•K} + 8.314 \times 10^{-3} \text{ kJ/mol•K} \times \ln 0.10} = 992 \text{ K} = 719°\text{C}$$

Solutions To Final Exercises

19.1 The pressure-volume (PV) work done by a system is the work done by the gaseous products when they expand and push back their surroundings. Useful work is all work other than PV work. PV work is not referred to as useful work, since in most reactions studied PV work simply pushes back the atmosphere and has no useful purpose.

The main example of a system where useful work is done is the battery; here chemical potential energy is converted into electrical energy and electrical work is performed.

19.3 Since the first law of thermodynamics requires that the total energy change be constant, the greater the useful work done, the less heat is evolved, and vice versa.

19.5 (a) $\Delta G = 0$ (b) ΔG is negative ($\Delta G < 0$) (c) ΔG is positive ($\Delta G > 0$)

19.7

$$\underset{2.00\text{ mol}}{H_2(g)} + \underset{1.00\text{ mol}}{1/2O_2(g)} \rightarrow \underset{2.00\text{ mol}}{H_2O(l)} \qquad \Delta H_f^\circ = -285.8\text{ kJ};\ \Delta G_f^\circ = -237.1\text{ kJ}$$

(a) maximum useful work $= 2.00\text{ mol} \times (-\Delta G_f^\circ) = 2.00\text{ mol} \times [-(-237.1\text{ kJ/mol})] = 474\text{ kJ}$
(b) heat evolved = (–enthalpy change) – work done $= 2.00\text{ mol} \times 285.8\text{ kJ/mol} - 474\text{ kJ} = 98\text{ kJ}$

19.9 (a) ΔG is the free energy change for the reaction when the reactants and products are not all in their standard states. ΔG° is the free energy change when all the reactants and products are in their standard states.
(b) $\Delta G = \Delta G^\circ + RT \ln Q$

19.11 (a) ΔG decreases with increasing reactant concentrations.
(b) ΔG increases with increasing product concentrations.

19.13 (a) $NH_3(g) \rightarrow 1/2N_2(g) + 3/2H_2(g)$; $\Delta G^\circ = +16.45$ kJ/mol; stable

(b) $HNO_3(l) \rightarrow 1/2H_2(g) + 1/2N_2(g) + 3/2O_2(g)$; $\Delta G^\circ = +80.71$ kJ/mol; stable

(c) $N_2O(g) \rightarrow N_2(g) + 1/2O_2(g)$; $\Delta G^\circ = -104.20$ kJ/mol; unstable

(d) $NO(g) \rightarrow 1/2N_2(g) + 1/2O_2(g)$; $\Delta G^\circ = -86.55$ kJ/mol; unstable

(e) $NO_2(g) \rightarrow 1/2N_2(g) + O_2(g)$; $\Delta G^\circ = -51.31$ kJ/mol; unstable

HNO_3 is the most stable and N_2O is the least stable.

Comment: An alternate approach is: Compounds for which ΔG_f° is negative are stable with respect to their elements under standard state conditions. These include (a) NH_3 and (b) HNO_3. Compounds for which ΔG_f° is positive are unstable with respect to their elements under standard state conditions. These include (c) N_2O, (d) NO, and (e) NO_2. HNO_3 has the lowest ΔG_f° (–80.71 kJ/mol) and is therefore the most stable. N_2O has the highest ΔG_f° (+104.20 kJ/mol) and is therefore the least stable.

19.15 (a)

$$2SO_2(g) + O_2(g) \rightleftharpoons 2SO_3(g)$$

ΔG_f°(kJ): 2×-300.19 0 2×-371.06

$\Delta G^\circ = 2 \times -371.06\text{ kJ} - (2 \times -300.19\text{ kJ} + 0\text{ kJ}) = -141.74\text{ kJ}$

Since $\Delta G^\circ < 0$, the spontaneous direction is the forward reaction.

(b) $Hg_2Cl_2(s) \rightleftharpoons HgCl_2(s) + Hg(l)$

ΔG°_f(kJ): −210.7 −178.6 0

$\Delta G^\circ = -178.6\text{ kJ} + 0\text{ kJ} - (-210.7\text{ kJ}) = +32.1\text{ kJ}$

Since $\Delta G^\circ > 0$, the spontaneous direction is the reverse reaction.

19.17 $Zn(s) + HgO(s) \rightarrow ZnO(s) + Hg(l)$

ΔG°_f(kJ): 0 −58.54 −318.30 0

$\Delta G^\circ = -318.30\text{ kJ} + 0\text{ kJ} - (0\text{ kJ} - 58.54\text{ kJ}) = -259.76\text{ kJ}$

The limiting reagent is HgO. The moles of HgO are

$$100\text{ mg HgO} \times \frac{1\text{ g}}{1000\text{ mg}} \times \frac{1\text{ mol HgO}}{216.59\text{ g HgO}} = 4.62 \times 10^{-4}\text{ mol HgO}$$

The maximum electrical work is

maximum electrical work $= -(4.62 \times 10^{-4}\text{ mol} \times -259.76\text{ kJ/mol}) = 0.120\text{ kJ}$

19.19 $2SO_2(g) + O_2(g) \rightleftharpoons 2SO_3(g)$

$\Delta G^\circ = 2 \times \Delta G^\circ_f(SO_3) - 2 \times \Delta G^\circ_f(SO_2) - \Delta G^\circ_f(O_2) = 2 \times -371.06\text{ kJ} - 2 \times -300.19\text{ kJ} - 0\text{ kJ} = -141.74\text{ kJ}$

$$\Delta G = \Delta G^\circ + RT\ln Q = \Delta G^\circ + RT\ln \frac{P^2_{SO_3}}{P^2_{SO_2}P_{O_2}}$$

$$\Delta G = -141{,}740\text{ J/mol} + 8.314\text{ J/mol•K} \times 298\text{ K} \times \ln \frac{(0.10)^2}{(0.00010)^2(0.20)} = -104{,}000\text{ J/mol or } -104\text{ kJ/mol}$$

Since $\Delta G <$ O, the forward direction is spontaneous under these conditions.

19.21 $H_2(g) + Cl_2(g) \rightleftharpoons 2HCl(g)$

$\Delta G^\circ = 2 \times \Delta G^\circ_f(HCl) - \Delta G^\circ_f(H_2) - \Delta G^\circ_f(Cl_2) = 2 \times -95.30\text{ kJ} - 0\text{ kJ} - 0\text{ kJ} = -190.60\text{ kJ} = -190{,}600\text{ J}$

$$\Delta G = \Delta G^\circ + RT\ln Q = \Delta G^\circ + RT\ln \frac{P^2_{HCl}}{P_{H_2}P_{Cl_2}}$$

$$\Delta G = -190{,}600\text{ J/mol} + 8.314\text{ J/mol•K} \times 298\text{ K} \times \ln \frac{(0.50)^2}{(0.50)(0.50)} = -190{,}600\text{ J/mol}$$

The net driving force under these conditions is the same as that for standard state conditions. Therefore, the reacting system is equally close to or far from equilibrium under both sets of conditions.

19.23 Refer to the derivation in Section 19.2 of the textbook.

19.25 (a) Substituting $K_p = 2.7 \times 10^{-2}$, $T = 773$ K, and $R = 8.314$ J/mol•K into Equation 19.4 gives

$\Delta G^\circ = -RT\ln K_p = -8.314\text{ J/mol•K} \times 773\text{ K} \times \ln(2.7 \times 10^{-2}) = +23{,}200\text{ J/mol or } 23.2\text{ kJ/mol}$

(b) $$\Delta G = \Delta G^\circ + RT\ln Q = \Delta G^\circ + RT\ln \frac{P^2_{NO}P_{Cl_2}}{P^2_{NOCl}}$$

$$\Delta G = +23{,}200\text{ J/mol} + 8.314\text{ J/mol•K} \times 773\text{ K} \times \ln \frac{(0.50)^2(0.50)}{(0.50)^2} = +18{,}700\text{ J/mol or } 18.7\text{ kJ/mol}$$

(c) $\Delta G > 0$; therefore the reverse direction is spontaneous under the conditions of part (b).

19.27 (a) $\Delta G° = 2 \times \Delta G_f^{\circ}(NO) - \Delta G_f^{\circ}(N_2) - \Delta G_f^{\circ}(O_2) = 2 \times 86.55\text{ kJ} - 0\text{ kJ} - 0\text{ kJ} = +173.1\text{ kJ}$

$$\ln K_p = \frac{\Delta G°}{-RT} = \frac{173{,}100\text{ J/mol}}{-8.314\text{ J/mol}\bullet\text{K} \times 298\text{ K}} = -69.9$$

and $K_p = \text{antilog}(-69.9) = e^{-69.9} = 4 \times 10^{-31}$

(b) $\Delta G° = 2 \times \Delta G_f^{\circ}(O_3) - 3 \times \Delta G_f^{\circ}(O_2) = 2 \times 163.2\text{ kJ} - 3 \times 0\text{ kJ} = +326.4\text{ kJ}$

$$\ln K_p = \frac{\Delta G°}{-RT} = \frac{326{,}400\text{ J/mol}}{-8.314\text{ J/mol}\bullet\text{K} \times 298\text{ K}} = -132 \text{ and } K_p = \text{antilog}(-132) = e^{-132} = 10^{-58}$$

(c) $\Delta G° = \Delta G_f^{\circ}(C_2H_6) - \Delta G_f^{\circ}(C_2H_4) - \Delta G_f^{\circ}(H_2) = -32.82\text{ kJ} - 68.15\text{ kJ} - 0\text{ kJ} = -100.97\text{ kJ}$

$$\ln K_p = \frac{\Delta G°}{-RT} = \frac{-100{,}970\text{ J/mol}}{-8.314\text{ J/mol}\bullet\text{K} \times 298\text{ K}} = +40.8$$

and $K_p = \text{antilog}(+40.8) = e^{+40.8} = 5 \times 10^{17}$

19.29 (a) $H_2O(l) \rightleftharpoons H_2O(g)$

$K_p = P_{H_2O} = 17.5\text{ torr}/760\text{ torr} = 2.30 \times 10^{-2}$

$\Delta G° = -RT \ln K_p = -8.314\text{ J/mol}\bullet\text{K} \times 293\text{ K} \times \ln(2.30 \times 10^{-2}) = 9190\text{ J/mol} = +9.19\text{ kJ/mol}$

(b) The vapor pressure at the normal boiling point of water is 760 torr = 1 atm.

$K_p = P_{H_2O} = 1$; $\Delta G° = -RT \ln K_p = -RT \ln 1 = 0$

19.31 (a) $K_a = 1.76 \times 10^{-5}$

$\Delta G° = -RT \ln K_a = -8.314\text{ J/mol}\bullet\text{K} \times 298\text{ K} \times \ln(1.76 \times 10^{-5}) = 27{,}100\text{ J/mol} = +27.1\text{ kJ/mol}$

(b) $K_f = 1.2 \times 10^2$

$\Delta G° = -RT \ln K_f = -8.314\text{ J/mol}\bullet\text{K} \times 298\text{ K} \times \ln(1.2 \times 10^2) = -11{,}900\text{ J/mol} = -11.9\text{ kJ/mol}$

(c) $K_b = 1.77 \times 10^{-5}$

$\Delta G° = -RT \ln K_b = -8.314\text{ J/mol}\bullet\text{K} \times 298\text{ K} \times \ln(1.77 \times 10^{-5}) = 27{,}100\text{ J/mol} = +27.1\text{ kJ/mol}$

19.33 The second law of thermodynamics states that for all spontaneous changes or processes the entropy of the universe, the total entropy of a system and its surroundings, increases. So, the second law means essentially that changes occur spontaneously in the direction of increasing overall randomness.

Examples are: 1. Aromas, cooking food, perfume, etc., spread through all the rooms, since the molecules become more randomly distributed. 2. Wet clothes dry on the wash line, because water vapor is more random than liquid water. 3. Batteries discharge, because heat energy is more random than stored potential energy. 4. Sugar dissolves in drinks, because mixed things are more random than separated things.

19.35 The entropy decreases. Separating a mixture into its components increases order.

19.37 (a) The kinetic energy of the gas increases when it is warmed and, as a consequence, the movements of the gas molecules become more chaotic.

(b) The volume of the gas increases when it expands, the gas molecules spread out and, as a consequence, the orderliness of the gas molecules decreases and the randomness increases.

19.39 Negative. A solid substance has a greater order, lower entropy, than a solution of the substance. This is especially true if the substance dissociates, which increases disorder, in solution.

19.41

	System	Universe
(a)	+	+
(b)	–	+
(c)	–	+
(d)	–	+
(e)	+	+

19.43 (a) The entropy increases. A gas is evolved.
(b) The entropy decreases. A gas is consumed.

19.45 The third law of thermodynamics states that the entropy of a pure, perfectly crystalline substance is zero at absolute zero (0 K).

Since all forms of motion, particularly translational, vibrational, and rotational, approach a minimum as zero kelvin is approached, a pure and perfect crystal should have nearly perfect order and, therefore, zero entropy at zero kelvin.

19.47 Refer to the derivation in the *Digging Deeper* in Section 19.4 of the textbook.

19.49 The heat absorbed by the surroundings (q_{surr}) is equal in magnitude but opposite in sign to the heat absorbed by the system (q_{sys}); $q_{surr} = -q_{sys}$.

19.51 The melting of benzene at 5.4°C and 1 atm is an equilibrium process; therefore Equation 19.8 applies. The change in enthalpy (the melting takes place under standard state conditions) is:

$$\Delta H° = 1.00 \text{ mol} \times \frac{78.11 \text{ g}}{1 \text{ mol}} \times \frac{126 \text{ J}}{1 \text{ g}} = 9840 \text{ J}$$

Substituting $\Delta H° = 9840$ J and T = 278.6 K into Equation 19.8 gives

$$\Delta S° = \frac{\Delta H°}{T} = \frac{9840 \text{ J}}{278.6 \text{ K}} = 35.3 \text{ J/K}$$

19.53 The vaporization of liquid ethanol at 78.5°C and 1 atm is an equilibrium process; therefore Equation 19.8 applies. The change in enthalpy (the vaporization takes place under standard state conditions) is:

$$\Delta H° = 45.0 \text{ g} \times \frac{1 \text{ mol}}{46.07 \text{ g}} \times \frac{38.6 \text{ kJ}}{1 \text{ mol}} = 37.7 \text{ kJ}$$

Substituting $\Delta H° = 37.7$ kJ and T = 351.6 K into Equation 19.8 gives

$$\Delta S° = \frac{\Delta H°}{T} = \frac{37.7 \text{ kJ}}{351.6 \text{ K}} = 0.107 \text{ kJ/K} = 107 \text{ J/K}$$

19.55 The $\Delta G°$ values come from Exercise 19.27; T = 298 K for each reaction.

(a) $N_2(g) + O_2(g) \rightleftharpoons 2NO(g)$; $\Delta G° = +173.1$ kJ

$$\Delta H° = 2 \times \Delta H^\circ_f(NO) - \Delta H^\circ_f(N_2) - \Delta H^\circ_f(O_2) = 2 \times 90.25 \text{ kJ} - 0 \text{ kJ} - 0 \text{ kJ} = +180.5 \text{ kJ}$$

$$\Delta S° = \frac{\Delta H° - \Delta G°}{T} = \frac{180.5 \text{ kJ} - 173.1 \text{ kJ}}{298 \text{ K}} = 0.0248 \text{ kJ/K} = +24.8 \text{ J/K}$$

(b) $3O_2(g) \rightleftharpoons 2O_3(g)$; $\Delta G° = +326.4$ kJ

$\Delta H° = 2 \times \Delta H_f^\circ(O_3) - 3 \times \Delta H_f^\circ(O_2) = 2 \times 142.7\text{ kJ} - 3 \times 0\text{ kJ} = +285.4\text{ kJ}$

$$\Delta S° = \frac{\Delta H° - \Delta G°}{T} = \frac{285.4\text{ kJ} - 326.4\text{ kJ}}{298\text{ K}} = -0.138\text{ kJ/K} = -138\text{ J/K}$$

(c) $C_2H_4(g) + H_2(g) \rightleftharpoons C_2H_6(g)$; $\Delta G° = -100.97$ kJ

$\Delta H° = \Delta H_f^\circ(C_2H_6) - \Delta H_f^\circ(C_2H_4) - \Delta H_f^\circ(H_2) = -84.68\text{ kJ} - 52.26\text{ kJ} - 0\text{ kJ} = -136.94\text{ kJ}$

$$\Delta S° = \frac{\Delta H° - \Delta G°}{T} = \frac{-136.94\text{ kJ} + 100.97\text{ kJ}}{298\text{ K}} = -0.121\text{ kJ/K} = -121\text{ J/K}$$

19.57 T = 298 K for each reaction.

(a) $\Delta G_f^\circ(CO_2) = -394.359$ kJ and $\Delta H_f^\circ(CO_2) = -393.509$ kJ

$$\Delta S_f^\circ = \frac{\Delta H_f^\circ - \Delta G_f^\circ}{T} = \frac{-393.509\text{ kJ} + 394.359\text{ kJ}}{298\text{ K}} = 2.85 \times 10^{-3}\text{ kJ/K} = +2.85\text{ J/K}$$

(b) $\Delta G_f^\circ(Hg(g)) = 31.82$ kJ and $\Delta H_f^\circ(Hg(g)) = 61.32$ kJ

$$\Delta S_f^\circ = \frac{61.32\text{ kJ} - 31.82\text{ kJ}}{298\text{ K}} = 0.0990\text{ kJ/K} = +99.0\text{ J/K}$$

19.59 (a)

	$2NaI(s)$	+	$Cl_2(g)$	→	$2NaCl(s)$	+	$I_2(s)$
S°(J/K):	2 × 98.53		223.07		2 × 72.13		116.135

$\Delta S° = 2 \times 72.13\text{ J/K} + 116.135\text{ J/K} - 2 \times 98.53\text{ J/K} - 223.07\text{ J/K} = -159.74\text{ J/K}$

(b)

	$N_2(g)$	+	$3H_2(g)$	→	$2NH_3(g)$
S°(J/K):	191.61		3 × 130.684		2 × 192.45

$\Delta S° = 2 \times 192.45\text{ J/K} - 191.61\text{ J/K} - 3 \times 130.684\text{ J/K} = -198.76\text{ J/K}$

19.61 (a)

	$N_2(g)$	+	$O_2(g)$	→	$2NO(g)$
S°(J/K):	191.61		205.138		2 × 210.76

$\Delta S° = 2 \times 210.76\text{ J/K} - 191.61\text{ J/K} - 205.138\text{ J/K} = +24.77\text{ J/K}$

(b)

	$3O_2(g)$	→	$2O_3(g)$
S°(J/K):	3 × 205.138		2 × 238.93

$\Delta S° = 2 \times 238.93\text{ J/K} - 3 \times 205.138\text{ J/K} = -137.55\text{ J/K}$

(c)

	$C_2H_4(g)$	+	$H_2(g)$	→	$C_2H_6(g)$
S°(J/K):	219.56		130.684		229.60

$\Delta S° = 229.60\text{ J/K} - 219.56\text{ J/K} - 130.684\text{ J/K} = -120.64\text{ J/K}$

19.63 Refer to Figure 19.8 in the textbook. (a) yes (b) yes (c) no

19.65 The $\Delta H°$ values come from Exercise 19.55 and the $\Delta S°$ values from Exercise 19.61. Assume that $\Delta H°$ and $\Delta S°$ do not change with temperature. T = 573 K for each reaction.

(a) $\Delta G° = \Delta H° - T\Delta S° = 180.5\text{ kJ} - 573\text{ K} \times 0.02477\text{ kJ/K} = +166.3\text{ kJ}$

(b) $\Delta G° = 285.4\text{ kJ} - 573\text{ K} \times -0.13755\text{ kJ/K} = +364.2\text{ kJ}$

(c) $\Delta G° = -136.94\text{ kJ} - 573\text{ K} \times -0.12064\text{ kJ/K} = -67.8\text{ kJ}$

19.67 $N_2O_4(g) \rightleftharpoons 2NO_2(g)$

At 298 K,

$\Delta H^\circ = 2 \times \Delta H^\circ_f(NO_2) - \Delta H^\circ_f(N_2O_4) = 2 \times 33.18 \text{ kJ} - 9.16 \text{ kJ} = +57.20 \text{ kJ}$

$\Delta S^\circ = 2 \times S^\circ(NO_2) - S^\circ(N_2O_4) = 2 \times 240.06 \text{ J/K} - 304.29 \text{ J/K} = +175.83 \text{ J/K} = +0.17583 \text{ kJ/K}$

When each gas has a partial pressure of 1 atm, then $\ln Q = \ln 1 = 0$ and $\Delta G = \Delta G^\circ$. In other words, the reaction is taking place under standard state conditions. We want to find the temperature at which $\Delta G^\circ = 0$. Assuming that ΔH° and ΔS° do not change with temperature and substituting their values and $\Delta G^\circ = 0$ into Equation 19.7 gives

$\Delta G^\circ = \Delta H^\circ - T\Delta S^\circ$

$0 \text{ kJ} = +57.20 \text{ kJ} - T \times 0.17583 \text{ kJ/K}$ and $T = 325 \text{ K} = 52°\text{C}$

19.69 (a)

	$Ag_2O(s)$	$\rightarrow$	$2Ag(s)$	+	$1/2O_2(g)$
ΔG°_f(kJ):	-11.20		0		0

$\Delta G^\circ = 0 \text{ kJ} + 0 \text{ kJ} - (-11.20 \text{ kJ}) = +11.20 \text{ kJ}$

Since $\Delta G > 0$, the spontaneous direction is the reverse reaction. Therefore, under these conditions of temperature (25°C) and O_2 partial pressure (1.00 atm), Ag_2O is thermodynamically stable with respect to its elements.

(b) At 298 K,

$\Delta H^\circ = 2 \times \Delta H^\circ_f(Ag) + 1/2 \times \Delta H^\circ_f(O_2) - \Delta H^\circ_f(Ag_2O) = 2 \times 0 \text{ kJ} + 1/2 \times 0 \text{ kJ} - (-31.05 \text{ kJ}) = +31.05 \text{ kJ}$

$\Delta S^\circ = 2 \times S^\circ(Ag) + 1/2 \times S^\circ(O_2) - S^\circ(Ag_2O) = 2 \times 42.55 \text{ J/K} + 1/2 \times 205.138 \text{ J/K} - 121.3 \text{ J/K} = +66.4 \text{ J/K}$

A value for ΔG° at 180°C can be obtained from Equation 19.7 if we assume that ΔH° and ΔS° do not change with temperature. Substituting $\Delta H^\circ = 31.05$ kJ, $\Delta S^\circ = 0.0644$ kJ/K, and $T = 453$ K into Equation 19.7 gives

$\Delta G^\circ = \Delta H^\circ - T\Delta S^\circ = 31.05 \text{ kJ} - 453 \text{ K} \times 0.0664 \text{ kJ/K} = +0.97 \text{ kJ}$

Since $\Delta G > 0$, the spontaneous direction is the reverse reaction. Therefore, under these conditions of temperature (180°C) and O_2 partial pressure (1.00 atm), Ag_2O is thermodynamically stable with respect to its elements.

19.71 The internal energy E of a substance is the total kinetic and potential energy contained by that substance. The Gibbs free energy, G, of a substance is a thermodynamic property that is related to the amount of potential energy contained by that substance and to its potential for change. Also, refer to the Chapter 6 and 19 definitions.

19.73 (a) The performance of useful work decreases the amount of heat evolved during a reaction.
(b) $\Delta G = -$maximum useful work

Since this an equilibrium process, $\Delta G = 0$ and none of the evolved energy can be obtained as useful work. Therefore, the answer is zero.

19.75 Since all four molecules can be in one flask, each flask must have a minimum of four places where the molecules can be. For simplicity, we will assume that each flask has only four places. The molecules are indistinguishable. We have to find the number of different ways of arranging four indistinguishable molecules among the available locations as specified in each part. Only arrangements we could physically see (observable arrangements), if we were capable, count.

(a) Two possible arrangements exist, one with all four molecules in flask 1 (F1) and none in flask 2 (F2), and vice versa.

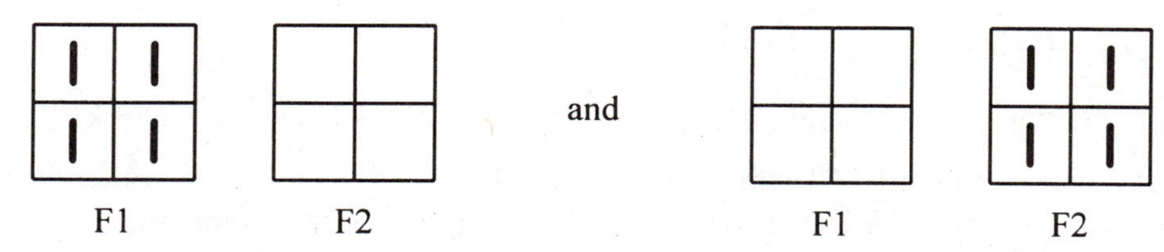

(b) Thirty-two possible arrangements exist when three molecules are in one flask and one molecule is in the other flask. The sixteen possible arrangements when three molecules are in F1 and one molecule is in F2 are:

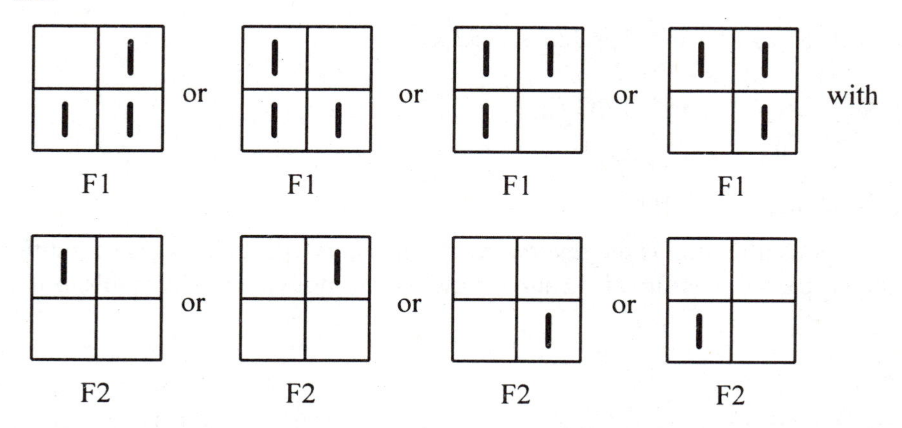

Another sixteen arrangements exist with one molecule in F1 and three molecules in F2.

(c) Thirty-six possible arrangements exist when two molecules are in each flask. The thirty-six possible arrangements are:

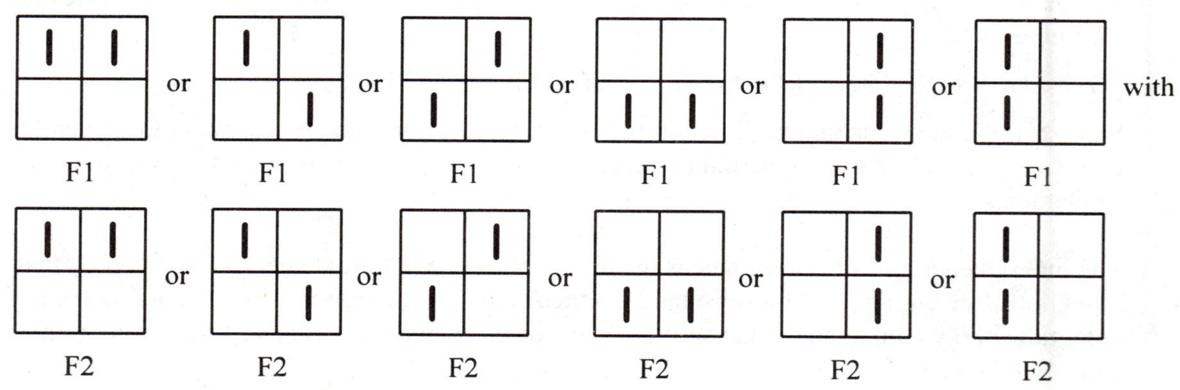

Answer: (a) two ways (b) thirty-two ways (c) thirty-six ways

Arrangement (c) is the most likely to occur, if the flasks are connected, and will have the highest entropy.

19.77 C(graphite) $\rightarrow$ C(diamond)

For this reaction at T = 298 K,

$\Delta H^\circ = \Delta H_f^\circ(\text{diamond}) - \Delta H_f^\circ(\text{graphite}) = 1.895 \text{ kJ} - 0 \text{ kJ} = +1.895 \text{ kJ}$

$\Delta S^\circ = S^\circ(\text{diamond}) - S^\circ(\text{graphite}) = 2.377 \text{ J/K} - 5.740 \text{ J/K} = -3.363 \text{ J/K}$

and $\Delta G^\circ = \Delta H^\circ - T\Delta S^\circ = +1.895 \text{ kJ} - 298 \text{ K} \times -3.363 \times 10^{-3} \text{ kJ/K} = +0.888 \text{ kJ}$

Therefore, the reaction is not spontaneous at 298 K, and since $\Delta H°$ is positive and $\Delta S°$ is negative, $\Delta G°$ will be positive at all temperatures (see Figure 19.8 in the textbook). Therefore, the reaction at 1 atm pressure is not spontaneous at any temperature and, consequently, diamond cannot be made from graphite at 1 atm pressure.

19.79 $ATP^{4-}(aq) + H_2O(l) \rightleftharpoons ADP^{3-}(aq) + HPO_4^{2-}(aq) + H^+(aq)$

(a) Substituting $\Delta G° = -31$ kJ/mol, T = 298 K, and R = 8.314×10^{-3} kJ/mol•K into Equation 19.4 gives

$$\ln K_H = \frac{\Delta G°}{-RT} = \frac{-31\ \text{kJ}}{-8.314 \times 10^{-3}\ \text{kJ/mol•K} \times 298\ \text{K}} = 12.5 \text{ (carry extra significant figure)}$$

and $K_H = \text{antilog } 12.5 = e^{+12.5} = 3 \times 10^5$

(b) $$\Delta G = \Delta G° + RT \ln Q = \Delta G° + RT \ln \frac{[ADP^{3-}][HPO_4^{2-}][H^+]}{[ATP^{4-}]}$$

$$\Delta G = -31\ \text{kJ/mol} + 8.314 \times 10^{-3}\ \text{kJ/mol•K} \times 298\ \text{K} \times \ln \frac{(0.00025)(0.00165)(1.0 \times 10^{-7})}{(0.00225)} = -92\ \text{kJ/mol}$$

Since $\Delta G < 0$, the forward reaction is spontaneous and, therefore, capable of providing useful work.

19.81 The van't Hoff equation is

$$\ln \frac{K_2}{K_1} = \frac{-\Delta H°}{R}\left(\frac{1}{T_2} - \frac{1}{T_1}\right) = \frac{\Delta H°}{R}\left(\frac{T_2 - T_1}{T_1 T_2}\right)$$

$\Delta G° = -RT \ln K$ (19.4) and $\Delta G = \Delta H - T\Delta S$ (19.7)

So, for T_1 and T_2 (and assuming $\Delta H°$ and $\Delta S°$ are constant over the temperature range) we have

$\Delta G_1° = \Delta H° - T_1\Delta S°$ (1) $\qquad$ $\Delta G_2° = \Delta H° - T_2\Delta S°$ (2)

$\ln K_1 = \Delta G_1°/-RT_1$ (3) $\qquad$ $\ln K_2 = \Delta G_2°/-RT_2$ (4)

Subtracting (3) from (4) we get

$$\ln K_2 - \ln K_1 = \frac{\Delta G_2°}{-RT_2} - \frac{\Delta G_1°}{-RT_1} = \frac{T_1\Delta G_2° - T_2\Delta G_1°}{-RT_1T_2} \quad (5)$$

Substituting the values of $\Delta G_1°$ and $\Delta G_2°$ from (1) and (2) into (5) gives

$$\ln \frac{K_2}{K_1} = \frac{T_1(\Delta H° - T_2\Delta S°) - T_2(\Delta H° - T_1\Delta S°)}{-RT_1T_2} = \frac{\Delta H°}{R}\left(\frac{T_2 - T_1}{T_1T_2}\right)$$

19.83 (a) (1)

	$4HCl(g)$	+	$O_2(g)$	→	$2Cl_2(g)$	+	$2H_2O(g)$
$\Delta G_f°$ (kJ):	4×-95.30		0		0		2×-228.572

$\Delta G° = 0\ \text{kJ} + 2 \times -228.572\ \text{kJ} - (4 \times -95.30\ \text{kJ} + 0\ \text{kJ}) = -75.9\ \text{kJ}$

(2) At 298 K,

$\Delta H° = 2 \times \Delta H_f°(Cl_2) + 2 \times \Delta H_f°(H_2O) - 4 \times \Delta H_f°(HCl) - \Delta H_f°(O_2)$

$= 2 \times 0\ \text{kJ} + 2 \times -241.818\ \text{kJ} - 4 \times -92.31\ \text{kJ} - 0\ \text{kJ} = -114.4\ \text{kJ}$

$\Delta S° = 2 \times S°(Cl_2) + 2 \times S°(H_2O) - 4 \times S°(HCl) - S°(O_2)$

$= 2 \times 223.07 \text{ J/K} + 2 \times 188.825 \text{ J/K} - 4 \times 186.91 \text{ J/K} - 205.138 \text{ J/K} = -129.0 \text{ J/K}$

A value for $\Delta G°$ at 400°C can be obtained using Equation 19.7 if we assume that $\Delta H°$ and $\Delta S°$ do not change with temperature. Substituting $\Delta H° = -114.4$ kJ, $\Delta S° = -0.1290$ kJ/K, and T = 673 K into Equation 19.7 gives

$\Delta G° = \Delta H° - T\Delta S° = -114.4 \text{ kJ} - 673 \text{ K} \times -0.1290 \text{ kJ/K} = -27.6 \text{ kJ}$

(b) (1) Substituting $\Delta G° = -75.9$ kJ/mol, T = 298 K, and R = 8.314×10^{-3} kJ/mol•K into Equation 19.4 gives

$$\ln K_p = \frac{\Delta G°}{-RT} = \frac{-75.9 \text{ kJ/mol}}{-8.314 \times 10^{-3} \text{kJ/mol•K} \times 298 \text{ K}} = 30.6$$

and K_p = antilog 30.6 = $e^{+30.6} = 2 \times 10^{13}$

(2) Substituting $\Delta G° = -27.6$ kJ/mol, T = 673 K, and R = 8.314×10^{-3} kJ/mol•K into Equation 19.4 gives

$$\ln K_p = \frac{\Delta G°}{-RT} = \frac{-27.6 \text{ kJ/mol}}{-8.314 \times 10^{-3} \text{kJ/mol•K} \times 673 \text{ K}} = 4.93$$

and K_p = antilog 4.93 = $e^{+4.93} = 140$

Using the van't Hoff equation gives $K_p = 100$. The difference in the answers is due to significant figure round off.

(c) The reaction at 25°C must be very slow compared to that at 400°C. In other words, at 25°C it would take too long to obtain a substantial amount of product, while at 400°C a substantial amount of product can be obtained in a reasonable amount of time, even though the equilibrium conditions are many orders of magnitude less favorable.

CHAPTER 20

ELECTROCHEMISTRY

Solutions To Practice Exercises

PE 20.1

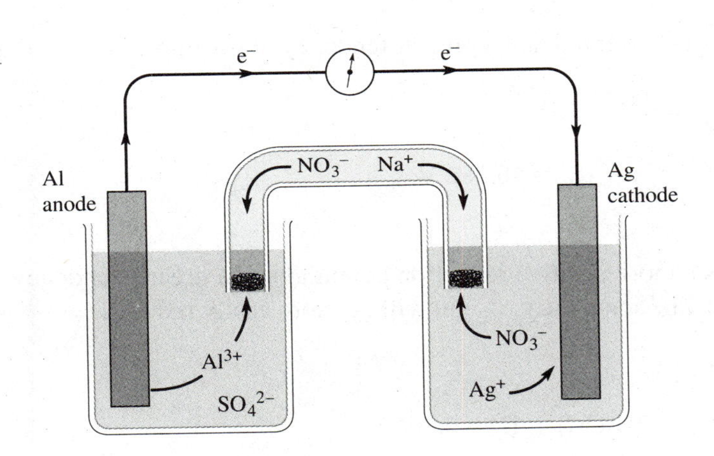

The half-reactions are

$Al(s) \rightarrow Al^{3+}(aq) + 3e^-$ anode, oxidation

$Ag^+(aq) + e^- \rightarrow Ag(s)$ cathode, reduction

The complete cell reaction is

$Al(s) + 3Ag^+(aq) \rightarrow Al^{3+}(aq) + 3Ag(s)$

PE 20.2 The half-reactions are

$Zn(s) \rightarrow Zn^{2+}(aq) + 2e^-$ anode, oxidation

$Br_2(l) + 2e^- \rightarrow 2Br^-(aq)$ cathode, reduction

The cell notation is

$Zn(s) \mid Zn^{2+}(aq) \parallel Br^-(aq) \mid Br_2(l) \mid C(gr)$

PE 20.3 $Zn(s) \mid Zn^{2+}$ (1 M) || Br^{-} (1 M) | $Br_2(l) \mid C(gr)$

PE 20.4 Substituting $Q = 1.00 \times 10^{-7}$ C/s $\times$ 600 s $= 6.00 \times 10^{-5}$ C and E = 1.35 V = 1.35 J/C into Equation 20.1 gives

maximum electrical work $= 6.00 \times 10^{-5}$ C $\times$ 1.35 J/C $= 8.10 \times 10^{-5}$ J

PE 20.5 The oxidation of 2 mol of Al atoms by Cu^{2+} ions requires the transfer of 6 mol of electrons. Substituting $n = 6$ mol e^{-}, F = 96,485 C/mol e^{-}, and $\Delta G° = -1166$ kJ into Equation 20.5b gives

$$E° = \frac{\Delta G°}{-nF} = \frac{-1166 \text{ kJ}}{-6 \text{ mol } e^{-} \times 96{,}485 \text{ C/mol } e^{-}} = 2.01 \times 10^{-3} \text{ kJ/C} = 2.01 \text{ J/C} = 2.01 \text{ V}$$

PE 20.6 $E^{\circ}_{oxid} = -E^{\circ}_{red} = -(-2.869 \text{ V}) = +2.869 \text{ V}$

PE 20.7 (a) The standard oxidation potentials are:

$2Cl^{-}(aq) \rightarrow Cl_2(g) + 2e^{-}$ $E^{\circ}_{oxid} = -1.360$ V

$2I^{-}(aq) \rightarrow I_2(s) + 2e^{-}$ $E^{\circ}_{oxid} = -0.535$ V

I^{-} has a higher (more positive) oxidation potential and a greater tendency to give up electrons than Cl^{-}; I^{-} is more easily oxidized.

(b) The standard reduction potentials are:

$Co^{2+}(aq) + 2e^{-} \rightarrow Co(s)$ $E° = -0.282$ V

$Cr^{3+}(aq) + 3e^{-} \rightarrow Cr(s)$ $E° = -0.74$ V

An aqueous solution of Co^{2+} ions has a more positive reduction potential and a greater tendency to accept electrons than an aqueous solution of Cr^{3+} ions; the Co^{2+} ion will be more easily reduced.

PE 20.8 One of the reactions listed in Table 20.1 is

$NO_3^{-}(aq) + 4H^{+}(aq) + 3e^{-} \rightleftharpoons NO(g) + 2H_2O(l)$ $E° = 0.964$ V

(a) No. From Table 20.1 we have

$Br_2(l) + 2e^{-} \rightleftharpoons 2Br^{-}(aq)$ $E° = 1.078$ V

Since the aqueous nitric acid reduction potential of 0.964 V is lower than the Br_2 reduction potential of 1.078 V, aqueous nitric acid cannot force the bromine reaction to go in reverse and, consequently, aqueous nitric acid cannot oxidize bromide ions.

(b) Yes. From Table 20.1 we have

$I_2(s) + 2e^{-} \rightleftharpoons 2I^{-}(aq)$ $E° = 0.535$ V

Since the $HNO_3(aq)$ reduction potential of 0.964 V is greater than the I_2 reduction potential of 0.535 V, $HNO_3(aq)$ can force the iodine reaction to go in reverse and, consequently, $HNO_3(aq)$ can oxidize iodide ions.

PE 20.9 (a) The reduction potentials from Table 20.1 are

$MnO_4^{-}(aq) + 8H^{+}(aq) + 5e^{-} \rightleftharpoons Mn^{2+}(aq) + 4H_2O(l)$ $E° = 1.512$ V

$I_2(s) + 2e^{-} \rightleftharpoons 2I^{-}(aq)$ $E° = 0.535$ V

The permanganate ion half-reaction has the more positive reduction potential; it will go forward as a reduction and the iodine half-reaction will go in reverse as an oxidation. The half-reactions are

oxidation: $2I^-(aq) \rightarrow I_2(s) + 2e^-$ $\qquad E^\circ_{oxid} = -0.535\ V$

reduction: $MnO_4^-(aq) + 8H^+(aq) + 5e^- \rightarrow Mn^{2+}(aq) + 4H_2O(l)$ $\qquad E^\circ_{red} = 1.512\ V$

(b) The cell voltage is obtained by substituting $E^\circ_{oxid} = -0.535$ V and $E^\circ_{red} = 1.512$ V into Equation 20.6:

$$E^\circ = E^\circ_{oxid} + E^\circ_{red} = -0.535\ V + 1.512\ V = 0.977\ V$$

PE 20.10 The reaction is the sum of the following half-reactions:

$Cu(s) \rightarrow Cu^{2+}(1\ M) + 2e^-$ $\qquad E^\circ_{oxid} = -E^\circ_{red} = -0.339\ V$

$2Ag^+(1\ M) + 2e^- \rightarrow 2Ag(s)$ $\qquad E^\circ_{red} = 0.799\ V$

The total voltage is

$E^\circ = -0.339\ V + 0.799\ V = 0.460\ V = 0.460\ J/C$

Two moles of electrons are transferred per mole of Cu oxidized; n = 2 mol e^-. Substituting in Equation 20.5b gives

$$\Delta G^\circ = -nFE^\circ = -2\ mol\ e^- \times \frac{96{,}485\ C}{1\ mol\ e^-} \times \frac{0.460\ J}{1\ C} = -8.88 \times 10^4\ J = -88.8\ kJ$$

PE 20.11 The cell reaction and standard voltage are

$Br_2(l) + 2Fe^{2+}(aq) \rightarrow 2Fe^{3+}(aq) + 2Br^-(aq)$ $\qquad E^\circ = 0.31\ V$

The reaction quotient is

$$Q = \frac{[Fe^{3+}]^2[Br^-]^2}{[Fe^{2+}]^2} = \frac{(0.10)^2(0.050)^2}{(0.10)^2} = 2.5 \times 10^{-3}$$

Two moles of electrons are transferred per mole of Br_2 reduced; therefore n = 2. Substituting the values for Q, n, and E° into the Nernst equation gives

$$E = 0.31\ V - \frac{0.0592\ V}{2} \times \log(2.5 \times 10^{-3}) = 0.39\ V$$

PE 20.12 The half-reaction and standard reduction potential are

$2H^+(aq) + 2e^- \rightarrow H_2(g)$ $\qquad E^\circ = 0.000\ V$

The reaction quotient is

$$Q = \frac{P_{H_2}}{[H^+]^2} = \frac{1.00}{(10^{-7})^2} = 1.00 \times 10^{+14}$$

Two moles of electrons are transferred per mole of H_2 formed; therefore n = 2. Substituting the values for Q, n, and E° into the Nernst equation gives

$$E = 0.000\ V - \frac{0.0592\ V}{2} \times \log(1.00 \times 10^{+14}) = -0.414\ V$$

$$\overset{+1}{2Cu^{+}(aq)} \rightleftharpoons \overset{0}{Cu(s)} + \overset{+2}{Cu^{2+}(aq)}$$

PE 20.13 (a) (1e⁻ lost: Cu⁺ → Cu²⁺; 1e⁻ gained: Cu⁺ → Cu)

One mole of electrons are transferred, hence n = 1. Substituting the values for n and K into Equation 20.8b gives

$$E° = \frac{0.0592\ V}{1} \times \log(1.1 \times 10^6) = 0.358\ V$$

(b) $\Delta G° = -nFE° = -1\ mol\ e^- \times \frac{96{,}485\ C}{1\ mol\ e^-} \times \frac{0.358\ J}{1\ C} = -34{,}500\ J = -34.5\ kJ$

PE 20.14 $Zn^{2+}(aq) + 4NH_3(aq) \rightleftharpoons Zn(NH_3)_4^{2+}(aq)$

This is not a redox reaction, but it can be written as the sum of two half-reactions whose potentials can be found in Table 20.1

$Zn(s) + 4NH_3(aq) \rightarrow Zn(NH_3)_4^{2+}(aq) + 2e^-$ $\quad E°_{oxid} = -E°_{red} = 1.015\ V$

$Zn^{2+}(aq) + 2e^- \rightarrow Zn(s)$ $\quad E°_{red} = -0.762\ V$

The sum of the half-cell voltages is

$$E° = E°_{oxid} + E°_{red} = 1.015\ V - 0.762\ V = 0.253\ V$$

Substituting n = 2 and E° = 0.253 V into Equation 20.8b gives

$$0.253\ V = \frac{0.0592\ V}{2} \times \log K$$

log K = 8.55 and K = antilog 8.55 = $10^{8.55} = 3.5 \times 10^8$

PE 20.15 anode: $2Cl^- \rightarrow Cl_2(g) + 2e^-$

cathode: $Mg^{2+} + 2e^- \rightarrow Mg(l)$

PE 20.16 $AuCl_4^-(aq) + 3e^- \rightarrow Au(s) + 4Cl^-(aq)$

The equation for the electrode reaction shows that 1 mol (196.97 g) of gold metal is deposited by 3 mol of electrons, or 3 × 96,485 C. The steps in the calculation are: 5000 C → mol e⁻ → mol Au → g Au.

$$5000\ C \times \frac{1\ mol\ e^-}{96{,}485\ C} \times \frac{1\ mol\ Au}{3\ mol\ e^-} \times \frac{196.97\ g\ Au}{1\ mol\ Au} = 3.40\ g\ Au$$

PE 20.17 The number of coulombs that passes through the circuit is the same as the number of coulombs produced by the zinc electrode. The electrode reaction, $Zn(s) \rightarrow Zn^{2+}(aq) + 2e^-$, shows that 1 mol (65.39 g) of zinc dissolves producing 2 mol of electrons, or 2 × 96,485 C. The steps in the calculation are: 350 mg Zn → mol Zn → mol e⁻ → C.

$$350\ mg\ Zn \times \frac{1\ g}{1000\ mg} \times \frac{1\ mol\ Zn}{65.39\ g\ Zn} \times \frac{2\ mol\ e^-}{1\ mol\ Zn} \times \frac{96{,}485\ C}{1\ mol\ e^-} = 1.03 \times 10^3\ C$$

PE 20.18 $Q = I \times t = 0.556\ C/s \times 25.0\ min \times 60\ s/min = 834\ C$

PE 20.19 The current is 15.0 A = 15.0 C/s. The electrode equation, $2Cl^- \rightarrow Cl_2(g) + 2e^-$, shows that 2 mol of electrons, or 2 × 96,485 C, produces 1 mol of chlorine. At STP (1.00 atm and 0°C), 1 mol of gas occupies approximately 22.4 L. The steps in the calculation are:
$C/s \rightarrow C/min \rightarrow mol\ e^-/min \rightarrow mol\ Cl_2/min \rightarrow L/min$.

$$\frac{15.0\ C}{1\ s} \times \frac{60\ s}{1\ min} \times \frac{1\ mol\ e^-}{96{,}485\ C} \times \frac{1\ mol\ Cl_2}{2\ mol\ e^-} \times \frac{22.4\ L\ Cl_2}{1\ mol\ Cl_2} = 0.104\ L\ Cl_2/min$$

Solutions To Final Exercises

20.1 (a) negative (b) positive (c) From the anode to the cathode.

20.3 (a) The salt bridge allows the flow of ions so that electrical neutrality is maintained in each half-cell, while keeping the different cell solutions from mixing. A direct chemical reaction would take place, negating the half-cells, if the solutions were to mix. Also, for the electrical circuit to be complete a pathway must be provided for ions to flow within the cell.
(b) If the flow of ions did not occur, an opposing charge difference would build up between the compartments and the reaction would stop.

20.5 (a)

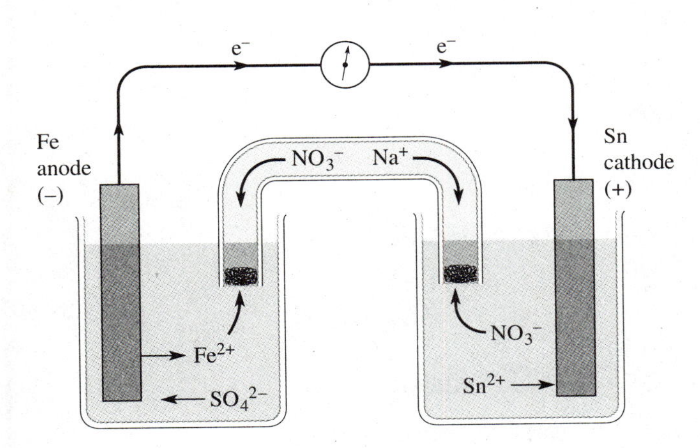

The half-reactions are

$Fe(s) \rightarrow Fe^{2+}(aq) + 2e^-$ anode, oxidation

$Sn^{2+}(aq) + 2e^- \rightarrow Sn(s)$ cathode, reduction

(b)

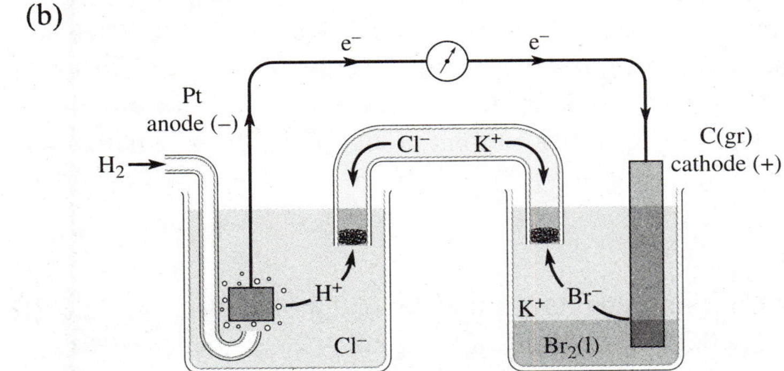

The half-reactions are

$H_2(g) \rightarrow 2H^+(aq) + 2e^-$ anode, oxidation

$Br_2(l) + 2e^- \rightarrow 2Br^-(aq)$ cathode, reduction

(c)

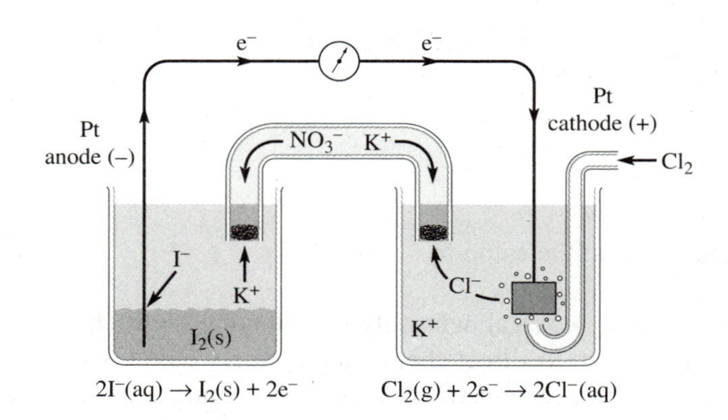

The half-reactions are

$2I^-(aq) \rightarrow I_2(s) + 2e^-$ anode, oxidation

$Cl_2(g) + 2e^- \rightarrow 2Cl^-(aq)$ cathode, reduction

20.7 (a) $Fe(s) \mid Fe^{2+}(aq) \parallel Sn^{2+}(aq) \mid Sn(s)$

(b) $Pt \mid H_2(g) \mid H^+(aq) \parallel Br^-(aq) \mid Br_2(l) \mid C(gr)$

(c) $Pt \mid I_2(s) \mid I^-(aq) \parallel Cl^-(aq) \mid Cl_2(g) \mid Pt$

20.9 (a) $3Ag(s) + NO_3^-(aq) + 4H^+(aq) \rightarrow 3Ag^+(aq) + NO(g) + 2H_2O(l)$

$Ag(s) \mid Ag^+ (1\ M) \parallel H^+ (1\ M), NO_3^- (1\ M) \mid NO (1\ atm) \mid Pt$

(b) The half-reactions are

$Zn(s) + 4NH_3(aq) \rightarrow Zn(NH_3)_4^{2+}(aq) + 2e^-$ anode, oxidation

$Ag(NH_3)_2^+(aq) + e^- \rightarrow Ag(s) + 2NH_3(aq)$ cathode, reduction

$Zn(s) \mid Zn(NH_3)_4^{2+} (1\ M), NH_3 (1\ M) \parallel NH_3 (1\ M), Ag(NH_3)_2^+ (1\ M) \mid Ag(s)$

20.11 (a) coulomb (C) (b) volts (V); 1 V = 1 J/C

20.13 In each separated half-cell compartment, a redox half-reaction equilibrium exists that imparts either a positive or negative electric charge to the electrode in that compartment. The difference in the accumulated charges on the two electrodes gives rise to the galvanic cell's potential difference. The electrode that takes on a negative charge relative to the other electrode is by definition the anode. When the circuit is closed, electrons flow from the anode to the cathode. The electric current produced can then be harnessed to perform electrical work.

20.15 $Zn(s) + Cu^{2+}(aq) \rightarrow Cu(s) + Zn^{2+}(aq)$

The oxidation of 1 mol of Zn atoms by Cu^{2+} ions requires the transfer of 2 mol of electrons. Substituting $n = 2$ mol e^-, $F = 96{,}485$ C/mol e^-, and $E = 0.35$ V $= 0.35$ J/C into Equation 20.5a gives

$$\Delta G = -nFE = -2 \text{ mol } e^- \times \frac{96{,}485 \text{ C}}{1 \text{ mol } e^-} \times \frac{0.35 \text{ J}}{1 \text{ C}} = -6.8 \times 10^4 \text{ J} = -68 \text{ kJ}$$

20.17 $Fe^{3+}(aq) + Ag(s) + Cl^{-}(aq) \rightarrow Fe^{2+}(aq) + AgCl(s)$

The reduction of 1 mol of Fe^{3+} ions to 1 mol of Fe^{2+} ions requires the transfer of 1 mol of electrons. Substituting n = 1 mol e^-, F = 96,485 C/mol e^-, and E° = 0.547 V = 0.547 J/C into Equation 20.5b gives

$$\Delta G° = -nFE° = -1 \text{ mol } e^- \times \frac{96{,}485 \text{ C}}{1 \text{ mol } e^-} \times \frac{0.547 \text{ J}}{1 \text{ C}} = -5.28 \times 10^4 \text{ J} = -52.8 \text{ kJ}$$

20.19 Refer to "Standard Electrode Potentials," in Section 20.3, in the textbook and Figure 20.7.

20.21 Any half-cell that has a greater standard reduction potential than hydrogen will undergo reduction and, therefore, will be the cathode and have a positive polarity. Any half-cell that has a lower standard reduction potential than hydrogen will undergo oxidation and, therefore, will be the anode and have a negative polarity. Since the potential of the standard hydrogen electrode is defined to be zero, the sign of the electrode polarity and the standard reduction potential are the same for this type of cell.

20.23 (a) Ni^{2+}. The higher the reduction potential, the greater the tendency to accept electrons.
(b) Cd. Since the reduction potential for Cd^{2+} ion is lower than that for Ni^{2+} ion, the oxidation potential for Cd metal is higher than that for Ni metal, and the higher the oxidation potential, the more easily the oxidation takes place.

20.25 The standard reduction potentials from Table 20.1 are:

$Cl_2(g) + 2e^- \rightarrow 2Cl^-(aq)$ E° = 1.360 V

$MnO_4^-(aq) + 8H^+(aq) + 5e^- \rightarrow Mn^{2+}(aq) + 4H_2O(l)$ E° = 1.512 V

An acidified solution of potassium permanganate has a more positive reduction potential than chlorine water; therefore the permanganate solution will be the stronger oxidizing agent.

20.27 Only acidified $KMnO_4$. From Table 20.1 we have

$MnO_2(s) + 4H^+(aq) + 2e^- \rightarrow Mn^{2+}(aq) + 2H_2O(l)$ E° = 1.229 V

$MnO_4^-(aq) + 2H_2O(l) + 3e^- \rightarrow MnO_2(s) + 4OH^-(aq)$ E° = 0.597 V

$S(s) + 2H^+(aq) + 2e^- \rightarrow H_2S(aq)$ E° = 0.144 V

$Hg^{2+}(aq) + 2e^- \rightarrow 2Hg(l)$ E° = 0.796 V

$Fe^{3+}(aq) + e^- \rightarrow Fe^{2+}(aq)$ E° = 0.769 V

$MnO_4^-(aq) + 8H^+(aq) + 5e^- \rightarrow Mn^{2+}(aq) + 4H_2O(l)$ E° = 1.512 V

To chemically oxidize Mn^{2+} to MnO_2, we need an oxidizing agent with a reduction potential greater than that of MnO_2, whose E° = 1.229 V. H_2S can be eliminated immediately, since H_2S is not an oxidizing agent; the sulfur atom in H_2S is already reduced. Hg^{2+}, Fe^{3+}, and MnO_4^- by itself are all oxidizing agents, but their standard reduction potentials are all below that of MnO_2. Only an acidified solution of $KMnO_4$ has a greater standard reduction potential than that of MnO_2. Therefore, only the acidified $KMnO_4$ is worth trying.

20.29 (a) 1. $Zn^{2+}(aq) + 2e^- \rightarrow Zn(s)$ E° = −0.762 V

$2H^+(aq) + 2e^- \rightarrow H_2(g)$ E° = 0.000 V

The hydrogen half-reaction has the more positive reduction potential; therefore hydrochloric acid will oxidize zinc metal. The equation for the reaction is

$Zn(s) + 2H^+(aq) \rightarrow Zn^{2+}(aq) + H_2(g)$

2. $Cu^{2+}(aq) + 2e^- \rightarrow Cu(s)$ $E° = 0.339$ V

$2H^+(aq) + 2e^- \rightarrow H_2(g)$ $E° = 0.000$ V

The copper half-reaction has the more positive reduction potential; therefore no reaction occurs.

(b) 1. $Ag^+(aq) + e^- \rightarrow Ag(s)$ $E° = 0.799$ V

$NO_3^-(aq) + 4H^+(aq) + 3e^- \rightarrow NO(g) + 2H_2O(l)$ $E° = 0.964$ V

The nitric acid half-reaction has the more positive reduction potential; therefore nitric acid will oxidize silver metal. The equation for the reaction is

$3Ag(s) + NO_3^-(aq) + 4H^+(aq) \rightarrow 3Ag^+(aq) + NO(g) + 2H_2O(l)$

2. $Au^+(aq) + e^- \rightarrow Au(s)$ $E° = 1.691$ V

or

$Au^{3+}(aq) + 3e^- \rightarrow Au(s)$ $E° = 1.498$ V

$NO_3^-(aq) + 4H^+(aq) + 3e^- \rightarrow NO(g) + 2H_2O(l)$ $E° = 0.964$ V

The gold half-reaction has the more positive potential; therefore no reaction occurs.

(c) $Cl_2(g) + 2e^- \rightarrow 2Cl^-(aq)$ $E° = 1.360$ V

$F_2(g) + 2e^- \rightarrow 2F^-(aq)$ $E° = 2.889$ V

The fluorine half-reaction has the more positive reduction potential; therefore no reaction occurs.

20.31 (a) oxidation: $Fe(s) \rightarrow Fe^{2+}(aq) + 2e^-$ $E^\circ_{oxid} = 0.409$ V

reduction: $Sn^{2+}(aq) + 2e^- \rightarrow Sn(s)$ $E^\circ_{red} = -0.14$ V

$E° = E^\circ_{oxid} + E^\circ_{red} = 0.409 \text{ V} - 0.14 \text{ V} = 0.27 \text{ V}$

Substituting n = 2 mol e^-, F = 96,485 C/mol e^-, and E° = 0.27 J/C into Equation 20.5b gives

$$\Delta G° = -nFE° = -2 \text{ mol } e^- \times \frac{96{,}485 \text{ C}}{1 \text{ mol } e^-} \times \frac{0.27 \text{ J}}{1 \text{ C}} = -5.2 \times 10^4 \text{ J} = -52 \text{ kJ}$$

(b) oxidation: $H_2(g) \rightarrow 2H^+(aq) + 2e^-$ $E^\circ_{oxid} = 0.000$ V

reduction: $Br_2(l) + 2e^- \rightarrow 2Br^-(aq)$ $E^\circ_{red} = 1.078$ V

$E° = E^\circ_{oxid} + E^\circ_{red} = 0.000 \text{ V} + 1.078 \text{ V} = 1.078 \text{ V}$

Substituting n = 2 mol e^-, F = 96,485 C/mol e^-, and E° = 1.078 J/C into Equation 20.5b gives

$$\Delta G° = -nFE° = -2 \text{ mol } e^- \times \frac{96{,}485 \text{ C}}{1 \text{ mol } e^-} \times \frac{1.078 \text{ J}}{1 \text{ C}} = -2.08 \times 10^5 \text{ J} = -208 \text{ kJ}$$

(c) oxidation: $2I^-(aq) \rightarrow I_2(s) + 2e^-$ $E^\circ_{oxid} = -0.535$ V

reduction: $Cl_2(g) + 2e^- \rightarrow 2Cl^-(aq)$ $E^\circ_{red} = 1.360$ V

$E° = E^\circ_{oxid} + E^\circ_{red} = -0.535 \text{ V} + 1.360 \text{ V} = 0.825 \text{ V}$

$$\Delta G° = -nFE° = -2 \text{ mol } e^- \times \frac{96{,}485 \text{ C}}{1 \text{ mol } e^-} \times \frac{0.825 \text{ J}}{1 \text{ C}} = -1.59 \times 10^5 \text{ J} = -159 \text{ kJ}$$

20.33 (a) oxidation: $Zn(s) \rightarrow Zn^{2+}(aq) + 2e^-$ $E^\circ_{oxid} = 0.762$ V

reduction: $2H^+(aq) + 2e^- \rightarrow H_2(g)$ $E^\circ_{red} = 0.000$ V

$E^\circ = E^\circ_{oxid} + E^\circ_{red} = 0.762 \text{ V} + 0.000 \text{ V} = 0.762 \text{ V}$

Substituting n = 2 mol e⁻, F = 96,485 C/mol e⁻, and E° = 0.762 J/C into Equation 20.5b gives

$$\Delta G^\circ = -nFE^\circ = -2 \text{ mol e}^- \times \frac{96{,}485 \text{ C}}{1 \text{ mol e}^-} \times \frac{0.762 \text{ J}}{1 \text{ C}} = -1.47 \times 10^5 \text{ J} = -147 \text{ kJ}$$

(b) oxidation: $Ag(s) \rightarrow Ag^+(aq) + e^-$ $E^\circ_{oxid} = -0.799$ V

reduction: $NO_3^-(aq) + 4H^+(aq) + 3e^- \rightarrow NO(g) + 2H_2O(l)$ $E^\circ_{red} = 0.964$ V

$E^\circ = E^\circ_{oxid} + E^\circ_{red} = -0.799 \text{ V} + 0.964 \text{ V} = 0.165 \text{ V}$

$$\Delta G^\circ = -nFE^\circ = -3 \text{ mol e}^- \times \frac{96{,}485 \text{ C}}{1 \text{ mol e}^-} \times \frac{0.165 \text{ J}}{1 \text{ C}} = -4.78 \times 10^4 \text{ J} = -47.8 \text{ kJ}$$

(c) No reaction occurs.

20.35 As the reaction proceeds, the reactant concentrations decrease and the product concentrations increase. Therefore, the free energy change of the cell reaction continuously decreases in absolute magnitude, becomes more positive or less negative, as the reaction proceeds and, consequently, the cell voltage continuously decreases as the reaction proceeds. Remember, the change in free energy is a quantitative measure of the driving force of the reaction. It is the maximum energy available to do useful work. The cell voltage or EMF is also a quantitative measure of the driving force of the reaction or tendency of the cell reaction to occur.

The cell voltage equals zero when equilibrium is reached, i.e., when Q = K. At equilibrium, ΔG = 0, $E_{cell} = 0$, Q = K, and the reaction has no net driving force.

Comment: Some galvanic cells have a constant cell voltage for most their operating lifetime, because Q is independent of the electrolyte concentration(s). The alkaline dry cell and mercury cell are two examples of such cells.

20.37 Refer to the derivation in Section 20.4 in the textbook.

20.39 The half-cell in which $[Cu^{2+}] = 0.10$ M is the anode. The half-cell with the higher (more positive) reduction potential will be the cathode. Since the half-cell reduction potential decreases with decreasing $[Cu^{2+}]$, the standard Cu | Cu^{2+} half-cell will have the higher reduction potential and, therefore, will be the cathode.

20.41 (a) The cell reaction and standard voltage are

$Cu(s) + 2Ag^{2+}(aq) \rightarrow Cu^{2+}(aq) + 2Ag(s)$

$E^\circ = E^\circ_{oxid} + E^\circ_{red} = -0.339 \text{ V} + 0.799 \text{ V} = 0.460 \text{ V}$

The reaction quotient is

$$Q = \frac{[Cu^{2+}]}{[Ag^+]^2} = \frac{1.5}{(1.0)^2} = 1.5$$

Substituting Q = 1.5, n = 2, and E° = 0.460 V into Equation 20.7b gives

$$E = 0.460 \text{ V} - \frac{0.0592 \text{ V}}{2} \times \log 1.5 = 0.45 \text{ V}$$

(b) This is an example of a concentration cell, a cell in which there is no net chemical reaction, only an equalization of the concentrations in the two cell compartments. Concentration cells harness the spontaneous mixing process. When the concentrations become equal, $\Delta G = 0$ and the current stops. The cell reaction and standard voltage are

oxidation:	$Cd(s) \rightarrow Cd^{2+}(0.0030\ M) + 2e^-$	$E^\circ_{oxid} = 0.402\ V$
reduction:	$Cd^{2+}(1.2\ M) + 2e^- \rightarrow Cd(s)$	$E^\circ_{red} = -0.402\ V$
net reaction:	$Cd^{2+}(1.2\ M) \rightarrow Cd^{2+}(0.0030\ M)$	$E^\circ = 0.000\ V$

The reaction quotient is

$$Q = \frac{[Cd^{2+}]_{\text{anode compartment}}}{[Cd^{2+}]_{\text{cathode compartment}}} = \frac{0.0030}{1.2} = 0.0025$$

Substituting $Q = 0.0025$, $n = 2$, and $E^\circ = 0.000\ V$ into Equation 20.7b gives

$$E = 0.000\ V - \frac{0.0592\ V}{2} \times \log 0.0025 = 0.077\ V$$

20.43 (a) $$Q = \frac{[Cr^{3+}]^2}{[Cr_2O_7^{2-}][H^+]^{14}} = \frac{(0.50)^2}{(0.30)(1)^{14}} = 0.8$$

Substituting $n = 6$, $E^\circ = 1.33\ V$, and $Q = 0.8$ into the Nernst equation gives

$$E = 1.33\ V - \frac{0.0592\ V}{6} \times \log 0.8 = 1.33\ V$$

(b) $$Q = \frac{1}{P_{O_2}[H^+]^4} = \frac{1}{(1)(1 \times 10^{-7})^4} = 1 \times 10^{28}$$

Substituting $n = 4$, $E^\circ = 1.229\ V$, and $Q = 1 \times 10^{28}$ into the Nernst equation gives

$$E = 1.229\ V - \frac{0.0592\ V}{4} \times \log(1 \times 10^{28}) = 0.815\ V$$

20.45 The cell reaction and standard voltage are

$8H^+(1\ M) \rightarrow 8H^+(0.01\ M)$ $\qquad E^\circ = 0.000\ V$

The reaction quotient is

$$Q = \frac{[H^+]^8_{\text{anode compartment}}}{[H^+]^8_{\text{cathode compartment}}} = \frac{(0.01)^8}{(1)^8} = 1 \times 10^{-16}$$

Substituting $n = 5$, $E^\circ = 0.000\ V$, and $Q = 1 \times 10^{-16}$ into the Nernst equation gives

$$E = 0.000\ V - \frac{0.0592\ V}{5} \times \log(1 \times 10^{-16}) = 0.189\ V$$

20.47 The $Ni \mid Ni^{2+}$ half-cell with the unknown $[Ni^{2+}]$ is the anode.

$Ni(s) \rightarrow Ni^{2+}(aq) + 2e^-$; $E_{oxid} = E - E_{red} = 0.054\ V - (-0.236\ V) = 0.290\ V$

$$E_{oxid} = E^\circ_{oxid} - \frac{0.0592}{n} \log Q_{oxid}$$

$$0.290 = 0.236 - \frac{0.0592}{2}\log [Ni^{2+}]$$

$\log [Ni^{2+}] = -1.82$; $[Ni^{2+}] = \text{antilog}(-1.82) = 10^{-1.82} = 0.015$ M

Comment: The Nernst equation for the overall cell reaction could also be used to solve for the unknown $[Ni^{2+}]$; with E = 0.054 V, E° = 0.000 V, n = 2, and Q = [Ni(?)]/[Ni(1 M)] = $x/(1) = x$.

20.49 (a) $2I^-(aq) \rightarrow I_2(s) + 2e^-$ $\quad E^\circ_{oxid} = -0.535$ V

$O_2(g) + 4H^+(aq) + 4e^- \rightarrow 2H_2O(l)$ $\quad E^\circ_{red} = 1.229$ V

$E^\circ = E^\circ_{oxid} + E^\circ_{red} = -0.535\text{ V} + 1.229\text{ V} = 0.694$ V

Substituting E° = 0.694 V and n = 4 into Equation 20.8b gives

$$\log K = \frac{nE^\circ}{0.0592} = \frac{4 \times 0.694\text{ V}}{0.0592\text{ V}} = 46.9 \text{ and } K = \text{antilog } 46.9 = 10^{46.9} = 8 \times 10^{46}$$

(b) $Zn(s) \rightarrow Zn^{2+}(aq) + 2e^-$ $\quad E^\circ_{oxid} = 0.762$ V

$Zn(NH_3)_4^{2+}(aq) + 2e^- \rightarrow Zn(s) + 4NH_3(aq)$ $\quad E^\circ_{red} = -1.015$ V

$E^\circ = E^\circ_{oxid} + E^\circ_{red} = 0.762\text{ V} - 1.015\text{ V} = -0.253$ V

Substituting E° = –0.253 V and n = 2 into Equation 20.8b gives

$$\log K = \frac{nE^\circ}{0.0592} = \frac{2(-0.253\text{ V})}{0.0592\text{ V}} = -8.55$$

and $K = \text{antilog}(-8.55) = 10^{-8.55} = 2.8 \times 10^{-9}$

20.51 The K_{sp} is the equilibrium constant for the reaction

$Hg_2Cl_2(s) \rightleftharpoons Hg_2^{2+}(aq) + 2Cl^-(aq)$

From Table 20.1 we have

$2Hg(l) \rightarrow Hg_2^{2+}(aq) + 2e^-$ $\quad E^\circ_{oxid} = -0.796$ V

$Hg_2Cl_2(s) + 2e^- \rightarrow 2Hg(l) + 2Cl^-(aq)$ $\quad E^\circ_{red} = 0.268$ V

and $E^\circ = E^\circ_{oxid} + E^\circ_{red} = -0.796\text{ V} + 0.268\text{ V} = -0.528$ V

Substituting n = 2 and E° = –0.528 V into Equation 20.8b gives

$$\log K_{sp} = \frac{nE^\circ}{0.0592} = \frac{2 \times (-0.528\text{ V})}{0.0592\text{ V}} = -17.8$$

and $K_{sp} = \text{antilog}(-17.8) = 10^{-17.8} = 2 \times 10^{-18}$

20.53 First, we have to find the equilibrium constant for the reaction:

$Cu(s) + 2Ag^+(aq) \rightarrow Cu^{2+}(aq) + 2Ag(s)$

From Table 20.1 we have

$Cu(s) \rightarrow Cu^{2+}(aq) + 2e^-$ $\quad E^\circ_{oxid} = -0.339$ V

$Ag^+(aq) + e^- \rightarrow Ag(s)$ $\quad E^\circ_{red} = 0.799$ V

and $E^\circ = E^\circ_{oxid} + E^\circ_{red} = -0.339\text{ V} + 0.799\text{ V} = 0.460$ V

Substituting E° = 0.460 V and n = 2 into Equation 20.8b gives

$$\log K = \frac{nE°}{0.0592} = \frac{2 \times 0.460\ V}{0.0592\ V} = 15.5 \text{ and } K = \text{antilog } 15.5 = 10^{15.5} = 3 \times 10^{15}$$

Since K is so large, virtually all of the Ag^+ will be converted to Ag, so long as there is a sufficient amount of copper metal present to react with all the Ag^+ ions present. Therefore, the $[Cu^{2+}] = 0.005$ M and the $[Ag^+]$ is calculated as follows:

$$\frac{[Cu^{2+}]}{[Ag^+]^2} = \frac{0.005}{x^2} = 3 \times 10^{15};\ x = [Ag^+] = 1 \times 10^{-9}\ M$$

20.55 Refer to Figures 20.9 and 20.12 and Section 20.5 in the textbook.

20.57 Refer to "Batteries and Fuel Cells" (Section 20.5), the section introduction, and the definition in the Glossary.

20.59 (a) From the half-cell reactions we see that 5 mol of ions are consumed and 1 mol of ions is produced, leading to an overall reduction of 4 mol of ions for every 2 mol of electrons that pass through the circuit. Therefore, the molality of the electrolyte in a discharged lead storage battery is considerably lower than that in a fully charged one and, consequently, the electrolyte in a discharged battery freezes at a higher temperature.

(b) Shaking and sudden shock can dislodge the Pb, PbO_2, and solid lead sulfate from the electrodes. Once dislodged, they fall to the bottom of the cell and the $PbSO_4$ can no longer be converted back to lead and lead dioxide. This diminishes the amount of reactants and can cause a short circuit if enough of this sludge builds up so that the sheets of lead and lead dioxide come in contact with one another.

20.61 A fuel cell is a galvanic cell in which the reactants are continuously supplied and the products continuously removed.

Fuel cells do not store electrical energy, while galvanic cells do. Also, the reactants are not initially present and the products do not remain in the cell. In other words, the reactants are continuously fed in and the products are continuously removed.

20.63 It's galvanic because it occurs as half-reactions at anodic and cathodic sites between which a potential difference develops and a current flows.

20.65 Metallic corrosion is a galvanic process that spontaneously occurs when metals are exposed to their environment. Since corrosion is galvanic in nature, it involves a cathodic process and, therefore, something in the environment has to be reduced. Oxygen is involved, because it is the most available (being both plentiful and a gas), readily reducible substance in the environment. Water is necessary for dissolving the cations produced by the oxidation reaction at the anode. The water is also necessary for carrying the ion current needed to complete the cell circuit, and for the ion flow or movement needed to maintain neutrality.

20.67 All metallic corrosion involves an anodic process, the oxidation of the metal to its ions with the release of electrons into the metal, and a cathodic process, a reaction that consumes the released electrons. Controlling corrosion consists of interfering with the anode or cathode reaction.

1. Cover the iron with a protective layer or coating. Coatings such as oil, porcelain, paint, and less active or corrosion-resistant metals, such as tin, chromium, and nickel, are used. This prevents moisture and oxygen from coming in contact with the iron. Protective coatings must remain intact in order to be effective.

2. Galvanize the iron. This places a coating of zinc on the iron. Galvanizing is a form of cathodic protection. The more active metal, zinc, is preferentially oxidized, thus protecting the less active iron from oxidation. The iron becomes the cathode and the reduction half-reaction takes place on its surface. An intact zinc coating also protects the iron from moisture and oxygen.
3. Use of a sacrificial anode. This is also a form of cathodic protection, in which a block of a more active metal, usually zinc or magnesium, is (electrically) connected to the less active iron. The sacrificial anode is preferentially oxidized, leaving the iron intact.
4. Keep the iron dry. The corrosion process requires the presence of moisture (water). Maintaining a dry environment will prevent rusting.

20.69 Galvanized iron is iron that has been coated with a layer of zinc. The zinc can be applied by several different methods. Dipping the iron in molten zinc, spraying on the zinc, and electroplating on the zinc are the three most commonly used methods.

Galvanizing confers cathodic protection to the iron. In any galvanic situation that arises, the zinc will become the anode and dissolve, while the iron will become the inert cathode and, therefore, will be preserved.

20.71 Cathodic protection is an anticorrosion technique that involves putting a more active metal such as zinc or magnesium in contact with a less active metal that is to be protected. The replaceable more active metal, the so-called sacrificial anode, has a higher oxidation potential and is thus preferentially oxidized.

Attach or electrically connect a zinc block to the underground oil storage tank. The zinc block will be a sacrificial anode and the iron storage tank an inert cathode, since zinc has a higher oxidation potential than iron, and will thus be preferentially oxidized. Hence, the less active iron storage tank will be left intact. Periodic replacement of the corroded zinc block with a new one is necessary. The wet soil forms the elecrolyte and ion migration in the wet soil completes the circuit.

20.73 (a) anode: positive
(b) cathode: negative
(c) They are the opposite of those in a galvanic cell.

20.75 The charge in the electrolytic cell is carried by the sodium and chloride ions. Therefore, the Na^+ and Cl^- ions must be mobile, which means that the NaCl must be in the molten state in order for the cell to work.

20.77 (a) $K^+ + e^- \rightarrow K(l)$ cathode, reduction

$2Cl^- \rightarrow Cl_2(g) + 2e^-$ anode, oxidation

The complete cell reaction is

$2K^+ + 2Cl^- \rightarrow 2K(l) + Cl_2(g)$ or $2KCl(l) \rightarrow 2K(l) + Cl_2(g)$

(b) $Ca^{2+} + 2e^- \rightarrow Ca(l)$ cathode, reduction

$2Cl^- \rightarrow Cl_2(g) + 2e^-$ anode, oxidation

The complete cell reaction is

$Ca^{2+} + 2Cl^- \rightarrow Ca(l) + Cl_2(g)$ or $CaCl_2(l) \rightarrow Ca(l) + Cl_2(g)$

20.79 (a) Since the reduction potential of water is considerably higher than that of potassium ion, the water would be reduced in preference to the potassium ion. This holds for other active metals as well.
(b) If water gets into the cell, hydrogen would be produced at the cathode instead of the active metal.

20.81 Refer to Figure 20.20 and Section 20.7 in the textbook.

20.83 (a) $2Cl^-(aq) \rightarrow Cl_2(g) + 2e^-$ anode

$2H_2O(l) + 2e^- \rightarrow H_2(g) + 2OH^-(aq)$ cathode

The net equation for the electrolysis is

$2H_2O(l) + 2Cl^-(aq) \rightarrow Cl_2(g) + H_2(g) + 2OH^-(aq)$

(b) $2Cl^-(aq) \rightarrow Cl_2(g) + 2e^-$ anode

$2H_2O(l) + 2e^- \rightarrow H_2(g) + 2OH^-(aq)$ cathode

The net equation for the electrolysis is

$2H_2O(l) + 2Cl^-(aq) \rightarrow Cl_2(g) + H_2(g) + 2OH^-(aq)$

Potassium and calcium are not produced, because the potassium and calcium ions are more difficult to reduce than water.

20.85 Refer to Figure 20.22 and "The Electrolysis of Water" in Section 20.7 in the textbook. No electrolysis takes place if sodium sulfate is not present. The sodium sulfate is needed to carry the current and maintain electrical neutrality.

20.87 $2H_2O(l) + 2e^- \rightarrow H_2(g) + 2OH^-(aq)$

(a) The equation for the electrode reaction shows that 2 mol (2×18.02 g = 36.04 g) of water are decomposed by the passage of 2 mol of electrons, or $2 \times 96{,}485$ C. The steps in the calculation are: 15,000 C $\rightarrow$ mol e^- $\rightarrow$ mol H_2O $\rightarrow$ g H_2O.

$$15{,}000\text{ C} \times \frac{1\text{ mol e}^-}{96{,}485\text{ C}} \times \frac{2\text{ mol H}_2\text{O}}{2\text{ mol e}^-} \times \frac{18.02\text{ g H}_2\text{O}}{1\text{ mol H}_2\text{O}} = 2.80\text{ g H}_2\text{O}$$

(b) The equation for the electrode reaction shows that 1 mol of hydrogen is produced by the passage of 2 mol of electrons, or 2 faradays (F). The steps in the calculation are: 1.25 F $\rightarrow$ mol H_2 $\rightarrow$ L H_2.

$$1.25\text{ F} \times \frac{1\text{ mol H}_2}{2\text{ F}} = 0.625\text{ mol H}_2$$

Substituting n = 0.625 mol, P = 1.00 atm, T = 298 K, and R = 0.0821 L•atm/mol•K into the ideal gas equation gives

$$V = \frac{nRT}{P} = \frac{0.625\text{ mol} \times 0.0821\text{ L·atm/mol·K} \times 298\text{ K}}{1.00\text{ atm}} = 15.3\text{ L}$$

20.89 $Pb^{2+}(aq) + 2e^- \rightarrow Pb(s)$

The equation for the electrode reaction shows that 1 mol (207.2 g) of lead metal is deposited by 2 mol of electrons, or $2 \times 96{,}485$ C. The steps in the calculation are: $I \times t \rightarrow$ C $\rightarrow$ mol e^- $\rightarrow$ mol Pb $\rightarrow$ g Pb.

$$\frac{0.15\text{ C}}{1\text{ s}} \times 3600\text{ s} \times \frac{1\text{ mol e}^-}{96{,}485\text{ C}} \times \frac{1\text{ mol Pb}}{2\text{ mol e}^-} \times \frac{207.2\text{ g Pb}}{1\text{ mol Pb}} = 0.58\text{ g Pb}$$

20.91 $Al^{3+} + 3e^- \rightarrow Al(l)$ cathode, reduction

$2O^{2-} \rightarrow O_2(g) + 4e^-$ anode, oxidation

(a) The equation for the electrode reaction shows that 1 mol (26.98 g) of aluminum metal is produced by the passage of 3 mol of electrons, or $3 \times 96{,}485$ C. The steps in the calculation are: $I \times t \rightarrow$ C $\rightarrow$ mol e^- $\rightarrow$ mol Al $\rightarrow$ g Al.

$$\frac{75.0\text{ C}}{1\text{ s}} \times 24.0\text{ h} \times \frac{3600\text{ s}}{1\text{ h}} \times \frac{1\text{ mol e}^-}{96{,}485\text{ C}} \times \frac{1\text{ mol Al}}{3\text{ mol e}^-} \times \frac{26.98\text{ g Al}}{1\text{ mol Al}} = 604\text{ g Al}$$

(b) The equation for the electrode reaction shows that 1 mol of oxygen is produced by the passage of 4 mol of electrons, or 4 × 96,485 C. The steps in the calculation are: $I \times t \rightarrow C \rightarrow \text{mol e}^- \rightarrow \text{mol } O_2 \rightarrow \text{L } O_2$.

$$\frac{75.0\text{ C}}{1\text{ s}} \times 24.0\text{ h} \times \frac{3600\text{ s}}{1\text{ h}} \times \frac{1\text{ mol e}^-}{96{,}485\text{ C}} \times \frac{1\text{ mol } O_2}{4\text{ mol e}^-} = 16.8\text{ mol } O_2$$

Substituting n = 16.8 mol, P = 800 torr/760 torr = 1.05 atm, T = 298 K, and R = 0.0821 L•atm/mol•K into the ideal gas equation gives

$$V = \frac{nRT}{P} = \frac{16.8\text{ mol} \times 0.0821\text{ L•atm/mol•K} \times 298\text{ K}}{1.05\text{ atm}} = 391\text{ L}$$

20.93 $Cu^{n+} + ne^- \rightarrow Cu(s)$

n is equal to the number of faradays required to produce 1 mol of Cu. The steps in the calculation are: $I \times t \rightarrow C \rightarrow C/\text{g Cu} \rightarrow C/\text{mol Cu} \rightarrow F/\text{mol Cu} \rightarrow n$.

$$\frac{1.10\text{ C}}{1\text{ s}} \times 30.0\text{ min} \times \frac{60\text{ s}}{1\text{ min}} \times \frac{1}{0.650\text{ g Cu}} \times \frac{63.55\text{ g Cu}}{1\text{ mol Cu}} \times \frac{1\text{ F}}{96{,}485\text{ C}} = 2\text{ F/mol Cu}$$

and n = 2

20.95 (a) $AuCl_4^-(aq) + 3e^- \rightarrow Au(s) + 4Cl^-(aq)$

The equation for the electrode reaction shows that 1 mol of gold metal is produced by the passage of 3 mol of electrons, or 3 × 96,485 C. Therefore, 3.00 F are required to deposit 1.00 mol of gold from this electrolyte.

(b) $Au(CN)_2^-(aq) + e^- \rightarrow Au(s) + 2CN^-(aq)$

The electrode equation shows that 1 mol of electrons, or 96,485 C, will deposit 1 mol of gold. Therefore, 1.00 F is required to deposit 1.00 mol of gold from this electrolyte.

20.97 The electrode reactions are

$Ag^+(aq) + e^- \rightarrow Ag(s)$

$Cu^{2+}(aq) + 2e^- \rightarrow Cu(s)$

$Ni^{2+}(aq) + 2e^- \rightarrow Ni(s)$

The number of coulombs that passes through each cell is the same as the number of coulombs that passes through the coulometer. Therefore, from the electrode reactions we see that for every mole of silver metal deposited, a half mole of copper metal and a half mole of nickel metal are produced.

(a) The steps in the calculation are: g Ag → mol Ag → mol Cu → g Cu.

$$0.156\text{ g Ag} \times \frac{1\text{ mol Ag}}{107.9\text{ g Ag}} \times \frac{0.500\text{ mol Cu}}{1\text{ mol Ag}} \times \frac{63.55\text{ g Cu}}{1\text{ mol Cu}} = 4.59 \times 10^{-2}\text{ g Cu}$$

Answer: The mass of the copper electrode increases by 4.59×10^{-2} g.

(b) The steps in the calculation are: g Ag → mol Ag → mol Ni → g Ni.

$$0.156 \text{ g Ag} \times \frac{1 \text{ mol Ag}}{107.9 \text{ g Ag}} \times \frac{0.500 \text{ mol Ni}}{1 \text{ mol Ag}} \times \frac{58.69 \text{ g Ni}}{1 \text{ mol Ni}} = 4.24 \times 10^{-2} \text{ g Ni}$$

Answer: The mass of the nickel electrode increases by 4.24×10^{-2} g.

20.99 (a) The net reaction and cell voltage is

$H_2(g, 1 \text{ atm}) + Cu^{2+}(1 \text{ M}) \rightarrow 2H^+(1 \text{ M}) + Cu(s) \qquad E° = 0.34 \text{ V}$

$E° = E^\circ_{oxid} + E^\circ_{red} = 0.34 \text{ V}$

If E°_{red} is defined to be zero, then the relative standard oxidation potential of the hydrogen electrode will be $E^\circ_{oxid} = 0.34$ V. Therefore, the standard reduction potential for the hydrogen half-cell would now be

$E^\circ_{red} = -E^\circ_{oxid} = -0.34 \text{ V}$

(b) The standard reduction potentials for all the other half-cells would have to be lowered by 0.34 V.

(c) 1. None of the negative potentials would become positive.
2. All the positive potentials below the copper half-reaction; i.e., all positive $E° < 0.34$ V, would become negative.

20.101 The use of strong reducing agents like Li, K, Ba, Ca, Na, etc., and strong oxidizing agents like F_2, H_2O_2, $KMnO_4$, Cl_2, etc., are not practical, because their highly reactive nature makes them difficult to control. For instance, the alkali metals, some of the alkaline earth metals, and fluorine cannot be allowed to come in direct contact with water, since they react with water. Subtracting the E° for a given half-reaction from 1.5 V gives the standard oxidization potential of the other half-reaction needed to make up the cell. Some combinations that could be used are:

1. $Hg_2^{2+}(aq) + 2e^- \rightarrow 2Hg(l) \qquad E^\circ_{red} = 0.796 \text{ V}$
   $Cr(s) \rightarrow Cr^{3+}(aq) + 3e^- \qquad E^\circ_{oxid} = 0.74 \text{ V}$
2. $Cu^{2+}(aq) + 2e^- \rightarrow Cu(s) \qquad E^\circ_{red} = 0.339 \text{ V}$
   $Mn(s) \rightarrow Mn^{2+}(aq) + 2e^- \qquad E^\circ_{oxid} = 1.182 \text{ V}$
3. $Fe^{3+}(aq) + e^- \rightarrow Fe^{2+}(aq) \qquad E^\circ_{red} = 0.769 \text{ V}$
   $Zn(s) \rightarrow Zn^{2+}(aq) + 2e^- \qquad E^\circ_{oxid} = 0.762 \text{ V}$

20.103 $$\text{maximum electrical work} = QE = ItE = \frac{0.030 \text{ C}}{1 \text{ s}} \times 2.0 \text{ s} \times \frac{1.5 \text{ J}}{1 \text{ C}} = 0.090 \text{ J}$$

20.105

$$ClO_4^-(aq) + ClO_2^-(aq) \rightleftharpoons 2ClO_3^-(aq)$$

(+2) × 1 [Cl in ClO_2^- → ClO_3^-]
(−2) × 1 [Cl in ClO_4^- → ClO_3^-]

(a) Substituting n = 2 and K = 1.9×10^4 into Equation 20.8b gives

$$E° = \frac{0.0592 \text{ V}}{2} \log(1.9 \times 10^4) = 0.127 \text{ V}$$

(b) $$\Delta G° = -nFE° = -2 \text{ mol } e^- \times \frac{96{,}485 \text{ C}}{1 \text{ mol } e^-} \times \frac{0.127 \text{ J}}{1 \text{ C}} = -24{,}500 \text{ J} = -24.5 \text{ kJ}$$

20.107 (a) $SO_3^{2-}(aq) + 2OH^-(aq) \rightarrow SO_4^{2-}(aq) + H_2O(l) + 2e^-$ $\quad E^\circ_{oxid} = 0.936\ V$

$MnO_4^-(aq) + 2H_2O(l) + 3e^- \rightarrow MnO_2(s) + 4OH^-(aq)$ $\quad E^\circ_{red} = 0.597\ V$

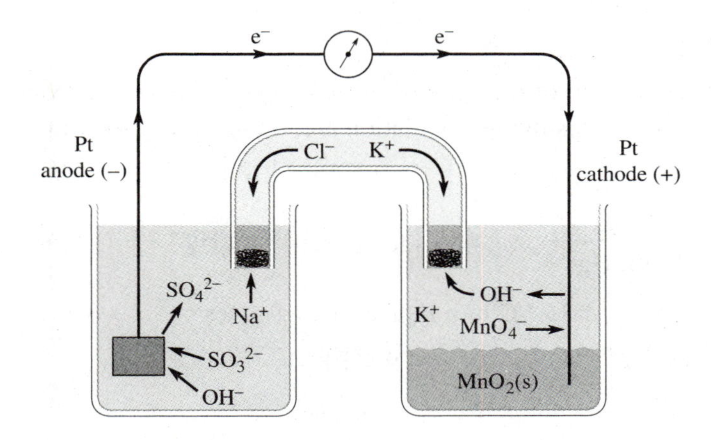

(b) $E^\circ = E^\circ_{oxid} + E^\circ_{red} = 0.936\ V + 0.597\ V = 1.533\ V$

Substituting $E^\circ = 1.533\ V = 1.533\ J/C$, $n = 6\ mol\ e^-$, and $F = 96{,}485\ C/mol\ e^-$ in Equation 20.5b gives

$$\Delta G^\circ = -nFE^\circ = -6\ mol\ e^- \times \frac{96{,}485\ C}{1\ mol\ e^-} \times \frac{1.533\ J}{1\ C} = -8.87 \times 10^5\ J = -887\ kJ$$

Substituting $E^\circ = 1.533\ V$ and $n = 6$ into Equation 20.8b gives

$$\log K = \frac{nE^\circ}{0.0592} = \frac{6 \times 1.533\ V}{0.0592\ V} = 155 \text{ and } K = \text{antilog } 155 = 10^{155}$$

(c) $$Q = \frac{[SO_4^{2-}]^3[OH^-]^2}{[MnO_4^-]^2[SO_3^{2-}]^3} = \frac{(0.10)^3(0.10)^2}{(0.10)^2(0.10)^3} = 1$$

Since $Q = 1$, $E = E^\circ = 1.533\ V$

20.109 (a) $Sn^{2+}(aq) \rightarrow Sn^{4+}(aq) + 2e^-$ $\quad E^\circ_{oxid} = -0.154\ V$

$Sn^{2+}(aq) + 2e^- \rightarrow Sn(s)$ $\quad E^\circ_{red} = -0.14\ V$

$E^\circ = E^\circ_{oxid} + E^\circ_{red} = -0.154\ V - 0.14\ V = -0.29\ V$

Substituting $E^\circ = -0.29\ V$ and $n = 2$ into Equation 20.8b gives

$$\log K = \frac{nE^\circ}{0.0592} = \frac{2(-0.29\ V)}{0.0592\ V} = -9.8 \text{ and } K = \text{antilog } (-9.8) = 10^{-9.8} = 2 \times 10^{-10}$$

(b) $$K = \frac{[Sn^{4+}]}{[Sn^{2+}]^2} = 2 \times 10^{-10} \text{ and } [Sn^{4+}] = 2 \times 10^{-10}[Sn^{2+}]^2$$

Since K is very small we can assume that the concentration of Sn^{2+} doesn't change; so $[Sn^{2+}] = 1.0\ M$. Hence,

$[Sn^{4+}] = 2 \times 10^{-10} \times (1.0)^2 = 2 \times 10^{-10}\ M$ and the fraction of Sn^{2+} converted to Sn^{4+} is

$$\frac{2 \times 10^{-10}\ M}{1.0\ M} = 2 \times 10^{-10}$$

20.111 (a) The current I = 0.10 μA = 1.0×10^{-7} A = 1.0×10^{-7} C/s. The time t = 1.00 h = 3600 s. Substituting I and t into Equation 20.9 gives

$$Q = I \times t = 1.0 \times 10^{-7}\ \text{C/s} \times 3600\ \text{s} = 3.6 \times 10^{-4}\ \text{C}$$

(b) $HgO(s) + H_2O(l) + 2e^- \rightarrow Hg(l) + 2OH^-(aq)$

The electrode equation shows that 1 mol (200.6 g) of mercury is formed when 2 mol of electrons, or $2 \times 96{,}485$ C, passes through the circuit. The steps in the calculation are: C → mol e^- → mol Hg → g Hg → μg Hg.

$$3.6 \times 10^{-4}\ \text{C} \times \frac{1\ \text{mol e}^-}{96{,}485\ \text{C}} \times \frac{1\ \text{mol Hg}}{2\ \text{mol e}^-} \times \frac{200.6\ \text{g Hg}}{1\ \text{mol Hg}} = 3.7 \times 10^{-7}\ \text{g Hg} = 0.37\ \mu\text{g Hg}$$

(c) Substituting Q = 3.6×10^{-4} C and E = 1.35 V = 1.35 J/C into Equation 20.1 gives

$$\text{maximum electrical work} = Q \times E = 3.6 \times 10^{-4}\ \text{C} \times 1.35\ \text{J/C} = 4.9 \times 10^{-4}\ \text{J}$$

20.113 $Zn(s) \rightarrow Zn^{2+}(aq) + 2e^-$

$2MnO_2(s) + 2NH_4^+(aq) + 2e^- \rightarrow Mn_2O_3(s) + 2NH_3(aq) + H_2O(l)$

(a) The current is 0.50 A = 0.50 C/s. The electrode equation shows that when 2 mol of electrons, or $2 \times 96{,}485$ C, passes through the circuit, 1 mol (65.39 g) of zinc metal is consumed. The steps in the calculation are: I × t → C → mol e^- → mol Zn → g Zn.

$$\frac{0.50\ \text{C}}{1\ \text{s}} \times 5.0\ \text{min} \times \frac{60\ \text{s}}{1\ \text{min}} \times \frac{1\ \text{mol e}^-}{96{,}485\ \text{C}} \times \frac{1\ \text{mol Zn}}{2\ \text{mol e}^-} \times \frac{65.39\ \text{g Zn}}{1\ \text{mol Zn}} = 0.051\ \text{g Zn}$$

(b) The electrode equation shows that when 2 mol of electrons, or $2 \times 96{,}485$ C, passes through the circuit, 2 mol of ammonia are produced. The steps in the calculation are: I × t → C → mol e^- → mol NH_3.

$$\frac{0.50\ \text{C}}{1\ \text{s}} \times 5.0\ \text{min} \times \frac{60\ \text{s}}{1\ \text{min}} \times \frac{1\ \text{mol e}^-}{96{,}485\ \text{C}} \times \frac{2\ \text{mol NH}_3}{2\ \text{mol e}^-} = 1.6 \times 10^{-3}\ \text{mol NH}_3$$

(c) $$Q = I \times t = \frac{0.50\ \text{C}}{1\ \text{s}} \times 5.0\ \text{min} \times \frac{60\ \text{s}}{1\ \text{min}} = 150\ \text{C}$$

$$\text{maximum electrical work} = Q \times E = 150\ \text{C} \times \frac{1.5\ \text{J}}{1\ \text{C}} = 225\ \text{J}$$

$$\text{power output} = \frac{225\ \text{J}}{300\ \text{s}} = 0.75\ \text{W}$$

20.115 $xF^- \rightarrow F_x(g) + xe^-$

x equals the number of faradays required to produce 1 mol of fluorine.

The number of moles of fluorine produced by the passage of 1.00 F can be obtained by substituting P = 1.00 atm, V = 12.8 L, T = 313 K, and R = 0.0821 L•atm/mol•K into the ideal gas equation

$$n = \frac{PV}{RT} = \frac{1.00\ \text{atm} \times 12.8\ \text{L}}{0.0821\ \text{K•atm/mol•K} \times 313\ \text{K}} = 0.498\ \text{mol}$$

Since 1 mol F_x × 1.00 F/0.498 mol F_x = 2.01 F are required to obtain 1 mol of fluorine, $x = 2$.

CHAPTER 21

METALS AND COORDINATION CHEMISTRY

Solutions To Practice Exercises

PE 21.1 VCl_3. In VCl_3 vanadium has a +3 oxidation state, while in VCl_2 it is only +2.

PE 21.2 (a) In_2O_3. Indium lies just below gallium in Group 3A; therefore, In_2O_3 should be more basic than Ga_2O_3.
(b) $Fe(OH)_2$. In $Fe(OH)_2$ iron has a +2 oxidation state, while in $Fe(OH)_3$ it has a +3 oxidation state.

PE 21.3 The complex ion is the bracketed portion, $Co(CN)_5NO^{2-}$. (The charge is 2– because the two Na^+ ions have a total charge of 2+.) Therefore, the sum of the oxidation numbers in the complex is –2. Each CN^- has a charge of 1– and Co(III) has an oxidation number of +3, so

$+3 + 5 \times (-1) + x_{NO} = -2$ and $x_{NO} = 0$

The charge on the nitrosyl ligand, NO, is zero.

PE 21.4 The AgCl precipitation reaction suggests that there is 1 mol of Cl^- ions per mole of compound. A formula consistent with the observations is $[Co(NH_3)_4Cl_2]Cl$.

PE 21.5 (a) Carbon monoxide coordinates through carbon (Table 21.5).

CO
|
Ni
OC CO
CO

(b) The cyanide ion coordinates through carbon (Table 21.5).

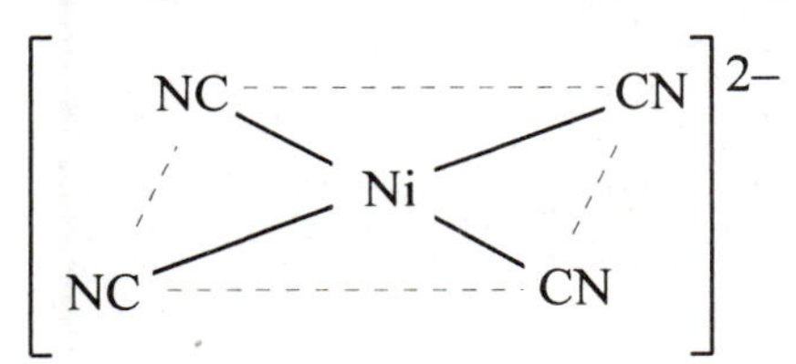

PE 21.6 Three. An octahedral complex has a coordination number of 6 and ethylenediamine is a bidentate ligand.

PE 21.7 (a) The complex ion is the bracketed portion, $Co(NO_2)_6^{3-}$. The six nitro ions are hexanitro. The complex is a negative ion, therefore the suffix -*ate* is added to cobalt. The cobalt is in the +3 oxidation state. The name is sodium hexanitrocobaltate(III).

(b) Ethylenediaminetetraacetato is EDTA. The oxidation numbers are –4 for EDTA and +2 for the plumbate(II), for a total of –2. The formula is $Na_2[Pb(EDTA)]$.

PE 21.8 (a)

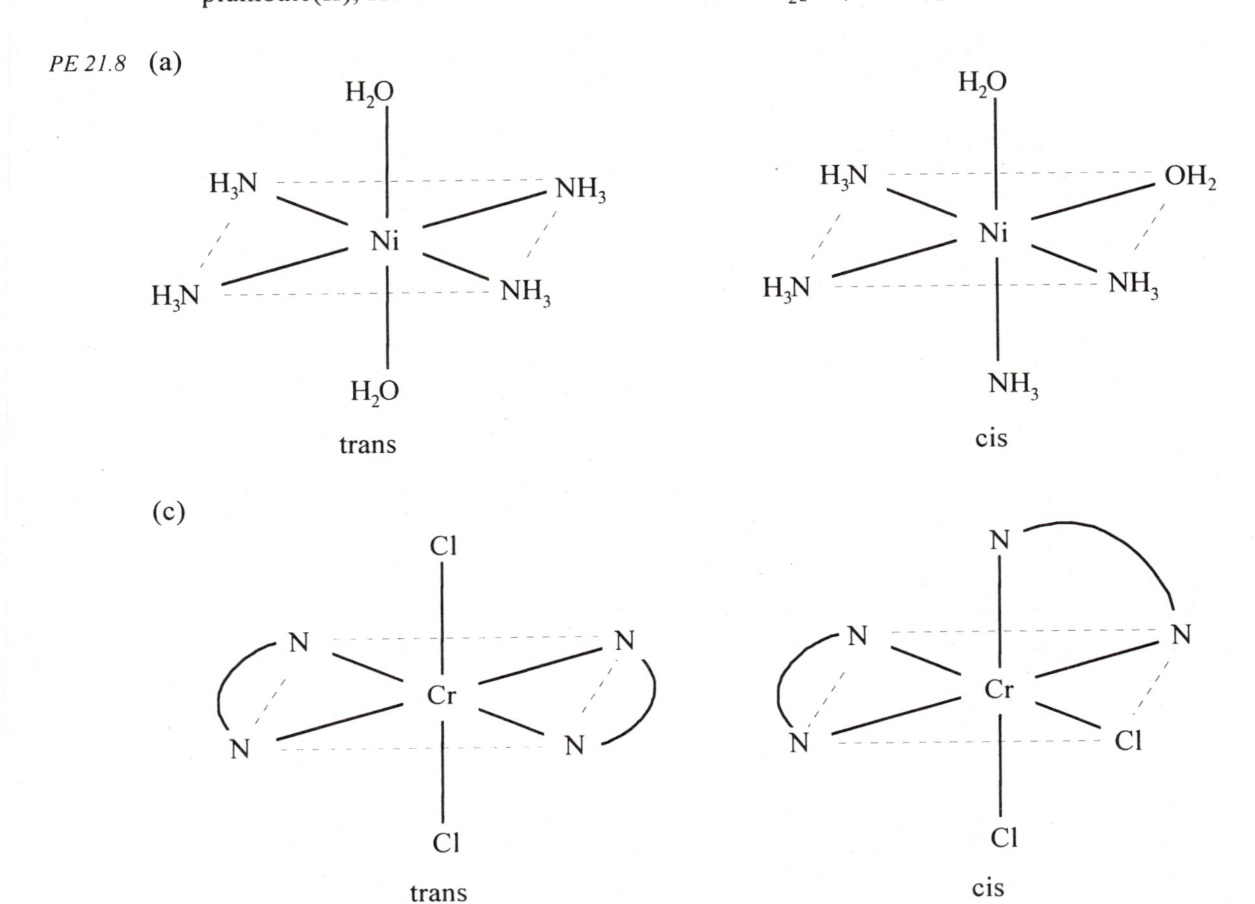

PE 21.9

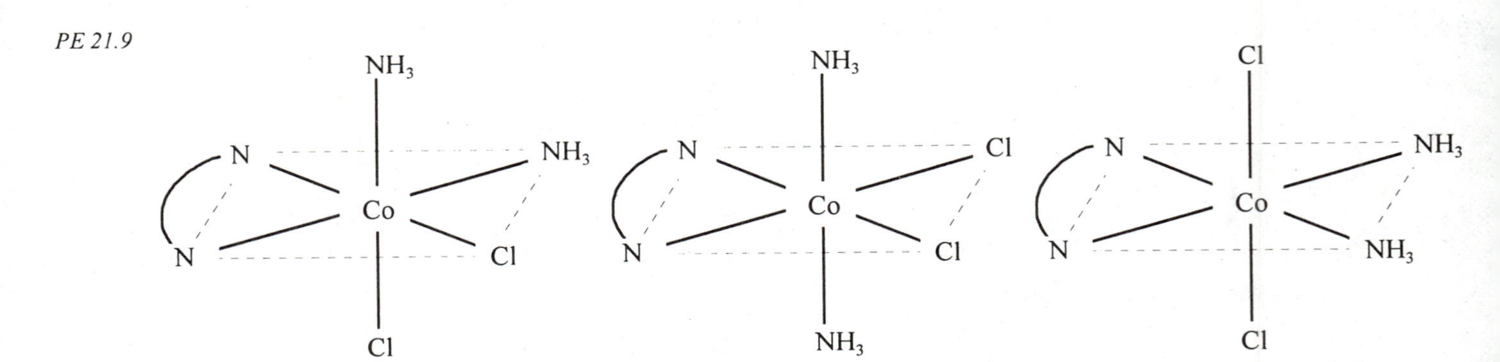

(a) Chiral (optically active).

(b) Each of the above *trans* isomers is achiral (optically inactive).

PE 21.10 $Ni(en)_3^{2+}$. Ethylenediamine (en) is a stronger ligand than NH_3 (see Table 21.8). Therefore, the orbital splitting energy, Δ, should be greater in $Ni(en)_3^{2+}$ and $Ni(en)_3^{2+}$ should absorb higher frequency light.

PE 21.11 Fe^{2+} has 6 d electrons. The high- and low-spin configurations are

high spin	low spin
1 1	— —
1⇂ 1 1	1⇂ 1⇂ 1⇂

Fe^{3+} has 5 d electrons. The high- and low-spin configurations are

high spin	low spin
1 1	— —
1 1 1	1⇂ 1⇂ 1

The only configuration that has one unpaired electron is the Fe^{3+} low-spin one; therefore the oxidation state of the iron is +3.

Solutions To Final Exercises

21.1 (a) Good electrical conductivity, good heat conductivity, and lustrous appearance of freshly cut surface.
(b) High density, hardness, high melting points, malleability, and ductility.
(c) Ca and Bi are brittle, Na and K are soft, Hg is liquid, and Ga has a low melting point (29.8°C).

21.3 (a) Metallic conduction is the transport of electric charge by the mobile valence electrons. Also refer to the definition in the Glossary.
(b) Increasing the temperature increases the random motions of the metal atoms that compose the metallic lattice and these increased random motions impede the movement of the valence electrons.

21.5 The hardness, melting point, and boiling point of a metal depend on the strength of the metallic bond, which decreases with increasing atomic size and increases with the number of valence electrons. Group 1A metals have one valence electron per atom compared to two valence electrons per atom for the Group 2A metals, and the Group 1A metals have larger atoms than the Group 2A metals that follow them in the same period. Hence, the metallic bond in the Group 1A metals is weaker than that in the Group 2A metals and, therefore, they are softer and have lower melting and boiling points than the Group 2A metals.

21.7 The number of electrons available for bonding is roughly similar for all lanthanides because each new electron in the series is added to an inner shell ($n = 4$). The atomic radius, however, decreases steadily going across the series; thus the metallic bond strength increases and with it the melting point.

21.9 (a) The absence of negative oxidation states. (Some metals, most notably nickel, do exhibit negative oxidation states in some of their complexes.)
(b) 1. The ability to form positive ions and ionic bonds. 2. The formation of basic oxides.

21.11 (a) 1. Separate the individual atoms from the solid metal. 2. Ionize the gaseous atoms to gaseous ions. 3. Change the gaseous ions into aqueous, hydrated ions.
(b) Steps 1 and 2 are endothermic. Step 3 is exothermic.

21.13 (a) In the d-block, f-block, and p-block. In the transition metals (d- and f-blocks) multiple oxidation states are attributed to the ability of the metal to employ electrons from more than one shell in bonding. In the p-block representative metals it is attributed to having valence electrons in both the s- and p-subshells of the valence shell.

(b) The electrons below the valence shell in the Group 1A metals are not used in bonding because their energies are too low. The single electron in the outer shell of the Group 1A metals is very easily removed, as can be seen from their ionization energies. However, removal of an electron from the next lower shell is not possible by chemical means, as can be seen from their much greater second ionization energies. Therefore, the Group 1A metals only part with one electron, and only have the +1 oxidation state.

Another way of viewing this is as follows: A 2+ alkali metal ion would be such a strong oxidizing agent, since it has a tremendous tendency to acquire electrons because of its high energy state, that it couldn't exist in a normal chemical environment. Therefore, the Group 1A metals only have or exhibit the +1 oxidation state, which corresponds to the removal of its single outer shell electron.

21.15 In general, it becomes more and more difficult to remove each subsequent electron from an atom, and ions with more than a 3+ charge are very rare. Consequently, atoms in a higher oxidation state tend to retain the electrons by sharing them to some degree so that the bonds becomes less ionic and more covalent.

21.17 (a) Thallium(I) hydroxide: TlOH; thallium(III) hydroxide: $Tl(OH)_3$. The strong and soluble base should be TlOH, since the lower oxidation state of thallium has the more ionic bonds.

(b) Tl(I) chloride: TlCl; Tl(III) chloride: $TlCl_3$. The bonds in $TlCl_3$ are more covalent than those in TlCl and, therefore, $TlCl_3$ should be more molecular than TlCl. Therefore, $TlCl_3$ should be the lower melting chloride, the 25°C melting chloride, and TlCl the higher melting chloride, the 430°C melting chloride.

21.19 (a) Above.

(b) They vary. Most of the first row of transition metals are more active than hydrogen. Most of the heavier transition metals are less active than hydrogen.

21.21 (a) 1. Small compact atoms. 2. Most are paramagnetic. 3. Variable oxidation states. 4. The activity of the transition metals usually decreases from top to bottom in a group. 5. Most of the transition metal compounds absorb visible light.

(b) The highest oxidation states of Zn, Cd, and Hg are the +2 states; that is, Zn, Cd, and Hg only part with their s valence electrons and their underlying d orbitals are always completely filled. Some chemists consider transition elements to be only those elements that have partially filled d orbitals in at least one observed oxidation state.

21.23 $Mg(s) \rightarrow Mg^{2+}(aq) + 2e^-$ $\qquad \Delta H^\circ_{oxid} = -467\text{ kJ}$

(a) $E^\circ_{oxid} = 2.357$ V (Table 20.1). Substituting n = 2 mol e^-, F = 96,485 C/mol e^-, and $E^\circ_{oxid} = 2.357$ V = 2.357 J/C into Equation 20.5b gives

$$\Delta G^\circ_{oxid} = -nFE^\circ_{oxid} = -2\text{ mol e}^- \times \frac{96{,}485\text{ C}}{1\text{ mol e}^-} \times \frac{2.357\text{ J}}{1\text{ C}} = -4.548 \times 10^5\text{ J} = -454.8\text{ kJ}$$

(b) $\Delta G^\circ_{oxid} = \Delta H^\circ_{oxid} - T\Delta S^\circ_{oxid}$

$$\Delta S^\circ_{oxid} = \frac{\Delta H^\circ_{oxid} - \Delta G^\circ_{oxid}}{T} = \frac{-467\text{ kJ/mol} + 454.8\text{ kJ/mol}}{298\text{ K}} = -4.1 \times 10^{-2}\text{ kJ/K}\cdot\text{mol} = -41\text{ J/K}\cdot\text{mol}$$

21.25 (a) Molecular orbital theory predicts that when N atomic orbitals combine they produce N molecular orbitals with a range of energies. When N is very large, however, the spacing or energy gap between the various energy levels is extremely small and we get a band, which is a collection of very closely spaced energy levels; the energy difference between any level and the next highest level being negligibly small. The foregoing holds for crystal metallic conductors. For insulators, the theory predicts that the N molecular orbitals formed will split into two bands separated by an energy gap called the forbidden zone.

(b) 1. Metallic conductors have a band that is only partly filled, called the conduction band, or a filled valence band that overlaps a vacant conduction band, the two bands forming one larger, partly filled conduction band. In some cases, a partly filled conduction band also overlaps an empty conduction band forming one larger, partly filled conduction band. When a band is only partly filled, the electrons can be easily promoted into the unfilled levels of the band and thereby conduct electricity.
2. Insulators have completely filled bands that are separated from empty bands by a large energy gap called the forbidden zone. Electrons cannot move into empty molecular orbitals and can therefore not conduct electricity; i.e., there is no way for them to gain and transport energy.
3. In semiconductors the forbidden zone or energy gap is smaller than that in an insulator and some electrons from the filled lower band can be excited into the unfilled conduction band above. Semiconductors have a very low electrical conductivity compared to metals, because they have very few electrons in the conduction band.

21.27 (a) The electron configuration of magnesium is Mg: $1s^22s^22p^63s^2$.

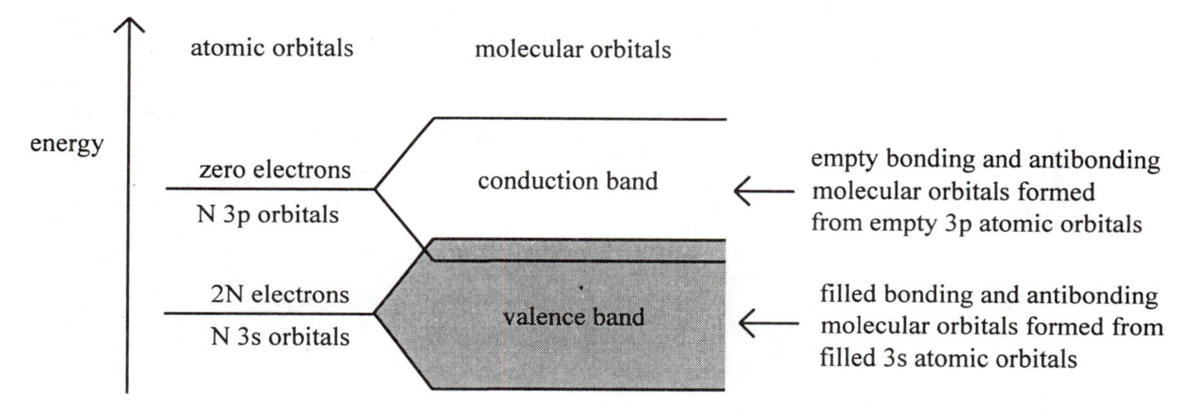

(b) The band formed from the 3s atomic orbitals is essentially full and the band formed from the 3p orbitals is essentially empty. If the 3s and 3p bands did not overlap, then Mg would have completely filled bands separated by a forbidden zone from an empty conduction band and would consequently be an insulator and not a metallic conductor. However, since the 3s and 3p bands in magnesium do overlap, there is no forbidden zone between them and excitation of electrons from the 3s valence band to the empty 3p conduction band occurs easily and, consequently, Mg is metallic in nature. (In reality, the 3s valence band and the empty 3p conduction band combine to form one conduction band in which the 3s electrons can freely move when excited.)

21.29 (a) The conductivity measurements suggest that there is 1 mol of Cl^- ion per mole of compound. A formula consistent with this is $[Cr(NH_3)_4Cl_2]Cl$.

(b) $[Cr(NH_3)_4Cl_2]Cl(s) \rightarrow Cr(NH_3)_4Cl_2^+(aq) + Cl^-(aq)$

21.31 (a) The AgBr precipitation reactions suggest that there is 1 mol of Br^- ion per mole compound A, but no Br^- ion in B. Formulas for A and B consistent with these observations are

A: $[Pt(NH_3)_3(NO_3)]Br$ and B: $[Pt(NH_3)_3Br]NO_3$

(b) $[Pt(NH_3)_3(NO_3)]Br(aq) + AgNO_3(aq) \rightarrow AgBr(s) + Pt(NH_3)_3(NO_3)^+(aq) + NO_3^-(aq)$

21.33 (a) $FeCl_6^{3-}$ (b) $Fe(C_2O_4)_3^{3-}$ or $Fe(ox)_3^{3-}$

21.35 (a) $x_{Co} = +3$ (b) $x_{Fe} = +3$ (c) $x_{Fe} = +2$ (d) $x_{Fe} = +3$

21.37 (a)

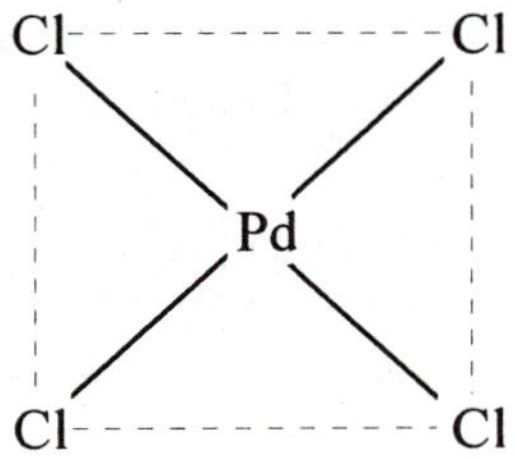

Cl Cl

Pd

Cl Cl

(c)

Cl–Ag–Cl

(b)

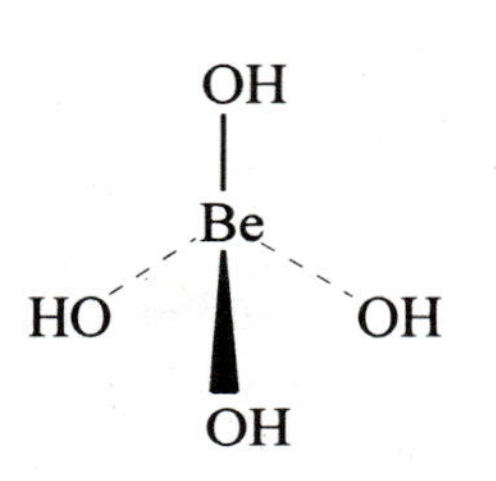

(d)

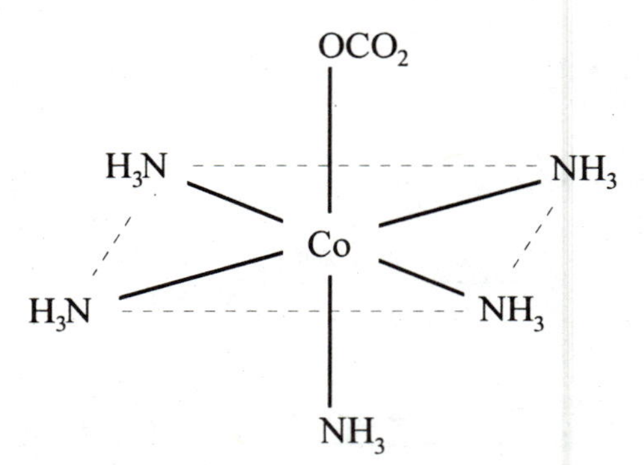

21.39 (a)

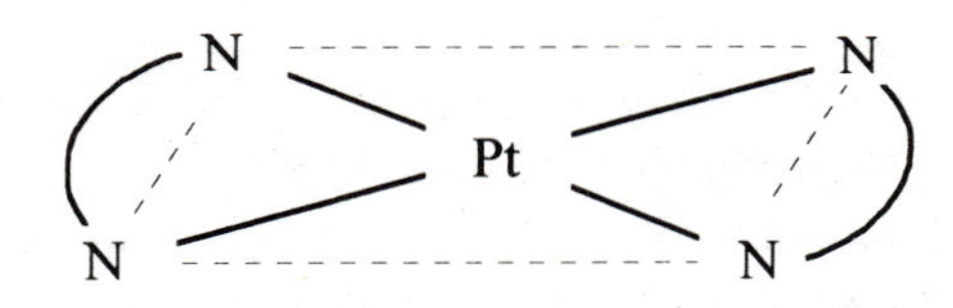

(b)

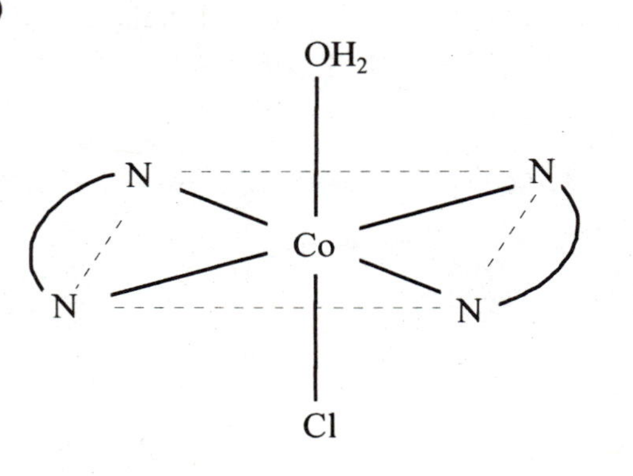

and

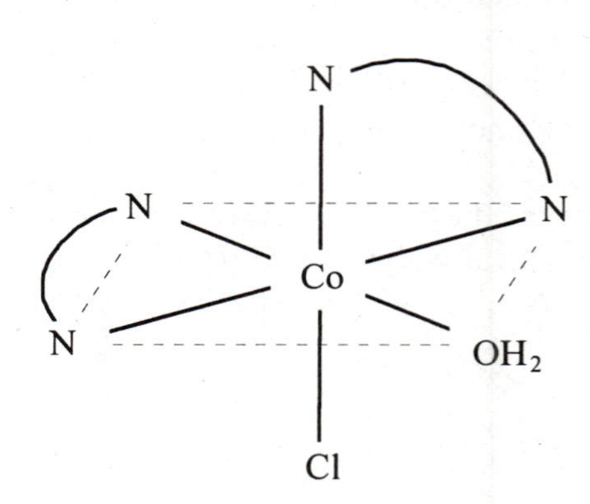

(c)

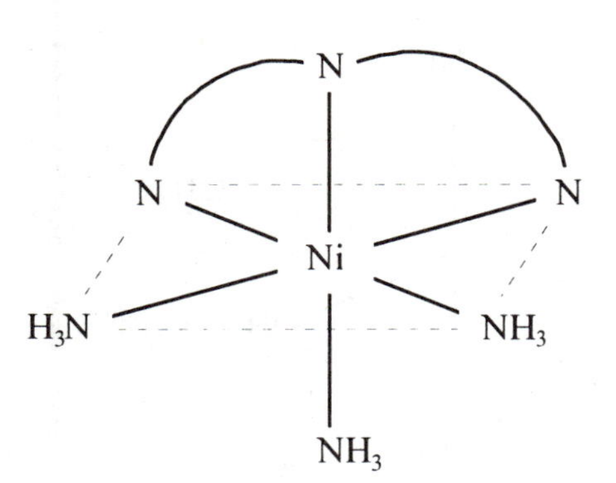

and

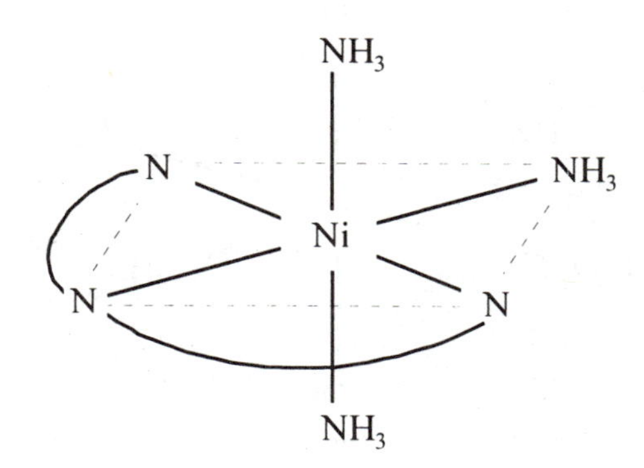

21.41 (a) $Ni(H_2O)_6^{2+}$ (b) $Cu(NH_3)_4^{2+}$ (c) $[Ir(NH_3)_3Cl_3]$ (d) $[Co(NH_3)_4Cl_2]$ (e) $Fe(H_2O)_6^{3+}$

21.43 (a) Exercise 21.35: (a) hexaamminecobalt(III)chloride; (b) potassium hexafluoroferrate(III); (c) potassium hexacyanoferrate(II); (d) potassium hexacyanoferrate(III).
(b) Exercise 21.37: (a) tetrachloropalladate(II) ion; (b) tetrahydroxoberyllate(II) ion; (c) dichloroargentate ion; (d) pentaamminecarbonatocobalt(III) ion.
(c) Exercise 21.39: (a) bis(ethylenediamine)platinum(II) ion; (b) aquachlorobis(ethylenediamine)cobalt(II) ion; (c) triamminediethylenetriaminenickel(II) ion.

21.45 Ionization isomerism is a form of structural isomerism exhibited by ionic coordination compounds in which each isomer has a different ion outside the complex. Ionization isomers have the same composition but yield different ions in solution. Some examples are: $[Co(NH_3)_4Cl_2]NO_2$ and $[Co(NH_3)_4Cl(NO_2)]Cl$; $[Co(NH_3)_5NO_2]SO_4$ and $[Co(NH_3)_5SO_4]NO_2$; and $[Pt(NH_3)_4Cl_2]Br_2$ and $[Pt(NH_3)_4Br_2]Cl_2$.

21.47 Linkage isomerism is a form of structural isomerism in which the isomers differ in the atom of a ligand, the coordinating atom of the ligand, that is bonded to the central metal atom. Linkage isomerism occurs with ligands capable of coordinating in more than one way. Examples are: $[Mn(CO)_5(SCN)]$ and $[Mn(CO)_5(NCS)]$; $[Co(en)_2(NO_2)_2]Cl$ and $[Co(en)_2(ONO)_2]Cl$; and $Co(NCS)_2Cl_2^{2-}$ and $Co(SCN)_2Cl_2^{2-}$.

21.49 Geometric isomers are stereoisomers that are not mirror images of each other. Enantiomers or optical isomers are stereoisomers that are nonsuperimposable mirror images of each other. Geometric isomers are also known as position isomers and cis-trans isomers.

Enantiomers are often called optical isomers because they have the ability to rotate the plane of polarized light.

21.51 (a)

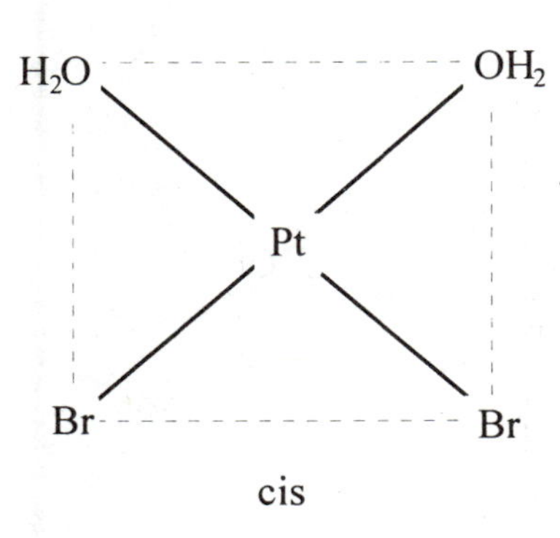

cis

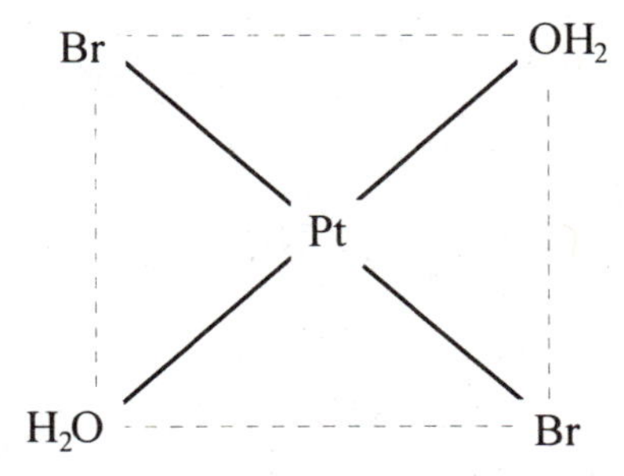

trans

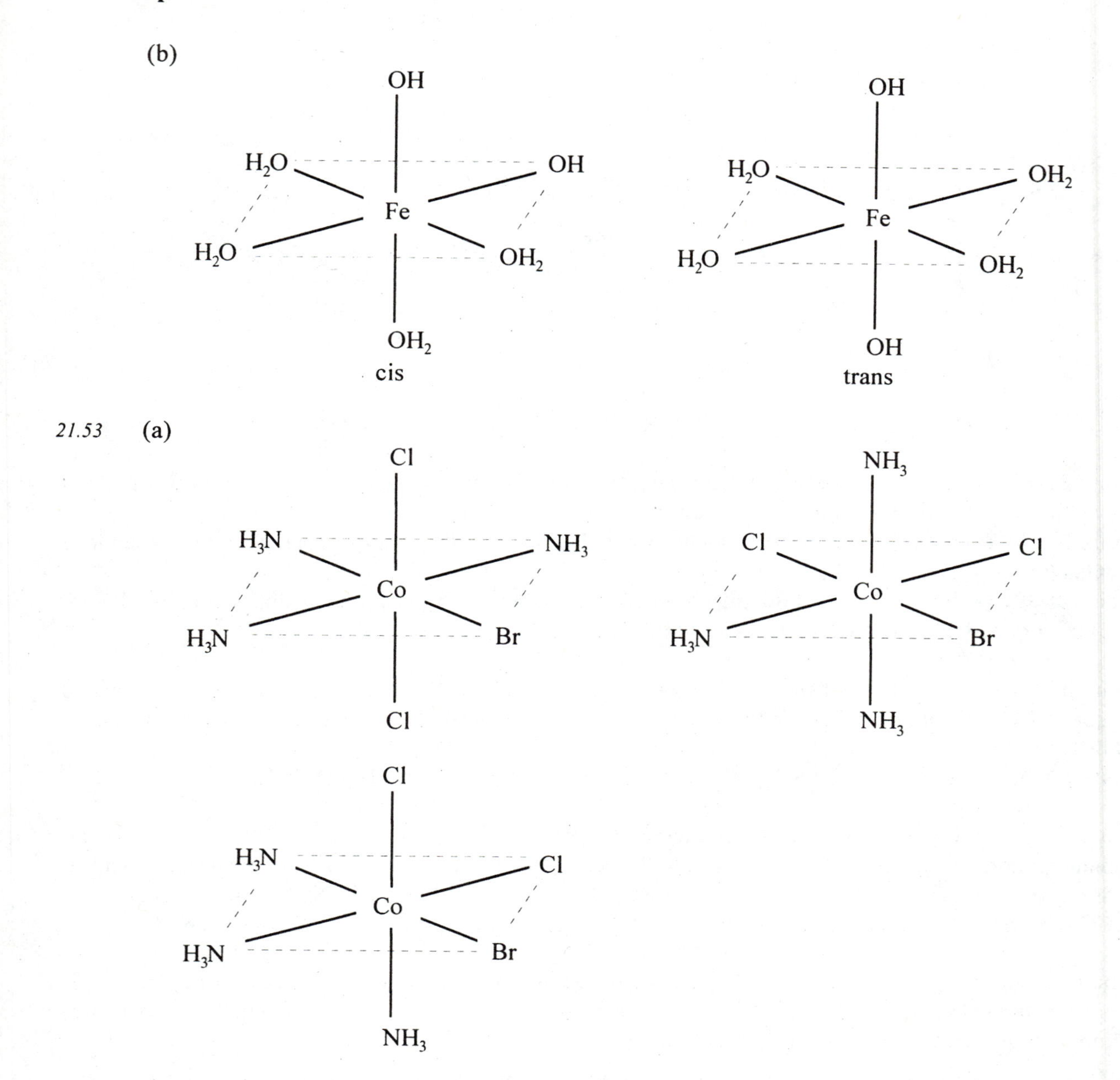

Note: Three similar ligands can either occupy the three corners of one face of the octahedron or they can occupy three corners of a square. An Ma_3d_2f octahedral complex has a total of 3 stereoisomers: 3 geometric isomers and no pairs of enantiomers.

(b)

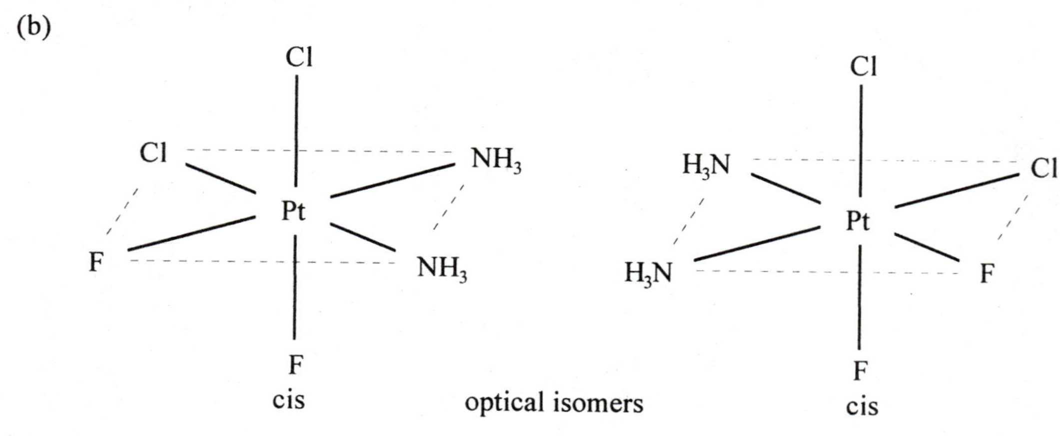

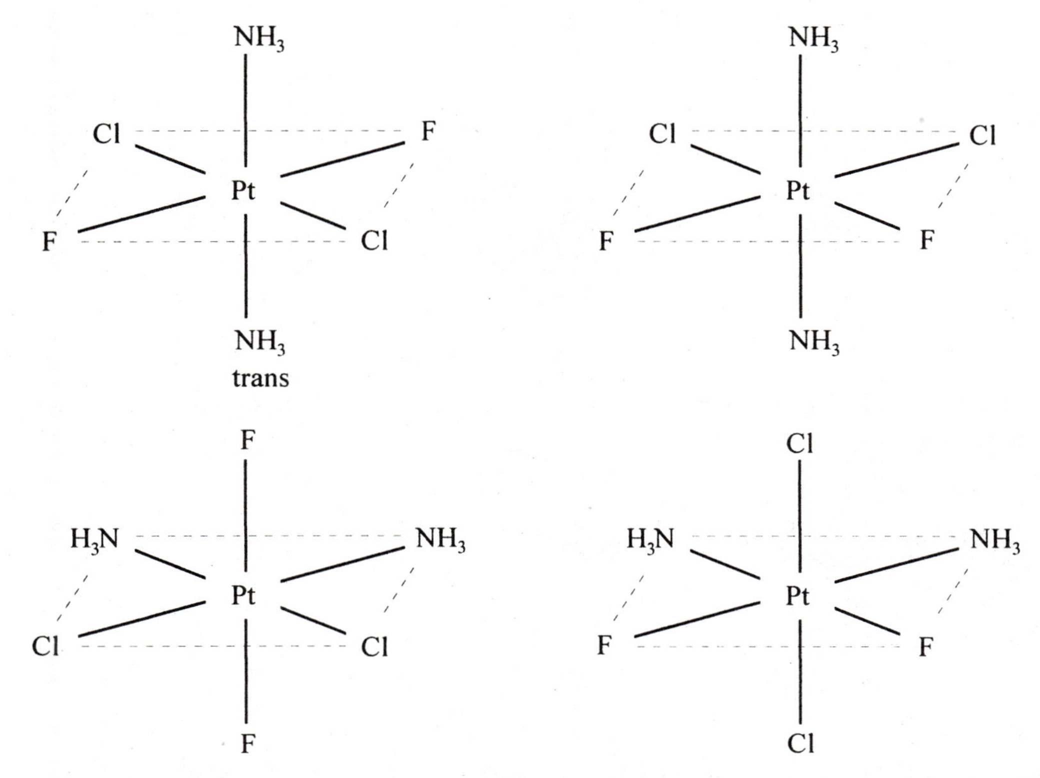

Note: An $Ma_2c_2e_2$ octahedral complex has a total of 6 stereoisomers: 5 geometric isomers and 1 pair of enantiomers.

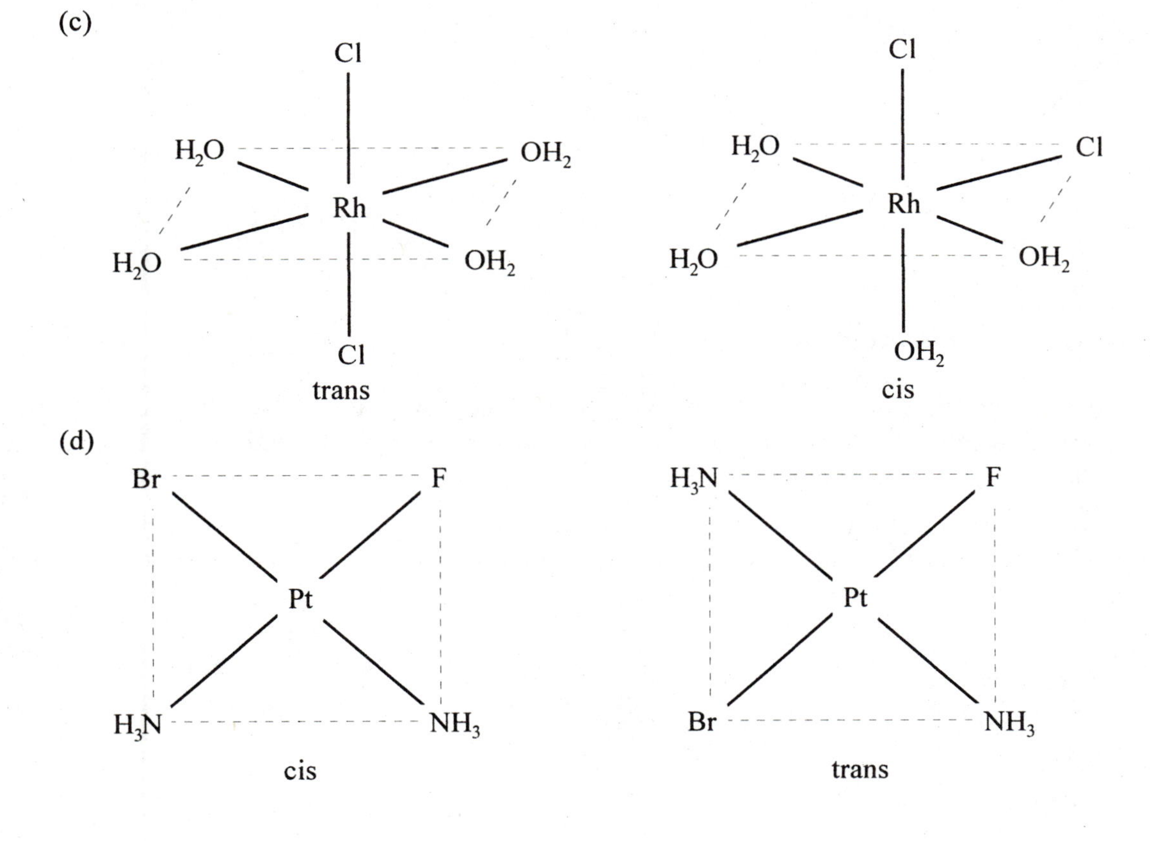

21.55 (a)

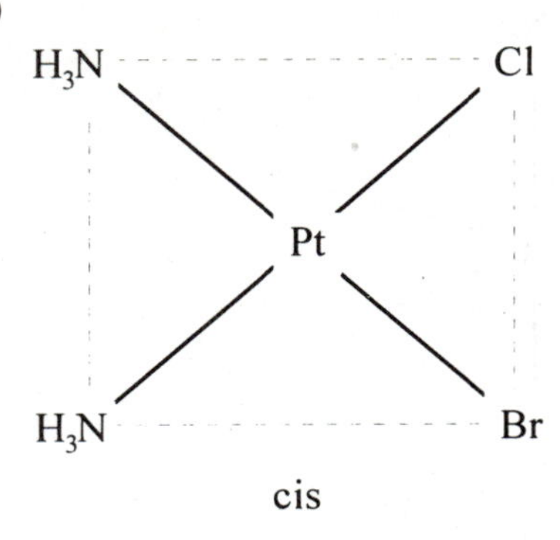

cis

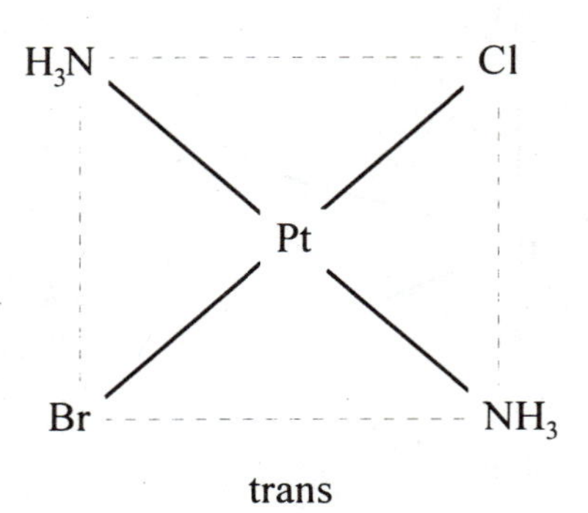

trans

All other sketches can be superimposed on one of these by just rotating through 90 or 180 degrees or by flipping the sketch.

(b)

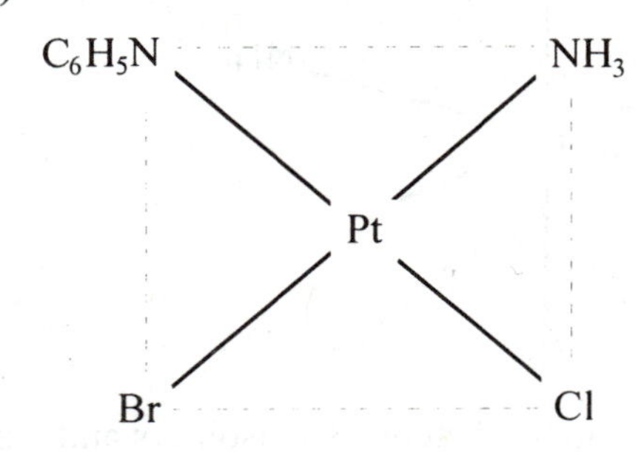

C$_6$H$_5$N Cl

Pt

Br NH$_3$

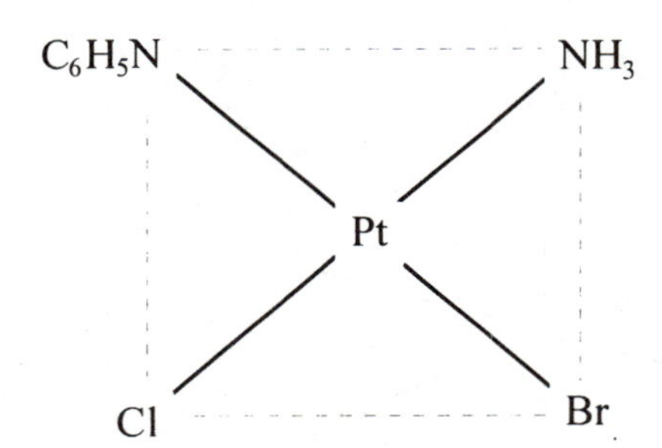

All other sketches can be superimposed on one of these.

21.57 In the electrostatic model, which deals with complexes containing a central metal ion, the metal-ligand bonding is treated as an ionic interaction or electrostatic attraction between the uniformly charged metal ion sphere and charged ligand ions or dipoles.

The strong points of the electrostatic model are its simplicity, its agreement with observations made on complexes of the representative metals, and its ability to explain the stability of many transition metal complexes.

Its shortcomings lie with its inability to explain some of the properties of the transition metal complexes, for example, their color and paramagnetism.

The electrostatic model works best for the complexes of the representative metals.

21.59 Refer to Figure 21.23 in the textbook.

21.61 Refer to Figure 21.25 in the textbook. Orbital splitting refers to the fact that the d electrons in the free atom are held in five equal energy d orbitals, while in the complex all the remaining d electrons fill into two sets of orbitals with different energies.

21.63 Refer to Figure 21.36 in the textbook.

21.65 Metal-H_2O bonds become stronger with increasing ionic charge and decreasing ionic radius, that is, with increasing charge density. And the stronger the bonds are, the larger the formation constant should be.

(a) $Be(H_2O)_4^{2+}$. Be^{2+} is much smaller than Cu^{2+}.
(b) $Fe(H_2O)_6^{3+}$. Fe^{3+} has a greater charge than Fe^{2+}.

21.67 $Cr(H_2O)_6^{3+}$. H_2O with its stronger field produces a larger orbital splitting energy (Δ) than Cl^-.

21.69

Compound	Observed Color	Approximate Absorbed Color	Approximate Absorbed Wavelength (nm)
$Co(NH_3)_6^{3+}$	orange yellow	violet blue	420
$Co(NH_3)_5(H_2O)^{3+}$	red	green	530
$Co(NH_3)_5Cl^{2+}$	purple	yellow green	560
$Co(NH_3)_5(NCS)^{2+}$	orange	blue	460

The lower the absorbed wavelength, the larger the orbital splitting energy (Δ) and the stronger the ligand. Therefore, in order of increasing field strength we have $Cl^- < H_2O < NCS^- < NH_3$. This agrees with the order in Table 21.8.

21.71 (a)

1 1

Ni^{2+}: $3d^8$ ⇅ ⇅ ⇅

(b)

1 1

Co^{2+}: $3d^7$ ⇅ ⇅ 1

(c)

1 1

Fe^{2+}: $3d^6$ ⇅ 1 1

(d)

1 1

Fe^{3+}: $3d^5$ 1 1 1

21.73 (a) two

    ↑ ↑

Ni^{2+}: $3d^8$ ↑↓ ↑↓ ↑↓

(b) none

    — —

Pt^{4+}: $5d^6$ ↑↓ ↑↓ ↑↓

(c) three

    — —

Mo^{3+}: $4d^3$ ↑ ↑ ↑

(d) none

    —

    ↑↓

    ↑↓

Pt^{2+}: $5d^8$ ↑↓ ↑↓

(e) one

    — —

Mn^{2+}: $3d^5$ ↑↓ ↑↓ ↑

(f) one

    — —

Ru^{3+}: $4d^5$ ↑↓ ↑↓ ↑

21.75 $Mn(H_2O)_6^{3+}$: ↑ — ; four unpaired electrons

$Mn^{3+}(3d^4)$ ↑ ↑ ↑
high-spin

$Mn(CN)_6^{3-}$: — — ; two unpaired electrons

$Mn^{3+}(3d^4)$ ↑↓ ↑ ↑
low-spin

21.77 Using the heavy line in the periodic table to separate approximately the metallic elements (left of the line) from the nonmetallic elements we get 22 main group metals and 22 main group nonmetals. (Metalloids have been counted as metals or nonmetals depending on which side of the line they fall on.) Therefore, there is an equal number of main group metals and nonmetals; i.e., there is a 1:1 ratio.

21.79 The atomic radius of gold is 0.144 nm (Figure 8.9).

(a) In a simple cubic lattice the atoms are touching one another in straight rows piled one atop the other and we only have to calculate the number of gold diameters that fit into 100 nm in order to determine the number of layers of atoms. The number of layers of atoms is therefore

$$\frac{100 \text{ nm}}{0.288 \text{ nm/layer}} = 347 \text{ layers}$$

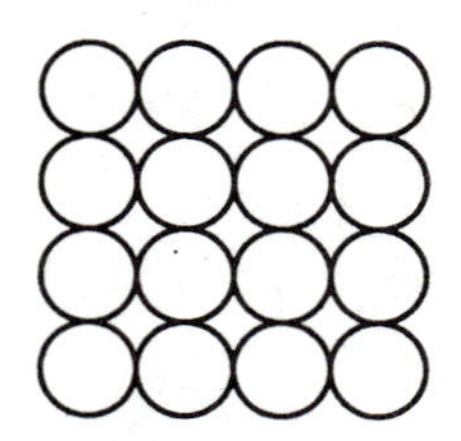

side view of simple unit packing

(b) Gold crystallizes in the face-centered cubic lattice system. In a face-centered cubic unit cell, the atoms that lie along an edge do not touch. Therefore, we will first calculate the width of three layers of atoms and then the number of layers in 100 nm.

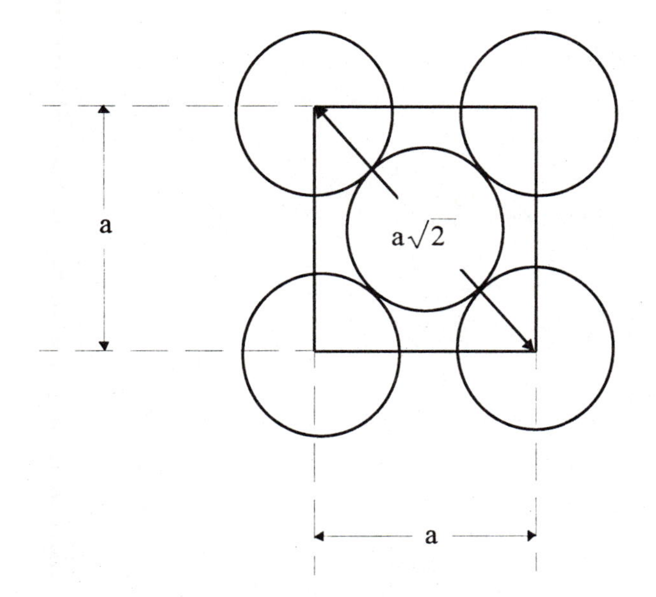

side view of a face-centered cubic unit cell

For a face-centered cubic unit cell we have

$$4r = a\sqrt{2} \text{ and } a = \frac{4r}{\sqrt{2}} = \frac{4 \times 0.144 \text{ nm}}{\sqrt{2}} = 0.407 \text{ nm}$$

The width of three layers is therefore

width three layers = a + 2r = 0.407 nm + 2 × 0.144 nm = 0.695 nm

The number of layers of atoms is therefore

$$\frac{100 \text{ nm}}{0.695 \text{ nm/3 layers}} = 432 \text{ layers}$$

21.81 The energy of a single energy gap is

$$\frac{1.06 \times 10^5 \text{J}}{1 \text{ mol}} \times \frac{1 \text{ mol}}{6.022 \times 10^{23} \text{ energy gaps}} = 1.76 \times 10^{-19} \text{ J/energy gap}$$

The lowest radiation frequency that would excite an electron across the gap is

$$\nu = \frac{E}{h} = \frac{1.76 \times 10^{-19} \text{J}}{6.63 \times 10^{-34} \text{J}\bullet\text{s}} = 2.65 \times 10^{14} \text{ Hz}$$

From Figure 7.3 we see that radiation of this frequency lies in the infrared region of the spectrum that borders on the visible portion of the spectrum. Therefore, the higher frequency visible light can excite or promote electrons across the energy gap, which increases the number of electrons in the conduction band and, hence, increases the conductivity.

21.83 Add barium nitrate to an aqueous solution of the compound. The first compound will form a precipitate of $BaSO_4$, while the second will not form a precipitate. Alternatively, add silver nitrate to an aqueous solution of the compound. The first compound will not form a precipitate, but the second will form a precipitate of AgBr.

21.85 (a) Purple compound: The steps in the calculation are: g AgCl → mol AgCl → mol Cl^- ions → mol Cl^- ions/mol compound → formula.

$$1.06 \text{ g AgCl} \times \frac{1 \text{ mol AgCl}}{143.32 \text{ g AgCl}} \times \frac{1 \text{ mol Cl}^- \text{ ions}}{1 \text{ mol AgCl}} = 7.40 \times 10^{-3} \text{ mol Cl}^- \text{ ions}$$

The moles of Cl^- ions per mole of compound is

$$\frac{7.40 \times 10^{-3} \text{ mol Cl}^- \text{ ions}}{0.100 \text{ L} \times 0.037 \text{ mol compound/L}} = 2 \text{ mol Cl}^- \text{ ions/mol compound}$$

A tentative formula is $[Co(NH_3)_5Cl]Cl_2{\bullet}H_2O$.

(b) Brick red compound:

$$4.3 \text{ g AgCl} \times \frac{1 \text{ mol AgCl}}{143.32 \text{ g AgCl}} \times \frac{1 \text{ mol Cl}^- \text{ ions}}{1 \text{ mol AgCl}} = 0.030 \text{ mol Cl}^- \text{ ions}$$

The moles of Cl^- ions per mole of compound is

$$\frac{0.030 \text{ mol Cl}^- \text{ ions}}{0.100 \text{ L} \times 0.10 \text{ mol compound/L}} = 3 \text{ mol Cl}^- \text{ ions/mol compound}$$

A tentative formula is $[Co(NH_3)_5H_2O]Cl_3$.

21.87 No.

Cl
Be
HO OH
Cl

The tetrahedral symmetry properties will only allow for stereoisomerism when all four groups attached to the central atom are different, and then only of the optical kind.

Comment: Optical isomerism exists in tetrahedral complexes of the type M(AB)(AB), where (AB) is an unsymmetrical bidentate ligand. However, tetrahedral complexes with four different unidentate ligands are generally too unstable to be isolated.

21.89 If the aqueous solution of $CrCl_3$ only contained $Cr(H_2O)_6^{3+}$ ions, then the solution would appear violet, since $Cr(H_2O)_6^{3+}$ has a violet color. Therefore, the green solution of $CrCl_3$ must also contain complex ions in which some of the water is replaced by chloride ion, for example, $Cr(H_2O)_5Cl^{2+}$ and $Cr(H_2O)_4Cl_2^+$. The observed color is the complementary color of the absorbed light and the wavelength of the absorbed light depends on the size of Δ, the orbital splitting energy. From the spectrochemical series (Table 21.8), we see that Cl^- is a weaker bonding ligand than H_2O. Consequently, Δ should decrease, which means that the wavelength of the absorption should increase, when H_2O is replaced by Cl^-. From the color wheel (Figure 21.28) we see that a shift to higher wavelength absorption shifts the observed color from violet toward green. Therefore, it is likely that the aquachlorochromium(III) complex ions are green and their presence makes the solution appear green.

21.91 (a) CoF_6^{3-}: Since the orbital splitting energy produced by F^- is less than the pairing energy, CoF_6^{3-} should be a high-spin complex.

↑ ↑

Co^{3+}: $3d^6$ ↑↓ ↑ ↑

$Co(NH_3)_6^{3+}$: Since the orbital splitting energy produced by NH_3 is greater than the pairing energy, $Co(NH_3)_6^{3+}$ should be a low-spin complex.

— —

Co^{3+}: $3d^6$ ↑↓ ↑↓ ↑↓

(b) CoF_6^{3-} is paramagnetic and $Co(NH_3)_6^{3+}$ is not.

SURVEY OF THE ELEMENTS 4
METALS AND METALLURGY

Solutions To Practice Exercises

PE S4.1 $Ni_2S_3(s) + 4O_2(g) \xrightarrow{heat} 2NiO(s) + 3SO_2(g)$

PE S4.2 Oxygen is the oxidizing agent:

$O_2(g) + 2H_2O(l) + 4e^- \rightarrow 4OH^-(aq)$ reduction

Gold is the reducing agent:

$Au(s) + 2CN^-(aq) \rightarrow Au(CN)_2^-(aq) + e^-$ oxidation

PE S4.3 $2Al(H_2O)_6^{3+}(aq) + 3S^{2-}(aq) \rightarrow 2Al(H_2O)_3(OH)_3(s) + 3H_2S(g)$

PE S4.4 (a) Yes. The standard oxidation potentials are:

$Sn^{2+}(aq) \rightarrow Sn^{4+}(aq) + 2e^-$ $E^\circ_{oxid} = -0.154$ V

$Hg(l) \rightarrow Hg^{2+}(aq) + 2e^-$ $E^\circ_{oxid} = -0.852$ V

Sn^{2+} has a higher (less negative) oxidation potential and a greater tendency to give up electrons than Hg; therefore, $Sn(NO_3)_2$ will reduce Hg^{2+} to Hg.

(b) No. The standard oxidation potentials are:

$Ag(s) + Br^-(aq) \rightarrow AgBr(s) + e^-$ $E^\circ_{oxid} = -0.0732$ V

$Sn^{2+}(aq) \rightarrow Sn^{4+}(aq) + 2e^-$ $E^\circ_{oxid} = -0.154$ V

The silver half-reaction has a higher (less negative) oxidation potential and a greater tendency to give up electrons than Sn^{2+}; therefore $Sn(NO_3)_2$ will not reduce AgBr to Ag.

PE S4.5 The boiling points (Table S4.5) of zinc and cadmium are 907°C and 767°C, respectively. Cadmium will volatilize at 767°C.

PE S4.6

$K_{sp} = [Hg_2^{2+}][Cl^-]^2$

$1.3 \times 10^{-18} = (x)(2x)^2 = 4x^3$ and $x = 6.9 \times 10^{-7}$ M $= [Hg_2^{2+}]$

The equilibrium concentration of Hg_2^{2+}, 6.9×10^{-7} mol/L, is also the molar solubility of Hg_2Cl_2.

PE S4.7

Refer to the solution of Exercise 20.100(c). Substituting n = 1 mol e^-, F = 96,485 C/mol e^-, and E° = 0.357 V = 0.357 J/C into Equation 20.5b gives

$$\Delta G° = -nFE° = -1 \text{ mol } e^- \times \frac{96{,}485 \text{ C}}{1 \text{ mol } e^-} \times \frac{0.357 \text{ J}}{1 \text{ C}} = -3.44 \times 10^4 \text{ J} = -34.4 \text{ kJ}$$

PE S4.8

$Cu(s) \rightarrow Cu^{2+}(aq) + 2e^-$ oxidation

$NO_3^-(aq) + 2H^+(aq) + e^- \rightarrow NO_2(g) + H_2O(l)$ reduction

The net ionic equation for the reaction is

$Cu(s) + 2NO_3^-(aq) + 4H^+(aq) \rightarrow Cu^{2+}(aq) + 2NO_2(g) + 2H_2O(l)$

PE S4.9

(a) Cl^- is –1; therefore the oxidation state of cobalt is +2. Hence, the central cobalt atom has 7 3d electrons.
(b) NO_2^- is –1; therefore the oxidation state of cobalt is +3. Hence, the central cobalt atom has 6 3d electrons.
(c) NH_3 is zero; therefore the oxidation state of nickel is +2. Hence, the central nickel atom has 8 3d electrons.

Solutions To Final Exercises

S4.1

An alloy is a mixture, homogeneous or heterogeneous, of metals, or a compound of metals, or some combination of both of these possessing metallic properties; some alloys contain nonmetals as well.

All of the various steels are ferrous alloys.

See Table S4.1 in the textbook for examples of nonferrous alloys.

S4.3

The three principal steps in obtaining a pure metal from its ore are: (1) concentrating the ore, (2) extracting (obtaining) the metal, and (3) refining (purifying) the crude metal.

S4.5

Pretreating an ore refers to the process where a hard-to-reduce mineral is converted to a form that is easy to reduce.

The roasting of sulfides to reducible oxides and the conversion of an oxide, hydroxide, or carbonate to a lower melting chloride before electrolytic reduction are examples of pretreatment.

S4.7

The Bayer process for obtaining pure aluminum oxide from bauxite is a chemical separation. The steps in the Bayer process are:

1. Bauxite, an aluminum ore consisting principally of alumina ($Al_2O_3 \cdot xH_2O$), is treated with hot 30% aqueous sodium hydroxide under pressure. The amphoteric aluminum oxide dissolves:

$$Al_2O_3(s) + 2OH^-(aq, 30\%) + 3H_2O(l) \xrightarrow[\text{to } 220°C]{150°C} 2Al(OH)_4^-(aq)$$

2. The resulting solution is filtered, cooled, and diluted to reduce the OH^- concentration. Aluminum hydroxide then precipitates:

$$Al(OH)_4^-(aq) \xrightarrow{\text{dilution}} Al(OH)_3(s) + OH^-(aq)$$

3. The pure anhydrous oxide is obtained by heating:

$$2Al(OH)_3(s) \xrightarrow{\text{heat}} Al_2O_3(s) + 3H_2O(g)$$

S4.9 In the electrorefining of copper, a bar of the crude copper to be refined is made the anode of an electrolytic cell. The cathode of the cell is a thin sheet of pure copper, and the electrolyte is an aqueous solution of copper sulfate. A voltage is then applied to the electrodes that is just high enough to oxidize the copper atoms in the anode:

$$Cu(s) \rightarrow Cu^{2+}(aq) + 2e^- \qquad \text{(at anode)}$$

Simultaneously, pure copper metal plates out at the cathode:

$$Cu^{2+}(aq) + 2e^- \rightarrow Cu(s) \qquad \text{(at cathode)}$$

More active impurities also dissolve from the anode, but stay in solution, while less active impurities are not oxidized and wind up in the anode sludge.

S4.11 First, crushed rock containing the native gold is aerated for several days in a solution of sodium cyanide. The gold dissolves; the cyanide, acting as a complexing agent, stabilizes the oxidized gold:

$$4Au(s) + 8CN^-(aq) + O_2(g) + 2H_2O(l) \rightarrow 4Au(CN)_2^-(aq) + 4OH^-(aq)$$

Then powdered zinc, used as a reducing agent, is added to the solution to displace gold from the complex:

$$2Au(CN)_2^-(aq) + Zn(s) \rightarrow 2Au(s) + Zn(CN)_4^{2-}(aq)$$

S4.13 (a) Aeration is the forcing of air through a solution or medium in order to bring about an oxidation by the oxygen in the air. In the cyanide extraction of gold, the oxygen of the air oxidizes the metallic gold to the gold(I) ion, Au^+.
(b) Cyanide is added as a complexing agent to stabilize the oxidized gold.

S4.15

$$Al(OH)_4^-(aq) \xrightarrow{\text{dilution}} Al(OH)_3(s) + OH^-(aq)$$

$$\text{or } Al(H_2O)_2(OH)_4^-(aq) + H_2O(l) \rightleftharpoons Al(H_2O)_3(OH)_3(s) + OH^-(aq)$$

$$Al(OH)_4^-(aq) + CO_2(aq) \rightarrow Al(OH)_3(s) + HCO_3^-(aq)$$

S4.17

(a) $2Bi_2S_3(s) + 9O_2(g) \xrightarrow{\text{heat}} 2Bi_2O_3(s) + 6SO_2(g)$

(b) $2MoS_2(s) + 7O_2(g) \xrightarrow{\text{heat}} 2MoO_3(s) + 4SO_2(g)$

S4.19

(a) $MnO_2(s) + 2CO(g) \xrightarrow{\text{heat}} Mn(l) + 2CO_2(g)$

(b) $FeCr_2O_4(s) + 4C(s) \xrightarrow{\text{heat}} Fe(l) + 2Cr(l) + 4CO(g)$

(c) $GeO_2(s) + 2H_2(g) \rightarrow Ge(s) + 2H_2O(g)$

S4.21 (a) Group 3A:
1. Physical properties: The trends are toward greater atomic radii, higher densities, and generally lower melting and boiling points in going down the group. The Group 3A elements are all solids.
2. Oxidation states: The +3 oxidation state is important for all five Group 3A elements. Gallium, indium, and thallium also show +1 oxidation states, and the +1 oxidation state is important for thallium. The trend is toward greater importance of the +1 oxidation state in going down the group.

3. Chemical activity (aqueous environment): For the metallic elements (Al, Ga, In, and Tl) the trend is toward a decrease in activity as you go down the group.
4. Oxides: The trend is from acidic oxides $\rightarrow$ amphoteric oxides $\rightarrow$ basic oxides in going down the group.

(b) Group 4A:
1. Physical properties: The trends are toward greater atomic radii, generally higher densities, and generally lower melting and boiling points in going down the group. The Group 4A elements are all solids.
2. Oxidation states: The +4 oxidation state is important for all five Group 4A elements. The +2 oxidation states of germanium, tin, and lead are also important. The trend is toward greater importance of the +2 oxidation state toward the bottom of the group.
3. Chemical activity (aqueous environment): The chemical activity of the metallic elements (Sn and Pb) decreases slightly in going down the group.
4. Oxides: The trend is from acidic oxides to amphoteric oxides in going down the group.

S4.23 (a) Refer to "Concentrating the Ore" in Section S4.1 in the textbook.

(b) $SnO_2(s) + 2C(s) \xrightarrow{\text{heat}} Sn(l) + 2CO(g)$
cassiterite

(c) $2PbS(s) + 3O_2(g) \xrightarrow{\text{heat}} 2PbO(s) + 2SO_2(g)$
galena

$PbO(s) + C(s) \xrightarrow{\text{heat}} Pb(l) + CO(g)$

S4.25 Aluminum is an active metal. It is a good reducing agent, and should react with water, liberating hydrogen. However, the film of Al_2O_3 that forms when aluminum is initially exposed to the atmosphere is a tough oxide coating and does not allow air (oxygen) and moisture (water) to come in contact with the aluminum metal underneath. Therefore, the Al_2O_3 film protects the Al from further atmospheric corrosion. Aluminum's widespread use as a structural metal is an important consequence of this.

Anodized aluminum is aluminum that is deliberately given an expecially thick protective oxide coating by having the aluminum be the anode in an electrolytic cell.

S4.27 Refer to Section S4.2 in the textbook.

S4.29 (a) $SnCl_2$. In general, as the oxidation state increases, the bonding becomes more covalent.
(b) $SnCl_4$. Since the bonding in $SnCl_4$ is more covalent, $SnCl_4$ has more of a molecular nature and, therefore, a lower boiling point.

S4.31 (a) $Sn(s) + O_2(g) \xrightarrow{\text{heat}} SnO_2(s)$
(b) $2Pb(s) + O_2(g) \xrightarrow{\text{heat}} 2PbO(s)$
(c) $4Bi(s) + 3O_2(g) \xrightarrow{\text{heat}} 2Bi_2O_3(s)$

S4.33 (a) $2Al(s) + 6HCl(aq) \rightarrow 2AlCl_3(aq) + 3H_2(g)$

(b) $Al_2O_3(s) + 6HCl(aq) \rightarrow 2AlCl_3(aq) + 3H_2O(l)$

(c) $SnO(s) + 2HCl(aq) \rightarrow SnCl_2(aq) + H_2O(l)$

(d) $SnO_2(s) + 6HCl(aq) \rightarrow 2H^+(aq) + SnCl_6^{2-}(aq) + 2H_2O(l)$

(e) $PbO(s) + 2HCl(aq) \rightarrow PbCl_2(s) + H_2O(l)$

(f) $Bi_2O_3(s) + 2HCl(dil) \rightarrow 2BiOCl(s) + H_2O(l)$

$BiOCl(s) + 2HCl(conc) \rightarrow BiCl_3(aq) + H_2O(l)$

Note: A form of Al_2O_3 called alumina (α-Al_2O_3) is a very inert, very hard, high-melting material used in the manufacture of refractories and abrasives, and is resistant to attack by acids.

S4.35 Treat a piece of the foil with concentrated nitric acid; if it is tin, a precipitate of SnO_2 will form. If no precipitate forms, the foil is aluminum. In fact, aluminum is made passive by HNO_3. Therefore, if the foil is aluminum, no reaction will occur.

S4.37 Zinc, cadmium, and mercury and the Group 2A elements that precede them all have just two electrons in the outermost shell and have no partially filled underlying shells. Therefore, they all have a characteristic +2 oxidation state.

Zinc, cadmium, and mercury have filled underlying d subshells (d^{10}). Since a filled subshell is very stable, the underlying d electrons are not used in bonding. Therefore, zinc, cadmium, and mercury only have the same two s electrons for use in bonding as the Group 2A elements, Ca, Sr, and Ba, which precede them in their periods.

S4.39
1. Physical properties: The trends are toward greater atomic radii, lower melting points, lower boiling points, and higher densities as you proceed down the group.
2. Oxidation states: The +2 oxidation state is important for all the members of the zinc group. The +1 oxidation state of mercury is also important.
3. Chemical activity (aqueous environment): The trend is toward a decrease in chemical activity as you proceed down the group.
4. Oxides: The trend is from amphoteric oxides to basic oxides as you proceed down the group.

S4.41 Zn: $2ZnS(s) + 3O_2(g) \xrightarrow{\text{heat}} 2ZnO(s) + 2SO_2(g)$

$ZnO(s) + C(s) \xrightarrow{\text{heat}} Zn(g) + CO(g)$

Cd: $2CdS(s) + 3O_2(g) \xrightarrow{\text{heat}} 2CdO(s) + 2SO_2(g)$

$CdO(s) + C(s) \xrightarrow{\text{heat}} Cd(g) + CO(g)$

Hg: $HgS(s) + O_2(g) \xrightarrow{\text{heat}} Hg(g) + SO_2(g)$

S4.43 (a) $4Zn(s) + 10HNO_3(6\ M) \rightarrow NH_4NO_3(aq) + 4Zn(NO_3)_2(aq) + 3H_2O(l)$

$4Cd(s) + 10HNO_3(6\ M) \rightarrow NH_4NO_3(aq) + 4Cd(NO_3)_2(aq) + 3H_2O(l)$

$3Hg(l) + 8HNO_3(6\ M) \xrightarrow{\text{excess acid}} 3Hg(NO_3)_2(aq) + 2NO(g) + 4H_2O(l)$

or $6Hg(l) + 8HNO_3(6\ M) \xrightarrow{\text{excess mercury}} 3Hg_2(NO_3)_2(aq) + 2NO(g) + 4H_2O(l)$

(b) $Zn(s) + 4HNO_3(12\ M) \rightarrow Zn(NO_3)_2(aq) + 2NO_2(g) + 2H_2O(l)$

$Cd(s) + 4HNO_3(12\ M) \rightarrow Cd(NO_3)_2(aq) + 2NO_2(g) + 2H_2O(l)$

$3Hg(l) + 8HNO_3(12\ M) \xrightarrow{\text{excess acid}} 3Hg(NO_3)_2(aq) + 2NO(g) + 4H_2O(l)$

or $6Hg(l) + 8HNO_3(12\ M) \xrightarrow{\text{excess mercury}} 3Hg_2(NO_3)_2(aq) + 2NO(g) + 4H_2O(l)$

S4.45 (a) Treat the zinc oxide with nitric acid:

$ZnO(s) + 2HNO_3(aq) \rightarrow Zn(NO_3)_2(aq) + H_2O(l)$

(b) Treat the zinc metal with potassium hydroxide:

$Zn(s) + 2KOH(aq) + 2H_2O(l) \rightarrow K_2Zn(OH)_4(aq) + H_2(g)$

(c) First treat the mercury with an excess of hot concentrated sulfuric acid to form $HgSO_4$. Then heat a dry mixture of mercuric sulfate and sodium chloride to form mercury(II) chloride, which sublimes.

$Hg(l) + 2H_2SO_4(aq) \xrightarrow{\text{excess acid}} HgSO_4(aq) + SO_2(g) + 2H_2O(l)$

$HgSO_4(s) + 2NaCl(s) \xrightarrow{\text{heat}} Na_2SO_4(s) + HgCl_2(g)$

An alternate method is:

$Hg(l) \xrightarrow[\text{excess}]{HNO_3} Hg(NO_3)_2(aq) \xrightarrow{OH^-} HgO(s) \xrightarrow{HCl} HgCl_2(aq)$

Also, $Hg(l) + Cl_2(g) \xrightarrow{\Delta} HgCl_2(s)$

(d) First dissolve excess mercury in concentrated nitric acid to form $Hg_2(NO_3)_2$, a soluble mercurous salt. Then add chloride ion to precipitate mercury(I) chloride, which is only slightly soluble.

$6Hg(l) + 8HNO_3(aq) \xrightarrow{\text{excess mercury}} 3Hg_2(NO_3)_2(aq) + 2NO(g) + 4H_2O(l)$

$Hg_2(NO_3)_2(aq) + 2NaCl(aq \text{ or } s) \rightarrow Hg_2Cl_2(s) + 2NaNO_3(aq)$

S4.47

1. Physical properties: The trend is toward higher densities as you proceed down the group. The melting points, boiling points, and atomic radii do not show a consistent trend.
2. Oxidation states: The +1 oxidation state is important for all three members of the copper group. The +2 oxidation state of copper and the +3 oxidation state of gold are also important.
3. Chemical activity (aqueous environment): The trend is toward a decrease in chemical activity as you proceed down the group.
4. Oxides: The trend is from stable basic oxides to unstable oxides in going down the group.

S4.49 The steps in the production of blister copper from its sulfide ore are:

1. The ore is concentrated to about 30% copper by flotation.
2. Impurities, especially FeS, are converted to a silicate slag by roasting and smelting the ore with SiO_2.

$2FeS(s) + 3O_2(g) \xrightarrow{\text{heat}} 2FeO(s) + 2SO_2(g)$

$FeO(s) + SiO_2(s) \xrightarrow{\text{heat}} FeSiO_3(s)$ (slag)

The slag, which floats on top of the molten copper sulfide, is poured off.

3. Air is blown through the remaining molten copper sulfide, reducing it to copper metal.

$Cu_2S(l) + O_2(g) \rightarrow 2Cu(l) + SO_2(g)$

S4.51 Refer to "Reactions and Compounds" in Section S4.4 in the textbook.

S4.53 The basic steps of the black-and-white photographic process are:

1. Exposure. Grains of silver bromide contained in the photographic film are exposed to light. They become "sensitized," which means they become more susceptible to decomposition. The sensitized grain possibly contains a small nucleus or seed of silver atoms caused by the light decomposing some of the silver bromide.

$$\underset{\text{grain}}{AgBr(s)} \xrightarrow{h\nu} \underset{\text{sensitized grain}}{AgBr(s)}$$

$$2AgBr(s) \xrightarrow{h\nu} 2Ag(s) + Br_2(g) \quad \text{(small amount)}$$

2. Developing. The exposed film is placed in the solution of a mild reducing agent. The sensitized grains are rapidly reduced to black grains of silver metal.

$$\underset{\text{sensitized grain}}{AgBr(s)} + e^- \rightarrow \underset{\text{black grain}}{Ag(s)} + Br^-(aq)$$

3. Fixing. The photographic film is then washed in a solution of sodium thiosulfate. The unreduced silver bromide is dissolved out of the film; the thiosulfate ion complexes the silver ion:

$$AgBr(s) + 2S_2O_3^{2-}(aq) \rightarrow Ag(S_2O_3)_2^{3-}(aq) + Br^-(aq)$$

The role of the developing solution is to reduce the sensitized silver bromide grains to elemental silver grains, which are black because the silver metal produced is finely divided.

The role of the fixing solution is to remove the unexposed silver bromide, since silver bromide slowly turns black when exposed to light and, consequently, the whole film would eventually turn black if it wasn't removed.

S4.55 (a) $Cu(s) + Cl_2(g) \rightarrow CuCl_2(s)$

$2Ag(s) + Cl_2(g) \rightarrow 2AgCl(s)$

$2Au(s) + 3Cl_2(g) \rightarrow 2AuCl_3(s)$

(b) $2Cu(s) + 2H_2S(g) + O_2(g) \rightarrow 2CuS(s) + 2H_2O(g)$

$4Ag(s) + 2H_2S(g) + O_2(g) \rightarrow 2Ag_2S(s) + 2H_2O(g)$

(c) Concentrated HNO_3:

$Cu(s) + 4H^+(aq) + 2NO_3^-(aq) \rightarrow Cu^{2+}(aq) + 2NO_2(g) + 2H_2O(l)$

$Ag(s) + 2H^+(aq) + NO_3^-(aq) \rightarrow Ag^+(aq) + NO_2(g) + H_2O(l)$

S4.57 (a) $Cu^{2+}(aq) + 2OH^-(aq) \rightarrow Cu(OH)_2(s)$

(b) $2Ag^+(aq) + 2OH^-(aq) \rightarrow Ag_2O(s) + H_2O(l)$

S4.59 Refer to Figure S4.2 and "Sources and Metallurgy" in Section S4.5 in the textbook.

S4.61 Refer to "Sources and Metallurgy" in Section S4.5 in the textbook.

S4.63 (a) Add potassium thiocyanate (KSCN) to the solution. If Fe^{3+} is present, the solution will turn deep red.

$$Fe^{3+}(aq) + SCN^-(aq) \rightarrow \underset{\text{deep red}}{Fe(SCN)^{2+}(aq)}$$

Fe^{3+} can also be identified in solution by adding potassium ferrocyanide ($K_4[Fe(CN)_6]$) to form a dark suspension of Prussian blue.

$$Fe^{3+}(aq) + K^+(aq) + Fe(CN)_6^{4-}(aq) \rightarrow \underset{\text{dark blue}}{KFe[Fe(CN)_6](s)}$$

(b) Add potassium nitrite (KNO_2) to the solution. If Co^{2+} is present, a yellow precipitate will form.

$$Co^{2+}(aq) + 7KNO_2(aq) + H_2O(l) \rightarrow \underset{\text{yellow}}{K_3[Co(NO_2)_6](s)} + 4K^+(aq) + NO(g) + 2OH^-(aq)$$

or in acid solution (pH close to neutral)

$$Co^{2+}(aq) + 7KNO_2(aq) + 2H^+(aq) \rightarrow K_3[Co(NO_2)_6](s) + 4K^+(aq) + NO(g) + H_2O(l)$$

(c) Add dimethylglyoxime to the solution. If Ni^{2+} is present, an insoluble bright red complex will form.

S4.65 (a) $3Fe(s) + 2O_2(g) \rightarrow Fe_3O_4(s)$ and/or $4Fe(s) + 3O_2(g) \rightarrow 2Fe_2O_3(s)$

$3Co(s) + 2O_2(g) \rightarrow Co_3O_4(s)$

$2Ni(s) + O_2(g) \rightarrow 2NiO(s)$

(b) $Fe(s) + 2HCl(aq) \rightarrow FeCl_2(aq) + H_2(g)$

$Co(s) + 2HCl(aq) \rightarrow CoCl_2(aq) + H_2(g)$

$Ni(s) + 2HCl(aq) \rightarrow NiCl_2(aq) + H_2(g)$

S4.67 (a) The high positive charge of Fe^{3+} attracts and strongly binds water molecules in aqueous solution:

$Fe^{3+}(aq) + 6H_2O(l) \rightarrow Fe(H_2O)_6{}^{3+}(aq)$

The positive central iron ion in the hydrated cation draws negative charge to it and, in so doing, increases the O–H bond polarity in the bound water molecules and makes them proton donors. Up to three protons can be removed from the hydrated cation.

$Fe(H_2O)_6{}^{3+}(aq) + H_2O(l) \rightleftharpoons Fe(H_2O)_5(OH)^{2+}(aq) + H_3O^+(aq)$

(b) Fe^{3+}. Fe^{3+} should be more acidic because it has a higher positive charge and, therefore, should be capable of causing a greater polarization of the O–H bond than Fe^{2+}.

S4.69 Iron. The oxidation potentials of iron and nickel (Table 20.1) are:

$Fe(s) \rightarrow Fe^{2+}(aq) + 2e^-$ $\quad E^\circ_{oxid} = 0.409\ V$

$Ni(s) \rightarrow Ni^{2+}(aq) + 2e^-$ $\quad E^\circ_{oxid} = 0.236\ V$

Iron has a higher (more positive) oxidation potential and a greater tendency to give up electrons than nickel; iron should be a better reducing agent.

S4.71 $Fe^{2+}(aq) \rightarrow Fe^{3+}(aq) + e^-$ $\quad$ oxidation

$O_2(g) + 4H^+(aq) + 4e^- \rightarrow H_2O(l)$ $\quad$ reduction

$4Fe^{2+}(aq) + O_2(g) + 4H^+(aq) \rightarrow 4Fe^{3+}(aq) + 2H_2O(l)$

Note: This is the net ionic equation for the reaction in acid solution.

S4.73 Refer to "Chromium" and "Manganese" in Section S4.6 in the textbook.

S4.75 (a) Similarities:

1. Both chromium and aluminum form oxide films that render them resistant to corrosion.
2. Their +3 oxides (Cr_2O_3 and Al_2O_3) and hydroxides ($Cr(OH)_3$ and $Al(OH)_3$) are both amphoteric.
3. The aqueous Cr^{3+} chemistry resembles that of aqueous Al^{3+}.
4. Both metals have high boiling points.

(b) Differences:
1. Chromium, being a transition metal, exhibits several oxidation states (+2, +3, and +6 being the important ones), while aluminum just has a +3 oxidation state.
2. Chromium is considerably denser than aluminum and has a much higher melting point.
3. Chromium forms many colored compounds and complexes.

S4.77 (a) $FeCr_2O_4(s) + 4C(s) \rightarrow Fe(s) + 2Cr(s) + 4CO(g)$
chromite ferrochrome

(b) $MnO_2(s) + 2C(s) \rightarrow Mn(s) + 2CO(g)$
pyrolusite

with some $Fe_2O_3(s) + 3C(s) \rightarrow 2Fe(s) + 3CO(g)$

and $Fe_2O_3(s) + 3CO(g) \rightarrow 2Fe(s) + 3CO_2(g)$

Comment: Manganese combines with carbon at high temperatures to form the carbide Mn_3C, similar to the reaction of iron and carbon that occurs at the high temperatures encountered in the blast furnace.

$3Mn(s) + C(s) \rightleftharpoons Mn_3C(s)$

(c) $Cr_2O_3(s) + 2Al(s) \rightarrow 2Cr(l) + Al_2O_3(l)$

Ferrochrome is an iron-chromium alloy used in the production of stainless steel.

S4.79 $Cr^{3+}(aq) + 3OH^-(aq) \rightarrow Cr(OH)_3(s)$ (amphoteric hydroxide)

$Cr^{3+}(aq) + 4OH^-(aq) \xrightarrow{\text{excess base}} Cr(OH)_4^-(aq)$ (chromite ion)

$Mn^{2+}(aq) + 2OH^-(aq) \rightarrow Mn(OH)_2(s)$

acid

S4.81 To answer this exercise we have to test the thermodynamic stability of the respective oxides at 500°C.

(a) Hg: $HgO(s) \xrightarrow{500°C} Hg(g) + 1/2O_2(g)$

At 298 K:

$\Delta H° = \Delta H_f°(Hg(g)) + 1/2 \times \Delta H_f°(O_2) - \Delta H_f°(HgO) = 61.3\text{ kJ} + 1/2 \times 0\text{ kJ} - (-90.8\text{ kJ}) = 152.1\text{ kJ}$

$\Delta S° = S°(Hg(g)) + 1/2 \times S°(O_2) - S°(HgO) = 175.0\text{ J/K} + 1/2 \times 205.1\text{ J/K} - 70.3\text{ J/K} = 207.2\text{ J/K}$

A value for $\Delta G°$ at 500°C can be obtained from Equation 19.7 if we assume that $\Delta H°$ and $\Delta S°$ do not change significantly with temperature. Substituting $\Delta H° = 152.1$ kJ, $\Delta S° = 0.2072$ kJ/K, and T = 773 K into Equation 19.7 gives

$\Delta G° = \Delta H° - T\Delta S° = 152.1\text{ kJ} - 773\text{ K} \times 0.2072\text{ kJ/K} = -8.1\text{ kJ}$

Since $\Delta G < 0$, HgO is thermodynamically unstable at 500°C and O_2 partial pressure 1 atm. Therefore, Hg will form from the roasting of HgS and not HgO.

Zn: $ZnO(s) \xrightarrow{500°C} Zn(l) + 1/2O_2(g)$

At 298 K: For Zn(l) use the values given for Zn(s) in Table 19.2.

$\Delta H° = \Delta H_f°(Zn) + 1/2 \times \Delta H_f°(O_2) - \Delta H_f°(ZnO) = 0\text{ kJ} + 1/2 \times 0\text{ kJ} - (-348.3\text{ kJ}) = +348.3\text{ kJ}$

$\Delta S° = S°(Zn) + 1/2 \times S°(O_2) - S°(ZnO) = 41.63\text{ J/K} + 1/2 \times 205.1\text{ J/K} - 43.6\text{ J/K} = 100.6\text{ J/K}$

At 773 K: $\Delta G° = \Delta H° - T\Delta S° = 348.3\text{ kJ} - 773\text{ K} \times 0.1006\text{ kJ/K} = 270.5\text{ kJ}$

Since $\Delta G > 0$, ZnO is thermodynamically stable at 500°C and O_2 partial pressure 1 atm. (The correction for atmospheric conditions, O_2 partial pressure 0.2 atm, is slight and we will ignore it.) Therefore, ZnO will form from the roasting of ZnS and not Zn.

(b) Ag: $Ag_2O(s) \xrightarrow{500°C} 2Ag(s) + 1/2O_2(g)$

At 298 K: $\Delta H° = 31.05$ kJ and $\Delta S° = 66.4$ J/K (See solution to Exercise 19.69.)

At 773 K: $\Delta G° = \Delta H° - T\Delta S° = 31.05 \text{ kJ} - 773 \text{ K} \times 0.0664 \text{ kJ/K} = -20.3 \text{ kJ}$

Since $\Delta G < 0$, Ag_2O is thermodynamically unstable at 500°C and O_2 partial pressure 1 atm. Therefore, Ag will form from the roasting of Ag_2S and not Ag_2O.

S4.83 (a) $Fe_2O_3(s) + 2Al(s) \rightarrow 2Fe(s) + Al_2O_3(s)$

$$\Delta H° = 2 \times \Delta H_f^\circ(Fe) + \Delta H_f^\circ(Al_2O_3) - \Delta H_f^\circ(Fe_2O_3) - 2 \times \Delta H_f^\circ(Al)$$
$$= 2 \times 0 \text{ kJ} - 1675.7 \text{ kJ} + 824.2 \text{ kJ} - 2 \times 0 \text{ kJ} = -851.5 \text{ kJ}$$

The $\Delta H°$ for the reduction of 1.00 mol of Fe_2O_3 with 2.00 mol of Al is 1.00 mol × (–851.5 kJ/mol) = –852 kJ.

(b) $\Delta H°$ for the reduction of 1.00 mol of Fe_2O_3 is –852 kJ. We want to calculate the temperature to which this amount of evolved heat can raise the products, which are 2.00 mol of iron and 1.00 mol of aluminum oxide.

$$q = n_{Fe} \times \frac{68 \text{ J}}{\text{mol}\bullet\text{K}} \times \Delta T + n_{Al_2O_3} \times \frac{124 \text{ J}}{\text{mol}\bullet\text{K}} \times \Delta T;\quad \Delta T = \frac{852 \times 10^3 \text{ J}}{(136 \text{ J/K} + 124 \text{ J/K})} = 3280 \text{ K}$$

and $T_{final} = 3280 \text{ K} + T_{initial} = 3280 \text{ K} + 298 \text{ K} = 3580 \text{ K}$

Answer: The final temperature will be 3580 K. In terms of degrees Celsius, a final temperature of 3300°C is obtained if the unrounded kelvin answer is used.

S4.85 (a) (1) $CuCl(s) \rightleftharpoons Cu^+(aq) + Cl^-(aq)$

$K_{sp} = [Cu^+][Cl^-]$; $1.7 \times 10^{-7} = (x)(x) = x^2$ and $x = 4.1 \times 10^{-4}$ M = $[Cu^+]$

(2) Concentration Summary:

Equation:	$Cu(CN)_2^-(aq) \rightleftharpoons$	$Cu^+(aq)$ +	$2CN^-(aq)$
Initial concentrations, mol/L	1.0	0	0
Concentration changes, mol/L	$-x$	$+x$	$+2x$
Equilibrium concentrations, mol/L	$1.0 - x$	x	$2x$
Approximate concentrations, mol/L	1.0	x	$2x$

Substituting the approximate concentrations and $K_d = 9.9 \times 10^{-25}$ into the dissociation constant expression gives

$$\frac{[Cu^+][CN^-]^2}{[Cu(CN)_2^-]} = K_d$$

$$\frac{(x)(2x)^2}{1.0} = 9.9 \times 10^{-25} \text{ and } x = 6.3 \times 10^{-9} \text{ M} = [Cu^+]$$

(b) (1) The net reaction for the disproportionation of CuCl is

$2CuCl(s) \rightleftharpoons Cu(s) + Cu^{2+}(aq) + 2Cl^-(aq)$

with the equilibrium constant

$K = [Cu^{2+}][Cl^-]^2$

The value of K can be obtained by breaking down the net reaction into a series of steps.

$$2CuCl(s) \rightleftharpoons 2Cu^+(aq) + 2Cl^-(aq) \qquad K_{sp}^2$$
$$2Cu^+(aq) \rightleftharpoons Cu(s) + Cu^{2+}(aq) \qquad K_c$$
$$2CuCl(s) \rightleftharpoons Cu(s) + Cu^{2+}(aq) + 2Cl^-(aq) \qquad K = K_{sp}^2 K_c$$

K_c is obtained from Equation 20.8b with $E° = 0.357$ V and $n = 1$ (see Exercise 20.100(c)).

$$\log K_c = \frac{nE°}{0.0592} = \frac{(1)(0.357\ V)}{0.0592\ V} = 6.03 \text{ and } K_c = 1.1 \times 10^6$$

The equilibrium constant for the net reaction is therefore

$$K = K_{sp}^2 K_c = (1.7 \times 10^{-7})^2(1.1 \times 10^6) = 3.2 \times 10^{-8}$$

The small value of K shows that very little Cu^{2+} will be present at equilibrium and that substantially no disproportionation occurs.

(2) The net reaction for the disproportionation of $Cu(CN)_2^-$ is

$$2Cu(CN)_2^-(aq) \rightleftharpoons Cu(s) + Cu^{2+}(aq) + 4CN^-(aq)$$

with the equilibrium constant

$$K = \frac{[Cu^{2+}][CN^-]^4}{[Cu(CN)_2^-]^2}$$

In a similar manner to that above, we get

$$K = K_d^2 K_c = (9.9 \times 10^{-25})^2(1.1 \times 10^6) = 1.1 \times 10^{-42}$$

The extremely low K value means that no disproportionation occurs.

S4.87 (a) $CuS(l) + O_2(g) \rightarrow Cu(l) + SO_2(g)$

$Cu_2S(l) + O_2(g) \rightarrow 2Cu(l) + SO_2(g)$

The mass of CuS (molar mass 95.61 g/mol) equals the mass of Cu_2S (molar mass 159.2 g/mol); let x equal this mass. Now the number of moles of Cu produced

$$1.00 \times 10^9 \text{ g Cu} \times \frac{1 \text{ mol Cu}}{63.55 \text{ g Cu}} = 1.57 \times 10^7 \text{ mol Cu}$$

is equal to the number of moles of CuS reduced plus twice the number of moles of Cu_2S reduced. Therefore,

$$\frac{x}{95.61 \text{ g/mol}} + \frac{2x}{159.2 \text{ g/mol}} = 1.57 \times 10^7 \text{ mol and } x = 6.82 \times 10^8 \text{ g}$$

The amount of SO_2 produced when the CuS is reduced is calculated as follows: g CuS → mol CuS → mol SO_2 → g SO_2.

$$6.82 \times 10^8 \text{ g CuS} \times \frac{1 \text{ mol CuS}}{95.61 \text{ g CuS}} \times \frac{1 \text{ mol } SO_2}{1 \text{ mol CuS}} \times \frac{64.06 \text{ g } SO_2}{1 \text{ mol } SO_2} = 4.57 \times 10^8 \text{ g } SO_2$$

The amount of SO_2 produced when the Cu_2S is reduced is calculated similarly

$$6.82 \times 10^8 \text{ g } Cu_2S \times \frac{1 \text{ mol } Cu_2S}{159.2 \text{ g } Cu_2S} \times \frac{1 \text{ mol } SO_2}{1 \text{ mol } Cu_2S} \times \frac{64.06 \text{ g } SO_2}{1 \text{ mol } SO_2} = 2.74 \times 10^8 \text{ g } SO_2$$

The total mass of SO_2 evolved during the production of 1000 metric tons of copper from this ore is therefore

$4.57 \times 10^8 \text{ g} + 2.74 \times 10^8 \text{ g} = 7.31 \times 10^8 \text{ g}$

(b) The equations needed are:

$2SO_2(g) + O_2(g) \rightarrow 2SO_3(g)$

$SO_3(g) + H_2O(l) \rightarrow H_2SO_4(aq)$

$CaCO_3(s) + H_2SO_4(aq) \rightarrow CaSO_4(aq) + CO_2(aq) + H_2O(l)$

The steps in the calculation are: g $SO_2 \rightarrow$ mol $SO_2 \rightarrow$ mol $H_2SO_4 \rightarrow$ mol $CaCO_3 \rightarrow$ g $CaCO_3 \rightarrow$ volume $CaCO_3$.

$$7.31 \times 10^8 \text{ g } SO_2 \times \frac{1 \text{ mol } SO_2}{64.06 \text{ g } SO_2} \times \frac{1 \text{ mol } H_2SO_4}{1 \text{ mol } SO_2} \times \frac{1 \text{ mol } CaCO_3}{1 \text{ mol } H_2SO_4} \times \frac{100.1 \text{ g } CaCO_3}{1 \text{ mol } CaCO_3} \times \frac{1 \text{ cm}^3 \ CaCO_3}{2.93 \text{ g } CaCO_3}$$

$= 3.90 \times 10^8 \text{ cm}^3 \ CaCO_3$ or $390 \text{ m}^3 \ CaCO_3$

S4.89 $[Hg - Hg]^{2+}$

S4.91 (a) $Fe_2O_3(s) + 3CO(g) \rightarrow 2Fe(s) + 3CO_2(g)$

At 25°C: $\Delta G° = 2 \times \Delta G_f^\circ(Fe) + 3 \times \Delta G_f^\circ(CO_2) - \Delta G_f^\circ(Fe_2O_3) - 3 \times \Delta G_f^\circ(CO)$

$= 2 \times 0 \text{ kJ} + 3 \times (-394.4 \text{ kJ}) + 742.2 \text{ kJ} - 3 \times (-137.2 \text{ kJ}) = -29.4 \text{ kJ}$

(b) At 25°C:

$\Delta H° = 2 \times \Delta H_f^\circ(Fe) + 3 \times \Delta H_f^\circ(CO_2) - \Delta H_f^\circ(Fe_2O_3) - 3 \times \Delta H_f^\circ(CO)$

$= 2 \times 0 \text{ kJ} + 3 \times (-393.5 \text{ kJ}) + 824.2 \text{ kJ} - 3 \times (-110.5 \text{ kJ}) = -24.8 \text{ kJ}$

$\Delta S° = 2 \times S°(Fe) + 3 \times S°(CO_2) - S°(Fe_2O_3) - 3 \times S°(CO)$

$= 2 \times 27.28 \text{ J/K} + 3 \times 213.74 \text{ J/K} - 87.40 \text{ J/K} - 3 \times 197.67 \text{ J/K} = 15.37 \text{ J/K} = 0.01537 \text{ kJ/K}$

We will assume that the values of $\Delta H°$ and $\Delta S°$ do not change significantly between 25°C and 1000°C (1273 K). Substituting these values and T = 1273 K into Equation 19.7 gives

$\Delta G° = \Delta H° - T\Delta S° = -24.8 \text{ kJ} - 1273 \text{ K} \times 0.01537 \text{ kJ/K} = -44.4 \text{ kJ}$

The standard free energy change is approximately –44.4 kJ per mole of reaction at 1000°C. The difference between $\Delta G°$ at 25°C and 1000°C is not that substantial. The main advantage of carrying out the smelting at high temperature is kinetic. At high temperatures the reaction proceeds very rapidly, while at low temperatures it takes too long to produce any substantial amount of product.

S4.93 For this reaction E° = 1.171 V and n = 2 (see Exercise 20.100 (b)). Substituting into Equation 20.8b gives

$$\log K_c = \frac{nE°}{0.0592} = \frac{2 \times 1.171 \text{ V}}{0.0592 \text{ V}} = 39.6$$

and $K_c = \text{antilog } 39.6 = 10^{39.6} = 4 \times 10^{39}$

CHAPTER 22

NUCLEAR CHEMISTRY

Solutions To Practice Exercises

PE 22.1 (a) The atomic number of mercury is 80. The alpha particle symbol is ${}^{4}_{2}He$, so

$${}^{187}_{80}Hg \rightarrow {}^{4}_{2}He + {}^{183}_{78}X$$

Element number 78 is Pt, and the complete equation is

$${}^{187}_{80}Hg \rightarrow {}^{4}_{2}He + {}^{183}_{78}Pt$$

Answer: mass number = 183; atomic number = 78; symbol is ${}^{183}_{78}Pt$

(b) The atomic numbers of aluminum and silicon are 13 and 14, respectively. So

$${}^{28}_{13}Al \rightarrow {}^{28}_{14}Si + {}^{0}_{-1}e$$

PE 22.2

$${}^{96}_{42}Mo + {}^{2}_{1}H \rightarrow {}^{97}_{43}Tc + {}^{1}_{0}n$$

$${}^{209}_{83}Bi + {}^{4}_{2}He \rightarrow {}^{211}_{85}At + 2{}^{1}_{0}n$$

PE 22.3 First calculate k using Equation 22.1. Substituting $N/N_0 = 0.908$ and $t = 48.0$ h into Equation 22.1 gives

$\ln 0.908 = -k \times 48.0\ h$ and $k = 2.01 \times 10^{-3}\ h^{-1}$

The half-life is obtained from k using Equation 22.2:

$$t_{1/2} = \frac{0.693}{k} = \frac{0.693}{2.01 \times 10^{-3}\ h^{-1}} = 345\ h$$

PE 22.4 Substituting $k = 1.21 \times 10^{-4}\ y^{-1}$ (see Example 22.4) and $N/N_0 = 1/8$ into Equation 22.1 gives

$$t = \frac{\ln (N/N_0)}{-k} = \frac{\ln (1/8)}{-1.21 \times 10^{-4}\ y^{-1}} = 17{,}200\ y$$

Or, since 1/8 of the radionuclides will remain after three half-lives; $t = 3 \times 5730\ y = 17{,}200\ y$.

PE 22.5 Substituting $N/N_0 = 10{,}000/10{,}056 = 0.99443$ and $k = \ln 2/t_{1/2} = \ln 2/(4.88 \times 10^{11}\ y)$ $= 1.42 \times 10^{-12}\ y^{-1}$ into Equation 22.1 gives

$$\ln 0.99443 = -1.42 \times 10^{-12}\ y^{-1} \times t \text{ and } t = 3.93 \times 10^9\ y$$

PE 22.6 The atomic number of rubidium is 37. Rubidium-95, with 37 protons and 58 neutrons, lies above the stability band of Figure 22.15. It will decay by a mode that decreases the n/p ratio. The most common mode for decreasing the n/p ratio is beta emission (neutron emission is rare):

$${}^{95}_{37}\text{Rb} \rightarrow {}^{95}_{38}\text{Sr} + {}^{0}_{-1}\text{e}$$

PE 22.7 $$96.5 \text{ KeV} \times \frac{1 \text{ MeV}}{10^3 \text{ KeV}} \times \frac{1 \text{ u}}{931.5 \text{ MeV}} = 1.04 \times 10^{-4} \text{ u}$$

PE 22.8 (a) The total mass of the separate nucleons is

28 protons × 1.00728 u/proton	=	28.20384 u
30 neutrons × 1.00866 u/neutron	=	30.25980 u
Total	=	58.46364 u

mass defect = mass of separate nucleons – nuclear mass of nickel-58
= 58.46364 u – 57.91999 u = 0.54365 u

(b) The binding energy is the energy equivalent of the mass defect or 0.54365 u × 931.47 MeV/u = 506.39 MeV.

(c) The binding energy per nucleon is 506.39 MeV/58 nucleons = 8.731 MeV/nucleon.

Solutions To Final Exercises

22.1 (a) Alpha particles are the nuclei of helium atoms, beta particles are electrons moving at very high speeds, and gamma rays consist of photons with very high frequencies.

(b) 1. The emission of an alpha particle reduces the mass number of a nucleus by four units and the atomic number by two units.
2. The emission of a beta particle raises the atomic number of a nucleus by one unit and leaves the mass number unchanged.
3. The emission of a gamma-ray photon does not affect the mass number or the atomic number of a nucleus.

22.3 The amount of deflection depends on the mass and speed of the particle. The greater the mass, the smaller the deflection. Since the beta particles are much less massive than the alpha particles, approximately 7300 times less massive, they undergo a greater deflection than the alpha particles.

22.5 (a) In order of increasing energy we have: beta particles (between 0.05 and 1 MeV) < gamma-ray photons (about 1 MeV) < alpha particles (about 5 MeV).

(b) In order of increasing penetrating ability we have: alpha particles < beta particles < gamma-ray photons.

(c) Alpha particles have a greater mass, lower speed, and higher charge than beta particles and this makes them more likely to be absorbed by molecules along their path. Therefore, alpha particles do not penetrate as far as beta particles. Gamma rays, on the other hand, have no mass, carry no charge, and move at the speed of light, and therefore have greater penetrating abilities than alpha or beta particles.

22.7 Refer to Figure 22.4 in the textbook.

22.9

(a) $^{228}_{90}Th \rightarrow ^{4}_{2}He + ^{224}_{88}Ra$

(b) $^{28}_{13}Al \rightarrow ^{0}_{-1}e + ^{28}_{14}Si$

(c) $^{82}_{37}Rb + ^{0}_{-1}e \rightarrow ^{82}_{36}Kr$ + x-ray photon

(d) $^{11}_{6}C \rightarrow ^{0}_{1}e + ^{11}_{5}B$

22.11

(1) $^{238}_{92}U \rightarrow ^{4}_{2}He + ^{234}_{90}Th$

(2) $^{234}_{90}Th \rightarrow ^{0}_{-1}e + ^{234}_{91}Pa$

(3) $^{234}_{91}Pa \rightarrow ^{0}_{-1}e + ^{234}_{92}U$

(4) $^{234}_{92}U \rightarrow ^{4}_{2}He + ^{230}_{90}Th$

(5) $^{230}_{90}Th \rightarrow ^{4}_{2}He + ^{226}_{88}Ra$

(6) $^{226}_{88}Ra \rightarrow ^{4}_{2}He + ^{222}_{86}Rn$

(7) $^{222}_{86}Rn \rightarrow ^{4}_{2}He + ^{218}_{84}Po$

(8) $^{218}_{84}Po \rightarrow ^{4}_{2}He + ^{214}_{82}Pb$

(9) $^{214}_{82}Pb \rightarrow ^{0}_{-1}e + ^{214}_{83}Bi$

(10) $^{214}_{83}Bi \rightarrow ^{0}_{-1}e + ^{214}_{84}Po$

(11) $^{214}_{84}Po \rightarrow ^{4}_{2}He + ^{210}_{82}Pb$

(12) $^{210}_{82}Pb \rightarrow ^{0}_{-1}e + ^{210}_{83}Bi$

(13) $^{210}_{83}Bi \rightarrow ^{0}_{-1}e + ^{210}_{84}Po$

(14) $^{210}_{84}Po \rightarrow ^{4}_{2}He + ^{206}_{82}Pb$

20.13 Refer to Figure 22.7 and "Digging Deeper: Particle Accelerators" in Section 22.3 in the textbook.

22.15

(a) $^{10}_{5}B + ^{4}_{2}He \rightarrow ^{13}_{7}N + ^{1}_{0}n$

(b) $^{8}_{3}Li + ^{1}_{0}n \rightarrow ^{9}_{3}Li$

(c) $^{238}_{92}U + ^{1}_{0}n \rightarrow ^{239}_{93}Np + ^{0}_{-1}e$

22.17 (a) $^{10}B(\alpha,n)^{13}N$ (b) $^{8}Li(n,0)^{9}Li$ (c) $^{238}U(n,\beta)^{239}Np$

22.19

$^{238}_{92}U + ^{2}_{1}H \rightarrow ^{238}_{93}Np + 2^{1}_{0}n$

$^{238}_{93}Np \rightarrow ^{238}_{94}Pu + ^{0}_{-1}e$

22.21

$^{209}_{83}Bi + ^{4}_{2}He \rightarrow ^{211}_{85}At + 2^{1}_{0}n$

22.23

$^{239}_{94}Pu + ^{1}_{0}n \rightarrow ^{240}_{94}Pu$

$^{240}_{94}Pu \rightarrow ^{240}_{95}Am + ^{0}_{-1}e$

$^{240}_{95}Am + ^{1}_{0}n \rightarrow ^{241}_{95}Am$

Or, alternatively

$^{239}_{94}Pu + 2^{1}_{0}n \rightarrow ^{241}_{94}Pu \rightarrow ^{241}_{95}Am + ^{0}_{-1}e$

22.25 No. Flint does not originate from previously living objects and, consequently, does not contain any carbon-14.

22.27

$$k = \frac{0.693}{t_{1/2}} = \frac{0.693}{1.83\ h} = 0.379\ h^{-1}$$

Substituting k and $N/N_0 = 0.750$ into Equation 22.1 gives

$$t = \frac{\ln (N/N_0)}{-k} = \frac{\ln 0.750}{-0.379\ h^{-1}} = 0.759\ h \text{ or } 45.5\ min$$

22.29 The rate constant is obtained from the half-life using Equation 22.2:

$$k = \frac{0.693}{t_{1/2}} = \frac{0.693}{28.1\ y} = 0.0247\ y^{-1}$$

(a) $\ln(N/N_0) = -0.0247\ y^{-1} \times 1\ y = -0.0247$ and $N/N_0 = e^{-0.0247} = 0.976$
$\ln(N/N_0) = -0.0247\ y^{-1} \times 10\ y = -0.247$ and $N/N_0 = e^{-0.247} = 0.781$

(b) $\ln(N/N_0) = -0.0247\ y^{-1} \times 55\ y = -1.36$ and $N/N_0 = e^{-1.36} = 0.26$

22.31 Each lead atom was once a uranium atom. The number of lead and uranium atoms in the rock are:

$$\text{Pb: } 85\ g\ Pb \times \frac{6.022 \times 10^{23}\ \text{Pb atoms}}{207.2\ g\ Pb} = 2.5 \times 10^{23}\ \text{Pb atoms}$$

$$\text{U: } 100\ g\ U \times \frac{6.022 \times 10^{23}\ \text{U atoms}}{238.0\ g\ U} = 2.53 \times 10^{23}\ \text{U atoms}$$

Therefore, for every 2.53×10^{23} U atoms there are now presently, there were 5.0×10^{23} U atoms in the rock originally, and the fraction of uranium-238 remaining in the rock is therefore

$N/N_0 = 2.53 \times 10^{23}/5.0 \times 10^{23} = 0.51$

Substituting this fraction and $k = 1.55 \times 10^{-10}\ y^{-1}$ (see Example 22.5) into Equation 22.1 gives

$\ln 0.51 = -1.55 \times 10^{-10}\ y^{-1} \times t$ and $t = 4.3 \times 10^9\ y$

The rock is about 4.3 billion years old.

22.33

$$k = \frac{0.693}{t_{1/2}} = \frac{0.693}{12.32\ y} = 0.0562\ y^{-1}$$

Substituting k and $N/N_0 = 2.26 \times 10^5/6.02 \times 10^5$ into Equation 22.1 gives

$$t = \frac{\ln(N/N_0)}{-k} = \frac{\ln(2.26 \times 10^5/6.02 \times 10^5)}{-0.0562\ y^{-1}} = 17.4\ y$$

22.35 The rem is the radiation unit used by the U.S. Environmental Protection Agency.

Average exposure from natural sources is about 0.15 rem per year.

Natural sources include cosmic rays, soil containing radium and its decay products, and potassium-40 in food.

22.37 $5.0 \times 3.7 \times 10^{-2}$ disintegrations/s•mL $\times$ 10 mL $\times$ 60 s/min = 110 disintegrations/min

22.39 (a)

$$k = \frac{0.693}{t_{1/2}} = \frac{0.693}{2.69\ \text{days}} = 0.258\ \text{days}^{-1}$$

Substituting k and $N/N_0 = 1/10$ into Equation 22.1 gives

$$t = \frac{\ln(N/N_0)}{-k} = \frac{\ln(1/10)}{-0.258\ \text{days}^{-1}} = 9\ \text{days}$$

(b) $\ln(N/N_0) = -0.258\ \text{days}^{-1} \times 14\ \text{days} = -3.61$; $N/N_0 = e^{-3.61} = 0.027$ and $N = 0.027N_0 = (0.027)(100\ \text{mCi}) = 2.7\ \text{mCi}$

22.41 No.

No.

The effectiveness of the nuclear force is limited to distances that do not exceed 10^{-15} m, the approximate diameter of a nucleus. It is also limited in the number of protons it can hold in a stable nuclear arrangement. The nuclear force can stabilize no more than 83 protons in a nucleus; above that number the nucleus becomes unstable.

22.43 This is explained by the tendency of the nucleons to pair up with each other. Paired nucleon arrangements are more stable than unpaired nucleon arrangements.

22.45 Alpha emission (Z > 65, usually), positron emission, electron capture, and proton emission (rare) are decay modes which result in an increased neutron/proton ratio.

Refer to the "Stability Band" in Section 22.5 in the textbook for illustrations.

22.47 The atomic number of strontium is 38. Strontuim-90, with 38 protons and 52 neutrons, lies above the stability band of Figure 22.15. It will decay by a mode that decreases the n/p ratio. The most common mode for decreasing the n/p ratio is beta emission (neutron emission is rare):

$${}^{90}_{38}\text{Sr} \rightarrow {}^{90}_{39}\text{Y} + {}^{0}_{-1}\text{e}$$

22.49 The atomic number of oxygen is 8. Oxygen-13, with 8 protons and 5 neutrons, oxygen-14, with 8 protons and 6 neutrons, and oxygen-15, with 8 protons and 7 neutrons, all lie below the stability band of Figure 22.15. They will all decay by a mode that increases the n/p ratio. Alpha emission is not common for elements with Z < 66, and proton emission rarely occurs. Therefore, the possible modes of decay are positron emission and electron capture. Since electron capture does not occur, ${}^{13}\text{O}$, ${}^{14}\text{O}$, and ${}^{15}\text{O}$ decay by positron emission:

(a) ${}^{13}_{8}\text{O} \rightarrow {}^{13}_{7}\text{N} + {}^{0}_{1}\text{e}$ (b) ${}^{14}_{8}\text{O} \rightarrow {}^{14}_{7}\text{N} + {}^{0}_{1}\text{e}$ (c) ${}^{15}_{8}\text{O} \rightarrow {}^{15}_{7}\text{N} + {}^{0}_{1}\text{e}$

Oxygen-19, with 8 protons and 11 neutrons, and oxygen-20, with 8 protons and 12 neutrons, both lie above the stability band of Figure 22.15. They will decay by a mode that decreases the n/p ratio. The most common mode for decreasing the n/p ratio is beta emission (neutron emission is rare):

(d) ${}^{19}_{8}\text{O} \rightarrow {}^{19}_{9}\text{F} + {}^{0}_{-1}\text{e}$ (e) ${}^{20}_{8}\text{O} \rightarrow {}^{20}_{9}\text{F} + {}^{0}_{-1}\text{e}$

22.51 (a) ${}^{45}_{22}\text{Ti} \rightarrow {}^{45}_{21}\text{Sc} + {}^{0}_{1}\text{e}$

${}^{45}_{22}\text{Ti} + {}^{0}_{-1}\text{e} \rightarrow {}^{45}_{21}\text{Sc}$

(b) ${}^{45}_{22}\text{Ti}$: n/p = 23/22 = 1.05; ${}^{45}_{21}\text{Sc}$: n/p = 24/21 = 1.14

(c) Below. Since the n/p ratio increases, titanium-45 is below the stability band.

22.53 The binding energy is the energy needed to decompose a nucleus into protons and neutrons.

Iron-56 has the highest binding energy per nucleon.

22.55 (a) $$\frac{1.0\text{ GeV}}{1\text{ particle}} \times \frac{10^{3}\text{ MeV}}{1\text{ GeV}} \times \frac{1.6022 \times 10^{-16}\text{ kJ}}{1\text{ MeV}} \times \frac{6.022 \times 10^{23}\text{ particles}}{1\text{ mol}} = 9.6 \times 10^{10}\text{ kJ/mol}$$

(b) $$\frac{1.0\text{ TeV}}{1\text{ particle}} \times \frac{10^{6}\text{ MeV}}{1\text{ TeV}} \times \frac{1.6022 \times 10^{-16}\text{ kJ}}{1\text{ MeV}} \times \frac{6.022 \times 10^{23}\text{ particles}}{1\text{ mol}} = 9.6 \times 10^{13}\text{ kJ/mol}$$

22.57 $4\,^{1}_{1}H \rightarrow \,^{4}_{2}He + 2\,^{0}_{1}e$

Reactant mass: $4\,^{1}H$ = 4.02912 u
Product masses: ^{4}He = 4.00150 u
$2\,^{0}e$ = 0.001097 u
Total product mass: 4.00260 u

The vanished mass is 4.02912 u – 4.00260 u = 0.02652 u.

(a) The energy equivalent of the vanished mass is 0.02652 u × 1.4924 × 10^{-10} J/u = 3.958 × 10^{-12} J. This is the energy released when one helium atom is produced.

(b) The atomic (isotopic) mass of helium-4 is 4.00150 u + 2(0.00054858 u) = 4.00260 u. The energy released in joules per gram of helium is

$$\frac{3.958 \times 10^{-12}\text{ J}}{4.00260\text{ u}} \times \frac{6.022 \times 10^{23}\text{ u}}{1\text{ g}} = 5.955 \times 10^{11}\text{ J/g}$$

22.59 $^{238}_{92}U \rightarrow \,^{234}_{90}Th + \,^{4}_{2}He$

Reactant mass: ^{238}U = 238.00032 u
Product masses: ^{234}Th = 233.99422 u
^{4}He = 4.00150 u
Total product mass: 237.99572 u

The vanished mass is 238.00032 u – 237.99572 u = 0.00460 u.

(a) The energy equivalent of the vanished mass in MeV is 0.00460 u × 931.5 MeV/u = 4.28 MeV. This is the energy released during the decay of one uranium atom.

(b) The atomic (isotopic) mass of uranium-238 is 238.00032 u + 92(0.00054858 u) = 238.05079 u. The energy released in joules per gram of uranium is

$$\frac{4.28\text{ MeV}}{238.05079\text{ u}} \times \frac{6.022 \times 10^{23}\text{ u}}{1\text{ g}} \times \frac{1.6022 \times 10^{-13}\text{ J}}{1\text{ MeV}} = 1.73 \times 10^{9}\text{ J/g}$$

22.61 (a) Beryllium-9 has 4 protons and 5 neutrons. The total mass of the separate nucleons is

4 protons × 1.00728 u/proton = 4.02912 u
5 neutrons × 1.00866 u/neutron = 5.04330 u
Total = 9.07242 u

mass defect = mass of separate nucleons – nuclear mass of beryllium-9 = 9.07242 u – 9.00999 u
= 0.06243 u

The binding energy is the energy equivalent of the mass defect, or 0.06243 u × 931.5 MeV/u = 58.15 MeV. The average binding energy per nucleon is 58.15 MeV/9 nucleons = 6.461 MeV/nucleon.

(b) Boron-10 has 5 protons and 5 neutrons. The total mass of the separate nucleons is

5 protons × 1.00728 u/proton = 5.03640 u
5 neutrons × 1.00866 u/neutron = 5.04330 u
Total = 10.07970 u

mass defect = mass of separate nucleons – nuclear mass of boron-10 = 10.07970 u – 10.01019 u
= 0.06951 u

The binding energy is the energy equivalent of the mass defect, or 0.06951 u × 931.5 MeV/u = 64.75 MeV. The average binding energy per nucleon is 64.75 MeV/10 nucleons = 6.475 MeV/nucleon.

(c) Silicon-28 has 14 protons and 14 neutrons. The total mass of the separate nucleons is

14 protons × 1.00728 u/proton = 14.10192 u
14 neutrons × 1.00866 u/neutron = 14.12124 u
Total = 28.22316 u

mass defect = mass of separate nucleons – nuclear mass of silicon-28 = 28.22316 u – 27.96925 u
= 0.25391 u

The binding energy is 0.25391 u × 931.5 MeV/u = 236.5 MeV. The average binding energy per nucleon is 236.5 MeV/28 nucleons = 8.446 MeV/nucleon.

The order of increasing stability is $^{9}Be < ^{10}B < ^{28}Si$.

22.63 1. You need the production of an unstable nucleus that undergoes fission, producing more neutrons than were consumed in initiating the reaction.
2. You have to have a critical mass of fissionable material; otherwise most of the neutrons generated will escape before reacting.
3. The sample shape and the concentration of fissionable atoms have to be properly adjusted.

22.65 Refer to Figure 22.21 and "How a Reactor Works" in Section 22.7 in the textbook.

22.67 The atomic bomb is needed to generate the high activation energy required to initiate the hydrogen bomb's fusion reaction.

22.69 (a) Advantages: 1. Attractive energy source; a lot of energy is produced from a small amount. 2. Relatively safe operation. 3. Saves hydrocarbon resources for better uses. 4. Cuts down on air pollution. 5. Reduces environmental damage caused by coal mining and oil spills.

Disadvantages: 1. Cost and difficulty of obtaining fuel. 2. Difficulty of reprocessing spent fuel. 3. Storage of the highly radioactive waste products generated. 4. Possibly disastrous radioactive leaks. 5. Limited fuel supply. 6. Possible thermal pollution of water adjacent to the plant.

(b) Advantages: 1. Attractive energy source. 2. Easily available supply of deuterium from seawater. 3. No long-lived radioactive waste products generated. 4. Saves hydrocarbon resources for better uses. 5. Cuts down on air pollution. 6. Reduces environmental damage caused by coal mining and oil spills. 7. No possibility of a disastrous radioactive leak.

Disadvantages: 1. Difficulty in obtaining required reaction temperature. 2. Problems related to the containment of the reaction.

The fusion plant, since it produces no toxic waste products and doesn't require extensive mining and refining operations.

22.71 (a) $^{238}_{92}U + ^{1}_{0}n \rightarrow ^{239}_{92}U$

$^{239}_{92}U \rightarrow ^{239}_{93}Np + ^{0}_{-1}e$

$^{239}_{93}Np \rightarrow ^{239}_{94}Pu + ^{0}_{-1}e$

(b) Advantages: Same as those of a fission reactor (see Exercise 22.69 (a)), plus conserves scarce nuclear fuel.

Disadvantages: Same as those of a fission reactor (see Exercise 22.69 (a)), plus 1. the plutonium fuel produced is an extremely toxic substance and 2. the plutonium fuel is much more difficult to handle, store, and transport than the uranium fuel.

22.73

(a) $^{235}_{92}U + ^{1}_{0}n \rightarrow ^{90}_{37}Rb + ^{144}_{55}Cs + 2^{1}_{0}n$

(b) $^{235}_{92}U + ^{1}_{0}n \rightarrow ^{90}_{38}Sr + ^{143}_{54}Xe + 3^{1}_{0}n$

(c) $^{235}_{92}U + ^{1}_{0}n \rightarrow ^{90}_{38}Sr + ^{144}_{54}Xe + 2^{1}_{0}n$

(d) $^{235}_{92}U + ^{1}_{0}n \rightarrow ^{97}_{39}Y + ^{137}_{53}I + 2^{1}_{0}n$

(e) $^{235}_{92}U + ^{1}_{0}n \rightarrow ^{137}_{52}Te + ^{97}_{40}Zr + 2^{1}_{0}n$

22.75 The atomic (isotopic) mass of uranium-235 is 234.99346 u + 92(0.00054858 u) = 235.04393 u. The energy released when 1 g of uranium-235 undergoes fission is

$$\frac{200\text{ MeV}}{235.04393\text{ u}} \times \frac{6.022 \times 10^{23}\text{ u}}{1\text{ g}} \times \frac{1.6022 \times 10^{-16}\text{ kJ}}{1\text{ MeV}} = 8.21 \times 10^{7}\text{ kJ/g}$$

The tons of TNT equivalent to this amount of energy is

$$8.21 \times 10^{7}\text{ kJ} \times \frac{1\text{ g}}{2.8\text{ kJ}} \times \frac{1\text{ ton}}{9.072 \times 10^{5}\text{ g}} = 32\text{ tons}$$

22.77 $^{2}_{1}H + ^{3}_{1}H \rightarrow ^{4}_{2}He + ^{1}_{0}n$

Reactant masses: ^{2}H = 2.01355 u
^{3}H = 3.01550 u

Total reactant mass: 5.02905 u

Product masses: ^{4}He = 4.00150 u
^{1}n = 1.00866 u

Total product mass: 5.01016 u

(a) The vanished mass is 5.02905 u − 5.01016 u = 0.01889 u. The energy equivalent of the vanished mass in MeV is 0.01889 u × 931.5 MeV/u = 17.60 MeV. This is the energy released during the production of one helium atom.

(b) The energy released in kilojoules per mole of product is

$$\frac{17.60\text{ MeV}}{1\text{ atom}} \times \frac{1.6022 \times 10^{-16}\text{ kJ}}{1\text{ MeV}} \times \frac{6.022 \times 10^{23}\text{ atoms}}{1\text{ mol}} = 1.698 \times 10^{9}\text{ kJ/mol}$$

22.79 At the time of Rutherford's experiments, neutrons were unknown and particle accelerators that could speed up electrons, protons, and other positive nuclei had not been invented yet. The only projectile particles at his disposal were alpha particles emitted by natural radionuclides.

22.81

(a) Iodine-131: used in the study of thyroid function and the detection of thyroid malignancies.
(b) Thallium-201: used for external heart scanning studies.
(c) Technetium-99m: used in medical diagnosis.
(d) Sodium-24: used in the study of blood circulation.
(e) Cobalt-60: used in cancer therapy.
(f) Plutonium-238: power source on Apollo lunar missions.
(g) Americium-241: used in smoke detectors.

22.83 $^{235}_{92}U$ decays to $^{207}_{82}Pb$

There are three natural decay series and each isotope in these series decays by either alpha particle or beta particle emission. Alpha particle emission lowers the mass number by four and the atomic number by two, but beta particle emission only raises the atomic number by one, so it does not affect the mass number at all. Therefore, the mass number is only affected by the alpha particle emissions, and the number of alpha particles emitted is equal to the difference in the mass numbers of uranium-235 and lead-207 divided by four:

$$\text{alpha particles emitted} = \frac{235-207}{4} = 7$$

The emission of seven alpha particles lowers the atomic number to $92 - 2 \times 7 = 78$. To bring the atomic number back up to 82, four beta particles must be emitted. Therefore, seven alpha particles and four beta particles are emitted during the sequence.

22.85 $^{68}_{29}Cu \rightarrow\, ^{68}_{30}Zn +\, ^{0}_{-1}e$

(a)

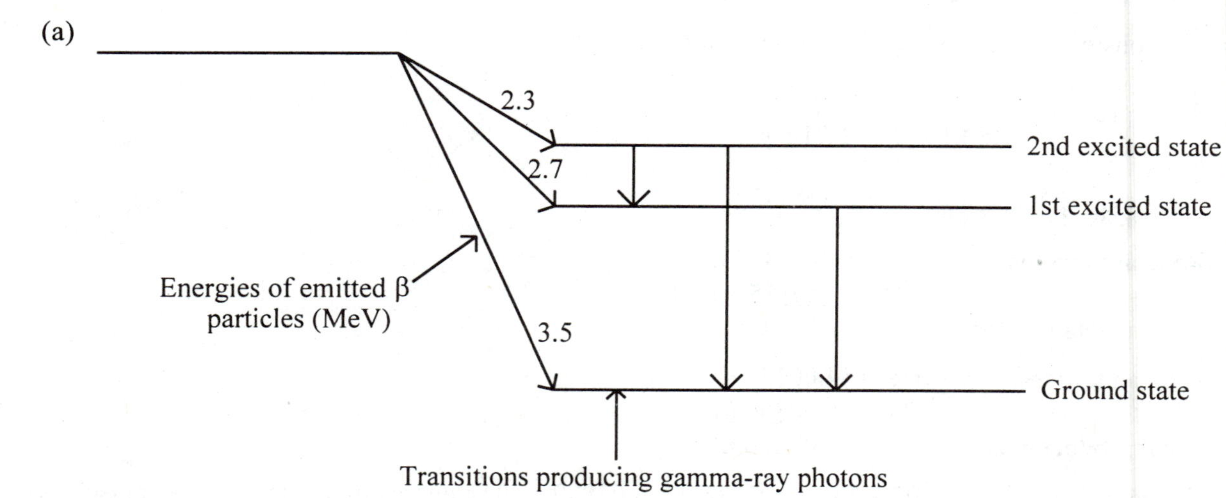

(b) Three different gamma-ray photons will be emitted with energies of $3.5 - 2.7 = 0.8$ MeV, $3.5 - 2.3 = 1.2$ MeV, and $2.7 - 2.3 = 0.4$ MeV.

22.87 (a) The mass equivalent of 39.35 kJ is

$$39.35\ \text{kJ} \times \frac{1\ \text{u}}{1.492 \times 10^{-13}\ \text{kJ}} \times \frac{1\ \text{g}}{6.022 \times 10^{23}\ \text{u}} = 4.380 \times 10^{-10}\ \text{g}$$

(b) The law of conservation of mass holds for chemical reactions because the mass change during a chemical reaction, on the order of 10^{-9} to 10^{-10} g/mol, is too small to be detected experimentally.

22.89 $^{6}Li + {}^{2}H \rightarrow 2\,{}^{4}He$

Reactant masses: ^{6}Li = 6.01348 u
^{2}H = 2.01355 u
Total reactant mass: 8.02703 u

Product mass: $2\,{}^{4}He$ = 8.00300 u

The vanished mass is 8.02703 u – 8.00300 u = 0.02403 u. The energy equivalent of the vanished mass in MeV is 0.02403 u × 931.5 MeV/u = 22.38 MeV. This is the energy released during the fusion of one lithium atom.

22.91 The energy released when 1 g of uranium-235 undergoes fission is 8.21×10^{7} kJ/g (see Exercise 22.75). The energy released when 1 g of deuterium fuses with tritium is 8.43×10^{8} kJ/g (see Exercise 22.78). Therefore, the energy released per gram of deuterium in the deuterium-tritium fusion is about ten times the energy released when 1 g of uranium-235 undergoes fission.

22.93 15.3 disintegrations/min•g × 1 min/60 s × $\frac{1 \text{ Ci}}{3.7 \times 10^{10} \text{ disintegrations/s}}$ = 6.9 × 10^{-12} Ci/g = 6.9 × 10^{-6} μCi/g

22.95 $^{40}_{19}K + ^{0}_{-1}e \rightarrow ^{40}_{18}Ar$ + x-ray photon

$^{40}_{19}K \rightarrow ^{40}_{20}Ca + ^{0}_{-1}e$

The rate constant is obtained from the half-life using Equation 22.2:

$$k = \frac{0.693}{t_{1/2}} = \frac{0.693}{1.26 \times 10^9 \text{ y}} = 5.50 \times 10^{-10} \text{ y}^{-1}$$

For every 1.00 mol of argon-40 that forms, 1.00/0.893 = 1.12 mol of potassium-40 must decay. So, N_0 = 1.5 + 1.12 = 2.6 mol per 1.5 mol of potassium-40 presently.

Substituting k and N/N_0 = 1.5/2.6 into Equation 22.1 gives

$$t = \frac{\ln (N/N_0)}{-k} = \frac{\ln (1.5/2.6)}{-5.50 \times 10^{-10} \text{ y}^{-1}} = 1.0 \times 10^9 \text{ y}$$

CHAPTER 23

ORGANIC CHEMISTRY AND THE CHEMICALS OF LIFE

Solutions To Practice Exercises

PE 23.1 4-ethyl-2-methylheptane

PE 23.2

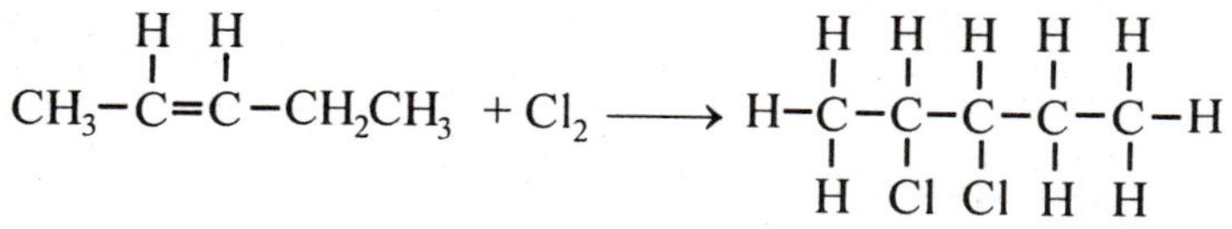

$$\mathrm{CH_3{-}\underset{}{\overset{H}{\overset{|}{C}}}{=}\overset{H}{\overset{|}{C}}{-}CH_2CH_3 + Cl_2 \longrightarrow H{-}\underset{\underset{H}{|}}{\overset{H}{\overset{|}{C}}}{-}\underset{\underset{Cl}{|}}{\overset{H}{\overset{|}{C}}}{-}\underset{\underset{Cl}{|}}{\overset{H}{\overset{|}{C}}}{-}\underset{\underset{H}{|}}{\overset{H}{\overset{|}{C}}}{-}\underset{\underset{H}{|}}{\overset{H}{\overset{|}{C}}}{-}H}$$

2-pentene 2,3-dichloropentane

PE 23.3

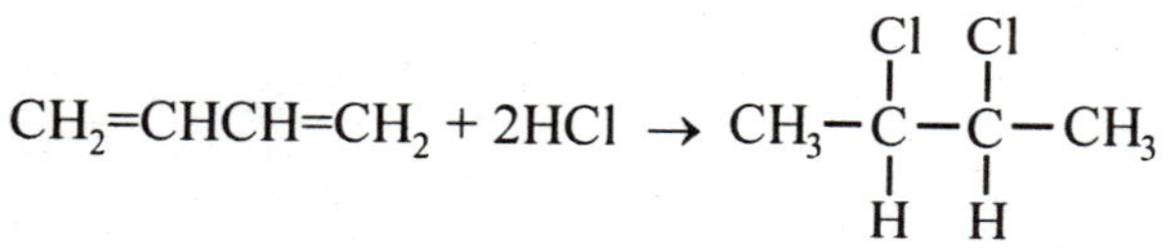

$$\mathrm{CH_2{=}CHCH{=}CH_2 + 2HCl \rightarrow CH_3{-}\underset{\underset{H}{|}}{\overset{Cl}{\overset{|}{C}}}{-}\underset{\underset{H}{|}}{\overset{Cl}{\overset{|}{C}}}{-}CH_3}$$

butadiene 2,3-dichlorobutane

PE 23.4 (a) ketone
(b) ether
(c) aldehyde
(d) phenol

PE 23.5 (a) glucose: $C_6H_{12}O_6$, 1:2
(b) glyceraldehyde: $C_3H_6O_3$, 1:2
(c) pyruvic acid: $C_3H_4O_3$, 3:4

They are all oxidations. (The formation of the second aldehyde group when glucose splits into two molecules of glyceraldehyde, is a redox process.)

PE 23.6

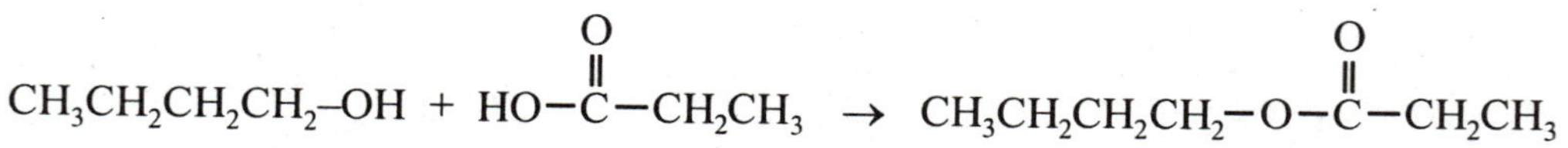

$$\mathrm{CH_3CH_2CH_2CH_2{-}OH + HO{-}\overset{O}{\overset{\|}{C}}{-}CH_2CH_3 \rightarrow CH_3CH_2CH_2CH_2{-}O{-}\overset{O}{\overset{\|}{C}}{-}CH_2CH_3}$$

butyl propionate

PE 23.7 (a) Not chiral. None of the carbon atoms has four different substituents.
(b) Chiral. The middle carbon atom has four different substituents; $CH_3{}^{*}CHClCH_2OH$.

PE 23.8

Solutions To Final Exercises

23.1 (a) A general formula is a formula that represents all the members of a particular group of organic compounds. Examples of general formulas are:
C_nH_{2n+2}: represents straight– or branched–chain alkanes
C_nH_{2n}: represents alkenes
C_nH_{2n-2}: represents alkynes

(b) A homologous series is a series of compounds having the same general formula in which one compound differs from a preceding one by a definite increment of number and kind of atoms. In most cases the increment is a $-CH_2-$ group, known as a methylene group. Homologous compounds have similar chemical properties, and the physical properties of homologous compunds change throughout the series in a regular way. Examples are the straight–chain alkanes and the straight–chain saturated carboxylic acids (also called fatty acids; general formula $C_nH_{2n+1}COOH$ or $C_nH_{2n}O_2$).

(c) A saturated hydrocarbon (alkane) is a hydrocarbon that contains only single bonds. In a saturated hydrocarbon, each carbon atom in the molecule is bonded to the maximum number of hydrogen atoms it can accomodate. Examples are methane (CH_4), ethane (C_2H_6), and propane (C_3H_8).

(d) An alkyl group is an alkane less one hydrogen atom. An alkyl group formed from a straight–chain alkane, is named by changing the suffix *-ane* to *-yl.* Examples are methyl group (CH_3-), ethyl group (CH_3CH_2-) and propyl group ($CH_3CH_2CH_2-$).

23.3 Refer to Figure 23.1 in the textbook.

23.5 No. Since each carbon atom is tetrahedrally bonded, the C–C–C bond angle is actually 109.5°, not 180°, and, therefore, the carbon chain is actually zig–zag in nature. Furthermore, free rotation about the C–C single bonds leads to a variety of convoluted chains or shapes.

23.7 (a) C_4H_{10}, cyclopentane

(b) C_6H_{12}, methylcyclopentane

23.9 (a)

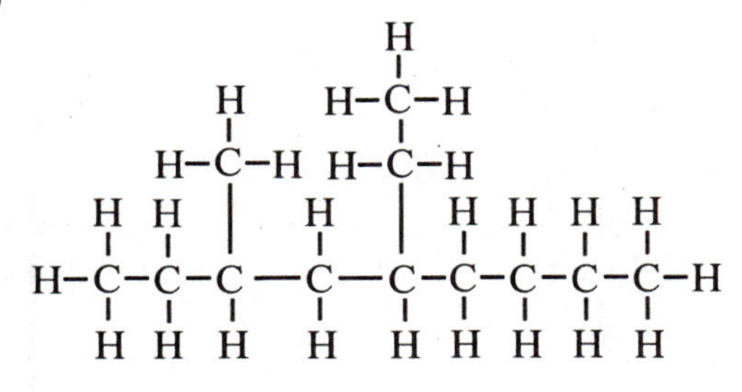

(b)

23.11 (a)

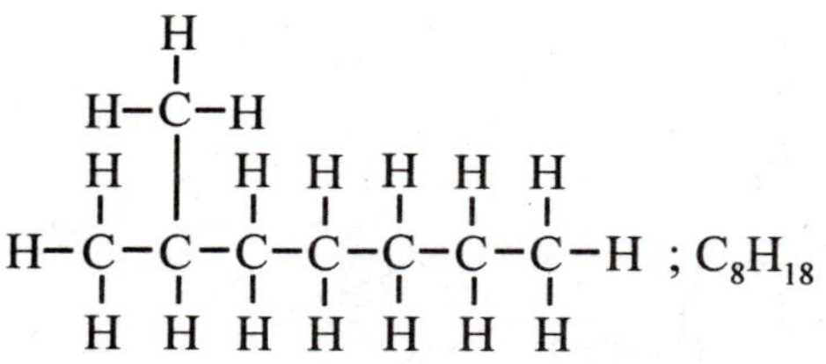

```
        H
        |
      H-C-H
    H   |   H H H H H
    |   |   | | | | |
  H-C - C - C-C-C-C-C-H ; C8H18
    |   |   | | | | |
    H   H   H H H H H
```

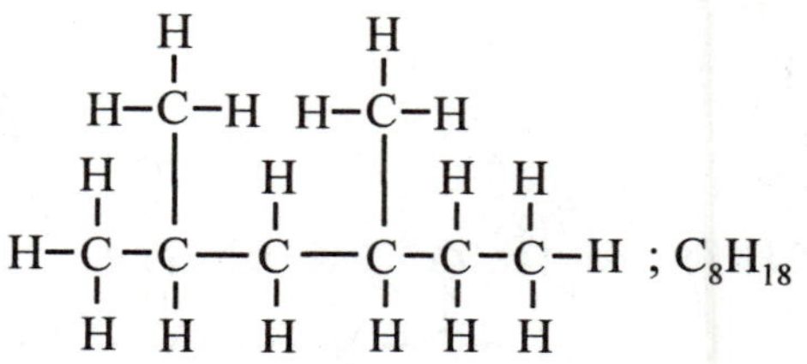

(b)

```
      H         H
      |         |
    H-C-H     H-C-H
  H   |     H   |   H H
  |   |     |   |   | |
H-C - C  -  C - C - C-C-H ; C8H18
  |   |     |   |   | |
  H   H     H   H   H H
```

(c)

```
          H
          |
        H-C-H
          |
        H-C-H
  H H     |   H H H
  | |     |   | | |
H-C-C  -  C - C-C-C-H ; C8H18
  | |     |   | | |
  H H     H   H H H
```

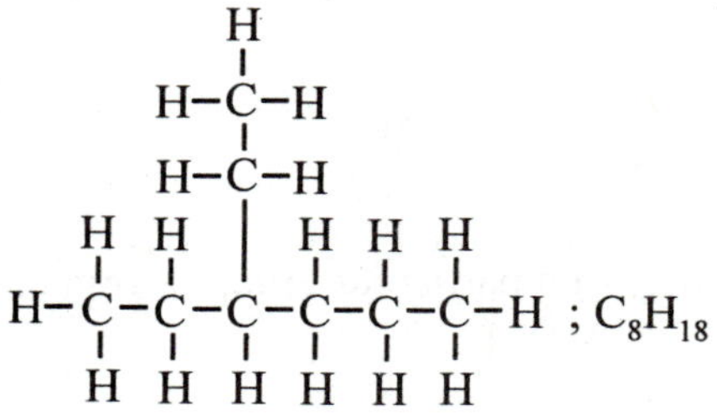

(d)

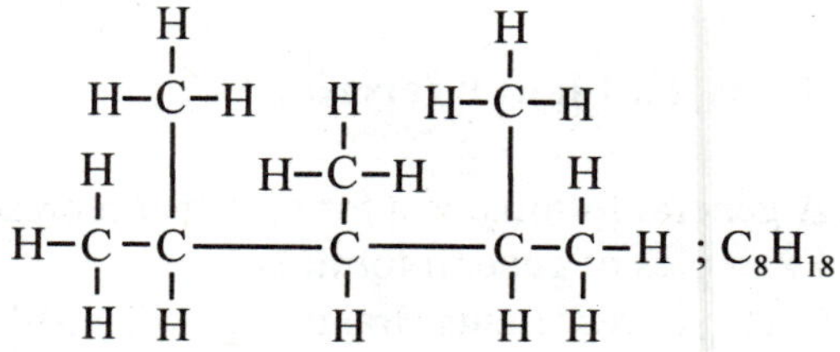

```
      H               H
      |               |
    H-C-H    H      H-C-H
  H   |    H-C-H      |   H
  |   |      |        |   |
H-C - C  ----C--------C - C-H ; C8H18
  |   |      |        |   |
  H   H      H        H   H
```

(e)

```
                 H
                 |
      H        H-C-H
      |          |
    H-C-H      H-C-H
  H   |          |   H H
  |   |          |   | |
H-C - C ---------C - C-C-H ; C8H18
  |   |          |   | |
  H   H          H   H H
```

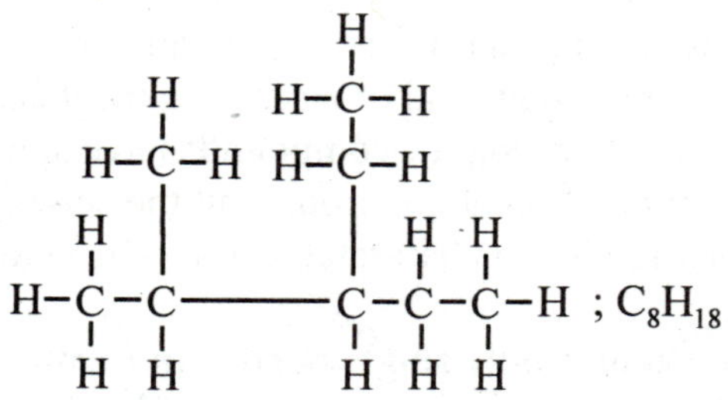

They are all structural isomers of the molecular formula C_8H_{18}.

23.13 (a) methylcyclobutane (b) 4-ethyl-5-methyloctane or 5-ethyl-4-methyloctane

23.15 (a) 17

(b)

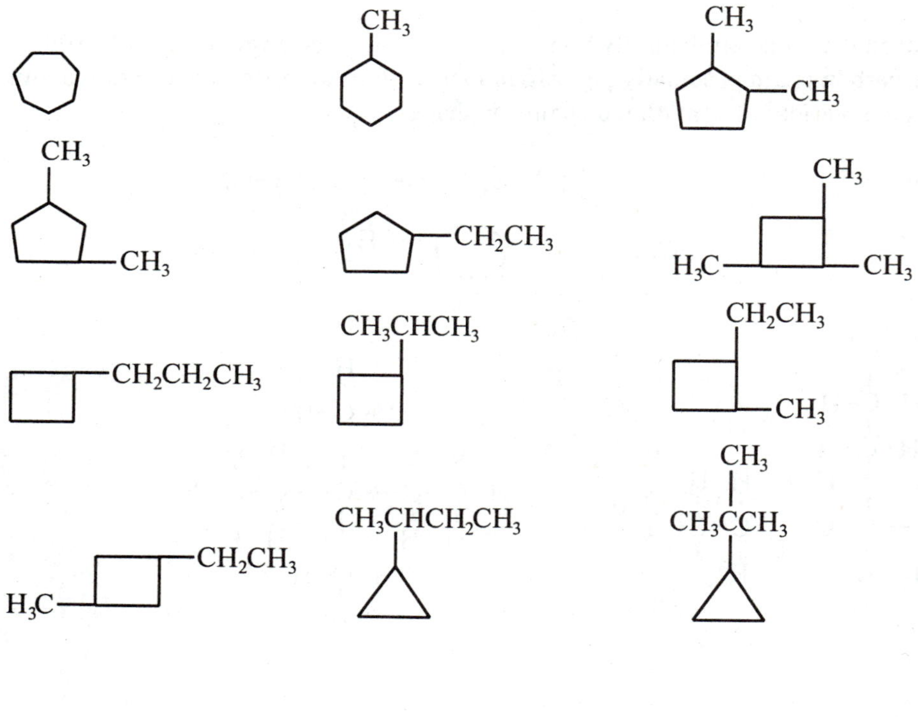

CH_3

CH_3 CH_3

CH_3 CH_3

CH_2CH_3

CH_3 H_3C CH_3

$CH_2CH_2CH_3$

CH_3CHCH_3

CH_2CH_3 CH_3

CH_3

H_3C CH_2CH_3

$CH_3CHCH_2CH_3$

CH_3CCH_3

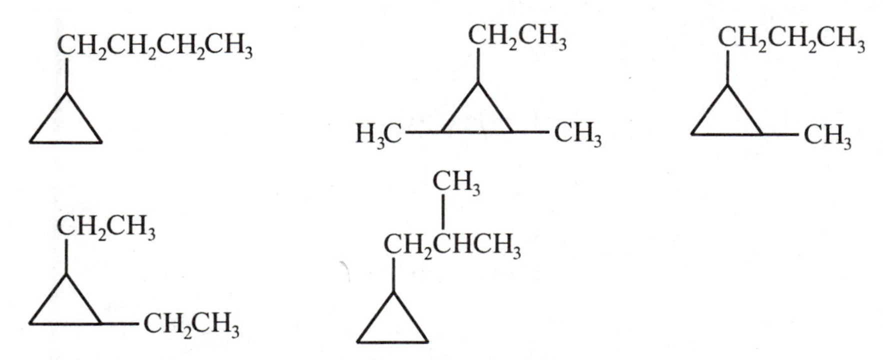

Note: Additional structural isomers exist in which two alkyl groups are attached to the same ring carbon. These structural isomers have been omitted. Also, only strucutral isomerism has been considered in this exercise.

23.17 $2C_4H_{10}(g) + 13O_2(g) \rightarrow 8CO_2(g) + 10H_2O(g)$

23.19 Step 1. $CH_4 + Cl_2 \xrightarrow{h\nu} CH_3Cl + HCl$

Step 2. $CH_3Cl + Cl_2 \xrightarrow{h\nu} CH_2Cl_2 + HCl$

Step 3. $CH_2Cl_2 + Cl_2 \xrightarrow{h\nu} CHCl_3 + HCl$

Step 4. $CHCl_3 + Cl_2 \xrightarrow{h\nu} CCl_4 + HCl$

The other product is hydrogen chloride.

23.21 (a) C_nH_{2n} (b) C_nH_{2n-2} (c) C_nH_{2n-4}

23.23

```
    H     H H            H  H  H  H
    |     | |            |  |  |  |
(a) C=C-C-C-H   (b) H-C-C=C-C-H     (c) H-C≡C-H
    | | | |              |        |
    H H H H              H        H
```

23.25 (a) No.

```
  H H H H H
  | | | | |
H-C=C-C-C-C-H
      | | |
      H H H
```

(b) No.

```
    H       H
    |       |
  H-C-H   H-C-H
H   |       |   H H
|   |       |   | |
H-C-C=======C-C-C-H
|               | |
H               H H
```

(c) No.

```
    H       H
    |       |
  H-C-H   H-C-H
H   |       |   H
|   |       |   |
H-C-C=======C-C-H
|               |
H               H
```

(d) Yes.

```
  H H H H H
  | | | | |
H-C-C=C-C-C-H
  |     | |
  H     H H
```

(e) No.

```
  H H H H
  | | | |
H-C=C-C=C-H
```

23.27

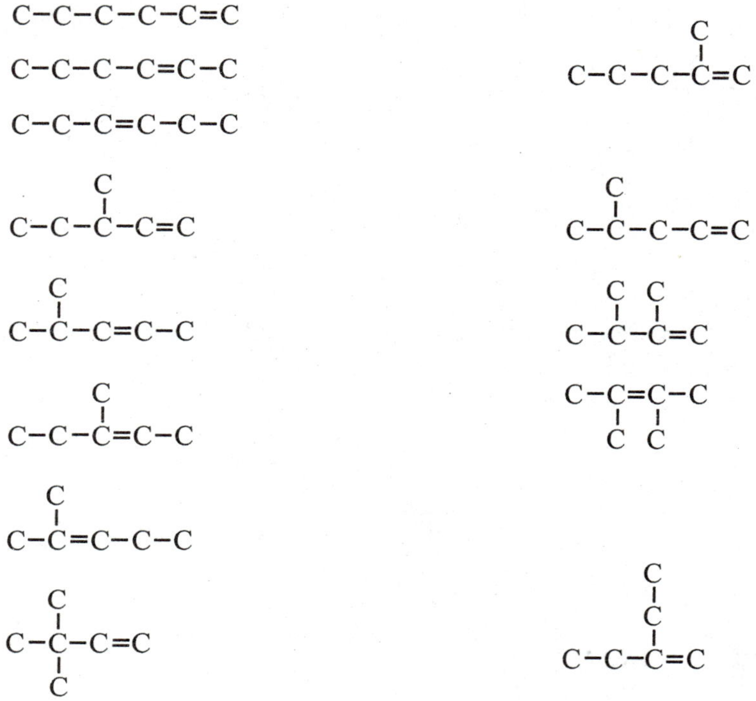

```
C−C−C−C−C=C                        C
                                   |
C−C−C−C=C−C                  C−C−C−C=C

C−C−C=C−C−C

      C                        C
      |                        |
C−C−C−C=C                    C−C−C−C=C

   C                           C C
   |                           | |
C−C−C=C−C                    C−C−C=C

      C                      C−C=C−C
      |                        | |
C−C−C=C−C                      C C

   C
   |
C−C=C−C−C
                                 C
   C                             |
   |                             C
C−C−C=C                          |
   |                         C−C−C=C
   C
```

23.29 *cis*-3-heptene

23.31 (a) $CH_3CH_2CH_2CH_3 + Br_2 \xrightarrow{h\nu} CH_3CHBrCH_2CH_3 + HBr$

butane → 2-bromobutane

(b)

```
                H  H
                |  |
CH3CH2−C=C−H  + HBr  →  CH3CH2CHBrCH3
```

butene → 2-bromobutane

The addition method is better, because it yields principally the desired product, while the substitution reaction is accompanied by a lot of unwanted side reactions, i.e., reactions that produce a product other than the one desired.

23.33

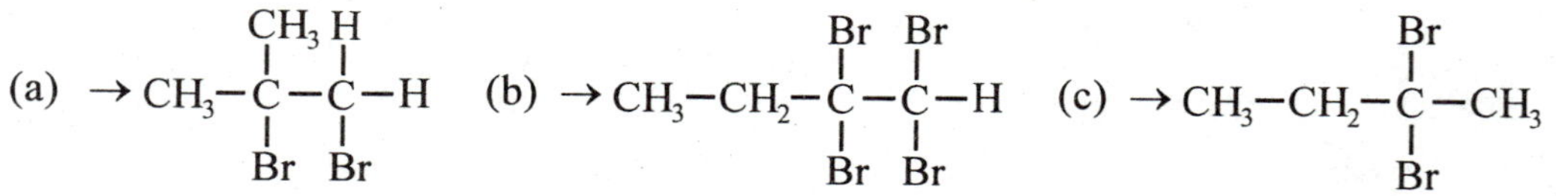

```
           CH3 H                      Br Br                      Br
           |   |                      |  |                       |
(a) → CH3−C−C−H   (b) → CH3−CH2−C−C−H   (c) → CH3−CH2−C−CH3
           |   |                      |  |                       |
           Br Br                      Br Br                      Br
```

23.35 (a) methylcyclohexane:

```
 CH3
  |
(cyclohexane ring)
```

(b)

```
     H   H   H   H   H   H            H   H
     |   |   |   |   |   |            |   |
   −C − C − C − C − C − C−    or   [ −C − C− ]
     |   |   |   |   |   |            |   |
     H  CH3  H  CH3  H  CH3          CH3  H   n
```

23.37 (a) cyclohexene $+ Cl_2 \rightarrow$ 1,2-dichlorocyclohexane (ring with −Cl, −Cl)   (b) cyclohexane $+ Cl_2 \xrightarrow{h\nu \text{ or heat}}$ chlorocyclohexane (ring−Cl) $+ HCl$

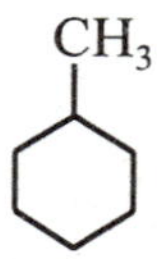

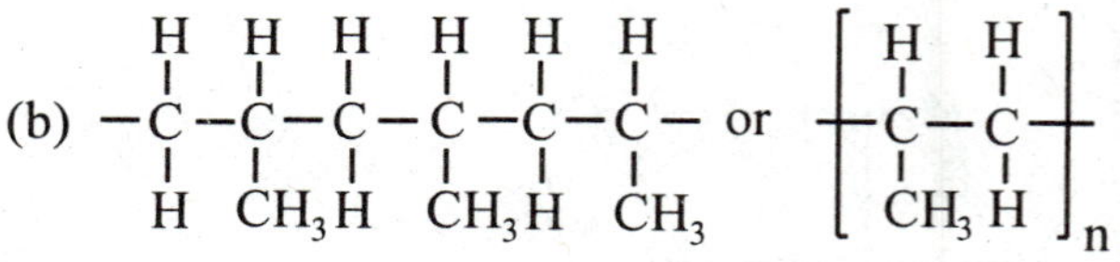

(c) $CH_3CH_2CH_2\text{–}C{\equiv}C\text{–}H + 2H_2 \xrightarrow{Pt}$ pentane

or

$H_2C{=}CHCH_2CH{=}CH_2 + 2H_2 \xrightarrow{Pt}$ pentane

(d) $H\text{–}C{\equiv}C\text{–}CH_2CH_2CH_3 + 2HBr \rightarrow$ 2,2-dibromopentane

23.39 $H_2C{=}CCl_2$

23.41 All aromatic compounds have delocalized pi bonds.

23.43 (a) 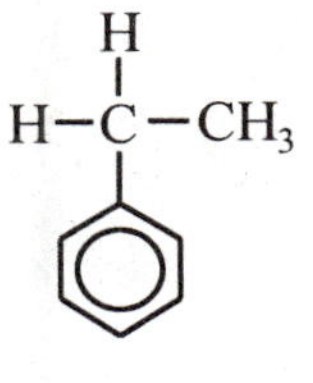 H–C(H)(C₆H₅)–CH₃ (b) Cl–C₆H₅ (c) branched chain with benzene ring

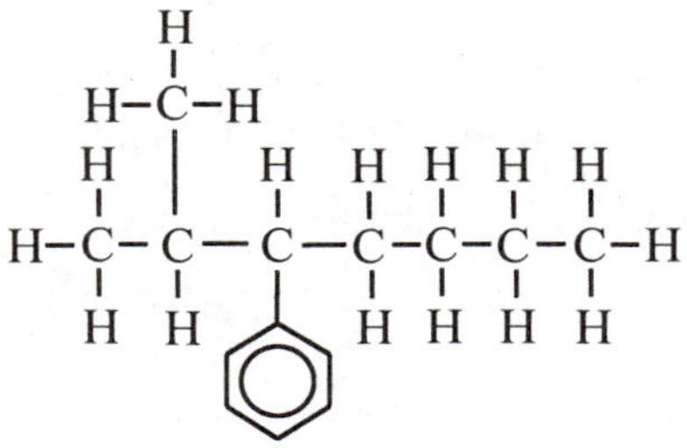

C_8H_{10} C_6H_5Cl $C_{14}H_{22}$

23.45 These structures are in Section 23.3 in the textbook.

23.47 (a) 1,2-dibromobenzene or o-dibromobenzene
(b) 1,3-dibromobenzene or m-dibromobenzene
(c) 1,4-dibromobenzene or p-dibromobenzene

23.49 (a) $C_6H_6 + 3H_2 \xrightarrow[200°C;\ 25\ atm]{Ni} C_6H_{12}$

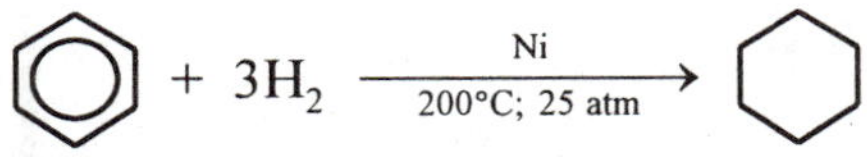

benzene $+ 3H_2 \xrightarrow[200°C;\ 25\ atm]{Ni}$ cyclohexane

(b) $C_6H_6 + Cl_2 \xrightarrow{FeCl_3} C_6H_5Cl + HCl$

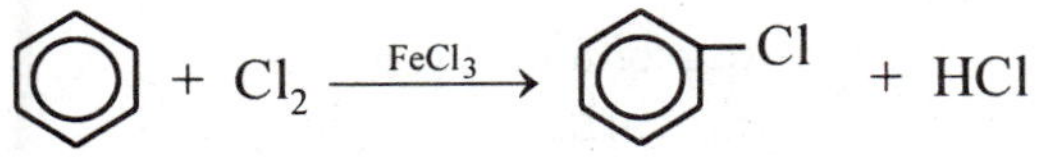

benzene $+ Cl_2 \xrightarrow{FeCl_3}$ chlorobenzene $+ HCl$

23.51 A functional group is a particular combination of atoms, situated in a portion of a molecule, that undergoes certain characteristic reactions; i.e., it is a reactive site or portion of a molecule. The unique chemical properties of a molecule are mostly the result of the properties of the functional groups present. Different molecules containing the same functional group normally have the same chemical properties, because the carbon skeleton generally remains unaffected during chemical reactions of the functional group. Therefore, general reactions, with R– used to represent the unreactive hydrocarbon portion of the molecule, are sufficient to illustrate the reactions that any member in that class, i.e., that any molecule containing that functional group, will undergo.

Refer to Table 23.4 in the textbook for examples of functional groups.

23.53 (a) R–Ö–R′ ; dimethyl ether (CH_3OCH_3) and diethylether ($CH_3CH_2OCH_2CH_3$); ether group, –Ö–.

(b) R–C(=O:)–R′ ; acetone (CH_3COCH_3) and diethyl ketone ($C_5H_5COC_2H_5$); carbonyl group, –C(=O:)–.

(c) $R-\ddot{\underset{\cdot\cdot}{O}}-H$; methanol (CH_3OH) and ethanol (C_2H_5OH); hydroxyl group, $-\ddot{\underset{\cdot\cdot}{O}}-H$.

(d) $R-\overset{:O:}{\overset{\|}{C}}-\ddot{\underset{\cdot\cdot}{O}}-H$; formic acid (HCO_2H) and acetic acid (CH_3CO_2H); carboxyl group, $-\overset{:O:}{\overset{\|}{C}}-\ddot{\underset{\cdot\cdot}{O}}-H$.

23.55 (a) 1. Catalytic hydration of ethylene.
2. Fermentation of carbohydrates.
3. Oxidation of ethane or reduction of acetaldehyde.

(b) Ethanol is used as a reagent in chemical synthesis and also as a fuel, solvent, antiseptic , and beverage ingredient.

23.57 (a) In phenol the aromatic ring withdraws electron density from the O–H bond and this makes the O–H bond more polar, while in ethanol the alkyl group does not.

(b) Aqueous hydroxides convert phenols into their salts.

$C_6H_5OH(l \text{ or } aq) + NaOH(aq) \rightarrow C_6H_5ONa(aq) + H_2O(l)$

23.59 Refer to Table 23.7, "Fats and Oils" in Section 23.5, and definitions in the Glossary.

23.61 (a) carbonyl group
(b) hydroxyl group
(c) carboxyl group
(d) two carboxyl groups
(e) halide and carboxyl groups
(f) carbonyl and carboxyl groups
(g) hydroxyl and carboxyl groups

23.63

```
  H H H H          H H OH H         H    H    H          H    OH   H
  | | | |          | |  | |         |    |    |          |    |    |
H-C-C-C-C-OH     H-C-C-C-C-H      H-C----C----C-H      H-C----C----C-H
  | | | |          | |  | |         |    |    |          |    |    |
  H H H H          H H  H H         H    |    H          H    |    H
                                       H-C-H                H-C-H
                                         |                    |
                                         OH                   H
```

23.65 Refer to Table 23.6 and appropriate sections in the textbook.

23.67 (a) Methanol. Hydrogen bonding decreases as the size of the hydrocarbon group increases.
(b) Formic acid. Hydrogen bonding decreases as the size of the hydrocarbon group increases.
(c) Oxalic acid. Oxalic acid has two carboxyl groups, while acetic acid has only one.
(d) Glycerol. Glycerol has three hydroxyl groups, while 2-propanol has only one.

23.69 (a) oxidation (b) reduction (c) reduction

Comment: For convenience, in organic chemistry oxidation is often defined as the addition of oxygen atoms to, or the removal of hydrogen atoms from, an organic compound. Reduction is defined as the addition of hydrogen atoms to, or the removal of oxygen atoms from, an organic compound. These definitions make it easy to recognize most organic redox reactions at a glance.

23.71 (a) oxidation (b) neither; this is an elimination reaction (c) reduction

See comment in Exercise 23.69.

23.73 (a) 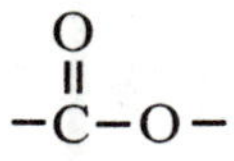 $-\overset{\overset{\displaystyle O}{\|}}{C}-O-$ ester linkage $R-\overset{\overset{\displaystyle O}{\|}}{C}-O-R'$ general formula for an ester

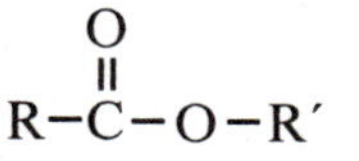

23.75 A fat is an ester of glycerol with three fatty acid molecules, i.e., a triglyceride.
(a) A saturated fat is one in which the fatty acids contain no double bonds.
(b) An unsaturated fat is a fat that contains double bonds.
(c) When the fat is liquid at room temperature, it is called an oil.

23.77 (a) 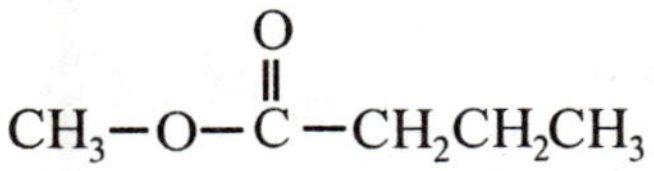 $CH_3-O-\overset{\overset{\displaystyle O}{\|}}{C}-CH_2CH_2CH_3$ (b) $C_6H_5-O-\overset{\overset{\displaystyle O}{\|}}{C}-CH_3$

(c)

$$\begin{array}{l} H_2C-O-\overset{\overset{\displaystyle O}{\|}}{C}-CH_2CH_3 \\ \ \ | \\ HC-O-\overset{\overset{\displaystyle O}{\|}}{C}-CH_2CH_3 \\ \ \ | \\ H_2C-O-\underset{\underset{\displaystyle O}{\|}}{C}-CH_2CH_3 \end{array}$$

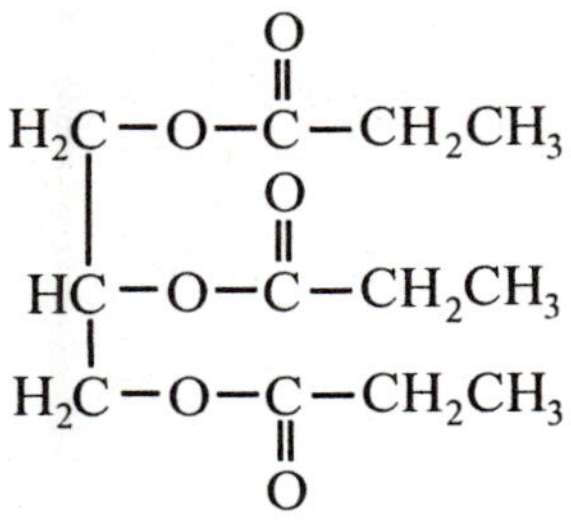

23.79

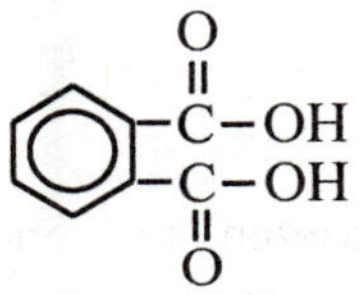

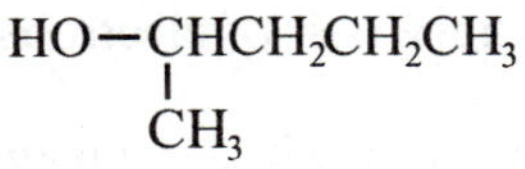

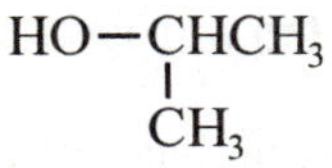

$C_6H_4(COOH)_2$ (phthalic acid, two $-\overset{\overset{\displaystyle O}{\|}}{C}-OH$ groups on the benzene ring) — acid

$HO-\underset{\underset{\displaystyle CH_3}{|}}{C}HCH_2CH_2CH_3$ — alcohol

$HO-\underset{\underset{\displaystyle CH_3}{|}}{C}HCH_3$ — alcohol

23.81 ethylene glycol: $HO–CH_2–CH_2–OH$; oxalic acid: $HOOC–COOH$

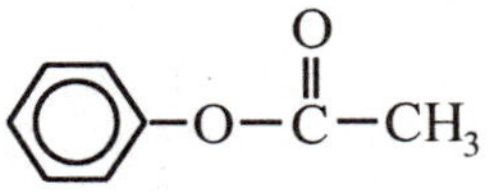

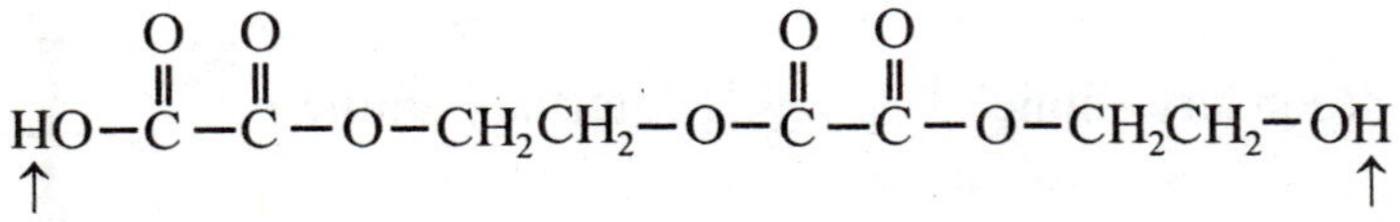

$$HO-\overset{\overset{\displaystyle O}{\|}}{C}-\overset{\overset{\displaystyle O}{\|}}{C}-O-CH_2CH_2-O-\overset{\overset{\displaystyle O}{\|}}{C}-\overset{\overset{\displaystyle O}{\|}}{C}-O-CH_2CH_2-OH$$

↑ another alcohol can condense at this end ↑ another acid can condense at this end

or

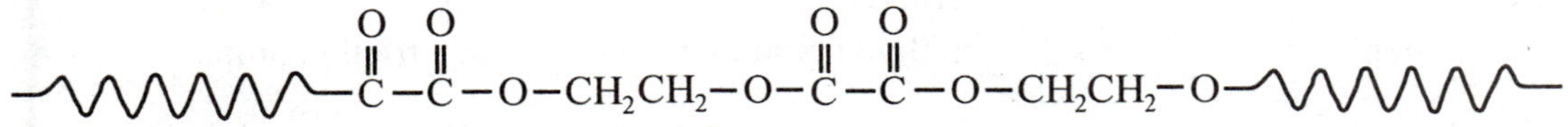

$$\sim\sim\sim\overset{\overset{\displaystyle O}{\|}}{C}-\overset{\overset{\displaystyle O}{\|}}{C}-O-CH_2CH_2-O-\overset{\overset{\displaystyle O}{\|}}{C}-\overset{\overset{\displaystyle O}{\|}}{C}-O-CH_2CH_2-O\sim\sim\sim$$

23.83 (a)

$$\begin{array}{l} H_2C-O-\overset{\overset{\displaystyle O}{\|}}{C}-(CH_2)_7CH{=}CH(CH_2)_7CH_3 \\ \ \ | \\ HC-O-\overset{\overset{\displaystyle O}{\|}}{C}-(CH_2)_7CH{=}CH(CH_2)_7CH_3 \\ \ \ | \\ H_2C-O-\underset{\underset{\displaystyle O}{\|}}{C}-(CH_2)_7CH{=}CH(CH_2)_7CH_3 \end{array}$$

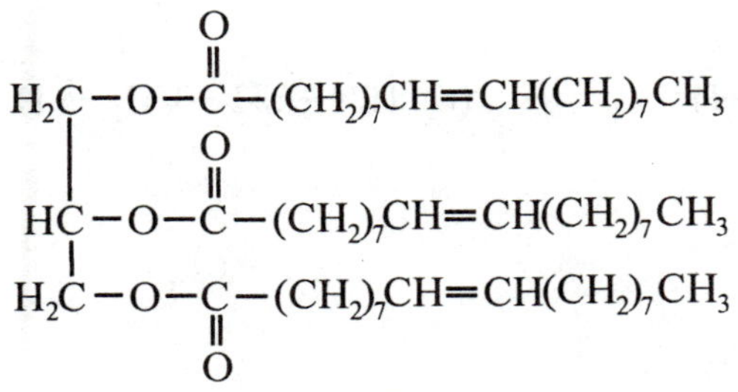

(b) 1.

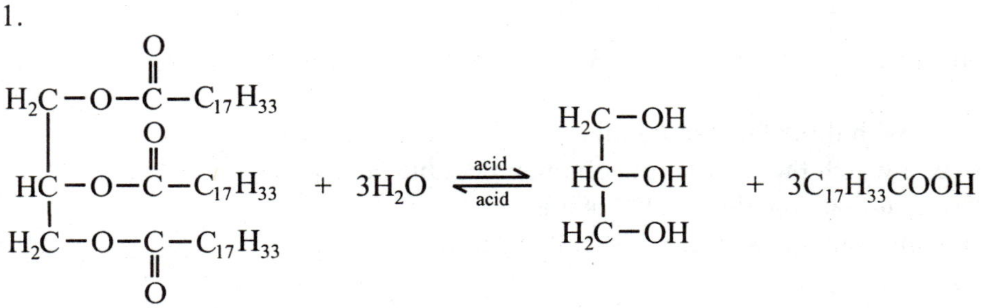

$$\begin{array}{l} H_2C-O-\overset{O}{\overset{\|}{C}}-C_{17}H_{33} \\ HC-O-\overset{O}{\overset{\|}{C}}-C_{17}H_{33} \\ H_2C-O-\underset{O}{\underset{\|}{C}}-C_{17}H_{33} \end{array} + 3H_2O \underset{acid}{\overset{acid}{\rightleftharpoons}} \begin{array}{l} H_2C-OH \\ HC-OH \\ H_2C-OH \end{array} + 3C_{17}H_{33}COOH$$

glycerol trioleate $\quad$ glycerol + oleic acid

2.

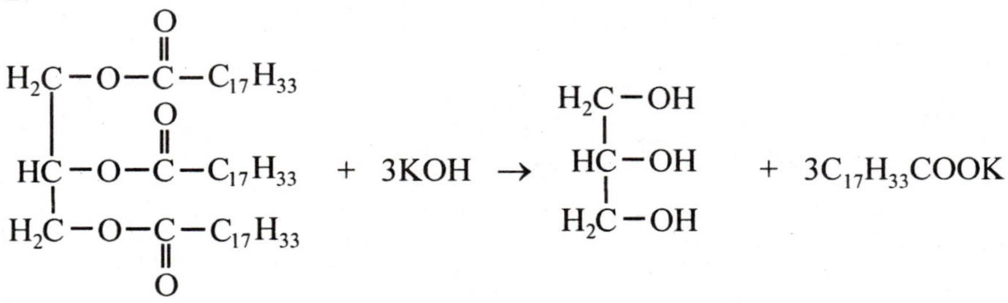

$$\begin{array}{l} H_2C-O-\overset{O}{\overset{\|}{C}}-C_{17}H_{33} \\ HC-O-\overset{O}{\overset{\|}{C}}-C_{17}H_{33} \\ H_2C-O-\underset{O}{\underset{\|}{C}}-C_{17}H_{33} \end{array} + 3KOH \rightarrow \begin{array}{l} H_2C-OH \\ HC-OH \\ H_2C-OH \end{array} + 3C_{17}H_{33}COOK$$

glycerol trioleate $\quad$ glycerol + potassium oleate, a soap

23.85 (a) 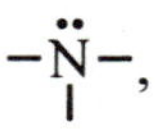 $-\ddot{N}-$ (with a third bond below), a bonded nitrogen atom with a lone pair of electrons.

(b) A primary amine has one alkyl group and two hydrogen atoms bonded to the nitrogen, a secondary amine has two alkyl groups and one hydrogen atom bonded to the nitrogen, and a tertiary amine has three alkyl groups bonded to the nitrogen. The general formulas are:

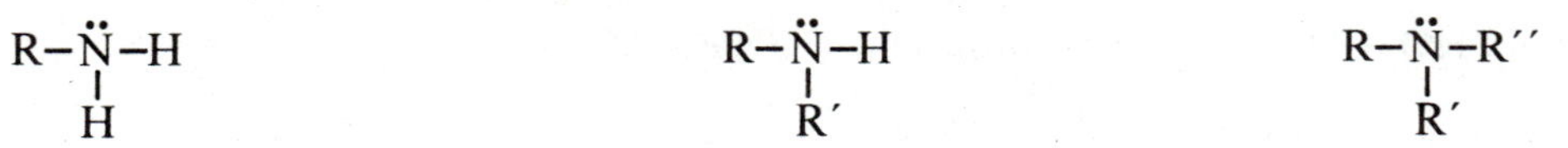

$$R-\underset{H}{\underset{|}{\ddot{N}}}-H \qquad R-\underset{R'}{\underset{|}{\ddot{N}}}-H \qquad R-\underset{R'}{\underset{|}{\ddot{N}}}-R''$$

primary amine $\quad$ secondary amine $\quad$ tertiary amine

Examples are:

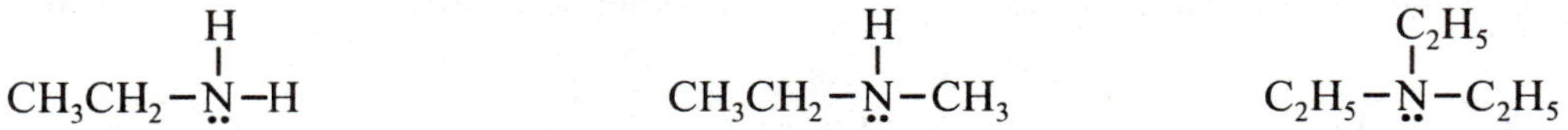

$$CH_3CH_2-\overset{H}{\overset{|}{\underset{\cdot\cdot}{N}}}-H \qquad CH_3CH_2-\overset{H}{\overset{|}{\underset{\cdot\cdot}{N}}}-CH_3 \qquad C_2H_5-\overset{C_2H_5}{\overset{|}{\underset{\cdot\cdot}{N}}}-C_2H_5$$

ethylamine $\quad$ methylethylamine $\quad$ triethylamine

23.87 An asymmetric carbon atom is a carbon atom with four different substituents. See examples in textbook.

23.89 (a) $CH_3-\overset{H}{\overset{|}{\underset{\cdot\cdot}{N}}}-H$ $\quad$ (b) $CH_3CH_2-\overset{H}{\overset{|}{\underset{\cdot\cdot}{N}}}-H$ $\quad$ (c) $H_2\ddot{N}-CH_2-CH_2-\ddot{N}H_2$

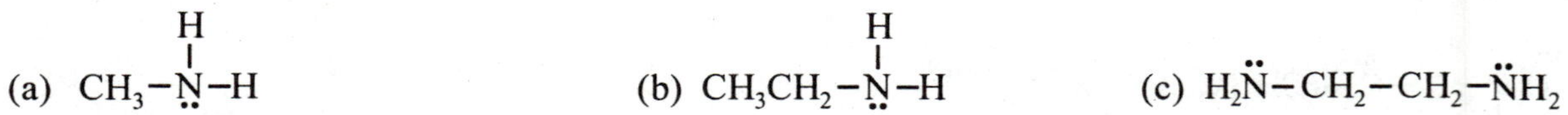

23.91 (a) 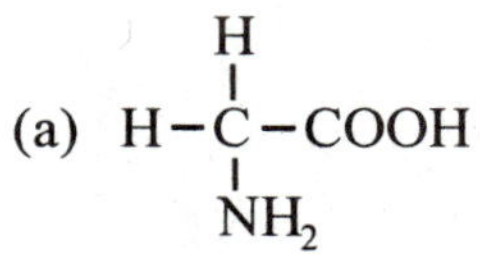 $H-\underset{NH_2}{\underset{|}{\overset{H}{\overset{|}{C}}}}-COOH$ $\quad$ (b) No

23.93 (a) 1-chloropropane: $CH_2ClCH_2CH_3$; No.
(b) 2-chloropentane: $CH_3CHClCH_2CH_2CH_3$; Yes, $CH_3\overset{*}{C}HClCH_2CH_2CH_3$.
(c) 3-bromopentane: $CH_3CH_2CHBrCH_2CH_3$; No.

(d) 1,3-dichlorobutane: $CH_2ClCH_2CHClCH_3$; Yes, $CH_2ClCH_2{}^{*}CHClCH_3$.
(e) Yes, $C_2H_5{}^{*}CHClCH_3$.

23.95 (a) $-\overset{\overset{\large O}{\|}}{C}-\overset{\overset{\large H}{|}}{N}-$ amide linkage (b) $R-\overset{\overset{\large O}{\|}}{C}-\overset{\overset{\large H}{|}}{N}-R'$ general formula for an amide

23.97 The primary structure of a protein refers to the sequence of the amino acid residues in the protein. The primary structure is held together or maintained by covalent bonds. The secondary structure refers to the protein's regular repeating conformations or arrangement. The secondary structure is held together by hydrogen bonding between the N–H and C=O groups along the backbone of the polypeptide chain.

The tertiary structure refers to the folds and bends of the protein that are not part of a repeating secondary structure. The tertiary structure is maintained by (1) hydrophilic interactions, (2) hydrogen bonding, (3) hydrophobic interactions, (4) ionic interactions, and (5) disulfide bridges. The quaternary structure refers to the spatial relationship of the polypeptide chains in proteins that are composed of two or more combined polypeptide chains, each with its own secondary and tertiary structure. The quaternary structure is maintained by hydrogen bonding, ionic interactions, and hydrophobic interactions.

(a) primary and secondary. (b) primary, secondary, and tertiary. (c) primary, secondary, tertiary, and quaternary.

23.99 (a) formamide: $H-\overset{\overset{\large O}{\|}}{C}-\overset{\overset{\large H}{|}}{N}-H$ (b) acetanilide: $H-\underset{\underset{\large H}{|}}{\overset{\overset{\large H}{|}}{C}}-\overset{\overset{\large O}{\|}}{C}-\overset{\overset{\large H}{|}}{N}-C_6H_5$

23.101 (a) $H-\overset{\overset{\large O}{\|}}{C}-\overset{\overset{\large H}{|}}{N}-H + H_2O \rightleftharpoons H-\overset{\overset{\large O}{\|}}{C}-OH + H-\overset{\overset{\large H}{|}}{N}-H$
formamide → formic acid + ammonia

$H-\underset{\underset{\large H}{|}}{\overset{\overset{\large H}{|}}{C}}-\overset{\overset{\large O}{\|}}{C}-\overset{\overset{\large H}{|}}{N}-C_6H_5 + H_2O \rightleftharpoons H-\underset{\underset{\large H}{|}}{\overset{\overset{\large H}{|}}{C}}-\overset{\overset{\large O}{\|}}{C}-OH + H_2N-C_6H_5$
acetanilide → acetic acid + aniline

(b) $H-\underset{\underset{\large H}{|}}{\overset{\overset{\large H}{|}}{C}}-\overset{\overset{\large O}{\|}}{C}-\overset{\overset{\large H}{|}}{N}-H + H_2O \rightleftharpoons H-\underset{\underset{\large H}{|}}{\overset{\overset{\large H}{|}}{C}}-\overset{\overset{\large O}{\|}}{C}-OH + H-\overset{\overset{\large H}{|}}{N}-H$
acetamide → acetic acid + ammonia

$CH_3-\overset{\overset{\large O}{\|}}{C}-\overset{\overset{\large H}{|}}{N}-C_6H_4-OH + H_2O \rightleftharpoons CH_3CO_2H + H_2N-C_6H_4-OH$
acetaminophen → acetic acid + p-aminophenol

23.103 (a) $n(HO-\overset{\overset{\large O}{\|}}{C}-R-\overset{\overset{\large O}{\|}}{C}-OH) + n(H_2N-R'-NH_2) \rightarrow \left[\overset{\overset{\large O}{\|}}{C}-R-\overset{\overset{\large O}{\|}}{C}-\overset{\overset{\large H}{|}}{N}-R'-\overset{\overset{\large H}{|}}{N}\right]_n + (2n-1)H_2O$

(b) Refer to "Polyamides" in Section 23.7 in the textbook for this equation.

23.105 (a)

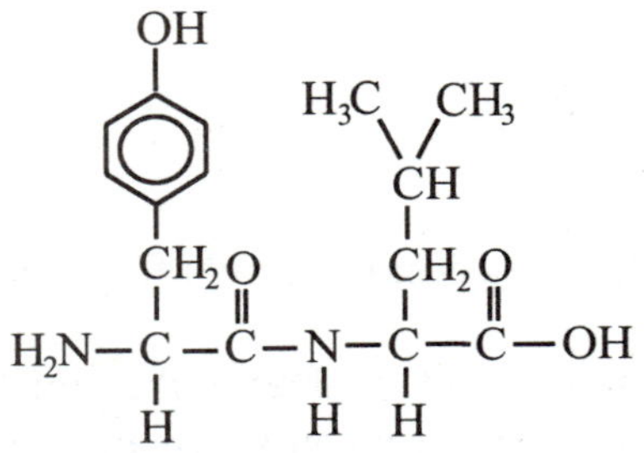

```
            OH
            |
          (ring)     H3C   CH3
            |           \ /
            |           CH
            |            |
           CH2 O        CH2 O
            |  ||        |  ||
      H2N—C—C—N—C—C—OH
            |      |  |
            H      H  H
```

Tyr-Leu

(b)

```
         H3C   CH3        OH
            \ /           |
            CH          (ring)
             |            |
            CH2 O        CH2 O
             |  ||        |  ||
       H2N—C—C—N—C—C—OH
             |      |  |
             H      H  H
```

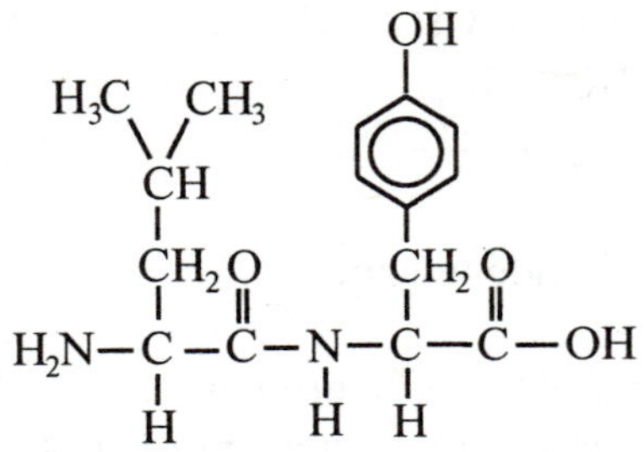

Leu-Tyr

23.107 (a) Hydrogen bonding: Ser, Gln, Tyr, Asn, and Thr.
(b) Ionic interactions: His, Asp, Glu, Lys, and Arg. These amino acids have charged R groups at physiological pH.
(c) Hydrophobic interactions: Gly, Ala, Val, Leu, Ile, Met, Phe, Pro, and Trp. These amino acids have nonpolar R groups.

23.109 (a) A hexose is a monosaccharide containing six carbon atoms, while a pentose is a monosaccharide containing five carbon atoms.
(b) An aldose is a monosaccharide containing an aldehyde functional group, while a ketose is a monosaccharide containing a ketone functional group.

Glucose is a hexose and an aldose, i.e., an aldohexose. Fructose is a hexose and a ketose, i.e., a ketohexose.

23.111
1. Sucrose: hydrolysis products are D-glucose and D-fructose.
2. Lactose: hydrolysis products are D-glucose and D-galactose.
3. Maltose: hydrolysis yields D-glucose.

23.113 (a) Starch, glycogen, and cellulose are all polymers of glucose.
(b) Starch is a condensation product of α-D-glucose, while cellulose is a condensation product of β-D-glucose. α-D-Glucose and β-D-glucose are the two different cyclic forms of open-chain D-glucose (see Figure 23.31). Glycogen is similar to starch but has more branching in the polymer chains.
(c) Starch and cellulose are found in plants. Glycogen is found in animals.

23.115 (a) The general structure of a nucleotide can be schematically represented as follows:

```
                 nitrogenous base
                        |
                        |
phosphate ———— sugar
```

See Figure 23.37 for an example.

(b) Hydrolysis of a nucleotide yields phosphoric acid, ribose or deoxyribose, and a nitrogenous base that is either adenine, quanine, thymine, cytosine, or uracil.

23.117 See Figure 23.35 in the textbook.

23.119 A nucleic acid is a polymer built up of nucleotide monomeric units. Refer to Figure 23.38 in the textbook.

23.121 The sequence of nucleotide bases in DNA holds the instructions for protein synthesis. The section of DNA containing the sequence or code needed for the synthesis or production of a specific protein is called a gene.

23.123 (a) A codon is a sequence of three nitrogenous bases or nucleotides in a nucleic acid that codes for a specific amino acid.

(b) Several codons. There are 64 codons and only 20 amino acids used in protein construction, and it turns out that each amino acid has more than one codon that codes for it (see Table 23.9 in the textbook).

(c) The "stop protein synthesis" codons are UAA, UAG, and UGA.

23.125 (a) Carbon has four valence electrons and with few exceptions forms four covalent bonds in compounds. Combining carbon's bonding capacity with its ability to form strong bonds with other carbon atoms allows for an unending array of straight and branched chains of varying lengths and complexity and rings (simple, multiple, and substituted) of different sizes. The small size of the carbon atom also permits its p orbitals to participate in π bonding, forming both double, C=C, and triple, C≡C, bonds with other carbon atoms, thus allowing for whole new categories of compounds containing these unsaturated bonds. In addition, carbon atoms bond with most of the elements in the periodic table.

(b) The bonds formed between the silicon atoms are not nearly as strong as those between carbon atoms. This severely limits the length and complexity of silicon chains and rings. Also, the larger size of the silicon atom doesn't allow it to participate in π bonding with its p orbitals.

23.127

```
             H  H             H  H
             |  |             |  |
1.  R•  +    C=C      →     R-C-C•                    chain-initiation
             |  |             |  |
             H  Cl            H  Cl

       H  H           H  H             H  H  H  H
       |  |           |  |             |  |  |  |
2.   R-C-C•     +     C=C      →     R-C-C-C-C•       chain-propagation
       |  |           |  |             |  |  |  |
       H  Cl          H  Cl            H  Cl H  Cl
```

3. Step 2 is repeated over and over again until the chain growth is terminated by combination with another free radical, either R• or another free radical chain.

23.129 Since each carbon can have no more than three hydrogen atoms, the formula for the hydrocarbon must be C_5H_{12}. The only saturated hydrocarbon with this formula that gives one monobromo derivative is

```
          H
          |
        H-C-H
   H      |      H                                        CH3 H
   |      |      |                                         |  |
 H-C------C------C-H   the monobromo derivative is  H3C-C----C-Br
   |      |      |                                         |  |
   H      |      H                                        CH3 H
        H-C-H
          |
          H
```

23.131 (a) Glyceryl trioleate has three double bonds per molecule or 3 mol of double bonds per mole of fat. Each mole of double bonds consumes 1 mol of iodine (I_2). The molar mass of glyceryl trioleate ($C_{57}H_{104}O_6$) is 885.4 g/mol. The road map is: 100 g fat → mol fat → mol double bonds → mol I_2 → g I_2 → iodine number.

$$100 \text{ g fat} \times \frac{1 \text{ mol fat}}{885.4 \text{ g fat}} \times \frac{3 \text{ mol double bonds}}{1 \text{ mol fat}} \times \frac{1 \text{ mol } I_2}{1 \text{ mol double bonds}} \times \frac{253.8 \text{ g } I_2}{1 \text{ mol } I_2} = 86 \text{ g } I_2$$

Answer: The iodine number is 86.

(b) Glyceryl trilinoleate has six double bonds per molecule or 6 mol of double bonds per mole of fat. Its molar mass ($C_{57}H_{98}O_6$) is 879.4 g/mol.

$$100 \text{ g fat} \times \frac{1 \text{ mol fat}}{879.4 \text{ g fat}} \times \frac{6 \text{ mol double bonds}}{1 \text{ mol fat}} \times \frac{1 \text{ mol } I_2}{1 \text{ mol double bonds}} \times \frac{253.8 \text{ g } I_2}{1 \text{ mol } I_2} = 173 \text{ g } I_2$$

Answer: The iodine number is 173.

(c) Stearin contains no double bonds. Therefore, its iodine number is zero.

23.133 Enzyme active sites can only accommodate molecules that have the correct shape. The glycosidic linkage in cellulose, which is a condensation product of β-D-glucose, has a different shape than the glycosidic linkage in starch, which is a condensation product of α-D-glucose. The enzyme active sites in humans cannot accommodate the shape of the glycosidic linkage in cellulose, only that in starch, and, consequently, humans cannot digest cellulose.